A HISTORY OF CHEMISTRY

A HISTORY OF CHEMISTRY

FROM EARLIEST TIMES TO THE PRESENT DAY

BEING ALSO

AN INTRODUCTION TO THE STUDY OF THE SCIENCE

BY

ERNST VON MEYER, PH.D.

PROFESSOR OF CHEMISTRY IN THE TECHNICAL HIGH SCHOOL, DRESDEN

TRANSLATED WITH THE AUTHOR'S SANCTION

BY

GEORGE McGOWAN, PH.D.

SECOND ENGLISH EDITION, TRANSLATED FROM THE SECOND GERMAN EDITION, WITH NUMEROUS ADDITIONS AND ALTERATIONS

London

MACMILLAN AND CO., LIMITED

NEW YORK: THE MACMILLAN COMPANY

1898

RICHARD CLAY AND SONS, LIMITED,
LONDON AND BUNGAY.

First Edition 1891.
Second Edition 1898.

PREFACE TO THE FIRST GERMAN EDITION

NEARLY five decades have passed by since Hermann Kopp's classical *Geschichte der Chemie*[1] began to appear, and it is now fifteen years since this was followed by the same indefatigable author's *Entwickelung der Chemie in der neueren Zeit.*[2]

The publication of these comprehensive works, in conjunction with which Höfer's *Histoire de la Chimie* must be named, and the further descriptions of the growth of chemistry within particular periods given both by Kopp himself and by other writers, might lead one to suppose that there was no pressing need for further work in the same direction at the present time.

This point can, the author thinks, be best discussed by his making a few remarks here with respect to the aim and plan of the present volume.

In this *History of Chemistry* the attempt has been made to describe within short compass the development of chemical knowledge, and especially of the general doctrines of chemistry which have thus been gradually evolved, from their earliest beginnings up to the present day. After a

[1] "History of Chemistry."

[2] "The Development of Chemistry in Recent Times."

general account of the main directions followed by chemistry in the various ages, the growth of particular branches of the science has been more or less minutely detailed.

In the general descriptions great emphasis has been laid upon the genesis of particular ideas, and their expansion into important dogmas or comprehensive theories. At the same time, in order that a vivid picture of the various periods and their distinguishing characteristics might be presented to the reader, short accounts have been given of the works, and in some cases of the lives, of the men who originated and developed such views.

In the special sections, on the other hand, the attempt has been made to collect together fundamental facts, which have been sifted and relegated to their proper branch of the science, and thus to offer as clear a description as possible of the state of chemical knowledge at the time in question.

That neither in this nor in the history of the development of theoretical views could completeness be thus achieved, hardly requires to be stated. But the author has at all events endeavoured to give a fair synopsis of the most important theories and facts which constitute the foundation of chemistry as we now know it.

The growth of chemical knowledge during recent times, since Boyle, and especially since Lavoisier, naturally forms the principal subject of the following chapters. The author is fully aware of the many difficulties which have to be met here, difficulties which increase in extent the nearer we approach to the history of our own period. We stand too close to the development of the theoretical views of these latter days to feel certain of always preserving the unbiassed temperament which is essential to the true historian. But, notwithstanding this, the author has ventured the attempt to carry the record of the history of chemistry up to the present day.

In this task he has done his best to preserve throughout an objective attitude; and he has further been guided by the earnest desire to contribute effectively towards shedding a clear light upon the opposing views held with respect to the development and the importance of the chemical doctrines of to-day. It has also been his duty as an historian to endeavour to apply to the services rendered by eminent investigators of quite recent years a calmer and juster criticism than has hitherto in many cases been meted out to them.

ERNST VON MEYER.

LEIPZIG, *7th October*, 1888.

TRANSLATOR'S PREFACE TO THE FIRST ENGLISH EDITION

THE author, in his preface to the original German edition, discusses the question whether there is any necessity for a new history of chemistry in his own language at the present day. That there is full room for one in this country will be admitted upon all hands. It is therefore hoped that the appended *history* will prove not only useful to the student, but also interesting to the general reader who is desirous of gaining some idea of the development of chemical science.

The translator has done his best to reproduce clearly the sense of the German original. And, since Professor von Meyer has been so kind as to read over the first corrected proofs, as well as to answer a great many queries, it is hoped that this has been achieved.

A considerable number of small alterations and additions have been made for this edition, most of them by the author, but some by the translator with the author's concurrence. While these may reasonably be supposed to have improved the book, they have not altered its character in the slightest degree. The translator has further added a number of duplicate references to English journals (to such papers as were published both in German and English), and also a few new ones, for the greater convenience of English readers.

In conclusion, he would express his indebtedness to the various gentlemen who have been kind enough to give him the benefit of their criticism and advice upon different points, with regard to which his own special knowledge was insufficient, and also to those others who have assisted him in the matter of references, etc.

University College of N. Wales, Bangor,
March, 1891.

AUTHOR'S NOTE TO THE FIRST ENGLISH EDITION

It was a great satisfaction to me that the translation of this *history* was undertaken by my former pupil, Dr. McGowan, and I desire to express here my appreciation of the manner in which he has entered into the spirit of the work, and to offer him my hearty thanks for all his trouble in the matter.

May the book find many friends among the English-speaking peoples, and help to stimulate the interest of its readers in the development of our science.

ERNST VON MEYER.

Leipzig, *February*, 1891.

TRANSLATOR'S PREFACE TO THE SECOND ENGLISH EDITION

THE present edition is a translation of the second German edition (published in 1895), with a number of further additions and alterations, most of these latter having been made by the author, but a good many of them by myself, with his approval; and, as in the case of the previous edition, the proof sheets had the benefit of the author's revision after my own corrections were made.

In his preface to the second German edition Professor von Meyer expresses his gratification at the success of the English version, and then goes on to speak of the additional sources of information on subjects of historical chemistry which have during the last few years become available for reference. Among these are the Berzelius-Liebig and the Liebig-Wöhler *Letters*, the *Letters* and *Journals* of Scheele, Priestley's *Letters*, and the autobiographical fragment which Liebig left behind him. In addition, there are the recently published and valuable historical researches of Berthelot on the chemistry of the early Middle Ages, and the writings of Ladenburg, Schorlemmer, Thorpe, Grimaux, and others on the development of chemistry within certain definite periods, or on the life and work of particular chemists.

I may, perhaps, be permitted to add my word of appreciation to what the author has said with regard to the friendly reception of the first English edition both in this country and in America, and to express the hope that the present edition may be found at least equally acceptable.

GEORGE McGOWAN.

Ealing, London, W.
July, 1898.

TABLE OF CONTENTS

PAGE

LIST OF ABBREVIATIONS xxiii

INTRODUCTION 1

CHAPTER I

FROM THE EARLIEST TIMES TO THE BIRTH OF ALCHEMY . . 5

Theoretical Views upon the Composition of Substances, and especially upon the elements, 6. Aristotle's Elements, 7.

The Empirical Chemical Knowledge of the Ancients, 9. *Metallurgy of the Older nations*—Gold, 11; Silver, Copper, Iron, 13; Lead, Tin, etc., 14; Mercury, 15. The Manufacture of Glass, 16. Pottery, 17. The Manufacture of Soap, 17. Dyeing, 17. The beginnings of Pharmacy, 18.

CHAPTER II

THE AGE OF ALCHEMY 21

General History of Alchemy 23

Origin and First Signs of Alchemistic Efforts, 23. The Alexandrian Academy, 27. *The Alchemy of the Arabians*—Geber and his Disciples, 28-30. *Alchemy among the Western Nations*, 30. Albertus Magnus, Roger Bacon, 31. Arnaldus Villanovanus, 32. Raymundus Lullus, 33. Basilius Valentinus, 36.

Special History of Alchemy 37

Theories and Problems of the Alchemistic Period, 37. The pseudo-Geber, 39-41. Views of Basilius Valentinus, etc., 41. The Philosopher's Stone, 42.

Practical-Chemical Knowledge of the Alchemists, 45.

Technical Chemistry—Gold, 46. Silver, 46. Copper and other metals, 47. Pottery, Glass, Dyeing, 47-48. *Pharmaceutical Chemistry*, 48.

Knowledge of the Alchemists with regard to Chemical Compounds, 49. Alkalies, 50, Acids, 51. Salts, 52. Preparations of Antimony, etc., 54. Organic Compounds, 56.
The Fortunes of Alchemy during the last Four Centuries, 58. A Short Review of Alchemistic Efforts, 63.

CHAPTER III

PAGE
HISTORY OF THE IATRO-CHEMICAL PERIOD 65
General History of this Period 67
Paracelsus and his school, 67. The Iatro-chemical Doctrines of Paracelsus, 69. Turquet de Mayerne, 73. Libavius, 74. *Van Helmont and his Contemporaries*, 75. The work of van Helmont, 73. Sala and Sennert, 80. Sylvius and Tachenius, 80. *Georgius Agricola*, 80. Palissy, 85. Glauber, 86.
Special History of the Iatro-chemical Period 87
Technical Chemistry, 87. Metallurgy, 88. Pottery and Glass Manufacture, 89. Dyeing, etc., 90.
Development of Pharmacy and of the Knowledge of Chemical Preparations, 91. Inorganic Compounds, 91. Organic Compounds, 96.

CHAPTER IV

HISTORY OF THE PERIOD OF THE PHLOGISTON THEORY, FROM BOYLE TO LAVOISIER 100
INTRODUCTION 100
General History of the Phlogistic Period 103
Robert Boyle, 103. Mayow, 107. Lemery and Homberg, 107. Kunkel and Becher, 109. *Stahl and the Phlogiston Theory*, 110. Fr. Hoffmann and Boerhave, 113.
The development of Chemistry, and particularly of the Phlogiston Theory, after Stahl's Time, 115. Neumann, Eller, Pott, Marggraf, 115-116. Geoffroy, Duhamel de Monceau, Rouelle, Macquer, 117-119. Black, 119. Cavendish, 121. Priestley, 122. Bergman and Scheele, 124-125.
Special History of the Phlogistic Period 128
Pneumatic Chemistry and its Relations to the Doctrine of Phlogiston, 128. The Discovery of Oxygen and the Composition of Air, 130.
Development of Theoretical Views in the Phlogistic Period, 134, Views regarding Elements and Chemical Compounds, 135.

Views regarding Chemical Affinities and its Causes, 137. Geoffroy's Tables of Affinities, 138.

Practical Chemical Knowledge in the Phlogistic Age, 140. The Development of Analytical Chemistry, 141. Boyle, 141. Fr. Hoffmann, Marggraf, Scheele, 142-143. Bergman, 143. The beginnings of Gas Analysis, 145.

Technical Chemistry in the Phlogistic Age—Metallurgy, 146. The Ceramic Industry, Dyeing, 147.

Technico-chemical Preparations—Acids and Alkalies, 147. The Discovery of Elements, 149. Inorganic and Organic Compounds, 150-151.

Pharmaceutical Chemistry, 154.

Concluding Remarks upon this Period, 155.

CHAPTER V

	PAGE
History of the most Recent Period (from the Time of Lavoisier up to now)	158
Introduction	158
General History of Chemistry during this Period	160

Lavoisier and the Antiphlogistic Chemistry, 160. Lavoisier's Life and Work, 160 *et seq.* His Combustion Theory, 164-167. Triumph of the Antiphlogistic Chemistry, 168. Beginnings of a Rational Chemical Nomenclature, 170. Guyton de Morveau, 172. Berthollet, 173. Fourcroy, 174. Vauquelin, 176.

The State of Chemistry in Germany at the end of the Eighteenth Century, 177. Klaproth, 178. *The State of Chemistry in England, Scotland and Sweden*, 180.

Development of the Doctrine of Chemical proportions, 181. Richter, 182. His law of Neutralisation, 183. The Beginnings of Stöchiometry, 185. Proust, 185. His contest with Berthollet, 186. Recognition of Constant Combining Proportions, 187.

Dalton's Atomic Theory	188

Law of Multiple Proportions, 189-190. Dalton's Attempts to determine the relative Atomic Weights of the Elements, 191. His Atomic Weights and Chemical Symbols, 193-194.

Further Development of the Atomic Theory, 194. Thomas Thomson, 194. Wollaston, 195. Humphry Davy, his Life and most important Work, 195-199. Gay-Lussac, 199. His Law of Volumes and Work generally, 200-201. *Prout's Hypothesis and its Effects*, 201.

Berzelius—A Survey of his Work	203

Biographical Notice, 204. His Influence upon the Development of Analytical and Organic Chemistry, 205-207. His Experimental Researches, 205-207. Berzelius as a Teacher and Writer, 207-209. His general Character, 209.

Development of the Atomic Theory by Berzelius, 210. His Determinations of relative Atomic Weights, 211 *et seq.* His Oxygen Law, 212.

Influence of Gay-Lussac's Law of Volumes upon the Atomic Theory, 214. Avogadro's Hypothesis, 215. Application of the Law of Volumes by Berzelius, 216. The Position of the Atomic Theory in 1818, 217. *Dulong and Petit's Law,* 220. *Influence of the Doctrine of Isomorphism upon the Atomic Theory,* 221. Mitscherlich, 222.

The Atomic Weight System of Berzelius from 1821 to 1826, 223. *Dumas' Attempt to alter the Atomic Weights,* 225. Failure of this Attempt, 227. Faraday, 227. His Law of Definite Electrolytic Action, 228.

The Electro-Chemical Theories of Davy and Berzelius, 229 *et seq. The Dualistic System of Berzelius,* 233. His Chemical Nomenclature and Notation, 234-237.

Manifestations against Dualism, 237. Discovery of the Alkali Metals, 238. Recognition of the Elementary Nature of Chlorine, 240. *Theory of the Hydrogen Acids* (Davy and Dulong), 241. *Doctrine of the Polybasic Acids* (Liebig), 243. Graham, 244.

Development of the Dualistic Doctrine in the Domain of Organic Chemistry, 246. *The Growth of Organic Chemistry previous to* 1811, 246. *The Position of Berzelius with regard to Organic Chemistry,* 248. Development of Views respecting Radicals, 249.

Isomerism and its Influence on the Development of Organic Chemistry, 250. Observations of Liebig, Wöhler, Faraday, and Berzelius, 251. Clearer Definition of the terms *Isomerism, Polymerism,* and *Metamerism* by Berzelius, 252.

The older Radical Theory, 253. The Etherin Theory, (Dumas and Boullay), 253-254. Liebig and Wöhler's Work upon Benzoyl Compounds, 254. The Ethyl Theory of Berzelius and Liebig, 256. Position of the Radical Theory in 1837, 258. Definition of the term *Radical,* 260. Bunsen, 260. His Work upon the Cacodyl Compounds, 261. The significance of the Radical Theory, 261.

Liebig, Wöhler, and Dumas—A Survey of their more important Work, 262. Justus Liebig, his Life and Work, 262. Liebig as a Teacher, 265. His Literary Activity, 266. His experimental Researches, 267. Friedrich Wöhler, 270. Wöhler as a Teacher and Writer, 271. His services to Science, 272. Dumas, his Life and Work, 272-275.

The Development of Unitary Views in Organic Chemistry, 275.

Substitution Theories, 275. *Dumas' Laws of Substitution*, 276. *Laurent's Substitution or Nucleus Theory*, 278. Criticism of the same, 279. *Dumas' Type Theory*, 280. His Unitary System, 281. *The Overthrow of Berzelius' Dualistic Doctrine*, 282. Berzelius' Fight against the Substitution Theory and his Defeat, 282 *et seq.*

Fusion of the older Theory of Types with the Radical Theory by Laurent and Gerhardt, 286. Laurent and Gerhardt, a Sketch of their Lives, 286. *Gerhardt's Theory of Residues*, 287. His Law of Basicity, 289. *Gerhardt's first Classification of Organic Compounds*, 289. *His Reform of the Atomic Weight System*, 290. *The distinguishing between the terms Molecule, Atom, and Equivalent by Laurent and Gerhardt*, 293. *Work preparatory to the new Type Theory*—Wurtz and A. W. Hofmann, 295-298. Williamson's Experiments on the Formation of Ethers, 298. His Opinions with regard to the "Typical" View, 299. *Gerhardt's new Theory of Types*, 300. Work preparatory to this, 301. Derivation of Organic Compounds from Types, 303. Gerhardt's Views upon Chemical Constitution, 304. Criticisms upon his Type Theory, 306. *Extension of the Type Theory by Kekulé*, 307. Kekulé, 308. Mixed Types, 308. Marsh Gas as a Type, 309. Position of the Type Theory in 1858, 310.

Development of the Newer Radical Theory by Kolbe—A Survey of Kolbe's Life and Work, 311. *The Re-animation of the Radical Theory by him—Frankland's Co-operation*, 313. Copulated or Conjugate Compounds, 315. Setting aside of the Notion of Copulation by Frankland, 316. Kolbe's Carbonic Acid Theory, 317. The Derivation of Organic Compounds from Inorganic, 317. *Kolbe's most important Experimental Researches*, 319. *His Attitude towards the older and the newer Chemistry*, 320. Kolbe's real Types, 321.

PAGE

The founding of the Doctrine of the Saturation-Capacity of the Elements by Frankland 322

Preparatory steps towards this Doctrine, 322. Frankland's services here, 322 *et seq.* Assumption of a varying Saturation-Capacity, 325. Discussions on the Subject by Odling, Williamson, and Wurtz, 326-327.

The Recognition of the Valency of Carbon, 327. Kekulé's services here, 329. Kolbe and Frankland's share in the Matter, 329.

Development of Chemistry under the Influence of the Doctrine of Valency during the last Thirty Years . . 331

Beginnings of the Structure Theory—Kekulé and Couper, 332. Establishment of the true Atomic Weights by Cannizzaro, 335.

Discussions regarding the Nature of "Structure" by Butlerow and Erlenmeyer, 336.

Controversies respecting constant and varying Valency of the Elements, 337. Views upon varying Valency held by Frankland, Kolbe, etc., 337-338; by Erlenmeyer, Wurtz, and Naquet, 338. Kekulé's Theory of a Constant Valency, 338; Grounds for the Assumption of a varying Valency, 340 *et seq.*

The further Development of the Structure Theory—The chief Directions taken by Organic Chemistry during the last Thirty Years, 342. Views upon the linking of Atoms, 343. *Constitution of Organic Compounds according to the Structure Theory*, 344. Saturated and Unsaturated Compounds, 344. *Kekulé's Theory of the Aromatic Compounds*, 346. Modifications in this Theory proposed by Ladenburg, Claus, and Baeyer, 348-349. Constitution of pyridine, pyrrol, etc., 350-351. Victor Meyer's more precise Conception of the term *Aromatic Compounds*, 351. *Application of Structural-chemical Conceptions to the Investigation of Isomerism*, 351. Position-isomerism, 353. Tautomerism or Desmotropism, 354-355. Geometrical isomerism (Wislicenus), 356. Allo-isomerism (Michael), 356. The supposed Spacial Arrangement of atoms, 357 *et seq.* *The Development of Important Methods for investigating the Constitution of Organic Compounds*, 361. *Synthetic Methods* (Wöhler, Kolbe, Frankland, Baeyer, Kekulé, Ladenburg, Fittig, W. H. Perkin sen., and others), 361. Chemical Behaviour of Organic Compounds, 365.

The Main Currents in Inorganic and General Chemistry during the last Thirty Years, 367. Application of the Structure Theory to Inorganic Compounds, 368. Important Researches in Inorganic Chemistry, 369. *The Periodic System of the Elements* (Newlands, L. Meyer, Mendelejeff), 370. Crookes' Hypothesis of a Primary Material, 374. *General Significance of Physico-chemical Investigations*, 375. Ostwald, 377. Van 't Hoff, 377.

CHAPTER VI

PAGE

SPECIAL HISTORY OF THE VARIOUS BRANCHES OF CHEMISTRY FROM LAVOISIER TO THE PRESENT DAY . . . 379

Introduction 381

History of Analytical Chemistry 384

Qualitative Analysis of Inorganic Substances, 384. Use of the Spectroscope for this purpose, 385. *Quantitative Analysis of Inorganic Substances*, 386. Kaproth, Vauquelin, 386. Lavoisier, Proust, Berzelius, 387. Dumas, Erdmann and Marchand, Marignac, and Stas, 388. H. Rose, Wöhler, Fresenius, 389.

Docimacy, 390. *Volumetric Analysis*, 390. Its Development by Gay-Lussac, Bunsen, Mohr, etc., 391. *Development of Methods of Gas Analysis*, 392. *The Analysis of Organic Substances* (Lavoisier, Gay-Lussac and Thénard, Berzelius, Liebig), 393-397. Legal-chemical Analysis, 397. Technico-chemical Methods, 398.

PAGE

The Progress in Pure Chemistry from Lavoisier to the Present Time 400

Special History of Inorganic Chemistry 400

The Discovery of Elements, and the Determination of their Atomic Weights, 401. *Oxygen, Nitrogen and Hydrogen*, 401-402. The Halogens, 402. Selenium, Tellurium, etc., 403. Boron and Carbon, 404. Allotropy, 405. The Metals of the Alkalies and Alkaline Earths, 407-408. Beryllium, Cadmium, Thallium, Aluminium, Indium, Gallium, 408-409. Metals of the Cerium Group, 409. Nickel and Cobalt, 410. Chromium, Titanium, Germanium, etc., 410-411. Vanadium and allied Elements, 412. Metals of the Platinum Group, 413. Argon, Krypton, Metargon, Neon and Helium, 414-417. Rayleigh, 414. Ramsay, 415. Supposed new Elements, 417.

Inorganic Compounds, 418. *Hydrogen Compounds of the Halogens*, 418. *Oxygen Compounds of Hydrogen and of the Halogens*, 418. *Sulphur, Selenium and Tellurium Compounds*, 420. Compounds of Nitrogen, Phosphorus, etc., 421-424. Compounds of Boron, Silicon, and Carbon, 424. Compounds of the Alkali and Alkaline Earth Metals, 426. Compounds of the Metals of the Iron Group, etc., 427. Compounds of Tin, Vanadium, etc., 429. Compounds of Gold, Platinum, etc., 430.

Special History of Organic Chemistry in the Nineteenth Century 432

Hydrocarbons and their Derivatives, 433. *The Alcohols and Analogous Compounds*, 437. *Carboxylic Acids*, 441. *Acid Chlorides, Anhydrides, and Amides*, 444. *Oxy- and Amido-Acids*, 446. *Aldehydes*, 448. *Ketones and Ketonic Acids*, 451. *Carbohydrates and Glucosides*, 454. *Haloid Derivatives of the Hydrocarbons, etc.*, 457. *Nitro- and Nitroso-Compounds*, 460. *Sulphur Compounds*, 462. *Organic Nitrogen Compounds* (Amines, etc.), 465. Phosphines, Arsines, Stibines, 469. Azo-Compounds, 470. Diazo-Compounds, 470. Hydrazines, Cyanogen Compounds, 472-478. *Pyridine and Quinoline Bases*, 478. Their Relation to Vegetable Alkaloids, 482. *Pyrrol and Analogous Compounds*, 484. *Organo-metallic Compounds*, 486.

PAGE

History of Physical Chemistry in Recent Times . . . 488

Determination of Vapour Density and the Application of this, 490. *Dissociation*, 492. *The Liquefaction of Gases*, 492. *The Kinetic Theory of Gases*, 493. *Spectrum Analysis*, 494. *Atomic Volumes of Solids and Liquids*, 495. *Laws regulating the Boiling Temperature*, 496. *Specific Heat of Solid Bodies*, 497. *Optical Behaviour of Solids and Liquids* (Refraction, Circular Polarisation), 498. *Diffusion, etc.*, 500. *Theory of Solution; Electrolytic Dissociation*, 501. *The Electrolysis of liquid or of dissolved Substances*, 503. *Isomorphism, etc.*, 505. *Thermo-Chemistry*, 507. Julius Thomsen; Berthelot, 508. *Photo-Chemistry*, 509.

Development of the Doctrine of Affinity since the Time of Bergman, 512. *Bergman's Doctrine of Affinity*, 512. *Berthollet's Doctrine of Affinity*, 513. *The Supplanting of Berthollet's Opinions by other Doctrines*, 515. *The Revival of Berthollet's Doctrines*, 517. *The latest Development of the Doctrine of Affinity*, 519.

Sketch of the History of Mineralogical Chemistry during the last Hundred Years 522

Its Earlier History, 522. The Chemical Mineral System of Berzelius, 524. Other Mineral Systems, 525. The more recent Development of Mineral Chemistry, 525-526. *The Artificial Production of Minerals—Beginnings of Geological Chemistry*, 527.

Development of Agricultural and of Physiological Chemistry 530

Agricultural Chemistry and Vegetable Physiology, 531. The Humus Theory, 531. Reform of Agricultural Chemistry by Liebig, 532. Its further Development by Liebig and his School, 533. Nitrification and the Assimilation of free Nitrogen by plants, 534-535.

The Development of Phyto-Chemistry, 536. Important Phyto-Chemical Researches, 537-538.

The Development of Zoo-Chemistry, 539. Researches upon the Constituents of the Animal Body, 539. The Chemistry of the Animal Secretions—Saliva, Gastric Juice, Bile, Blood, 541; Milk, Urine, 542. Metabolism, 544.

Fermentation; Putrefaction, 546. Views regarding Fermentation, 546 *et seq.* Organised and Unorganised Ferments, 548. The Phenomena of Putrefaction, 548. The Ptomaïnes, 549.

The Relation of Chemistry to Pathology and Therapeutics, 549. Bacteriology, 550. Antiseptics, Anæsthetics, and Antipyretics, 550-551.

The Relation of Chemistry to Pharmacy, 552.

PAGE

History of Technical Chemistry during the last Hundred Years 554

Introduction, 554. Development of Technical Instruction, 556. Literature on Technical Chemistry, 556.

The Progress of Metallurgy, 557. Iron and Steel, 557. Nickel, Silver, the Galvano-Plastic Process, Aluminium, 558-559. Mineral Pigments, 560.

Development of the Great Chemical Industries, 561.

Sulphuric Acid, 561. The Soda Industry, 562-564. Hydrochloric Acid, Chlorine, and Bleaching Powder, 565-566. Bromine and Iodine, 566. Nitric Acid, Gunpowder, 566-567. Other Explosives, Matches, 568.

The Manufacture of Soap, etc., 569. Ultramarine, 570. Glass, Earthenware, and Pottery, 570, 571. Mortar, Paper, 571, 572. Starch, Beet-Sugar, 572-574.

Fermentation Processes, 574. The Manufacture of Spirits, 575. The Quick Vinegar Process, 575.

The Aniline Colours and other similar Dyes, 576. Phthaleïns, Azo-Dyes, 578. Alizarine, the Safranines, Indigo Blue, 579. *Dyeing*, 580. *Tanning*, 580.

Various Chemical Preparations, 581. *Various Products from Coal-tar*; *Illuminants*, 582. *Heating Materials*, 584.

The Growth of Chemical Instruction in the Nineteenth Century, more especially in Germany 586

The State of Education in Science at the end of the Eighteenth Century, 586. Experimental Lectures, 587. The Development of Practical Instruction (Berzelius, Liebig), 587-588. The Erection of Laboratories for General Instruction in Germany, 588 *et seq.* Erdmann, 590. The State of Scientific Education in France, Great Britain, etc., 591. Improvements in the Construction of Chemical Laboratories, 593.

Chemical Literature, 594. Text-books, 594. Larger Treatises and Encyclopedias, 595. Periodical Journals, 596. Yearly Reports (*Jahresberichte*), 597. The Necessity for Criticism in Chemical Literature, 598. The Study of Original Memoirs, 598.

INDEX OF AUTHORS' NAMES 601

INDEX OF SUBJECTS 613

ABBREVIATIONS OF THE NAMES OF MOST OF THE JOURNALS TO WHICH REFERENCE HAS BEEN MADE

Ann. Chem. . .	Liebig's Annalen der Chemie und Pharmacie (begun 1832).
Ann. Chim. . .	Annales de Chimie et de Physique (begun 1816; five series).
Ann. de Chimie .	The same journal from 1789 to 1815.
Ann. des Mines .	Annales des Mines.
Ann. of Philosophy	Annals of Philosophy (edited by Thomas Thomson, 1813-26). This journal was subsequently merged in the *Philosophical Magazine*.
Ann. Phys. . .	The new Series (*Neue Folge*) of Poggendorff's Annalen.
Archiv. Pharm. .	Archiv der Pharmacie (begun 1832).
Bayer. Akad. . .	Sitzungsberichte der Bayerischen Akademie der Wissenschaften.
Ber.	Berichte der Deutschen chemischen Gesellschaft (begun 1868).
Bull. Soc. Chim. .	Bulletin de la Société Chimique de Paris (begun 1864).
Chem. Centr. . .	Chemisches Centralblatt (begun 1848).
Chem. News . .	Chemical News (begun 1860).
Compt. Rend. . .	Comptes Rendus des Séances de l'Académie des Sciences (begun 1835).
Crell's Ann. . .	Chemische Annalen von L. v. Crell (1784-1805).
Dingl. Journ. . .	Dingler's Polytechnisches Journal (begun 1820).
Gazz. Chim. Ital. .	Gazzetta Chimica Italiana (begun 1871).
Gilb. Ann. . . .	Annalen der Physik von Gilbert und Gren (1798-1824).
Hofmann's Bericht, etc. .	Bericht über die Entwickelung der Chemischen Industrie während des letzten Jahrzehnts von Hofmann (began 1875, but ceased after the publication of two volumes).
Jahres. Berz. . .	Jahresberichte über die Fortschritte der Chemie und Mineralogie von Berzelius (1821-47).
Jahres. d. Chemie	Jahresberichte über die Fortschritte der Chemie von Liebig und anderen (begun 1847).

Journ. Chem. Ind. Journal of the Society of Chemical Industry (begun 1882).

Journ. Chem. Soc. Journal of the Chemical Society (Memoirs and Proceedings, vols. i.-iii., 1841-47; Journal begun 1848).

Journ. de Phys. . Journal de Physique (1778-94; 1798-1823).

Journ. pr. Chem. Journal für praktische Chemie (begun 1834; the new series begun 1870).

Mon. Scient. . . Moniteur Scientifique (edited by Quesneville, begun 1857).

Phil. Mag. . . Philosophical Magazine (begun 1798).

Phil. Trans. . . Philosophical Transactions of the Royal Society (begun 1666).

Phil. Trans. E. . Philosophical Transactions of the Royal Society of Edinburgh (begun 1788).

Pogg. Ann. . . Annalen der Physik und Chemie von Poggendorff (begun 1824; new series begun 1877).

Proc. R. S. . . Proceedings of the Royal Society [begun 1800. Vols. i.-iv. (1800-1843) are entitled "Abstracts of the Papers printed in the Philosophical Transactions of the Royal Society of London," and vols. v., vi. (1843-1854) "Abstracts of Papers communicated to the Royal Society." The final form of title, "Proceedings of the Royal Society of London," begins with vol. vii., published in 1856].

Proc. R. S. E. . Proceedings of the Royal Society of Edinburgh (begun 1845).

Rec. Trav. Chim. . Recueil des Travaux Chimiques (begun 1882).

Schweigg. Journ. Journal für Chemie und Physik von Schweigger (1811-33).

Wagner's Jahresber. Jahresbericht über die Leistungen der chemischen Technologie von Wagner (begun 1856).

Wiener Monatshefte Monatshefte für Chemie und verwandte Theile anderer Wissenschaften (begun 1880).

Ztschr. anal. Chem. Zeitschrift für analytische Chemie von Fresenius (begun 1862).

Ztschr. Chem. . Zeitschrift für Chemie (1865-71); this was a continuation of the Kritische Zeitschrift (begun 1858).

Ztschr. phys. Chem. Zeitschrift für physikalische Chemie, Stöchiometrie, und Verwandtschaftslehre (edited by Ostwald and van 't Hoff; begun 1887).

ERRATA.

Page 27, line 18, from top, *for* "Synesius" *read* "Synesios."
,, 357, ,, 2, from top, *for* "Lebel" *read* "Le Bel."
,, 386, ,, 2, from top, *for* "Kirchoff" *read* "Kirchhoff."
,, 407, ,, 19, from top, *for* "Matthiessen" *read* "Mathiessen."
,, 446, ,, 17, from top, *for* "R. Hofmann" *read* "R. Hoffmann."
,, 481, ,, 5, from foot, *for* "Huisberg" *read* "Hinsberg."
,, 499, ,, 5, from foot, *for* "Lebel" *read* "Le Bel."
,, 502, ,, 11, from foot, *for* "Eykman" *read* "Eykmann."
,, 505, ,, 11, from foot, *for* "Scherer" *read* "Scheerer."
,, 537, ,, 7, from top, *for* "formic acid" *read* "formic aldehyde."
,, 543, ,, 11, from foot *for* "Scherer" *read* "Scheerer."
,, 558, ,, 7, from foot, *for* "Pattison" *read* "Pattinson."
,, 575, note 4, *for* "Schüzenbach" *read* "Schützenbach."

A HISTORY OF CHEMISTRY

INTRODUCTION

CHEMISTRY has for the last two hundred years or so been defined as the study of the composition of substances. Its first task, therefore, lies in ascertaining the constituents of which the material world surrounding us is composed, in reducing these constituents to their elements, and in building up new chemical compounds from the latter. Hand in hand with these analytic and synthetic problems there goes the further task of determining the laws which regulate the chemical combination of matter.

The problems just indicated occupy, in the widest sense of the word, the attention of chemists to-day. The problems of chemistry were, however, different in former times, and it is precisely these differences in aim which characterise the various epochs into which the history of the science may therefore be divided.

The oldest nations with regard to which we possess reliable information—the Egyptians, Phœnicians, Jews and others—did indeed possess a certain disjointed knowledge of chemical processes acquired accidentally; but these were applied for their practical results alone, and not with the object of deducing any comprehensive scientific explanation from them. We meet with similar conditions among the earliest cultured European nations, the Greeks and Romans, who owed most of their knowledge of chemical facts to the peoples just named. Nowhere do we find in antiquity the endeavour to gain an insight into chemical processes by means of definitely planned experiments. Although the Ancients were wholly without such data, furnished by exact research, as are nowadays held to be indispensable, this did not prevent them from speculating as to the nature of the universe; indeed, those theoretical views upon the nature of matter, on the "elements" of which the world was composed, have given to the earliest age of chemistry its own particular stamp. Some of these systems—especially Aristotle's system of the elements—continued to hold sway for many centuries, and influenced more especially the whole teaching of the Middle Ages.

From the above-mentioned doctrine of the nature of the elements was developed the theory of the transmutation of metals, or rather the fixed belief that one element can be transformed into another. Even so far back as the beginning of our own era, at first in Egypt, there began the striving to transmute the base metals into the noble, to "create" gold and silver.

The art by which this was to be achieved was termed *chemia* (χημεία), a name dating, so far as actual proof goes, from the fourth century, but in reality probably from an earlier period.[1]

There are many indications that this conception of the aim of chemistry and of the problems which it had to solve predominated for centuries following, *e.g.* it lies at the root of the definition given by Suidas, the author of an encyclopedia, who lived in the eleventh century: "Chemistry, the artificial preparation of silver and gold;" further, "χρυσοποιία" was a very commonly used designation for chemistry over a long period.

This task, the solution of which was the aim of the

[1] This word is of Egyptian origin and is probably founded on the North Egyptian word *Chēmi*, the name for Egypt. It also means, however, "black," and hence there is still some doubt whether the word χημεία of that period denotes Egyptian art or, as Hoffmann in the article "Chemie," in the *Dictionary of Chemistry* edited by A. Ladenburg, endeavours to prove, the employment of a black-coloured preparation valuable for alchemistical purposes. The mode of writing χυμεία, and the derivation of this word from χυμός, may be regarded as incorrect.

so-called Alchemy,[1] characterises the alchemistic period, a period extending from at least the fourth century of our era to the first half of the sixteenth. It is impossible to state with perfect exactitude the date at which alchemy took its rise, its origin being lost in the mists of the past. The labours of the alchemists, who strove by all imaginable methods to attain to the philosopher's stone (by the aid of which not only were the noble metals to be produced from the base, but also the life of man to be prolonged), had the effect of largely extending the area of the then existing knowledge of chemical facts.

In the first half of the sixteenth century, almost contemporaneously with the Reformation, *i.e.* with the birth of a new epoch in the world's history, chemistry began to develop in a new direction, without, however, losing all at once its alchemistic tendencies. Chemistry, which had already proved itself a valuable helpmeet to medicine in the preparation of active remedies, came to be looked upon as the basis of the whole medical art. Health and illness were reduced to chemical processes in the human body; only by means of chemical preparations could an unhealthy body be restored to its normal condition; in short, the absorption of medicine in chemistry, the fusion of both together, was the cry which emanated from Paracelsus. Van Helmont, de le Böe Sylvius, Tachenius and others were the chief exponents of this doctrine, which characterises the period of Medical or Iatro-Chemistry. The fact that technical chemistry was advanced at the same time, through the labours of individuals such as Georgius Agricola, was without influence on the prevailing tendency of the science of that age.

From the middle of the seventeenth century on, the iatro-chemical current gradually underwent substitution by another. After that date chemistry strove hard to become a self-supporting branch of natural science, quite independent of every other. Indeed, the history of chemistry proper begins with Robert Boyle, who taught, as its main object, the acquisition of a knowledge of the composition of bodies.

The conception of this aim marks the date from which

[1] This term with the Arabic prefix "al" became naturalised very early.

chemistry may be regarded as a science striving towards an ideal goal along the paths of exact research, without regard to practical results, and solely with the object of arriving at the truth.

The most important problem, whose solution occupied all the chemists of note at that day, was the question of the chemical reasons underlying the phenomena of combustion. Since Stahl's attempt to explain the latter, the hypothetical fire stuff Phlogiston—which was supposed to escape during every combustion—was regarded as the universal principle of combustibility. This doctrine held sway over chemists at the end of the seventeenth and during the greater part of the eighteenth centuries to such an extent that we are justified in characterising this period (after the death of iatro-chemistry) as the period of the Phlogiston Theory.

The fall of the latter, and its replacement by the anti-phlogistic system of Lavoisier, bring us to the commencement of the chemical era in which we are still living. For, upon the foundation laid by Lavoisier and his co-workers, and firmly fixed by Dalton, Berzelius and others, the structure of the new chemistry rises. The founding and developing of the chemical atomic theory, and its extension to all parts of chemical science, characterise this latest epoch, to which the period of Lavoisier's reform of chemistry was a necessary stepping-stone; it is, therefore, to be designated as the period of the Chemical Atomic Theory. An insight into the conditions which it involved being only possible by careful quantitative researches, the balance has been, since the time of Lavoisier; the most valuable instrument of the chemist. H. Kopp is, therefore, fully justified in naming the epoch which begins with the French *savant* the period of quantitative research. Of late years the first aim of chemistry, *i.e.* the exact determination of the composition of substances, has been accompanied by the investigation of the relations which exist between their physical properties and chemical composition. But the light of the atomic theory permeates the whole, so that one is forced to regard it as the guiding star of modern chemistry.

CHAPTER I

FROM THE EARLIEST TIMES TO THE BIRTH OF ALCHEMY

THE characteristics of this period, which have been already referred to, justify one in designating it the period of crude empiricism with regard to chemical facts. In sharp contrast with the disinclination of the Ancients towards experiment, through which alone the secrets of nature are to be unravelled, stood their great love of speculation, by means of which they did not hesitate to attempt an explanation of the ultimate reasons of all things. Aristotle, to whom the natural sciences owed the direction which they followed for a very long time, pointed to deduction as the road which should lead to the goal. Instead of drawing general conclusions from accurately observed facts, the Ancients preferred to advance from the general to the particular. The position of all the natural sciences in far-back times, especially that of chemistry, is sufficient to prove how the most mischievous errors crept in and became firmly established in consequence of following the purely deductive method.

The philosophical writings of the Ancients, especially those of the Greeks and Romans, give us a tolerably distinct idea of their theoretical views. Certain writings of Aristotle, and also the "περὶ λίθων" of his pupil Theophrastus, are of especial value for the criticism of the empirical chemical knowledge of these times. The works of Dioscorides on *Materia Medica* and particular chapters of the *Historia Naturalis* of the elder Pliny give us an exceptionally clear insight into the knowledge of the Ancients. Dioscorides, who was born about the middle of the first century at Anazarbos, enlarged his acquirements, already great, by experiences collected on long journeys. His fame as a physician holds good among the Turkish doctors to this day. The work of Pliny above-mentioned contains exceedingly valuable records of the state of scientific knowledge in his time; it also shows, however, that the author was by no means master of the immense amount of material which he had collected[1] from tradition, but which he had not really assimilated.

Theoretical Views upon the Composition of Substances, and especially upon the Elements.[2]

The question of the ultimate constituents of bodies, *i.e.*, of the elements which go to build up the world, occupied the minds of the oldest nations. To give an exhaustive description of their speculations on the point does not come within the scope of this work; what is wanted is rather to call special attention to those views which have exercised a permanent influence upon the chemical ideas of later times.

This applies in a particular degree to the doctrine of the elements, which originated with Empedocles, although it usually bears Aristotle's name; also, but to a much lesser extent, to the ideas of the older Greek philosophers regarding the original material of which the world, according to them, was built up. Views like that of Thales (in the sixth century B.C.), that water is the ground material, or those of Anaximenes and Heraclitus (in the same century), who ascribed to air and fire respectively the same *rôle*, have had no influence upon the development of chemical knowledge.

[1] Pliny the younger characterised the work of his uncle as "*opus diffusum, eruditum, nec minus varium, quam ipsa natura*," and similar admiration of it was expressed by other authors of the day. Our thanks are due to E. O. von Lippmann, who has recently published a memoir entitled *Die chemischen Kenntnisse des Plinius* ("Pliny's Knowledge of Chemistry"), in which the whole subject is treated in lucid style (*vide Mittheilungen aus dem Osterlande*, vol. v., p. 370).

[2] Cf. Kopp, *Geschichte der Chemie*, vol. i. p. 29; vol. ii. p. 267; also Höfer, *Histoire de la Chimie*, vol. i. p. 72.

Democritus (in the fifth century B.C.) also took a ground material as the basis of his speculations, but subdivided this further in that he imagined it to be made up of the smallest possible particles, of atoms, which differed from one another in form and size, but not in the nature of their substance. All the changes in the world consisted, according to him, in the separation and recombination of these atoms, which were supposed to be in a state of continual motion. This doctrine, which at first sight appears to accord with our present chemical atomic theory, but which in reality has nothing in common with the latter, was further developed by Epicurus, as may be well seen in the didactic poem of Lucretius, *De Rerum Natura.*

The four so-called "elements"—air, water, earth and fire—were regarded by that intellectually great philosopher, Empedocles of Agrigent (about 440 B.C.), as the basis of the world; but neither he himself nor Aristotle, who adopted these into his system of natural philosophy, looked upon them as different kinds of matter, but as different properties carried about by one original matter.[1] Their chief qualities (the *primæ qualitates* of the later scholastics) he held to be those apparent to the touch, viz., warm, cold, dry, and moist. Each of the four so-called elements is characterised by the possession of two of these properties, air being warm and moist, water moist and cold, earth cold and dry, and fire dry and warm. The differences in the material world were, therefore, to be ascribed to the properties inherent in matter. From the assumption that these latter can alter, there necessarily follows the immediate conviction that substances can be transformed, one into the other. It is easy to see how, when based upon speculations of this nature, the belief in the transformation of water into air should establish itself, for both have the property of moistness in common, while cold, the individual property of water, can be converted by the addition of heat into the second chief property of air. And it is not surprising that considerations of this kind on the

[1] Cf. the ingenious exposition by Th. Gomperz in his work "*Griechische Denker,*" p. 183 (Leipzig, Veit and Co.).

states of aggregation of matter should lead to the idea of transforming one kind of matter into another. It was doubtless by the generalisation of such views that the belief in the possibility of the transmutation of metals, which formed the chief feature of the alchemistic period, grew to the extent that it did.

Aristotle considered that his four elements were insufficient in themselves to explain the phenomena of nature; he therefore assumed a fifth one, termed *οὐσία*, which he imagined to possess an ethereal or immaterial nature and to permeate the whole world. As the "*quinta essentia*" this played an immense *rôle* among the followers of the Aristotelian doctrine in the Middle Ages, and gave rise to endless confusion, from the endeavours of many (who, unlike Aristotle, supposed it to be material) to isolate it.

There seems to be a high degree of probability in the assumption that Empedocles and Aristotle did not themselves deduce their theory of the elements, but derived it from other sources; thus the oldest writings of India teach that the world consists of the four elements mentioned above,[1] together with ether,[2] which last is most likely related to Aristotle's *οὐσία*.

It is unnecessary to point out how widely the above views of the Greek philosophers with regard to the elements deviate from the conceptions of modern chemistry.

With respect also to the meaning of the term "chemical combination," one meets, even if only occasionally, with opinions diametrically opposed to those obtaining at the present day; the formation of a substance by the interaction of others was looked upon as the creation of a new matter, and the destruction of the original substances from which it was produced was assumed. Everywhere men were contented with theoretical explanations, without attempting to prove their correctness by actual experiment. This want shows

[1] Instead of air, the element wind is given.

[2] So teaches Buddha (as Dr. Pfungst has been good enough to inform me); see the *Anguttara Nikâja*, vol. i. fol. c.e. Here consciousness is named as the sixth element.

itself very markedly in the manner in which the Ancients regarded the numerous chemical facts which they had learned by empirical methods, and mostly by accident.

The Empirical Chemical Knowledge of the Ancients.[1]

The Egyptians stand out prominently from among the earlier civilised nations as having usefully applied their knowledge of chemical processes, acquired by chance observations, to useful purposes; the needs of everyday life and the desire to make that life a comfortable one were the incentives.

Their country formed a kind of focus in which was concentrated the chemical knowledge of the time, if one may so designate an acquaintance with technical processes. The Egyptians already possessed at a very early date a large experience in the production of metals and alloys, in dyeing, in the manufacture of glass, and also in the making and application of pharmaceutical and antiseptic preparations. The chemical art proper, revered as "holy" (*ἅγια τέχνη*), was jealously guarded by the priesthood as a treasure at once precious and profitable. Only the elect might penetrate its mysteries. That laboratories, in which chemical operations of various kinds were carried out, actually formed adjuncts to the temples, is clearly proven by the inscriptions found in such chambers, *e.g.* at Dendera and Edfu.

There can scarcely be a doubt that the Phœnicians and Jews obtained their knowledge of the manufacture of important technical products from the Egyptians. In like manner, and to an even greater extent, was there a wealth of chemical experience laid open to the Greeks, and afterwards to the Romans, by reason of their close relations with the ancient country Chêmi (see p. 2, note 1). The writings of such eminent Greek philosophers as Solon, Pythagoras, Democritus and Plato, who succeeded in gaining the confidence of the Egyptian priesthood, contributed in no small degree to the spread of such practical knowledge.

[1] Cf. Kopp, *Gesch. d. Chemie*, vols. iii. and iv.; Höfer, *Hist.*, vol. i. p. 106 *et seq.*

But all the knowledge so gained was purely empirical; long ages were to pass before the various items of which it was composed were brought together under a general scientific standpoint. In this section of the book merely those portions of applied chemistry which were known to the Ancients will be treated of. That a people, so gifted as the Greeks were, should have failed to understand how to group together the numerous observations in those subjects which lay ready to their hand, and to draw conclusions from them, can only be explained by the whole tendency of their thought, and particularly by their undervaluing the inductive method. Aristotle's opinion that "industrial work tends to lower the standard of thought" was certainly of influence here. In accordance with this dictum the educated Greeks held aloof from the observation and practice of technical chemical processes; a theoretical explanation of the reactions involved in these lay outside their circle of interests. To this want of sympathy is certainly to be ascribed the fact that the discovery of even the most important chemical processes is but very seldom to be connected with the names of distinct historical persons; while, on the other hand, the old historians give detailed records of those men who advanced untenable opinions on the constitution of the world.

Before giving an account of the state of practical chemical knowledge in early times, it may be remarked in passing that much uncertainty often prevailed in consequence of different products being called by one and the same name. Substances were not distinguished according to their chemical behaviour, the investigation of which possessed no interest for the Ancients, but were classified according to their outward appearance and source, a confounding of similar or identification of dissimilar substances thus frequently resulting. Two samples of one and the same compound—soda, for instance—were looked upon as different, if the external appearance seemed to indicate a dissimilarity. Much discrimination has been found to be, and still is, requisite in order to clear up the indistinct points in the records of the old historians.

Metallurgy of the Old Nations.[1]

We find in the earliest records of the civilised nations (the Egyptians, Jews, Indians, etc.) an acquaintance with the working of different metals. By the younger of those nations mythical personages were held to have taught this art, *e.g.* Prometheus, Cadmus, etc., by the Greeks. If the translations of the Hebrew words in the Old Testament signifying "metals" are correct, then the Jews were acquainted with six, viz. gold, silver, copper, iron, lead and tin; this may be considered certain as regards the first four, which either occur native or are readily reduced from their ores. They are recorded in the Old Testament in the order just given.

The name "metals" is derived, according to Pliny, from the fact of their never occurring separately but in veins together, *μετ' ἄλλα.*[2] Even at that early period glance, ductility and hardness were held to be characteristics of a metal. With regard to the origin of metals and ores in the interior of the earth, the Ancients had formed the most extravagant conceptions; they believed, on the ground of Aristotle's weighty testimony, that they were produced by the penetration of air into the vitals of the earth, and consequently assumed that the amount of metal or ore increased as the mine proceeded inwards.

The Greeks, and especially the Romans, were intimately acquainted with many metallurgical processes; Dioscorides, Pliny and later historians give fairly exact data for the obtaining and smelting of ores, but not the slightest attempt is made to explain the chemical processes which this involves.

The noble metals gold and silver, whose stability in the

[1] The following works have been used for reference:—R. Andree, *Die Metalle bei den Naturvölkern* (Veit und Co., Leipzig, 1884); Beck, *Geschichte des Eisens* (Vieweg, Braunschweig, 1884; 2nd. ed. 1891); O. Schrader, *Sprachvergleichung und Urgeschichte* (Jena, 1883); and also various treatises by K. B. Hofmann, to whom the author is greatly indebted for much information on the subject.

[2] Herodotus gives *μέταλλον* as signifying a mine.

fire had not escaped the Ancients, were those earliest known (in prehistoric times), and were highly valued; the fact of their occurring native, and the ease with which they can be worked, afford a sufficient explanation of this.[1] The exceeding malleability of gold excited, the astonishment of the older nations in a high degree, and rendered possible the gilding of objects by covering them with thin plates of the metal. The later discovery of affixing a layer of gold by means of the amalgamation process was known considerably before the time of Pliny.

In the second century B.C. we meet with the first records[2] of a cupellation process, by which gold was freed from admixtures; in fact, an operation similar to the so-called lead process was then carried out, gold dust being melted with lead and salt for a number of days. The purification of gold by means of mercury was also well known in Pliny's time.

Silver, which the enterprising Phœnicians are supposed to have supplied to the other civilised nations from Armenia and Spain, where rich silver ores occur, was, according to the record of Strabo, *i.e.* at the begining of our era, purified in a precisely similar manner to gold, viz. by fusion with lead. The separation of silver from gold does not appear to have been known before our era, at any rate an extant record[3] states that Archimedes was not possessed

[1] The gold mines of Nubia (the Egyptian name *nub*, *i.e.* gold, is perhaps connected with the designation of that country) were worked very extensively by the Egyptians. According to the records of Agatharchides and of Diodorus Siculus, in which pity is expressed for the slaves employed in the work, the finely ground gold ore was washed out and the heavy residue melted. In the time of Rameses II. the mines yielded gold to the value of £125,000,000 sterling per annum. The gold-producing land of Ophir, from which the Phœnicians obtained the precious metal, is supposed to have been in India, Midian (Arabia), or on the east coast of Africa. The same energetic trading nation opened up for the Greeks the first gold mines on the island of Thasos.

[2] This record, which originated with Agatharchides, is to be found in Diodorus.

[3] Archimedes attempted to determine whether the crown of King Hiero contained silver, and, if so, how much; this problem he tried to solve by taking the specific gravity, not by chemical means.

of the means to accomplish this. From indications which Pliny gives, however, it appears that in his time a kind of cementation process was practised, which probably consisted in the treatment of silver containing gold with salt and alum shale. Moreover, an amalgam of gold and silver was regarded in ancient times as a particular individual metal, being termed *asem* by the Egyptians, and ἤλεκτρος by the Greeks (amber being distinguished as τὸ ἤλεκτρον). From this also it may be concluded that at that time no method was known of separating the metals.

The data concerning copper (termed χαλκός, *aes*[1]), which has been known from very primitive times (being first found in the neolithic stone age), frequently refer to its alloys with other metals, especially to bronze; the latter, as is well known, was very early used for making weapons, ornaments and utensils. Copper, which was universally employed in prehistoric times, was found native in many places (*e.g.* in Egypt), or was readily smelted from malachite or similar copper ores. All the civilised nations, which have been named, were acquainted with bronze before they had learnt to prepare its other constituent, metallic tin, no mention of which is made in old Egyptian records. With regard to the smelting processes by which the "*aes*" of the Ancients was obtained, nothing certain is known.

Iron, the extraction and working of which was not discovered till after that of copper and bronze, but which, nevertheless, goes back to very ancient times also,[2] was prepared in smelting furnaces; the old authors do not, however, give any particulars as to the actual process.[3] The ores used are

[1] The Roman *aes* has the same stem as the Sanscrit word *ayas*, signifying ore; the latter designation *cuprum* for copper is an abbreviation of *aes cyprium* (so called because of its occurrence in Cyprus).

[2] According to *Lepsius*, iron has been in use in Egypt for more than 5000 years, having served primarily for the manufacture of hard instruments, while utensils of all kinds were made from bronze.

[3] Old Roman smelting furnaces with their appurtenances have recently been excavated near Eisenberg in the Pfalz. The form of apparatus used by the Egyptians for the smelting of iron can be arrived at approximately from inscriptions, etc.; it is worthy of note that bellows of the same shape as those of Ancient Egypt are in use in the interior of Africa at the present day.

supposed to have been brown iron ore and magnetite; that meteoric iron was first employed is an improbable and unproven assumption. The tempering of iron was early learnt in Ancient Egypt; even in the time of Pliny the undesirable property of impure iron, which we now term brittleness, was known, and its capability of assuming the peculiarity of the magnet stone when brought into contact with the latter, was also observed.

Lead[1] was likewise known from very early times, having been prepared and turned to good account, by the Greeks and Romans more especially. Little, however, is known with regard to the smelting processes for it, Pliny's records on the subject being indistinct; but the smelting was probably carried out on a refining hearth. On the other hand we have many details as to the use of lead for making water pipes, writing tablets, coins, etc. Soldering with lead or with an alloy of lead and tin was also well known. Since cooking utensils were often made of lead, symptoms of lead poisoning occurred frequently; but notwithstanding this, the metal was prized as a medecine.

Recent discoveries in Egyptian tombs have brought to light the fact that tin also was prepared fairly pure in olden times, and that it found numerous applications. Among the Romans lead and tin were distinguished from one another as *plumbum nigrum* and *plumbum candidum*.[2] The alloy of the two together, *i.e.* solder, played, as already mentioned, an important part in technical work. Still older and of even greater significance was the use of bronze,[3] which one meets with among the most ancient civilised nations.

[1] Cf. K. B. Hofmann's *Das Blei bei den Völkern des Alterthums* (Berlin, 1885).

[2] The word *stannum*, which now denotes tin, appears in Pliny's time to have signified an alloy of tin and lead. Whether the *κασσίτερος* of the *Iliad* stood for tin is likewise highly problematical. It is equally uncertain from whence the Phœnicians obtained this metal (or an alloy of it); whether from India, with which they had commercial relations, or from Britain and Iberia. The similiarity between the Sanscrit word *kastira* and the Greek word *κασσίτερος* has been used as an argument in favour of the former assumption (cf. A. v. Humboldt, *Kosmos*, ii. § 409).

[3] K. B. Hofmann considers that the name bronze, with regard to the

Zinc,[1] as an individual metal, was certainly not known to the Ancients, but its alloys with copper (χαλκός, ὀρειχαλκος) found the widest application.

Brass, the first description of which is given by Aristotle as the "metal of the Mosynoeci" (from which the German word *Messing*, signifying brass, is undoubtedly derived), was for long regarded as copper which had been coloured yellow by fusing it with an earth (*cadmia*);[2] it was only recognised as an alloy at a much later date. The change in colour produced in copper by certain additions to it played—in the transmutation of metals—an important part in the alchemistic age.

The first records as to mercury are to be found in Theophrastus (about 300 B.C.), who gives its preparation from cinnabar by means of copper and vinegar, and terms it "liquid silver." Dioscorides describes the production of mercury, which he at first termed ὑδράργυρος, from cinnabar and iron, *i.e.* by a process of simple elective affinity, without, however, making the slightest attempt to explain the process. For the carrying out of this operation, an exceedingly imperfect distilling apparatus was used. Pliny makes mention both of the purification of the metal, by squeezing it through leather, and also of its poisonous nature. It did not escape the Ancients that other metals, gold in especial, were altered by mercury (cf. p. 12); indeed Vitruvius gives a minute recipe for the recovery of gold in worn-out sewn draperies by means of it.

An account will be given later on of several metallic compounds known in ancient times.

origin of which there has been much dispute, is probably derived from the word βρουκήσιον, meaning an alloy, a word possibly borrowed from the Persian. The view, held even so early as in Pliny's time, that "bronze" was derived from (*aes*) *Brundusinum*, has been proved untenable.

[1] Cf. K. B. Hofmann's paper in the *Zeitschrift fur Berg- und Hüttenwesen*, vol. xli. Nos. 46—51.

[2] Even so early as 300 B.C., "*cadmia*" was famous as a medicine. The word likewise means "tutty" (oxide of zinc), or also rich zinc ore. According to K. B. Hofmann, it is not improbable that *galmei* (cadmia, calamine) is derived from *cadmia*; *i.e.* those three terms appear to be synonymous.

Beginnings of Technical Chemistry among the Ancients.

The Manufacture of Glass.—The art of making vessels from glass originated in China and Egypt, and had for a long time its chief habitat in Thebes; from there it spread to the Phœnicians and other Eastern nations, the Greeks first acquiring it—so far as actual proof goes—in the fifth century B.C. Pliny is the first to give a distinct account of the preparation of glass by fusing sand and soda together.[1]

The artificial colouring of glass by metallic oxides, especially oxide of copper, was very early discovered. Many of the remains which have been found in Ancient Egypt indicate that the manufacture of glass must at that time have attained to a high degree of perfection, methods for producing enamels and artificial gems being then known. Pliny states that beryl, opal, sapphire, amethyst, etc., could be imitated, but that at the same time these imitations were distinguishable from the real stones through being softer and lighter.

The first preparation of glass presupposes in any case an acquaintance with soda or potash; the former of these was found as a natural product on the shores of certain lakes, *e.g.* in Macedonia and Egypt, while carbonate of potash was obtained from a very early period by lixiviating the ashes of plants, and also, according to Dioscorides, by igniting tartar. These two salts[2] were frequently mistaken for one another on account of their similar action. They were largely used for the preparation of soap, and also directly for washing clothes, cleansing the skin and the teeth (just as the ash of tobacco, which is rich in carbonate of potash, is often employed as a tooth-powder at the present

[1] The discovery of glass in Egypt was undoubtedly accidental, soda having been added as a flux to sand containing gold for the purpose of extracting the latter.

[2] The Hebrew *neter* probably denotes soda, while the Latin *nitrum* is employed by Pliny for both alkaline salts. The designation *alkali* came originally from the Arabs.

day), and also as ingredients of medicines. Lastly, the ashes of plants and saltpetre were much prized as effective manures.

To the art of pottery must be ascribed an age at least as great as that of the preparation of the nobler metals and of glass. Even the old Egyptians understood how to coat their originally simple earthen vessels with coloured enamel. At a later date the ceramic industry prospered among the Etruscans, and also in many towns of Southern Italy and Asia Minor. Porcelain, which was discovered and employed by the Chinese, remained entirely unknown to the older civilised European nations.

The Manufacture of Soap.—Of no slight interest is the fact that the saponification of fats by means of alkalies, with the object of preparing soap—that is to say, a complicated process of organic chemistry—, was already practised in ancient times. Pliny's records on the subject make it probable that in Germany and Gaul soap was prepared from animal fat and the aqueous extract of ashes, the latter being strengthened (rendered caustic) by the addition of lime. Further, there was a distinction drawn between soft and hard soaps, according as potash or soda (the latter being obtained from the ashes of shore plants in Gaul) was used in the preparation.[1]

Dyeing likewise belongs to the arts which the Egyptians, Lydians, Phœnicians and Jews greatly developed. They knew how to fix certain dyes on cloth by means of mordants, alum [2] playing an important part here; indeed the dyeing of purple had attained to a high state of perfection among the Phœnicians. Pliny mentions the application both of madder dye and of litmus (the gatulian purple). Indigo blue seems to have been more used at that time for painting than for

[1] From K. B. Hofmann's researches it appears to be doubtful whether the *sapo* of the Romans meant soap, and not rather a depilatory.

[2] Under στυπτηρία or *alumen* of the Ancients must be understood substances of astringent properties generally, although alum itself is what is usually meant; being prepared from alum shale, it contained green vitriol as an impurity.

dyeing, but with this exception mineral substances were employed as paints. The principal of these in Pliny's time were white lead, cinnabar, vermilion, smalt,[1] verdigris, red oxide of iron and soot. This last, mixed with gum, also served as ink. Numerous recent researches[2] have proved that sulphide of lead (galena) formed the basis of the much-used old Egyptian cosmetic *mesdem*,—and not native sulphide of antimony, as was at one time supposed. *Mesdem* was also a highly prized medicine. The use of preparations of antimony belongs to a later period.—The sulphides of arsenic, realgar and orpiment, served both as pigments and as medicines, although their poisonous action was known. In short, the Ancients had access to a considerable number of colouring chemical compounds, some of these being the earliest chemical preparations to be manufactured on a large scale.

As has been already indicated, the use of such artificially prepared products in medicine also extends to a period very far back, even although, in referring to this, one can only speak of the first beginnings of a pharmaceutical chemistry. But a connection between the chemical art and pharmacy established itself very early indeed, *e.g.* among the Egyptians, who were doubtless the first to employ actual chemical preparations for medicinal purposes. Thus verdigris, white lead, litharge, alum, soda and saltpetre served for the making of salves and other medicaments, while the preparation of lead plaister from litharge and oil was much practised in the time of Dioscorides. Iron rust was a very old medicine, its use being ascribed to Æsculapius, while sulphur and copper vitriol containing iron (*chalcanthum*) were valuable ingredients of the medical treasury

[1] Davy found cobalt in certain antique glasses, and assumed from this that smalt had been used in their manufacture. According to Fouqué (*Compt. Rend.*, vol. cviii. p. 325) Egyptian glass contained only oxide of copper as the colouring material; but vitrifiable pigment containing cobalt has been again found recently in small Egyptian statuary.

[2] Collected and critically examined by K. B. Hofmann in his paper:—*Ueber Mesdem* (*Mittheilungen des Vereins der Ärzte in Steiermark*, 1894, Nos. 1 and 2).

before our era; but the important preparations of antimony and mercury can be proved to have first come into notice in the alchemistic period.

Most of the officinal compounds just referred to were also used for other purposes, as has already been mentioned in a few cases. The combustion-product of sulphur, for instance, was employed for fumigation (*vide* Homer), for the purification of clothes, the conservation of wine, and for destroying impure colours (Pliny), while copper vitriol and alum were used in dyeing operations.—In closing this short account of the knowledge possessed by the Ancients with regard to chemical compounds, the following substances may be mentioned, substances whose practical application dates from a very early period. In ancient times lime was burnt, and after being slaked, was used for preparing mortar, and also, as already stated, for causticising soda (cf. p. 17). Of the acids, acetic acid [1] in the form of crude wine vinegar was the earliest known, its presence being assumed in all acid plant juices. The mineral acids, which are of such importance in technical chemistry, were only discovered in the succeeding epoch.

Other organic compounds known at the beginning of our era, and doubtless even before then, were sugar (from the sugar-cane), starch [2] (from wheat), many fatty oils (from seeds and fruits, the oil being extracted either by pressing or by boiling with water), petroleum, and oil of turpentine, which last was obtained by the distillation of pine resin in very imperfect apparatus.[3] Of the fatty oils, olive, almond and castor oils, etc., were known and used for a variety of purposes, the first-named—*e.g.*—for extracting perfumes from

[1] The Ancients had the most extravagant ideas with regard to the solvent power of vinegar upon mineral substances, as may be gathered from the concordant statemants of Livy and Plutarch that Hannibal, in his passage across the Alps, cleared the way of rocks by means of it. The story which Pliny tells of Cleopatra may also be recalled here,—how she, in fulfilment of her wager to consume a million sesterces at one meal, dissolved costly pearls in vinegar and drank the solution.

[2] ἄμυλον, so called from its being prepared without millstones, and the production of which is described by Dioscorides.

[3] Prof. K. B. Hofmann kindly tells me that the earliest account of a *destillatio per decensum* is to be found in Aëtius (Aldine Ed., fol. 10).

flowers, leaves, etc. Ethereal oils were also known and employed in large number.—The animal fats played an important part in medicine, and Pliny's mention of sheeps-wool grease, among other things, is noteworthy here, seeing that it has recently been brought into use again in the form of lanoline.—Pliny does not seem to have been acquainted with cane sugar; but one frequently comes across passages in his writings referring to the occurrence and remarkable actions of vegetable poisons (alkaloids).

Such compounds as spirits of wine, carbonic acid, etc., which are formed in many processes of fermentation, *e.g.* in the making of wine, beer and bread, remained unknown to the Ancients. It is true that they noticed in these cases and also in others—natural emanations of gas, for instance—the presence of a kind of air prejudicial to breathing and even under certain circumstances fatal to life, but it did not occur to them to recognise in this a gas different from atmospheric air.

This lack of the gift of observation, this disinclination to go to the root of any phenomenon, in fact, a certain indifference with regard to natural events, are characteristics of the attitude of the Ancients towards nature. Instead of experimenting with natural products, they infinitely preferred to call speculation to their aid, so that the most superficial observations gave rise to opinions which, when uttered by high authorities, attained to the dignity of dogmas. How otherwise than from an extreme lack of the desire of observation can one explain Aristotle's assertion that a vessel filled with ashes will contain as much water as one which is empty? A further instance of the credulity of that time is given in the conviction expressed by Pliny, and universally held, that air can be transformed into water, and *vice versa*, that earth is produced from water, and that rock-crystal also proceeds from the latter. The assumption that water can be transformed into earth has often come up again at later periods, having exercised the minds of people even in comparatively recent times; as it subsequently assumed the form of an important question of dispute, it will be referred to in detail later on.

CHAPTER II

THE AGE OF ALCHEMY

IN the introduction to this book Egypt is spoken of as the mother-land of Alchemy. The University of Alexandria was especially instrumental in the propagation of the latter during the first centuries of our era; it was the carrier and intermediary for the alchemistic doctrines, more particularly at the time of the fall of the Western Roman Empire.

The attempts to convert the base metals into the noble ones had their origin in superficial observations, which appeared to give a strong support to the belief in this transmutation. Among such accidental observations was that of the deposition of copper upon iron utensils left in copper mines from the waters which accumulated there. What more natural than to conclude that a transmutation of iron into copper had occurred? For the production of gold or silver from copper, the transformation of the latter into yellow or white alloys by means of earthy substances such as calamine or arsenic appeared to give warrant. Finally, the fact that a residue of gold or silver remained behind when an alloy with lead or an amalgam with mercury was strongly heated, indicated the generation of those noble metals.

To these considerations of a practical nature, which strengthened the conviction as to the transmutation of metals, but which inferred a gross self-deception on the part of the observer himself—to say nothing of their being turned to good account by crafty knaves—there came to be allied, in this epoch for the first time, the tendency to group together chemical facts from common points of view.

It was precisely in the mode in which it was attempted to explain the composition of the metals that there lay a powerful and ever-active charm, leading to the belief in the ennobling of the baser metals and to continually repeated efforts to achieve this. The first beginnings in an experimental direction, which we meet with early in the alchemistic period, although very incomplete, indicate nevertheless a distinct step in advance as compared with the deductive method which had hitherto reigned supreme, and whose fruits consisted, for the most part, in the setting up of mystic cosmogonies. The few observations which were made remained, however, isolated—that is, were not grouped together in a connected manner.

That the attempts to attain to a knowledge of the processes of nature by the inductive method were but slight at best in the alchemistic period, is explained by the supremacy of the Aristotelian doctrine, which, amalgamated with the Neo-Platonic philosophy, trammelled the minds of men throughout almost the whole of the Middle Ages. Even the Christian theology had to compromise with this system, the product of the joint work being scholasticism, which imprinted its stamp upon all the mental efforts of that time and prevented their free development. The relation of the alchemistic tendencies to the Aristotelian philosophy has been already indicated (p. 7).

The limitation of this epoch between the first appearance of alchemistic conceptions (in the fourth century) and the bold attempt of Paracelsus to call in chemistry to the aid of medicine (in the beginning of the sixteenth century) is thus a natural one, since, during the whole of this time, one and the same keynote runs through all the questions bearing upon chemistry, viz. the idea of ennobling the metals. People were so convinced of the practicability of this for many centuries, that almost every one who devoted himself to chemistry, and many others besides, strove hopefully towards this long-desired goal. The early mixing up of astrological and cabalistic nonsense with these alchemistic endeavours marks very distinctly the degeneration of the latter.

Alchemy by no means ceased to exist on the appearance of the new iatro-chemical doctrines, but gradually receded as chemistry became more of a science. True, its seductive problems are often seen to throw a weird lightning flash on the chemists' camp, and to exercise upon even the most eminent of them an undoubted influence; but upon the main lines which chemistry has followed ever since the time of Boyle, the phantasies of alchemy have had no appreciable effect. Notwithstanding, however, that this influence was but slight, a short account of the position of alchemy during the last four centuries cannot properly be omitted, and will therefore be added as an appendix to this section of the book.

GENERAL HISTORY OF ALCHEMY.[1]

Origin[2] *and First Signs of Alchemistic Efforts.*

The sources from which the belief in the practicability of the transmutation of metals was nourished, and which in the course of centuries gradually expanded into a broad stream of the most mischievous errors, have their origin in the gray mists of antiquity. No actual proof of these must be looked for: we depend, with regard to them, upon mythical and mystical traditions. The first historical sources, too, are small in number and very obscure. But we find among various nations distinct signs of alchemy having been pursued as a secret science and having been held in honour.

When one recalls to mind that Ancient Egypt was a

[1] Cf. Kopp, *Gesch. d. Chemie*, vol. i. p. 40, *et seq.*; also his work, *Die Alchemie in älterer und neuerer Zeit* (Heidelberg, 1886).

[2] Cf. particularly M. Berthelot's *Les Origines de l'Alchimie* (Paris, 1885) and his *Introduction à l'Étude de la Chimie des Anciens et du Moyen-âge* (Paris, 1889); also H. W. Schaefer's admirable treatise:—*Die Alchemie; ihr aegyptisch-griechischer Ursprung, &c.* (Fleusburg, 1887; School-calendar). M. Berthelot has indeed rendered signal service by his publication and critical revision of old alchemistic works, such as the Leyden papyrus, and Greek and Arabic MSS. Quite recently, in conjunction with certain philologists, he has given to the world the *Collection des Anciens Alchimistes Grecs* and *La Chimie en Moyen-âge.*

centre of the higher culture, and, especially, that it was a country where the chemical art was practised, one feels no surprise that the earliest reliable records of alchemy are to be found there. Egyptian sources, partly such as have been preserved to us by the Leyden papyrus (about 300 A.D.), and partly the writings of the Alexandrians from the third to the seventh century A.D., constitute the most valuable aids at our disposal for a historical proof of the origin of alchemy. The influence of the doctrines and practical recipes contained in these works upon the alchemy of the entire Middle Ages is easily demonstrable.

The tradition according to which, among other knowledge, the art of ennobling metals had been brought from heaven to earth by demons, was universally diffused in the first centuries of our era; Zosimos of Panopolis states that the mystical book from which this art was to be learned was termed, *χῆμευ* and the art itself *χημεία*. This myth doubtless sprang from one exactly similar which is to be found in the apocryphal book of Enoch; indeed indications of it are to be met with even in Genesis. The later alchemists were inclined to refer the origin of alchemy to the time before the flood, thinking that a special sanctity would accrue to their art from this great age. Moreover, they wrote down various biblical characters as alchemists, on the authority of certain passages in Holy Writ, for instance, Moses and his sister Miriam, and the Evangelist John. When legends such as these found credence in the Middle Ages, it is hardly surprising that the records as to the origin of this art, which remain to us from ancient times, should have upheld their authority over a very long period.

The first personality with which the origin of alchemy is associated is that of Hermes Trismegistos,[1] "the three times great," who was said to have been the author of books upon the holy art; he was, moreover, generally reverenced as the discoverer of all the arts and sciences. The then popular expressions "hermetic," "hermetic writings" and "hermetic

[1] This designation is probably first found in Tertullian (end of the second century of our era). Cf. Schaefer, p. 4, &c.

art"[1] recalled this undoubtedly mythical personage even so recently as in the present century. In Romish Egypt pillars were erected in honour of this Hermes, upon which alchemistic inscriptions were cut in hieroglyphics.

Who then was this Hermes? One has to seek in him, as ancient traditions indicate almost certainly, the personified idea of strength, *i.e.* the old Egyptian godhead Thot (or Theuth), which, when endowed with the serpent-staff as the symbol of wisdom, was compared by the Greeks with their Hermes, the latter designation being thus transferred to the Egyptian god.[2] Alchemy, as a holy and divine art, whose special task consisted in the preparation of the metals, was kept secret and fostered by the priesthood, the sons of kings alone being permitted to penetrate its mysteries. The estimation in which it was held rose in exact proportion with the belief that Egypt owed to alchemy its riches.

When and in what way the influence of other nations made itself felt upon the alchemy of the Egyptians, it is difficult to determine. The Babylonish astrologers, without doubt, undertook the fusion of astrology and magic; in particular, the mutual relations between the sun and planets and the metals, which were taken for granted for so many centuries, were of old Babylonish origin. According to the account of the Neo-Platonist Olympiodor (in the fifth century A.D.), gold corresponds to the sun, silver to the moon, copper to Venus, iron to Mars, tin to Mercury, and lead to Saturn.[3]

Certain passages in the works of Dioscorides, Pliny, and the Gnostics enable us to conclude that the transmutation of copper into silver and gold was regarded as an ascertained fact during the first centuries of our era.[4] The

[1] The designation "spagiric art" (from σπάω, to separate, and ἀγείρω, to unite) occurs for the first time in the sixteenth century.

[2] This identity is confirmed by the fact that, in the inscriptions on the temple of Dakke on the Nile dedicated to *Thot*, the three names *Thot*, *Hermes* and *Mercurius* occur, the first in hieroglyphics, the second in Greek, and the third in Latin (cf. Schaefer, p. 7).

[3] Even in Galen are to be found statements with regard to the influence of the planets upon the metals.

[4] The Chinese also busied themselves with alchemy at that time, the

"duplication of the metals," which is to be found in the writings of first-century authors, and which also plays a part in the Leyden papyrus, likewise refers to the transmutation of metals. The designation of this art as "Chemia" probably appears for the first time in an astrological treatise of Julius Firmicus (in the fourth century).

Berthelot has made a careful study of the Leyden papyrus (found in Thebes in the third century A.D.), and has compared it with later alchemistic writings. This has led him to the conclusion that an intimate connection existed between the industrial production of the noble metals, the dyeing of fabrics, and the colouration of glass (whence the frequent expressions:—*Tingiren der Metalle; Tincturen*, etc.). The alleged processes of transmutation, which were currently believed for hundreds of years, consisted in artifices for debasing the noble metals, but at the same time imitating their appearance as nearly as possible in less costly alloys. It is quite likely that, as time went on, the idea took possession of many kinds that the gold and silver were newly created by some supernatural aid. It would thus seem as if alchemy originated in the fraudulent practices of gold-workers.

The records of the study of alchemy go on increasing from the 4th century, much information regarding it being found in the writings of the Alexandrian *savants* of that time, especially in those of Zosimos, Synesios and Olympiodor. In addition to these, various pseudo-authors, especially pseudo-Democritus, are cited here as witnesses to the spread of alchemy; the philological-historical critic is not yet, however, in a position to fix the dates at which these works were written. In the Middle Ages people did not hesitate to accept the writings of the false Democritus, and also those of a pseudo-Aristotle, as originating from the ancient philosophers Democritus and Aristotle themselves. The later alchemists also fathered counterfeit writings upon Thales, Heraclitus and Plato, in order to make use of the great authority of those names for their own ends.

transformation of tin into silver, and of the latter into gold, being held to have been actually accomplished.

Zosimos of Panopolis, a voluminous author of the fifth century, who was looked upon as one of the greatest authorities among alchemists both of that date and of later times, is said to have written twenty-eight books treating of alchemy, of which, however, only small fragments remain. His mystical recipes are quite unintelligible, and yet he distinctly speaks of the fixation of mercury, of a tincture[1] which changes silver into gold, and also of a divine water (panacea). Reference is frequently made to the work of the pseudo-Democritus, φυσικὰ καὶ μυστικά. The graphic and mysterious language of Zosimos appears to have exercised a permanent influence upon the works of the later Alexandrians, and also, subsequently, upon those of the alchemists of the Middle Ages.

The end of the fourth century and the beginning of the fifth constitute, without doubt, the period in which the study of alchemy reached its zenith among the Alexandrians; but the works of Synesius upon alchemy and magic, and those of Olympiodor, who bore the surname of "ποιητής," *operator*, do not yield much certain information with regard to definite operations or to the knowledge of chemical facts. How many works which would have been valuable for the history of chemistry were lost through the destruction of the Serapeum, which marked the completion of the overthrow of Hellenic culture in Egypt, cannot at this distance of time be estimated. That all acquaintance with chemical operations, and chemical knowledge and skill generally, were not thereby quite exterminated was due to the relations which were before that developed between the Alexandrians and the Byzantine *savants*; for, from the sixth century on, applied chemistry, which may also be said to include alchemy, found a foothold at Byzantium. Even in Egypt itself the knowledge of chemistry was not completely extirpated by that catastrophe, but continued to exist by fostering certain branches of industry, which, without it, could never have been developed. Lastly, the conviction that metals could

[1] The term "*mercurius philosophorum*," which is often found in later writings, was first used by Synesius.

be transmuted had fixed its roots too deeply to allow of this art, by which endless riches were to be attained, dying a natural death.

The Alchemy of the Arabians.

The germs of chemical knowledge, which had lain hidden in the brains of a few philosophers, attained to a marvellous growth among the Arabians, who overran and conquered Egypt in the seventh century; it might have appeared much more likely that they would crush the arts and sciences rather than be the instruments of their resurrection. It was certainly curious that this people, originally strangers to science, should assume the care of it and cause it to flourish in an undreamt-of degree, at a time when culture remained at its lowest ebb in most European countries, and everything had to give way to the pressure of the conditions produced by the migration of the nations.[1]

The first appearance of the Arabians in Egypt, where they destroyed much priceless literary treasure by fire, did not seem to herald any such change of opinion. They very soon learnt, however, to assimilate the elements of the education of the conquered peoples,[2] so that we find (especially after the conquest of Spain, in the beginning of the eighth century) many cities of learning springing up; to these in the following centuries the European nations—especially France, Italy and Germany—sent crowds of earnest students,

[1] Alex. von Humboldt gives expression to this point in the following words: "The Arabians, an original Semitic stock, partially drove away the barbarism which had overwhelmed Europe for two centuries, convulsed as it had been by revolutions. They turned to the everlasting springs of Greek philosophy, and thereby assisted not only in preserving the culture of science, but in widening it and opening out new paths to the investigators of nature."

[2] Reference may just be made here to the important part played by the Nestorians in engrafting the scientific spirit upon the Arabians, and in enriching them with practical chemical knowledge. The latest researches of Berthelot and others leave no doubt that the Arabians derived from the Syrians much—if not indeed the greater part—of their knowledge of chemistry.

who applied themselves, for the most part, to the study of medicine, mathematics and optics. From the Arabian universities of Cordova and other Spanish cities, where alchemy was also ardently studied, it made its way to the other western nations, among which it attained to its full development in the thirteenth century.

A renown quite unexampled, and an authority which continued all through the Middle Ages, were attained by the physician and alchemist Dschafar, afterwards known to western nations by the name of Geber. About his life (he is supposed to have lived in the ninth and tenth centuries) nothing is known. It is possible, too, that Geber himself has been sometimes confused with his pupil Dschabir of Tharsis.

There can indeed be no dispute that with the name *Geber* was propagated the memory of a personality, with which the chemical knowledge of the time was bound up. But Berthelot's recent researches [1] have proved that the Latin writings hitherto ascribed to Geber cannot have come from him. The oldest of these—the celebrated *Summa Perfectionis Magisterii*—was not written before the middle of the fourteenth century; and the *De Investigatione Veritatis* and *De Investigatione Perfectionis Metallorum*, formerly regarded as genuine, belong to an even later date. In fact, the whole of what were supposed to be Geber's writings are apocryphal.

The Arabic MSS. of the real Geber, which Berthelot's investigations have now brought to light, prove that he did not really profess the knowledge and the opinions with which he has been credited. On the contrary, we find Geber adhering closely to the Græco-Alexandrian alchemists, and bringing forward many mystical views, *e.g.*, the belief in the influence of the planets upon the metals. There is no distinct indication in his writings of the theory of the metals hitherto ascribed to him (see below). Geber can therefore no longer be regarded as the author of the Latin treatises with which, up to now, his name has been associated. These

[1] See p. 23, Note 2. Cf. also two papers by Berthelot in the *Revue des deux Mondes*, Sept. 15th and Oct. 1st, 1893.

writings contain, in fact, the collected knowledge of the four or five centuries after his time.

The disciples of Geber, famous Arabian physicians like Maslema, Rhazes, Avicenna, Avenzoar, Abukases and Averrhoes, may possibly have exercised a retarding influence upon the development of medical science and of pharmacy. And it is extremely doubtful whether they advanced chemistry in any material degree. It is worthy of note that Rhazes distinctly assumes the transmutation of metals, while Avicenna disputes it.

Alchemy among the Christian Nations of the West during the Middle Ages.

The doctrines of the Egypto-Greek and Arabian alchemists gradually penetrated into France, Italy and Germany, certain Byzantine *savants*—Michael Psellus among them—also contributing to the spread of alchemistic ideas. Eastern influence is recognised distinctly for the first time in the earliest appearance—of which there is clear proof—of an alchemist in Germany at the court of Adalbert von Bremen (about 1063), as recorded by Adam von Bremen; a baptised Jew named Paul gave out that he had learnt in Greece the art of transmuting copper into gold, and he appears to have imposed upon the above-named ecclesiastical prince. The next certain records of alchemistic endeavours in Germany date from the thirteenth century, at which period alchemy was studied by men famous for their learning, and was consequently developed in a high degree.

The transformation of the base metals into the noble by means of the philosopher's stone formed at that date the cardinal point towards which all chemical knowledge was directed. Vinzenz of Beauvais[1] (in the first half of the thirteenth century) and, after him, men like Albertus Magnus, Roger Bacon, Arnaldus Villanovanus and Raymund Lully, whose chief works belong to the same century,

[1] Vincentius Bellovacensis.

regarded the transmutation of metals as an incontrovertible fact. These maintained that the philosopher's stone did exist, and was endowed with the most marvellous powers (see below), their dogmas being based upon those of the Aristotelians and of the Egypto-Greek alchemists. In addition to these, the most distinguished representatives of chemistry, all of whom belonged to the priestly class, must be mentioned the famous Thomas Aquinas; the latter did not indeed materially advance the knowledge of chemistry, but he stood up at various times for the truth of the doctrine of transmutation of metals.

The influence of the four men above-mentioned upon the history of chemistry renders biographical notices of them desirable; their views upon the alchemistic problem, and also their very considerable practical knowledge, will be treated of under special sections. Their writings have to be criticised with some caution, since many of the alchemistic treatises of later times were given out to the world under their names.

Albertus Magnus, or, more properly, Albert von Bollstädt, born at Lauingen on the Danube in 1193, taught philosophy, grammar, alchemy, etc., publicly as a Dominican in Hildesheim, Regensburg, Cologne and Paris, and became Bishop of Regensburg in 1260. He retired, however, to the cloister five years later, and died in the Dominican convent of Cologne after having devoted himself for fifteen years to scientific work. Albertus Magnus was held, both by his contemporaries and still more during the later Middle Ages, as a man of the greatest erudition and widest acquirements, the degrees of which are given by Tritheim, an author of the fifteenth century, in the following words: *Magnus in magia naturali, major in philosophia, maximus in theologia.* His noble character also earned for him the highest respect. Of his numerous memoirs, the two—*De Alchymia* and *De Rebus Metallicis et Mineralibus* are of the most value for adjudging his position with regard to alchemy.

Roger Bacon was born in Somersetshire in 1214, and

studied science, as well as theology, both at Oxford and Paris. The veneration felt by posterity for his marvellous and many-sided knowledge is shown by the title which it conferred upon him of *Doctor Mirabilis.* Since he did not hesitate to oppose in many points the orthodox beliefs of his day, he was subjected to bitter persecution and penalties. His death probably occurred in the year 1294.

His firm belief in the power of the philosopher's stone, not only to transform a million times its own weight of base metal into gold, but also to prolong life, seems to us incomprehensible when contrasted with the otherwise enlightened views which he held and propagated. This undisguised recognition of miracle-working, and this bias towards the marvellous, are directly opposed by the fact that Roger Bacon taught the working out of carefully devised experiments as a special kind of research, by which new data for the knowledge of nature should be acquired. He is to be regarded as the intellectual originator of experimental research, if the departure in this direction is to be coupled with any one name—a direction which, followed more and more as time went on, gave to the science its own particular stamp, and ensured its steady development. The most important works of Roger Bacon are the following:—*Opus Majus; Speculum Alchemiæ;* and *Breve Breviarium de Dono Dei.* He did not apparently do much towards the spread and development of practical chemical knowledge.

In the life and work of the two notable alchemists, Arnaldus Villanovanus and Raymundus Lullus, the alchemistic tendencies of their century are clearly reflected, although much uncertainty exists as to many points, especially in the life of the latter, and also with regard to the works ascribed to Lully. Both of them at all events were held in high esteem, not only during their lives, but also in the centuries following. Arnaldus Villanovanus, whose birthplace is uncertain, practised as a physician in Barcelona in the second half of the thirteenth century. His opinions, however, causing great offence to the priests, he was obliged to flee from there, and after vainly endeavouring to escape persecution in Paris and in various towns of Italy, he at last found an asylum in Sicily with King Frederick II. Summoned to Avignon by Pope Clement V., then seriously ill, he lost his life by shipwreck on the way thither, about the year 1313. He had special opinions of his own as to the nature and efficacy of the philosopher's stone, and also with regard to the noble metals obtained through its means. Among his writings may be mentioned: *Rosarius Philosophorum; De Vinis;* and *De Venenis.*

A similarly restless life was foreordained for Raymund Lully, a life which comprised in itself the greatest contradictions and eccentricities. Shortly after his death the object of a traditional glorification, Lully possessed among alchemists the fame of having attained to the highest which it was in the power of their art to achieve. The historical critic has a difficult task in dealing with him; for while, on the one hand, many of the writings ascribed to him are obviously counterfeit, there are, on the other, no sufficient data for deciding as to which of the remainder are really genuine. Thus there is very great uncertainty whether the alchemist Raymund Lully is identical with the famous grammarian and dialectician of the same name, who was called by his admirers *Doctor Illuminatissimus*; for this view, which has been held by many, is strongly opposed by the fact that criticisms of alchemy are to be found in many of the works of the latter.

Most of the records which we possess of the life of Raymund Lully agree in stating that he was descended from a noble Spanish family, and was born in the year 1235. After leading a dissipated life at the court of Aragon, he abjured the pleasures of the world in his thirtieth year and devoted himself to science. It was probably Bacon and Villanovanus who initiated him into the secrets of alchemy. When somewhat aged, he gave himself up to the conversion of the heathen, undertaking several journeys to Africa for this purpose; his reception there, however, was more than once of the worst, and he was at last stoned to death in the year 1315. Tradition has it that he lived for several years after that date in the unresting study of alchemy, but there can be no doubt as to the untenability of this report.

His alchemistic doctrines were very obscure; and still more incomprehensible and hidden in deep mystic darkness are his recipes for the ennobling of the metals. Certainly none of the alchemists who preceded him have ascribed to the philosopher's stone such powers as he did; for he was able to cry out presumptuously: "If the sea were of mercury, I would change it into gold."[1] And not only gold, but also all precious stones, and that highest good—health,—together with long life, were to be obtained through its means. Of the writings which are attributed to him, the *Testamentum*, *Codicillus seu Vademecum*, and *Experimenta* are regarded as genuine.

The earliest of the Latin writings formerly ascribed to Geber (*e.g.*, the *Summa*, mentioned on p. 29) may possibly have come into circulation soon after Lully's death. It is worthy of note, and also important for fixing with more or less accuracy the date at which they lived, that neither Albertus Magnus nor Raymund Lully refer to these writings, which grew in repute from the close of the fourteenth century onwards. The information which one finds in the works of the pseudo-Geber is by no means inconsiderable. Great progress is apparent in the recipes given for the making of preparations; in the use of apparatus such as the water-bath, the ash-bath, and improved furnaces; and in the description of chemical operations like sublimation, filtration, crystallisation, distillation, &c., &c. All this leaves no doubt on the mind as to the high standard of practical chemical knowledge which the pseudo-Geber possessed. The important question of the constitution of the metals out of mercury and sulphur will be discussed later on.

The history of alchemy in the fourteenth and first half of the fifteenth centuries contains no single name which will compare in eminence with those of the above-mentioned

[1] *Mare tingerem, si mercurius esset.*

philosophers, as the alchemists themselves preferred to be called.

This must not be taken as meaning that the supposed art of making gold had died out; on the contrary, it bore its strangest fruit during that period. If it be desired to connect specific names with the study of alchemy at that time, then the Frenchman Nicolas Flamel, Isaac Hollandus the elder and the younger, Count Bernardo da Trevigo, and Sir George Ripley may be mentioned as among those who were supposed to be in possession of the wonder-working philosopher's stone. These men did nothing, however, to materially advance the knowledge of chemistry.

Alchemy was at this time fostered and protected at many of the European courts, for nothing appeared to be more simple than to recuperate embarrassed finances by means of artificial gold. Many documents in the history of that century bear record to the frequent disappointments which were certain to come about sooner or later,—decrees against the practice of alchemy, threatenings of those who contravened these with the severest punishments, and accounts of discoveries of the most bare-faced impositions. Alchemy found especial protection at the court of Henry VI. of England, in spite of the fact that the kings preceding him had had to pay heavily for their leaning towards the hermetic art, and that a stringent law against it had been promulgated by Henry IV. The consequence of the favour shown to it by these monarchs was the production of large quantities of counterfeit gold which, in the form of coinage, inundated neighbouring countries. Charles VII. of France, who was then at war with England, was seduced by an alchemist, Le Cor, into a similar experiment, and thereby materially increased the debt of his country; to the alchemistic gold which he set in circulation were added the English "Rose nobles." Counterfeit coining, carried out on such a large scale, was hardly calculated to raise the reputation in which alchemy was held.

Chemistry, not being enriched during that time by any facts of importance, likewise suffered from this depreciation;

it first received new life from the work of Basil Valentine, whose acquirements in practical chemistry even now excite our wonder. This remarkable man was the real precursor of the iatro-chemical period, even although he was unable to free himself from the fetters of the alchemistic faith. Of his life practically nothing is known; from his writings we learn his name and the time, approximately, in which he lived, viz. the second half of the fifteenth century, and also that he was a Benedictine monk of Southern Germany. The most important of his works were published in the beginning of the seventeenth century by a city chamberlain Tölde in Frankenhausen, Thuringia; whether foreign matter has become mixed up with them cannot now be determined. So much is certain, that Basil Valentine was regarded as an oracle by alchemists so early as the end of the fifteenth and beginning of the sixteenth centuries, and was held in higher honour than Geber, higher even than Raymund Lully, besides being admired by many who had nothing to do with alchemy. His works were spread abroad by means of copies, and excited the interest of the Emperor Maximilian I. to such a degree that he caused a searching inquiry to be made in the year 1515 as to which Benedictine convent the famous author had dwelt in; but unfortunately his efforts in this direction were without result, as were also all later ones.

An account will be given further on both of his theoretical views and of his wide acquaintance with practical chemistry. Among the writings which are presumably his, and which are at the same time of the most importance, are the following:—*Triumphwagen des Antimonii* ("Triumphal Car of Antimony"); *Von dem grossen Stein der Uralten Weisen* ("On the Great Stone of the Ancient Philosophers"); *Offenbarung der verborgenen Handgriffe* ("Revelation of the Hidden Key"); *Letztes Testament* ("Last Testament"); *Schlussreden* ("Concluding Words").

In the first-mentioned work we possess what for that time was a marvellous description of an element and its compounds, the knowledge of these being due mainly to

Basil Valentine himself. The language which he employs is frequently obscured by mystical pictures and alchemistic ideas; but, while he thus appears as a visionary on the one hand, he excites on the other our highest admiration from the fulness of his temperate and conscientious observations, as well as from the rational views he takes of subjects which were then, for the most part, judged erroneously. The rich experiences in practical chemistry which he made his own cause him to stand out as the most distinguished chemist of the whole alchemistic period. His boldness, too, in proposing the use of chemical preparations for medicinal purposes led the way in a direction which soon after his time became the prevailing one, viz. the iatro-chemical, which dominated the succeeding age. Notwithstanding all this, Basil Valentine was an out-and-out alchemist, holding, as such, the most exaggerated ideas as to the power of the philosopher's stone; just as the tendency towards alchemy and the firm belief in the possibility of transmuting metals and of prolonging life continued engrained in many of the iatro-chemists.

SPECIAL HISTORY OF ALCHEMY.

Theories and Problems of the Alchemistic Period.

The alchemistic ideas, with the transmutation of metals as their leading principle, have been proved (as already mentioned) to have originated and to have been first systematically fostered in Egypt. The first attempt to explain this assumed transmutation, by a theoretical conception of the nature of metals, was made very early. From a similar endeavour, *i.e.* from regarding transmutation—then looked upon as an incontrovertible fact—as a consequence of the constitution of the metals, there sprang the doctrine contained in the works ascribed to Geber, which in its essentials predominated during the alchemistic period. It was thus always the metals which gave rise to the early chemical theories.

If we penetrate to the kernel of the doctrines of the Alexandrians through the veil of mysticism which envelops it, we see that these philosophers were permeated with the idea that the metals were alloys of varying composition. From this it necessarily followed that the transformation of one metal into another was possible, either through the addition of new metallic substances or the expulsion of some already present. Such transformations of similar substances into one another appear much less wonderful than those of dissimilar ones like air, water and earth, which were mutually convertible, according to the teaching of the Platonists and Aristotelians. The means for bringing about these changes in the metals, the substances which it was necessary to add to them, and the operations which had to be gone through, were either kept secret or obscured by indistinct figurative language. The various colours of the metals, and their alteration by melting them with others, played a prominent part in alchemistic processes; in imparting thereby the colour of a noble metal to a base one, much was supposed to have been attained. For the Alexandrians, therefore, and also for the alchemists of the Middle Ages, the colouring of metals was synonymous with their transmutation. The chief operations were the so-called *Xanthosis*, *Leukosis* and *Melanosis*, which were compared with the processes followed in the dyeing of cloth. The old designation of *tincturæ* for the media by which this transformation was brought about, gives expression to the idea that the latter consisted in a dyeing operation.

As may be imagined, no trace can be found of any distinct chemical conception, or of any knowledge of the actual operations which take place in these transmutations. At the root, however, of these endeavours of the Alexandrian alchemists to produce noble metals from base, lay speculations purely philosophical, which strongly excited and strengthened the belief in the transmutation of metals. These were partly taken from the writings of Plato, especially from his *Timæus*, which was highly esteemed by the Alexandrians, and partly from the philosophy of Aristotle. Both of these Greeks held the opinion that the (so-called) elements in general were capable of transformation into one another,[1] and an extension of this idea led to the assumption that the same applied to the metals. The observations of the supposed generation of noble metals from base, which have been already discussed, were looked upon as proofs of the correctness of this supposition.

In the writings hitherto attributed to Geber, but which, according to Berthelot, are not of earlier date than the fourteenth century, we find a specific chemical theory of the metals, a theory which, supported by the great authority of Geber's name, found universal recognition in the later Middle Ages. This theory looks upon classes of bodies from a chemical point of view, and seeks to explain the difference between the substances comprising these by assuming a peculiar chemical composition. The metals consist of sulphur and mercury, which are present in them in different proportions and in different degrees of purity.[2]

The transmutation of metals consists, according to him, in an arbitrary alteration of their composition; the ennobling of them, specially, in a purification and fixation of the mercury. The idea of creating a metal anew, which we find highly developed among Western alchemists, is not to be found in the pseudo-Geber's writings. This, together with the application of his theory, is proved by the following sentences, which comprise in themselves his theoretical and practical chemical programme: "To assert that one substance can be produced from another which does not contain it, is

[1] This idea comes out very clearly in the following passage from *Timæus*:—"We believe from observation that water becomes stone and earth by condensation, and wind and air by subdivision; ignited air becomes fire, but this, when condensed and extinguished, again takes the form of air, and the latter is then transformed into mist, which dissolves into water. From this, lastly, are produced rocks and earth."

[2] The pseudo-Geber sometimes added arsenic to the above-named constituents of the metals as a third possible one, without however laying emphasis upon this extension. Here and there, also, Aristotle's doctrine of the four different states of matter appears to get mixed up with his views upon the composition of the metals, the "four elements" being regarded to some extent as subsidiary constituents, sulphur and mercury being the principal ones.

folly. Since, however, all metals consist of sulphur and mercury, we can add to them the constituent in which they are deficient, or abstract the one which is present in excess. In order to achieve this, make use of the art: calcination, sublimation, decantation, solution, distillation, coagulation (crystallisation), and fixation. The active agents are the salts, alums, vitriols, borax, the strongest vinegar and fire."

The varying origin of the works hitherto ascribed to Geber explains why in many passages of these writings no distinction is drawn between the supposed two constituents of the metals and natural sulphur and mercury, while we frequently find him expressing, in others, the opinion that the former are not identical with the latter. The mercury and sulphur present in the metals were, in this second case, looked upon as being of an abstract nature; thus mercury conferred glance, malleability, fusibility, and what we consider metallic properties generally, while sulphur, on account of its combustibility, was regarded as being present because of the alteration of many metals in the fire. The noble metals, those which withstood the fire, therefore consisted of almost pure mercury, which however could not be identical with the ordinary substance of that name, since the latter was volatile; this property was ascribed to the fact of ordinary mercury containing sulphur. By means of these and similar assumptions, contradictions between theory and facts were easily set aside, the alchemists of later times especially distinguishing themselves in this way.

For the solution of the possible problem of the transmutation of metals—possible, that is, in the sense of the above theory,—so-called "medicines" are, according to the pseudo-Geber, requisite, these being distinguished as possessing different power and virtue. The medicines of the first order do indeed produce changes in the base metals, but these changes are not permanent. Those of the second order partially alter the properties of such metals into those of the noble ones,[1] but the transmutation proper is only effected by

[1] The *Particulare* of the later alchemists appear to have corresponded to medicines of the second order.

the medicine of the third order, which is variously designated as the *Philosopher's Stone, the Grand Elixir*, or the *Magisterium* (masterpiece).[1] The accounts which the pseudo-Geber gives of the preparation of the medicines of higher order are wholly unintelligible; it should, however, be emphasised that there is a wide difference between these and the incredible exaggerations of which other alchemists were guilty, when speaking of the efficacy of such secret preparations.

One cannot but feel surprised that the alchemists of the thirteenth and fourteenth centuries, possessing as they did a fairly extensive knowledge of chemistry, should have remained satisfied with such speculations as to the constitution of the metals, without actually trying to isolate the substances that they assumed as being present in these and other bodies. Instead of endeavouring to gain an insight into their composition by experiment, they brought forward fresh hypotheses to controvert obvious objections, *e.g.*, that the above-mentioned constituents (mercury, &c.) were not identical with the substances commonly so named.

The above theory of the metals underwent an extension by Basil Valentine, who assumed the presence in them of a third constituent, viz. salt.[2] By the term *salt* he did not mean a definite chemical compound, such as common salt, but rather the principle of solidification and power of withstanding fire, just as sulphur determined the combustibility or change in the fire and also the colour, and mercury the metallic character and volatility. Basil Valentine generalised his opinion in this way, that he assumed these three essential principles in all substances, an assumption which Paracelsus appropriated later on, and made the basis of his iatro-chemical doctrine.

Their views upon the composition of the elementary bodies being so very obscure and so utterly wrong, one sees how impossible it was for the alchemists to explain chemical

[1] At a later period the great elixir was distinguished from the small one, which only transmuted the base metals into silver.

[2] Isaac Hollandus had before this spoken of the saline principle of the metals.

processes rightly, connected as these are with the formation of compounds. Some very incomplete attempts were made to give a theoretical explanation of isolated observations, but these only led to gross errors creeping in; the calcination of the metals, for instance, was supposed to depend upon the escape of moisture or of some other constituent, an idea which reappeared in another form in the later theory of phlogiston. The above theory of the composition of metals is sufficient evidence of the small amount of trouble which was taken to find out the true chemical constituents of bodies.

We may safely say that scientific chemistry only really began with the fruitful endeavours to discover the real composition of substances. It is out of the question to speak of this as applying to a time when it was considered as proved that the formation of a chemical compound was identical with the annihilation of its original components, a new substance being created. This view was the almost sole predominating one during the later alchemistic period, although in the works of the pseudo-Geber we find some indications of more correct opinions on the composition of many chemical compounds (the recognition of mercury and sulphur, for instance, as constituents of cinnabar).

Contemporaneously with the holding of such theories based upon no facts whatever, the Western alchemists strove in every imaginable way to obtain the philosopher's stone,—*mercurius philosophorum*.[1] Those of them who were in happy possession of the means for transmuting metals, attributed to it the most astounding powers. In order to give some idea of the aberration of mind caused by the alchemistic problem, a few of the extraordinary assertions of well-known alchemists with regard to the preparation and efficacy of the philosopher's stone may be mentioned here.

For its preparation (we are now speaking more particularly of the thirteenth century onwards) a *materia prima* was requisite, to obtain which was the hardest task of all. The most incredible substances, natural products of every

[1] Cf. the Engler lecture:—*Der Stein der Weisen* (Carlsruhe, 1889).

kind, were taken as raw materials for the manufacture of this preparation, and worked up in every conceivable way. Those who laid claim to the possession of the philosopher's stone took very good care to keep the secret of their *materia prima* to themselves. They described all kinds of operations with it[1] in the most enigmatical recipes, employing at the same time mystical drawings, such as those of the dragon, the red or green lion, the lily, the white swan, &c., and well knew how to keep their imitators, of whom there were formerly shoals (isolated cases being found even in this century), in a state of continual tension. That this was possible is explained by the immovable and almost universal belief in the transmutation of metals, by means of the philosopher's stone, during the Middle Ages.

To the latter the greatest miracles were ascribed; thus, Roger Bacon does not hesitate to say that it was able to transform a million times its weight of base metal into gold (*millies millia et ultra*). Others, *e.g.* Arnaldus Villanovanus, were more modest in their estimate of its powers, stating that it could convert into gold one hundred times its weight of mercury. Others, again, surpassed even Bacon, as the following passage from the *Testamentum Novissimum*, ascribed to Lully, proves: "Take of this precious medicine a small piece, as large as a bean. Throw it upon a thousand ounces of mercury, and this will be changed into a red powder. Put one ounce of the latter upon one thousand ounces of mercury, which will thereby be transformed into a red powder. Of this, again, an ounce thrown upon a thousand ounces mercury, will convert it entirely into medicine. Throw an ounce of this on a thousand ounces of fresh mercury, and it will likewise turn into medicine. Of this last medicine, throw once more an ounce upon a thousand ounces of mercury, and this will be entirely changed into gold, which is better than gold from the mines." One sees clearly, from these and other fraudulent assertions, that the simple standpoint which the Egypto-Greek alchemists

[1] The process of fixation, a term which indicated the solidification of mercury by the transmutation, was of special importance.

assumed with regard to the question of the transmutation of metals, was departed from in the later Middle Ages.

In view of such excesses, which are an insult to the human understanding, it causes no surprise to find attributed to the philosopher's stone other results which are, if possible, even more incredible; being a universal medicine, health and life were to be preserved and ensured by it. Statements as to the power of prolonging life possessed by the elixir were also rife in the later Middle Ages, and it was no unusual assertion that adepts, the fortunate possessors of the panacea, had been able to prolong their lives to 400 years and more. The long lives of the patriarchs were explained by the assumption that they were acquainted with this universal medicine. In the time of the Arabian alchemists healing properties were ascribed to gold prepared artificially and brought into the potable form (*aurum potabile*), and from this the belief in the medicinal power of the philosopher's stone appears to have originated.

Alchemistic ideas produced their most absurd results towards the end of the Middle Ages and in still more recent times, the creation of living beings by means of the philosopher's stone being not merely held as possible, but being actually taught; this marks the acme of the mental aberration they induced.

The melancholy picture, which the condition of alchemy presents to us at various periods, becomes still more sombre and involved in deeper shadow from the fact that men did not hesitate to affirm the Divine assistance and to claim predestination, in order to explain the marvellous effects of the philosopher's stone. Gross abuse was made in this way of the name of the Deity, and also of prayers and biblical quotations, by the alchemists of the thirteenth century, and still more by their successors. There is no need to go into further details upon this point here, but it is necessary to mention it in order that the methods by which the problems of alchemy were treated at different periods may appear in their proper light.

Upon the development of chemistry as a science, the

alchemistic doctrines—especially the theories of the composition of metals—had only a slight and an indirect influence. The excesses to which they gave rise have—as aberrations of mind, enchaining a large portion of the educated—a higher value for the history of civilisation than for that of chemistry. The main significance of alchemy for the latter lies in this,—that the endeavours to solve the problem of the transmutation of metals were the cause of actual work with materials of every kind; and the result of this was a not inconsiderable increase in the knowledge of applied chemistry during the alchemistic age. The following section will be devoted to an account of the latter.

Practical-Chemical Knowledge in the Alchemistic Period.[1]

When one considers upon what superficial observations the conviction of the transmutability of metals was based, and how readily wholly untenable theories upon the composition of bodies were brought forward and accepted, one feels no surprise that comparatively little progress was made, during the succeeding epochs, towards explaining the numerous chemical processes already known to the Ancients. The acquirements in chemistry during these centuries themselves likewise remained, for the most part, empirical; it was but seldom that the composition of chemical compounds was even in some degree correctly indicated. The fantastic treatment of chemistry—a treatment wholly foreign to the exact sciences—has been sufficiently detailed in the preceding section. We must not omit to mention, however, that the addition of new facts to those already known, and the gain of experience in the fields of technical and pharmaceutical chemistry and in the manufacture of chemical preparations, were not inconsiderable.

[1] Cf. Kopp, *Gesch. d. Chemie*, vols. iii. and iv.; Höfer, *Histoire*, etc., vol. i. p. 317, *et seq.*; Gmelin, *Gesch. d. Chemie.*; and Berthelot, *La Transmission de la Science Antique du Moyen-âge.*

Technical Chemistry.—Metallurgy, upon which the infant powers of an early developed *technique* were expended, shows, upon the whole, but little progress. Towards the end of the alchemistic period certain other metals were indeed added to those already known, viz., the semi-metal antimony, together with bismuth and zinc; but these can only lay claim to a subordinate position in the circle of metallurgical processes generally. From the eleventh century on, mining increased among the Western nations, in Germany especially in the Harz, Nassau and Schlesien. So far as our present information goes, only trifling alterations were made in the preparation and purification of the metals.[1]

Gold was obtained and purified from other metals and admixtures by the old method of cupellation (working with lead), already accurately described by the pseudo-Geber. The latter knew that the desired result was ensured and its progress hastened by the addition of saltpetre, and, further, that copper and tin, but not silver, could be separated from gold in this way. In the fifteenth century there was added to this the process of purifying gold by fusing it with antimony trisulphide ore (*Spiessglanzerz*), a method which is given in detail by Basil Valentine. Alloys of gold were often fraudulently prepared of set purpose.

The extraction of silver from its ores was accomplished, as in Pliny's time, by fusion with lead, an operation first termed "*Aussaigern*" by Basil Valentine. The only means of separating gold from silver which was known up to a comparatively recent date, was the cementation process of the Ancients. The wet process with nitric acid appears to have been first successful in the time of Albertus Magnus, at least he is the earliest to indicate it; an absolutely

[1] The work entitled *Schedula Diversarum Artium*, which was written by Theophilus Presbyter, a Benedictine of the eleventh century, gives a true picture of the state of technical industry in his time, particularly of the working up of metals, something being also said about their production from the ores.—A tenth-century manuscript, *Mappæ Clavicula*, edited by Berthelot, contains an essay on the noble metals, and by its agreement with recipes found in the Leyden papyrus, conclusively shows the close connection with the Egypto-Greek alchemy.

certain acquaintance with the process is first to be found in Agricola.

From the importance which was attached to the successful working-up of gold and silver ores, one understands how the closest attention was given from an early period to the definite quantitative yield of the noble metals. Accurate balances came into use, their employment in cupellation and cementation processes being made obligatory by law; one thus meets here with the first beginnings of a docimacy.

With regard to the metallurgy of iron, lead, tin and copper in the alchemistic period, there are no particular improvements to record. Basil Valentine (fifteenth century) states that the last metal was also obtained by the wet process as the so-called cement copper, by precipitating a solution of copper vitriol with iron. The changes undergone by these metals on being heated and on treatment with chemical reagents, especially acids, were closely studied, and thereby the knowledge of metallic preparations decidedly enlarged. (See p. 52.)

Mercury, which played such an important part in the theoretical views of the alchemists, was prepared on a large scale for technical purposes by roasting quicksilver ores in improved furnaces, especially after the opening up of the rich Idrian mines in the fifteenth century. The preparation of the metal by distilling a mixture of sublimate and caustic lime was well known to Basil Valentine. For its purification he gives various processes, some of which had been already described by the pseudo-Geber. Mercury was much used, particularly for the extraction of gold and silver (by the so-called amalgamation process) and for gilding.

Metallic zinc and bismuth, and also cobalt ore, are likewise mentioned by Basil Valentine, but the metals themselves do not seem to have been employed technically; some preparations of zinc, however, were. A special place among chemical preparations is to be assigned to antimony and its compounds, the knowledge of which is due to Basil Valentine himself. (See p. 54.)

In pottery and glass manufacture, important improvements in single points were made during the alchemistic period; but it is also noticeable here that the interest in the chemical processes remains a purely external one, no attempt being made to give a scientific explanation of the facts empirically arrived at. The general use of glazes containing lead and tin for earthenware vessels is worthy of mention, as is also the burning of colours into glass (the whole mass having formerly been coloured by the addition of metallic oxides during fusion).

Dyeing remained stationary on the whole, so far as the chemical media for fixing the colour on the fibre were concerned; alum was universally employed as a mordant, being manufactured on a large scale in different places. The introduction of the kermes dye (cochineal) into European countries by the Arabians, that of orchilla, already known in ancient Rome (from the East in the thirteenth century), and, lastly, the gradual supplanting of the (blue) dye from woad by indigo, are the most important technico-chemical events in the domain of dyeing.

Condition of Pharmaceutical Chemistry.

Although the Arabians and the later Western *savants* busied themselves with chemical operations, and thereby arrived at preparations of the most various kinds, the pharmaceutical chemistry of that period only profited slightly by this; it was Basil Valentine who inaugurated a new era by his bold attempts to apply chemical preparations to medicinal purposes. The opening up of the intimate connection existing between chemistry and medicine, which led to the high development of pharmacy, was reserved for the period of iatrochemistry. The Arabians prepared their medicines strictly according to the recipes of Galen, Andromachus and others, which were transmitted to them, according to Leo Africanus, by the Nestorians.[1] Apothecaries' shops, in which the remedies were almost exclusively prepared from vegetable

[1] For their influence upon the Arabians see note 2, p. 28.

substances, sprang up at an early date. To the Arabians belongs the credit of having improved and rendered the process of distillation serviceable for this purpose; distilled water, ethereal oils, and other products (especially spirit of wine) obtained by distillation, to which the most wonderful results were ascribed, soon came into general use.

These apothecaries' shops with their fittings then spread into Spain, Southern Italy (into Salerno in the eleventh century) and, somewhat later, into Germany. The recipes of that time for the preparation of medicines, the imperfect pharmacopeias,[1] show that the doctrines and axioms of Galen and the Arabian physicians remained the standards up to the end of the fifteenth century. The position of the physician with regard to the apothecary was early fixed by legal statute, it being considered advisable to draw a sharp distinction between the man who had to prescribe the medicines and the man who had to make them.

With respect to chemical preparations proper, a few new ones were added to those already used in medicine, *e.g.* saltpetre, mercury in the form of grey ointment, and—towards the end of the fifteenth century, at the instigation of Basil Valentine—various mercurial and antimonial preparations (see p. 54). Almost all the physicians of that time took up, however, an antagonistic position with regard to the last of these, being of opinion that the undoubted poisonous properties of antimony compounds were incompatible with their internal use.

Knowledge of the Alchemists with regard to Chemical Compounds.

It has already been mentioned that the knowledge of the true composition of chemical compounds was but slightly extended during this period; we have therefore to deal here with the state of empirical knowledge as affecting substances prepared artificially, together with a few occurring naturally.

[1] The first German pharmacopeia (*Arzneibuch*) was drawn up by Ortholph von Baierland and appeared in 1477.

The tendency to group together observed facts under a common point of view showed itself at an early date with respect to salts, of which a large number were known. The pseudo-Geber regarded solubility in water as a general characteristic; later on the generic name *sal* was made to include a variety of substances, *e.g.* the vitriols, potash, soda, saltpetre, alum, etc. Other chemical compounds of totally different nature, viz. the alkalies and acids, were added to the class of salts by many alchemistic writers, the term *sal* being thus widely extended and distorted; it was reserved for a later century to fix it without any ambiguity. In addition to the common designation *sal* for a number of heterogeneous bodies, we find in the writings of that time the generic name *spiritus* for the volatile acids, *e.g. spiritus salis* for hydrochloric acid; also the name *spiritus urinæ* for volatile alkaline salt (carbonate of ammonia). The individual salts are distinguished by the word which follows *sal*, for instance, *sal petræ, sal maris*, etc.; for alkalies, such as caustic potash, the expression *nitrum alcalisatum* is frequently used. One seldom meets in the alchemistic age with a strict distinction between potash and soda, or between their carbonates, while, on the other hand, preparations of carbonate of potash obtained in different ways were regarded as dissimilar products.[1]

This acquaintance with the carbonates of soda and potash was accompanied by a knowledge of the lyes obtained from them by the addition of lime, the strongly alkaline and solvent power of these lyes being largely made use of, *e.g.* in the preparation of milk of sulphur. The name "alkali" is first met with in the writings ascribed to Geber, while the designation "caustic" had been already employed by Dioscorides for burnt lime, and at a later period for lyes. The question of the occurrence of alkalies in plants was frequently discussed among the alchemists; although it did not escape some of them that different amounts of ash and of alkali were found in different parts of a plant, only a few held the opinion that the alkali was really present in the plant itself, most of

[1] The salt from the ashes of plants was termed *sal vegetabile*, and that from tartar, *sal tartari*.

them believing that it was first produced during the incineration of the latter.

It was formerly taken for granted that the Arabians possessed a very considerable knowledge of the acids, in comparison with that of the Ancients, who were totally unacquainted with the mineral acids. This assumption was based upon the fact that in the treatise *De Inventione Veritatis*, attributed to Geber, he explained the method of obtaining nitric acid by distilling a mixture of saltpetre, copper vitriol, and alum in certain proportions; it was designated *aqua dissolutiva* or *aqua fortis*. We know now, however, that this manuscript does not date further back than the fourteenth century. The preparation of nitric acid from saltpetre and sulphuric acid was known to alchemists of a later date; and we find Basil Valentine speaking of it as a process which had been in operation for a long time.

Sulphuric acid was certainly obtained by the pseudo-Geber, for he mentions as noteworthy that when alum is strongly heated, a spirit distils over which possesses a high degree of solvent power; he does not, however, appear to have investigated its properties more closely. The writings of Basil Valentine show that the preparation of sulphuric acid by distilling a mixture of iron vitriol and pebbles, and also by setting fire to sulphur after the addition of saltpetre to it, was known not only to himself but also to his predecessors. An aqueous solution of sulphurous acid, the combustion product proper of sulphur, was frequently confounded with sulphuric acid.

Basil Valentine is the first to describe the preparation of aqueous hydrochloric acid, which he terms *spiritus salis*, by heating a mixture of common salt and green vitriol, and also its behaviour towards many of the metals and their oxides. He likewise knew that a mixture of this acid with *aqua fortis* was the so-called *aqua regis*, now termed *aqua regia*, which the pseudo-Geber had already made use of, obtaining it by the solution of salmiac in nitric acid.

Nitric acid and *aqua regia*[1] (so-called because it dissolved

[1] Albertus Magnus terms them respectively *aqua prima* and *aqua secunda*.

gold, the king of metals) were highly prized by the alchemists of the West. The observation that almost nothing was able to withstand this *aqua regia*, even sulphur being "consumed" by it, strengthened the conviction that in it they possessed a liquid which very nearly approximated to the long-sought-for "alkahest," the universal solvent. On the same grounds oil of vitriol was greatly valued, many indeed regarding it as the *sulphur philosophorum*, or, at least, as a substance which would lead to the acquirement of the *materia prima*.

Among the salts which were already known in Pliny's time, and whose properties were carefully investigated by the alchemists, alum and some of the vitriols deserve special mention, the former being obtained in various places from alum shale. The pseudo-Geber tells us how to purify it by recrystallisation from water, and terms it *alumen de rocca* (from the name of its chief source, the town Roccha), a term which long remained in vogue in France as *alun de roche*. The fact that alum contained an alkaline salt was overlooked, and its true composition remained unknown. Iron and copper vitriols were lagely employed in different chemical operations. The pseudo-Geber describes the preparation of the pure products by crystallisation, and Basil Valentine the production of iron vitriol by dissolving iron in sulphuric acid, a method which indicated the composition of the salt, although he did not explain this correctly.

The important salts, saltpetre, salmiac and carbonate of ammonia, first became known and used for chemical purposes in the alchemistic period. The author of the works ascribed to Geber was well acquainted with potash saltpetre, as it served him for the preparation of nitric acid; and there is every reason to suppose that it was used in even earlier times for the production of fire-works and such like things, after its property of deflagrating with red-hot carbon had been recognised. The oldest designations for it were *sal petræ* and *sal petrosum*. Raymund Lully also termed it *sal nitri*, but distinguished between it and *nitrum*, the fixed alkali of the older writers; in the sixteenth century this latter word was converted into *natron,* while the name *nitrum* was given to potash saltpetre.

The same applies to the term salmiac, *sal ammoniacum,* as to that of *nitrum,* in so far that both of them had originally a different meaning from what they now possess; for the *sal ammoniacum* of the Ancients was without doubt rock-salt. At the time when the pseudo-Geber's works were written, on the other hand, this name, which is also metamorphosed into *sal armeniacum* (Armenian salt), could only mean salmiac. The *sal armoniacum* of Basil Valentine led to the contraction salmiac. At first this salt appears to have been partly prepared from dung, and partly to have been found as a natural product of volcanic origin.

Carbonate of ammonia, well known to the alchemists of the thirteenth century as volatile alkaline salt (*spiritus urinæ*) was obtained by distilling putrefied urine. Basil Valentine taught how to prepare it from salmiac and fixed (carbonated) alkali, a method which led a long time afterwards to the proper recognition of the composition of the salt. The pharmaceutical use of these two ammonia compounds, just named, probably belongs to a later date.

The knowledge of the metallic salts was very decidedly increased during the alchemistic period. A special interest attached to a solution of gold in *aqua regia,* since from this *aurum potabile* the most wonderful medicinal effects were expected. The pseudo-Geber was the first to become acquainted with nitrate of silver in the crystalline state, and to observe the precipitation of its solution by one of common salt, a reaction which came to be applied as a test both for silver and for salt. The alchemists were also aware of the beautiful precipitation of metallic silver from a solution of its nitrate by means of mercury or copper.

Compounds of mercury early attracted the interest of those who carried out chemical operations. The pseudo-Geber described the preparation of mercuric oxide by calcining the metal, and that of sublimate (mercuric chloride) by heating a mixture of mercury, common salt, alum and saltpetre; he also taught how to prepare various amalgams.[1] Basil Valentine was acquainted with basic mercuric sulphate, and also with mercuric nitrate. Being an advocate of heroic treatment, he recommended the medicinal use both of the latter and of sublimate.

Preparations of zinc and bismuth (*e.g.* zinc vitriol) were well known towards the end of the fifteenth century, but detailed records are wanting both of their formation and their properties. Antimony and its compounds, on the other hand, were the object of unwearied labours on the part of Basil Valentine, as his treatise *Triumphwagen des Antimonii* ("Triumphal Car of Antimony") sufficiently testifies. He shows how to prepare antimony itself from the native sulphide (which was termed *antimonium* or *stibium* and was known to the Ancients), by fusing it with iron. In his treatise, *Wiederholung des grossen Steins der Uralten Weisen* ("Recovery of the Great Stone of the Ancient Philosophers") he writes: "If one adds some iron to the fused *Spiessglas*,[2] there is produced by a particular manipulation a curious star, which the wise men before me called the signet star of philosophy." Basil Valentine was well aware that antimony did not possess the properties of a metal in full degree, and so he regarded it as a variety of one, especially as a variety of lead; he sometimes talks of it as the *lead of antimony.* Even in his time antimony was employed for alloys, which served for the manufacture of printer's type, mirrors and bells. It did not escape him, either, that *Spiessglas* contained sulphur, and he was also acquainted with amorphous sulphide of antimony and *sulphur auratum* (a mixture of Sb_2S_3 and Sb_2S_5). He gives distinct recipes for the preparation of antimony trichloride (butter of antimony), of powder of algaroth (basic chloride of antimony), of antimony trioxide,

[1] This word is first found in the writings of Thomas Aquinas. The part played by amalgams in the transmutation of metals has been already considered.

[2] This designation of Basil Valentine's for native sulphide of antimony became altered later on into *Spiessglanz.*

and of potassic antimoniate, and there can hardly be a doubt that he recommended and applied those preparations for internal use. With regard to the composition of these, he only appears to have had a tolerably clear idea of that of the sulphide.

Arsenic, which is so closely allied chemically to antimony, and with whose sulphides the Ancients were acquainted, was first prepared by the Western alchemists in the thirteenth century; Basil Valentine regarded it as a "bastard metal" analogous to antimony. Arsenious acid is first distinctly spoken of by the pseudo-Geber, having been obtained by the roasting of realgar; it was known as white arsenic, in contradistinction to the red and yellow varieties (realgar and orpiment). Its occurrence in the smoke from pyrites burners was also noticed by the observant Basil Valentine. Mention has already been made of the important part which was played in alchemistic operations by the property which arsenic possessed of turning copper white;[1] indeed, this contributed materially to the belief in the possibility of the transmutation of copper into silver.

In addition to the metallic oxides which have been already spoken of (those of mercury, antimony, etc.), and the early known oxide of copper and oxides of lead (PbO and Pb_3O_4), oxide of zinc and peroxide of iron may be specially mentioned. The former of these, which separated in woolly flakes when zinc was burnt, and which was therefore termed *lana philosophica*, appears to have been known to Dioscorides, but it is in the alchemistic period that we first come across an intimate acquaintance with it. The alchemists of the Middle Ages were familiar with peroxide of iron in the different forms, red and yellow; the designation *colcothar*, for the ignited oxide, is to be found for the first time in Basil Valentine's writings.

The theoretical importance which, from early times, was ascribed to sulphur as a constituent of the metals, and also of other bodies, leads to the question—How was the actual know-

[1] On account of this behaviour, the pseudo-Geber calls arsenic *medicina Venerem dealbans*.

ledge of this element and of its compounds acquired? The property possessed by sulphur of dissolving in aqueous alkalies, and of being thrown down from such a solution as sulphur milk upon the addition of acids, is described by the pseudo-Geber in his treatise *De Inventione Veritatis;* the disappearance of sulphur, when acted upon by *aqua regia*, was likewise regarded as solution. Basil Valentine is the first to give definite details regarding flowers of sulphur, and also regarding the taking up of the element by many oils, a property upon which the preparation of sulphur balsam depended.

Mention has already been made of various sulphur compounds, the sulphides of mercury and antimony among others which were the most valuable materials for the production not only of sulphur itself, but also of other bodies. These had already been grouped together as forming a particular variety of compounds, under the name of *marcasitæ* (Albertus Magnus), zinc blende, galena, and iron and copper pyrites being included among them. The peculiarity, which these substances had in common, of giving off a product of such characteristic odour as sulphurous acid when roasted, may not unlikely have formed the main reason for thus gathering them into one group. It must not be forgotten, however, that the formation of several metallic sulphides from their components had been observed (*e.g.* that of cinnabar from quicksilver and sulphur), and this may be supposed to have contributed materially to a knowledge of their composition.

In spite of many unequivocal observations to the contrary, people still held to the assumption that the metals and almost all other substances contained sulphur. Organic bodies, too, had to conform to this hypothesis; their real constituents remained hidden, no sharp general distinction being drawn between them and inorganic compounds. The meagre attempts made to explain the formation of organic substances, *e.g.* in fermentation processes, only give evidence of confused and untenable views. The organic preparations which were known in the alchemistic age were but few in number. Among them spirit of wine[1] takes a prominent place, its manufacture being gradually simplified and improved after more perfect apparatus had been introduced by the Alexandrians. In accordance with its importance for medicinal and alchemistic purposes, it was usually termed *aqua vitæ*, the name alcohol being first met with in Libavius (end of the sixteenth century). The preparation of concentrated spirit of wine—as an excellent solvent for many things—by repeated distillation, and also by dehydration with fused potashes, was already known to Raymund Lully. To test its strength, Basil Valentine recommends that a portion be burnt, in order to see whether any water remains behind or not. The latter alchemist was also acquainted with various chemical transformations of alcohol, although he did not obtain the resulting compounds in a state of purity; among these were the production of common ether by the action of sulphuric acid, and of nitric and hydrochloric ethers by the action of nitric and hydrochloric acids respectively. By the "sweetening" (*Versüssung*) of alcohol is to be understood our term etherification. That alcohol is only formed during the various processes of fermentation, which yield wine, beer and spirits, was not perceived even by the most acute observers of that time; its pre-existence in unfermented materials was thus taken for granted.

Increasing attention was likewise paid to the product of the acetic fermentation. The alchemists of the later Middle Ages taught how to concentrate vinegar by distillation, and they also prepared various salts of acetic acid, *e.g.* basic acetate and sugar of lead. Other organic acids, too, were noticed in different plant juices, but they were mostly mistaken for acetic acid. The addition to the medical treasury of various resins and oils, especially ethereal oils, which were obtained from plants by distillation in improved apparatus,

[1] Berthelot (*Ann. Chim.*, (6), vol. xxiii., p. 433) has traced with great care the history of the discovery of spirit of wine, and has found that the preparation of alcohol by distilling wine was accurately known so far back as the time of Marcus Græcus (eighth century, A.D.).

is no evidence of scientific progress; this really begins for organic chemistry with the discovery of methods for arriving at the composition of organic compounds.

The Fortunes of Alchemy during the last Four Centuries.

After the labours of Basil Valentine, and especially after the beginning of the iatro-chemical period, alchemy gradually became separated from chemistry, which was raising itself to the rank of a science. Although, therefore, a record of the alchemistic aims or rather errors of the last few centuries does not properly come within the scope of a short history of chemistry, they cannot be passed over in complete silence; the justification for this lies in the relations in which the most eminent chemists of the sixteenth and seventeenth centuries stood with regard to alchemy. The support given by such men to the latter undoubtedly accounts to a large extent for the belief in the transmutation of metals as an incontrovertible fact being but seldom affected, and this, notwithstanding the great increase in chemical knowledge. Another effective means by which the life of alchemy was prolonged, consisted in the favour with which it was regarded by many princes; the seductive prospect of easily acquired treasure often rendered the latter a prey to designing alchemists.

The actual decay of alchemy, for which the numberless disappointments of honest workers and the exposure of numerous frauds paved the way, may be dated from the first half of the eighteenth century, when the conviction of the practicability of transmuting metals began to die out among most chemists. Even up to the present century, however, we find able and educated men in the thralls of alchemistic chimeras, and directly opposing the simplest rules of reason.

A distinction must be drawn during the iatro-chemical period between alchemists and chemists, inasmuch as the latter aimed at the solution of a scientific problem, viz. the knowledge of the relations between chemistry and medicine. At the same time this distinction must not be taken as meaning that the most eminent among the iatro-chemists were not firmly convinced that the ennobling of metals was a fact, indeed some of them maintained that they were in possession of the most powerful alchemistic specifics; it was but seldom, however, that chemists were at the same time practical alchemists.

Paracelsus, who was greatly given to romantic exaggerations, claimed for himself the widest knowledge of alchemy. Van Helmont, whose authority was especially weighty, went so far as to describe in detail the transmutation of mercury into gold and silver, as effected by himself with the aid of a very small quantity of a gold- and silver-producing philosopher's stone. The opinion held by the highly esteemed Libavius respecting alchemy and what it could effect is equally significant of the judgment of that period upon the subject; he regarded the transmutation of metals as an accomplished fact. Other influential physicians of the sixteenth century, such as Agricola—famed as an observant and accomplished metallurgist,—Sennert, and Angelus Sala, were more cautious in their assertions with respect to alchemy, but they never seriously contended against the possibility of transmutation. Tachenius alone, the last iatro-chemist of note, took up a sceptical position with regard to the alchemistic problem; he considered the evidence adduced in favour of the ennobling of metals as insufficient, notwithstanding that his famous teacher Sylvius had given himself up unreservedly to the belief in their transmutation.

The power of this belief was still so great at that time, when the phlogistic period was just beginning and chemistry was striving to develop itself independently, that it took firm root in the minds of even the most discerning men, with Boyle at their head. The latter was firmly convinced of the possibility of transmuting individual metals into one another, as were also many of his contemporaries and successors, *e.g.* Glauber, Homberg, Kunkel, Stahl and Boerhave, of whose earnest desire to arrive at the truth there can be no doubt whatever. That the wished-for goal was never reached in spite of the most unwearied efforts, did not shake their belief in the correctness of the assumptions of alchemy; Stahl alone began to doubt these towards the end of his life, and warned his brethren against alchemistic frauds. The vitality of the belief in transmutation depended chiefly on the theoretical opinions which these men held regarding the composition of metals; the primal error of the pseudo-Geber and his disciples was thus propagated for centuries through the alchemistic age.

Boerhave was the last distinguished chemist to support with his great authority some of the alchemistic views, while he failed to criticise others of the fraudulent assertions with sufficient sharpness. After his time no notable exponent of chemistry—which had now attained to the rank of a science—spoke in their favour; but all the greater was the number of cheats and swindlers who cultivated the lucrative field of gold-making even during the eighteenth century. The conviction of the impossibility of transmutation, which was at that time establishing itself among scientific chemists, made its way but slowly into outer circles. Credulity, and the hope of obtaining riches for nothing, were the means of leading many into very doubtful paths even so late as the end of last century and the beginning of this one.[1] The final echoes of the alchemistic problem, which had for so long a period of time held the cultured of every nation in a state of tension, and had even blinded eminent scientific men, only appear to die away during the last decades of our own century.

Seeing the marvellous results which alchemy produced, it is but natural to inquire more closely into the supposed evidence in favour of the ennobling of metals, and to ask what kind of observations led to this being regarded as a matter of fact. If most weight is to be laid upon the

[1] For details on these points, especially for an account of the interesting relations of the Rosicrucians to alchemy, and of secret alchemistic associations, etc., see H. Kopp's *Die Alchemie in älterer und neuerer Zeit*, a book which gives us a clear insight into the workings of the alchemists.

statements of men who had established their claim as practised observers, then first place must be given to the records of the great physician and chemist van Helmont (towards the middle of the seventeenth century), respecting transmutation as carried out by himself; these records afford the most remarkable testimony to the power of alchemistic illusions. Van Helmont had received from an unknown source a small specimen of the philosopher's stone, and with this he states that he transformed several portions of mercury into pure gold, giving the exact proportions by weight; one part of this preparation sufficed to transmute 2000 parts of mercury.

Soon after the death of van Helmont, Helvetius, body-physician to the Prince of Orange, published a detailed account of the transmutation of lead into gold, by means of a trifling quantity of a preparation which had come to him from the hand of a stranger. It appeared impossible to doubt the testimony of such men, who were held in high esteem by all the scientific investigators of that time.

More palpable proof of the actual transmutation of metals was held to be furnished by the coins or ornaments prepared from alchemistic gold up to and in the eighteenth century.[1] The evidence, which came for the most part too late, that these consisted of worthless alloys (*e.g.* bronze gilt over), was all too soon forgotten. The findings of courts of justice, too, in favour of alchemistic operations, were looked upon as proofs of transmutation having been actually accomplished.

As has been already mentioned, a large number of German princes gave unremitting support to the efforts of the alchemists, being induced to do so by the hope of large gains. Many of them worked zealously at transmutation themselves, among others John, Burgrave of Nürnberg, who received the surname of "the Alchemist"; the Emperor Rudolph II., the most powerful protector of the makers of gold; the Elector Augustus of Saxony, the Elector John George of Brandenburg, etc. etc. The courts of these princes were the field-grounds of adepts, who for long

[1] Cf. H. Kopp's *Alchemie*, vol. i., p. 90, *et seq.*

succeeded, by means of clever experiments, in maintaining a belief in their art among these Mæcenases, until, as usually happened, they were unmasked as cheats and generally severely punished, after having been the cause of excessive expenditure on the part of their patrons.

It is impossible to enter here into details of the romantic lives of alchemists like Leonhard Thurneysser, physician at the court of John George of Brandenburg, Sendivogius, Caëtano (on whom the title of Count was bestowed), St. Germain, Cagliostro, etc. The two last named lived at a time when chemistry was strong enough as a science to protect itself against the frauds of alchemy. The opposition to the latter which was raised in the course of the preceding century by chemists of repute, *e.g.* Geoffroy the elder (the earlier warnings of Erasmus of Rotterdam, Athanasius Kircher and Palissy having had no effect), led to its ultimate fall, which even the amalgamation of alchemistic aims with those of the secret societies (Rosicrucians, Illuminates, etc.) was powerless to retard. The belief in the possibility of the transmutation of metals received its actual deathblow from the new chemistry which began with Lavoisier.[1] At the same time, however, (*i.e.* about the year 1790), the *Hermetic Society* endeavoured to foster and maintain the alchemistic illusion in Germany. It has only recently come to light that the leaders of this undertaking were Kortum (the poet-author of the *Jobsiade* and a practising physician in Bochum, Westphalia) and a Dr. Bährens, a clergyman. But Wiegleb, a chemist and pharmacist of merit, combated those belated efforts with entire success.

The melancholy errors which arose from the introduction of the mystical religious element into alchemy can but be indicated here; the assertion, frequently made by adepts,

[1] Schmieder, who published a history of alchemy in 1832 (in Halle), did not hesitate to accept the transmutation of metals as having been actually accomplished by various adepts. He expresses himself with more caution regarding the assumed efficacy of the philosopher's stone as a medicine and a means of prolonging life. Even in quite recent times we find the study of alchemy carried on, ostensibly with result, *e.g.*, in Paris in 1844 (cf. Baudrimont, *Traité de Chimie*, vol. i.).

that the secret of making gold was revealed to them through the grace of God, only excites feelings of repugnance.[1] Other frauds, which were likewise the products of alchemistic effort during the eighteenth century, to go no further back, merely provoke satire; among these may be mentioned the endeavours to prepare from the air the so-called "substance of shooting stars" (the alga *Nostoc commune*, which is found in wet ground, was so regarded), and the *materia prima* from "air-salt."

The benefits which have accrued to chemistry during the last four centuries from the mania for producing gold from the base metals, can only be estimated as very slight. It was but very seldom that a discovery of technical importance, like that of the making of porcelain by Böttger, sprang from alchemistic work. On the other hand, it did a vast amount of harm during that period, for it crippled the usefulness of many able men who would undoubtedly have advanced science had they not been influenced by chimeras of an exciting nature; as it was, they were led away into the most tortuous paths.

We are thus forced to the above unfavourable criticism of the work of the alchemists on their problem of the transmutation of metals, in spite of the striking and seemingly incontestable evidence in favour of the latter; in spite, also, of a strong inclination at the present time to a belief in the mutual convertibility of elements chemically similar—a belief grounded upon speculations which do not seem to be without foundation. But in no single case, as yet, has there been any positive evidence brought forward in support of this idea.

If, therefore, we review the work of the alchemists during

[1] Had such misuse of the name of God and of the Bible been made in the time of Luther, as was later the case, or had he been aware of it, his opinion of alchemy would have been a much lower one; as a matter of fact he valued it because of its bearing upon religious feeling. In contradistinction to this stands Melanchthon's criticism of alchemy, a criticism which testifies to the sobriety of his judgment (he called it *imposturam quandam sophisticam*).

the last fifteen centuries, we arrive at the conclusion that it was based upon a series of falsely interpreted chemical problems. The expectation of the easy acquirement of boundless riches, the *auri sacra fames* to which it led, formed the powerful stimulus to the useless and yet continually renewed efforts of an unsatisfied mankind.

CHAPTER III

HISTORY OF THE IATRO-CHEMICAL PERIOD

INTRODUCTION.—Traditional belief, which dominated every branch of science during the Middle Ages, exercised its power not least in the domain of alchemy, for almost every one engaged in chemical pursuits was deluded by the idea that gold and other bodies could be artificially prepared. In the course of the fifteenth century, however, this yoke, which had hindered the development of free inquiry, was in many quarters cast off. The sciences, hitherto studied almost alone in the cloister, now found a foothold in the universities of France, England, Germany and other countries, which were then both increasing in number and expanding rapidly; the free interchange of ideas among these seats of learning rendered a development of the sciences possible, as it had never been before. That the discovery and spread of the art of printing contributed materially to this, hardly requires to be stated; for new ideas, which were opposed to those prevalent up till then, and which had hitherto been restricted to a narrow circle, became quickly disseminated by its aid. Any one could inform himself as to the range of any particular science by means of the encyclopedias and special memoirs which were being printed in increasing numbers. As a consequence of this, the capacity for independent criticism spread, one of the most effectual of remedies against the domination of the scholastics being thereby created. A further aid to controverting scholastic principles was found in the inductive method, then gradually forcing itself forward, by means of which the experimental sciences were called into life.

In addition to these impulses of a freer spirit, chemistry received a powerful impetus from the increase in scientific knowledge which resulted from the discovery of the New World and of the ocean route to the East Indies. All these events testified to the birth of a new era, which found its most powerful expression in the works of the Reformation.

At that time chemistry strove to free itself from the exclusive domination of the alchemistic idea. And even although the latter was not totally supplanted, still another aim came into prominence, an aim to which a scientific character could not be denied; the chemical knowledge of that day was, however, so very imperfect that a solution of this new problem was not to be expected. Chemistry was, in fact, to be intimately conjoined with medicine; each (so many opined) was to help the other. The chemist was to discover the medicines, prepare them carefully, and investigate them chemically, while the physician was to examine and explain their action; or, better still, both things were to be united in one person. The mutual interaction of chemistry and medicine is the main idea which runs through the iatro-chemical age, and which gives to the latter its own particular stamp.

What benefit, then, accrued to both of them from this? The answer is, a mutual enrichment, which did almost more for chemistry than for medicine; for the former was raised to a higher level through being transferred from the hands of laboratory workers, who were mostly uneducated, to those of men belonging to a learned profession and possessing a high degree of scientific culture. The iatro-chemical age thus formed an important period of preparation for chemistry, a period during which the latter so extended her province that she was enabled in the middle of the seventeenth century to stand forth as a young science by the side of her elder sister physics. That period was for chemistry an apprenticeship in the fullest sense of the word, during which she laboriously acquired the capacity to see that the iatro-chemical doctrines were untenable, and to apply herself to her true vocation.

GENERAL HISTORY OF THE IATRO-CHEMICAL PERIOD AND PARTICULARLY OF ITS THEORETICAL VIEWS.[1]

The main currents of the iatro-chemical age emanated from Paracelsus, van Helmont and de le Boë Sylvius, with whose name must be coupled that of his most distinguished pupil, Tachenius, their doctrines being spread by schools of greater or lesser importance. Besides these there were some men who worked independently, or who at least did not entirely subordinate themselves to their authority, of whom Libavius, Glauber and Sala may be mentioned. Other men like Agricola, Palissy, etc., employed their energies in a totally different direction, giving all their attention to technical chemistry.

Paracelsus and his School.[2]—Paracelsus was the man who, in the first half of the sixteenth century, opened out new paths for chemistry and medicine by joining them together. To him is undoubtedly due the merit of freeing chemistry from the restrictive fetters of alchemy, by a clear definition of scientific aims. He taught that "the object of chemistry is not to make gold but to prepare medicines." True, chemical remedies had been employed now and again before his time, Basil Valentine in particular having suggested their use; but Paracelsus differed from his predecessor in the theoretical motives which led him to employ them.

[1] Cf. Kopp, *Geschichte der Chemie*, vol. i. p. 84.

[2] The recent researches upon Paracelsus—more especially Fr. Mook's *Theophrastus Paracelsus* (Wurzburg, 1876); E. Schubert and K. Sudhoff's *Paracelsus-Forschungen* (Frankfurt, 1887–9); and Aberle's *Grabdenkmal, Schädel und Abbildungen des Theophrastus Paracelsus, etc.* (Salzburg, 1891) ("The Gravestone, Skull, and Portraits of Theophrastus Paracelsus, etc.") —have thrown much light upon the life and works of this truly eccentric man. They materially enhance our appreciation of the real services which he rendered.

He regarded the healthy human body as a combination of certain chemical matters; when these underwent change in any way, illnesses resulted, and the latter could therefore only be cured by means of chemical medicines. The foregoing sentence contains the quintessence of Paracelsus' doctrine; the principles of the old school of Galen were quite incompatible with it, these having—indeed—had nothing to do with chemistry.

Paracelsus entered the lists with great boldness and with a marvellous vigour, to combat the old doctrines long accepted by all physicians. Although his exaggerations here are distinctly to be condemned, still he effectually obviated by his action the growing stagnation of medicine, and partly carried through valuable innovations, partly incited others to do so.

His career was not calculated to raise him in the esteem of his opponents, that is, of nearly all the physicians of the time. Paracelsus (his full name was Philippus Aureolus Paracelsus Theophrastus Bombastus) was born at Einsiedeln[1] in Switzerland in 1493, and returned to his native country about 1525 as a physician celebrated for his wonderful cures, after an extremely unsettled life and the most romantic wanderings in almost every country in Europe. The chair of Medical Science (therapeutics) at Basle was conferred upon him, and this position, together with his fame as a doctor, he made use of to spread the iatro-chemical doctrine, and to fight against the old medical school with every possible dialectic weapon. He discredited the hitherto undisputed authority of Galen and Avicenna, and succeeded by means of popular lectures given in German, as well as by his rude originality, in gaining a large number of adherents. A quarrel with the Basle Municipal Council soon compelled him, however, to leave that town (in 1527), and after moving about restlessly in Alsace, Bavaria, Austria and Switzerland, he at last came to Salzburg in the Tyrol, where he died in 1541 in wretched circumstances. The

[1] The name *Eremita* (Hermit) which was given to him by many, recalled that of his native town (the verb *einsiedeln* means "to live like a hermit").

assertion that Paracelsus was done to death by the hirelings of physicians who were his enemies, has been proved to be unfounded (Cf. Aberle, *loc. cit.*).

There has at all times been much difference of opinion in criticising this gifted man, whose life offered such a rude contrast to his mental capacity. Rated too high and even extolled by his disciples,[1] and also by many who disapproved of his doctrines, he was, on the other hand, disparaged by his opponents and by chemists who criticised him as historians. The good to which he incited by his reforming labours seldom found the recognition it deserved, from its being so much mixed up with charlatanism and coarseness, while the overweening estimation in which he held himself helped to make him ridiculous in the eyes of thoughtful physicians.

At the root of his iatro-chemical doctrines, which he imagined were based upon ample experience, lay the idea already mentioned,—that the operations which go on in the human body are chemical ones, and that the state of health depends upon the composition of the organs and the juices. With respect to the constituents of organic bodies, Paracelsus adhered to Basil Valentine's assumption that the latter were composed of mercury, sulphur and salt. Indeed, in spite of many contradictions in the details of his theoretical views, this hypothesis forms the foundation of his whole system.[2] When one of these elements predominates or when it falls below its normal amount, illnesses ensue. This idea is expressed in the most fantastic manner in the writings of this strange man, as the following sentences show :—

An increase of the sulphur gives rise to fever and the plague, an increase of mercury to paralysis and depression, and an increase of salt to diarrhœa and dropsy. By the elimination of

[1] Cf. A. N. Scherer's memoir *Theophrastus Paracelsus* (St. Petersburg, 1821). Francis Bacon criticised him more reasonably, praising his endeavours to get at the truth through the light of experience.

[2] Medicine rests, according to the confused statement of Paracelsus, upon four pillars, of which chemistry forms one; the three others are philosophy, astronomy and virtue.

the sulphur, gout results, and by distilling it from one organ into another, delirium, and so on.—However unfounded such opinions are, it is possible to find a certain sense in them; on the other hand, his utterances upon the relations of the individual organs and the secretions of the human body to the metals and planets, to both of which he ascribes a mystical influence, are quite unintelligible. Not less incomprehensible is his assumption of a connection between the plague and shooting stars. He designates *tartarus* as the cause of various illnesses, meaning by this expression precipitates from juices which in the healthy state contain no solid particles. The deposition of concretionary matter, which he may have observed in the affected organs during many diseases (such as gout, stone in the kidneys and gall-stones), no doubt led him to this partially sound conclusion. The comparison of such secretions with known sediments, particularly with tartar, led to the general designation *tartarus*; the word had possibly also a double meaning, recalling the severe pains which people afflicted with these ailments had to endure.

While Paracelsus endeavoured in this semi-rational, if also fantastic, manner to reduce pathological processes to chemical causes, he assumed nevertheless for his iatro-chemical doctrine the action of particular forces in certain cases, which forces he, in his drastic manner, pictured to himself as personified. Digestion, in especial, was regulated by the action of *Archeus*, who—as a good genius—rendered the nutriment consumed digestible, effected the separation of indigestible matters, and provided generally for the preservation of a proper equilibrium. Diseases in the stomach were produced by *Archeus* becoming ill. In this interpretation of such a specific chemical process as digestion, Paracelsus was disloyal to his own principles. It fell to the later iatro-chemists to clear their doctrinal system from this incongruity.

Diseases were to be cured by medicines (*arcana*), the preparation of which, as we have already seen, was—according to Paracelsus—the aim of chemistry. Due recognition must be given here to the fact that this axiom infused new life into the effete medical doctrines. Paracelsus enriched medicine with a large number of valuable preparations. The manner in which he applied most of these must remain unknown to us; but it is certain that he effected numerous brilliant cures in cases of serious illness. With regard to the preparations which he employed, we know that he was the first to stamp copper vitriol, corrosive sublimate, sugar of lead and various antimony compounds as medicines, these metallic compounds having hitherto been looked upon with dread, on account of their poisonous properties. Further, he brought into use dilute sulphuric acid, "sweetened oil of vitriol" (sweetened by spirit of wine, and which was known at a later date as Haller's acid), tinctures of iron and iron saffron; and he also introduced better methods for preparing and utilising various essences and extracts. He appears to have attained great success by the judicious prescription of laudanum.

That Paracelsus gave a tremendous impetus to the higher development of the apothecary's calling by such generous additions to the medical treasury goes without saying; for before his time apothecaries' shops were nothing more than stores for roots, herbs, syrups and confections of every kind, the preparation of the latter being carried out exclusively in them. The making of new medicines presupposed an acquaintance with chemical facts and processes; pharmacists had therefore to be continually striving to attain to this knowledge, pharmacy, in the proper sense of the word, taking its beginning here. The service which Paracelsus rendered in instigating physicians and apothecaries to busy themselves with chemistry was a great one, but Scherer goes too far when he says that "pharmacy owes everything to Paracelsus."[1]

The trenchant innovations which Paracelsus strove to introduce gave rise to violent agitations among his contemporaries, agitations which were continually receiving new food from his numerous memoirs, circulated in various languages, and dating the most part from the time after

[1] *Loc. cit.*

his departure from Basle. These gave frequent opportunity for vehement contradictions on the part of the old medical school. So far as their composition goes, his writings stand upon a very low level, faithfully reflecting as they do the unsettled life and rude attitude of their author. Every one of them shows an illimitable self-conceit, many indeed being written in a manner quite unworthy of an educated man. His chemical knowledge and his views with regard to the origin of diseases are best seen in the following works:—*Archdioxa; De Tinctura Physicorum; De Morbis ex Tartaro Oriundis; Paramirum; Grosse Wundarznei.*

The results of the labours of Paracelsus were not long in manifesting themselves. His pupils, inspired by the new doctrines, glorified him as the reformer of medicine; while the adherents of the old school, on the other hand, resisted desperately the innovations and attacks which undermined their views. A violent contest ensued and continued for a long time, until it was decided, if not in favour of Paracelsus, at least in that of the more moderate iatro-chemists. It does not lie within the scope of this work to enter minutely into these controversies, sufficing as it does to indicate here the significance of the new medico-chemical views for the development of chemistry. But we may mention that the Swiss physician Erastus (whose German name was Lieber), who remained faithful to the doctrines of Galen, was Paracelsus's chief opponent, and was especially instrumental in exposing the mischievous contradictions which were accumulated in his later writings. The medical world was agitated during the sixteenth century by polemical writings on both sides. Of the disciples of Paracelsus, who, less gifted than their master, paraphrased his ideas and imitated his less amiable peculiarities, especially his charlatanism, but who fell short of him as scientists, Leonhard Thurneysser[1] (called *zum Thurm*) was

[1] A good account of Thurneysser's performances is to be found in Mochsen's admirable work, *Beiträge zur Geschichte der Wissenschaften in der Mark Brandenburg, etc.* (Berlin and Leipzig, 1783). Cf. also A. W. Hofmann's admirable lecture, *Berliner Alchemisten und Chemiker* (1882).

the best known. The latter achieved nothing of any note for chemistry, but his unsuccessful appearance as an adept ensures for him a place in the history of alchemy cf. (p. 62).

The acts of men of this calibre, who wrought immense mischief by the reckless use of poisonous preparations, render intelligible the attempts which were made to put a stop to their excesses by legal statute. This is seen, for instance, by the parliament of Paris prohibiting the prescription of antimonial preparations, and by the sentence of condemnation which the medical faculty of Paris hurled against every attempted innovation in the healing art.

But there belonged also to the school of Paracelsus men of scientific eminence who did not subscribe to all his doctrines, but rather regarded them from a critical point of view, and who endeavoured in a rational manner to extract the good which they contained. The most prominent of these at the end of the sixteenth and beginning of the seventeenth centuries were Turquet de Mayerne and Libavius, Oswald Croll and Adrian van Mynsicht. These were for some time contemporaries of van Helmont, and formed the connecting link between Paracelsus and that remarkable man. They greatly enriched not only medicine but also chemistry.

Turquet de Mayerne was born at Geneva in 1573, and became a noted physician in Paris. Holding, however, as he did, that the antimonial preparations now in ill-repute were necessary, and therefore prescribing them, he found it impossible to keep on good terms with his professional brethren in that city, and preferred to become body-physician to the King of England, in which country he died in 1655. His knowledge of chemistry was very highly developed for that age, as a consequence of which he laboured earnestly for the rational application of chemical remedies, without falling into the exaggerations of Paracelsus on the one hand, or rejecting all the medicines of the school of Galen on the other.

The physicians Croll and van Mynsicht busied themselves in a similar manner and at about the same time.

Having a good knowledge of chemistry, they brought into vogue many of the medicaments of Paracelsus, together with other new preparations; among the latter, Croll was the first to recommend the use of sulphate of potash and of volatile salt of amber (succinic acid), and van Mynsicht that of tartar emetic.

Andreas Libavius (Libau) attracts our attention in a high degree by the critical position which he took up with regard to the gross errors of the school of Paracelsus, and especially also by many new observations which he contributed to chemistry. He was the first chemist of note in Germany who stood up manfully against the excesses of Paracelsus, and who vigorously combated the defects in his doctrines, the obscurities in his writings, his phantasies and sophisms, and the employment of "secret remedies." Originally a physician, Libavius attained to a wide knowledge of chemistry, which he helped to extend, although latterly he devoted himself chiefly to historical and philological studies. He died in 1616 as director of the gymnasium at Coburg, having previously worked with great success as a physician and at the same time as head of the "Latin School" at Rothenburg on the Tauber from 1591 to 1607. Thanks to his medical knowledge and to his thorough general education, Libavius was able to appreciate better than his contemporaries the influence which chemistry ought to exercise upon medicine; he took up a position midway between those of Paracelsus and his opponents, the latter of whom wished nothing less than to banish chemistry from medical science. Notwithstanding his sound judgment, however, of which he gave many proofs, he could not quite free himself from the predilection of his time towards alchemy.

Libavius did chemistry a real service in writing his text-book, which was published in 1595 under the title *Alchymia*, and which contained all the most important facts and theories germane to the subject at that date. His other writings, in which he combated the weak points of the Paracelsian school (as indicated above), and also described new chemical observations, appeared in three volumes shortly before his death, under the title *Opera Omnia Medico-chymica.* We shall still have frequent occasion to refer to his practical chemical knowledge, which was attested by the discovery of important facts.

It is worthy of note that Libavius made a vigorous effort to establish chemical laboratories, in which scientific work should be carried out. From the proposals which he brought forward with this end in view, it is evident that he was desirous to provide plenty of accommodation in these laboratories, and to furnish them with fittings of the most varied kind.[1]

Johann Baptist van Helmont and his Contemporaries.

A distinguished place and a detailed notice in the history of the iatro-chemical period is due to van Helmont, one of the most eminent and independent chemists of his time. Endowed with rich acquirements and experiences in medicine and chemistry, he surpassed those of his contemporaries who worked in the same field. His life was for the most part that of a scholar working in quiet, although his brilliant outward circumstances (he belonged to a noble Brabantine family) were hardly in keeping with this. Born in Brussels in the year 1577, he applied himself at an unusually early age to the study of philosophy and theology; but finding no satisfaction in these, he renounced them to devote himself to medicine. At first an adherent of the old school of the Galenites, he soon recognised its deficiencies and turned to the doctrines of Paracelsus, accepting them, however, only in part. With a growing enthusiasm for his physician's calling, he fought against the old medical system, and materially contributed by his brilliant services in bringing about its fall. Without van Helmont, iatro-chemistry would never have attained to

[1] For an account of the life and work of Libavius, cf. Ottmann's lecture in the *Verhandlungen der Gesellschaft Deutscher Naturforscher, etc.*, 1894, vol. ii. p. 79.

the height to which it was raised later on by Sylvius and Tachenius. In addition, he enriched pure chemistry by a very great number of valuable observations. So attached did he become to his scientific pursuits that he declined the tempting offers of princes, preferring to investigate the secrets of nature in his laboratory at Brussels, in which city he died in 1644.

In van Helmont wonderful contradictions were united. In contrast with his gift of sharp and temperate observation, there was an intense inclination towards the supernatural,—possibly the result of his mystical and magical studies, to which, as well as to theology, he had applied himself. Thus this same man, who laid the foundation of the first knowledge of gases, and showed thereby a keenness of perception unapproached before his time by any other observer, defended the transmutation of the base metals into gold with the utmost vigour (cf. p. 61); his belief in this was grounded so firmly that illusions arose from it which are to us incomprehensible.

After this it is easy to understand that van Helmont was not free from fantastic ideas of a less questionable nature. His theoretical views upon the elements and his iatro-chemical doctrines yield many proofs of this; but, on the other hand, much of his knowledge was so sound, and he was able to expound it so much better than any of his predecessors, that the good service which he rendered far outweighed the bad effect of any of his mistakes.

Van Helmont had his own opinion with regard to the primary substances of which matter was composed; he neither accepted all the four Aristotelian elements[1] nor those which were assumed by Basil Valentine, but looked upon water as the chief constituent of all matter. That it was present in organic bodies he concluded from the fact of invariably finding it as a product of their combustion. He imagined that he contributed a strong proof of this by an

[1] With respect to air, it is uncertain whether van Helmont looked upon it as an element or not. He denied altogether that fire could be of a material nature, which is evidence of his extraordinary clearness of perception.

experiment which showed that plants could be made to grow luxuriantly in pure water alone, which, he believed, was their only nutriment under the circumstances. That he was thereby convinced of the transformation of water into earthy matter is therefore quite intelligible.

Whilst van Helmont thus subscribed to the same error that held possession of many minds both before and after his time, he nevertheless recognised much more clearly than his contemporaries the unchangeableness of matter in numerous instances; thus he contributed more than any one else to do away with the belief that the copper thrown down from a solution of copper vitriol by means of iron was newly created. He further showed that the same substance continued to exist in many of its compounds, *e.g.*, silver in its salts, and silica in water glass, the latter yielding, on decomposition with acids (according to his own memorable observations), the same amount of silicic acid as was originally used to prepare it. These were views and observations of the greatest moment; for, in place of the former obscure conceptions as to the formation of chemical compounds, he substituted the doctrine that the original substance, even after undergoing chemical changes, remains present in the new products. He had therefore clearly grasped the fundamental idea of the conservation of matter in particular cases.

Van Helmont thus stands out as unique in those ideas, which pointed out new paths to chemistry. The relations between chemistry and medicine too, the latter of which he also ardently fostered, led him to views which likewise possess a partial originality, since he endeavoured to decide theoretical questions by means of experiments with juices and other secretions of the animal body. The reactions which go on in the liquids of the body were in his opinion of especial importance, for, according as the latter were acid or neutral, they regulated its most important functions. Besides the chemical nature of the juices, fermentation was, according to him, the principal cause of the organic processes; but he expresses himself less clearly upon this point than upon the significance of the chemical reactions. Indeed, he could not quite free himself from the idea of *Archeus* governing digestion and the processes connected with it. On the other hand, he stood on solid ground in his explanation of vital processes, when he took into account the chemical nature of the juices. He held that the acid of the gastric juice brought about digestion, but this, if present in excess, gave rise to discomfort and illnesses, which were the more serious the more acid there was; and the latter could not then, as under normal conditions, be neutralised by the alkali of the bile, which mixes with the gastric juce in the duodenum. To cure any of the ailments produced in this way, van Helmont declared that medicines of an alkaline nature (alkaline salts) must be used; while those of an opposite kind, which arose from a deficiency of acid, were to be treated by medicines of an acid nature. He also recommended the latter in cases of gout, stone and similar diseases, which likewise originated (in his opinion) from an insufficient or irregular admixture of the juices. These views show a distinct advance upon those of Paracelsus. For, while the latter assumed the presence of arbitrary constituents—incapable of preparation—in organic matter, van Helmont searched for the actual substances themselves, and compared the interactions of the various juices which mingle with one another with similar reactions of solutions outside the organs; a procedure which laid the first foundation, however insecure, of chemical physiology.

Van Helmont proved himself an original investigator of the first rank, who opened out new ground for chemical science by his researches on gases—researches which constitute him the real founder of pneumatic chemistry; though this indeed only attained to a considerable development a century after his time, when the discoveries connected with it brought about the great reform of the science. If we consider that before van Helmont's time the most various gases, such as hydrogen, carbonic acid and sulphurous acid, were looked upon as not differing materially from ordinary air, and that he was the first to characterise gaseous substances as different, by investigating their properties, we gain some idea of the immense services which he rendered. He it was who gave to them the generic name of "gas,"[1] and he further distinguished them from vapours, in so far that the latter were condensed to liquids upon cooling, while the former were not.

Van Helmont specially examined carbonic acid, and showed how it was produced from limestone or potashes with acids, from burning coal, and in the fermentation of wine and beer; he also pointed out its presence in the stomach, and its occurrence in mineral waters and in many natural cavities in the earth. He usually termed it *gas sylvestre.*[2] To the want of suitable apparatus for collecting gases are to be ascribed the imperfections in many of his observations, and also the confounding of carbonic acid with other gases which were non-supporters of combustion like itself; nevertheless he described the two combustible gases —hydrogen and marsh gas—as peculiar varieties of air. His collected works were published in 1648 by his son under the title, *Ortus Medicinæ vel Opera et Opuscula Omnia.*

Van Helmont's influence upon his contemporaries and upon the development of the iatro-chemical doctrines must be rated very high. By his introduction of chemical ideas into medical science, the latter was advanced, because the use of chemical medicines seemed natural from thenceforth; moreover, in his *Pharmacopolium ac Dispensatorium Modernum,* he published suitable prescriptions for the preparation of medicines. The scientific spirit which he endeavoured to introduce into the healing art tended to its more healthy development, in contrast with the crude empiricism of the Paracelsian school.

In a similar manner, if in lesser degree, various other physicians of that time were also active. Well equipped

[1] In choosing this designation, van Helmont had Chaos in his mind,—possibly also the process of fermentation (the Dutch word for the verb "to ferment" is *gisten*).

[2] By the designation *sylvestre,* he doubtless meant to indicate the impossibility of condensing the gas; at least he says in one passage: *Gas sylvestre, sive incöercibile, quod in corpus cogi non potest visibile.*

with chemical knowledge, they pursued the practice of their calling, and were enabled by their clearness of vision to recognise and combat many evils, *e.g.* those which arose from the use of secret remedies; among them we must mention Angelus Sala and Daniel Sennert. Sala, who practised as body-physician at the Mecklenburg Court in the first half of the seventeenth century, awakens our surprise by his able criticisms both of the Paracelsian and of the old medical schools, and also by his (for that time) wide knowledge of chemistry. This knowledge, conjoined with his solid medical experience, was of the utmost value not only to pharmacy but also to pure chemistry; for he formed correct ideas with regard to the composition and reactions of many chemical compounds, such as had never been advanced before his time. Thus he tells us that salmiac consists of hydrochloric acid and carbonate of ammonia (*flüchtiges Laugensalz*), and he also knew that sulphuric acid was able to drive out nitric acid from its salts,—and so on.

Sennert, who taught as professor at Wittenberg in the first quarter of the seventeenth century, devoted his energies chiefly to proving to the medical world the wonderful efficacy of chemical remedies, when these were properly applied. It is true that he was never able to disentangle himself from many of the erroneous conceptions of Paracelsus, for instance, from the doctrine of the three primary elements; but he worked effectively against the serious abuses which had crept into medicine through the influence of the last-named, especially against the so-called universal remedies.

Sylvius and Tachenius.—F. de le Boë (Dubois) Sylvius was born at Hanau in 1614, and, after a thorough grounding in scientific and medical studies, practised with great success as a physician, and later on, until his death in 1672, was famous as professor of medical science in Leyden. In his knowledge of medicine he far surpassed most of his contemporaries. He was aware of the difference between arterial and venous blood, and ascribed the red colour of the former to the air absorbed in breathing. Combustion and respiration were in his view precisely similar phenomena. He directed all his efforts, as instanced in this latter case, to proving that the processes which go on in the human body —whether they be normal or pathological—were purely chemical ones. The spiritualistic element which was mingled with the doctrines of Paracelsus and van Helmont was to be entirely set aside. Digestion, for instance, which only appeared possible to the two latter by the intervention of a spirit (*Archeus*), was regarded by Sylvius as a chemical process in which the saliva primarily, but also the gastric and pancreatic juices and the bile, were the most important acting agents. To the acid, alkaline, or neutral reactions of the juices of the body he ascribed an equal, if not a higher, significance than van Helmont himself, following the latter in this as in similar questions. Sylvius had a predilection for comparing chemical with physiological and pathological processes, which frequently led him into error. Medicine as a whole, he considered, ought simply to be applied chemistry. That these one-sided endeavours were bound to miscarry, considering the state of chemical knowledge at that time, requires no demonstration. And it is equally easy to understand why his chemical doctrines brought less benefit to medicine than to chemistry, seeing that educated physicians, if they wished to comprehend them, were compelled to go minutely into the study of chemical questions. This applied in a very special degree to the new remedies, the preparation and rational application of which presupposed a knowledge of chemistry. Sylvius, addicted as he was to the use of heroic medicines, did not hesitate to prescribe *lapis infernalis* (nitrate of silver), sublimate and zinc vitriol for internal use; and he was particularly enthusiastic about antimonial and mercurial preparations.

While there are but few discoveries in pure chemistry by Sylvius himself to chronicle, his pupil Otto Tachenius proved himself an independent investigator, to whom the science is indebted both for extremely valuable observations and for speculations deduced from these. Of his life we

G

only know that he was born at Herford in Westphalia, and that, after moving about from place to place as an apothecary's assistant, he applied himself to the study of medicine in Italy towards the middle of the seventeenth century, and practised in Venice as a physician. Although he attached the greatest weight to clear relations between chemistry and medicine, he had no hesitation in working mischief with secret remedies. Tachenius was the last iatro-chemist of note who followed the doctrines of Sylvius with enthusiasm. In addition to him may be mentioned here the famous English physician Willis (*ob.* 1675), who likewise advocated similar views.

Tachenius, among his other valuable observations, contributed materially to elucidating that problem which Boyle considered the most important of all, viz. a knowledge of the composition of bodies. It was with him that the first serviceable definition of the term "salt," as a compound of an acid and an alkali, originated. His statements on the composition of various compounds show great acuteness, which is also seen in the value he attached to certain reactions as tests for different substances. While Tachenius thus laid the foundations of qualitative analysis in a more systematic manner than his predecessors, his attention was also directed to the quantitative proportions in which substances react chemically —a point to which hardly any attention had hitherto been paid; and this he exemplified with tolerable accuracy by noting the increase in weight which took place when lead was transformed into minium. His writings, and also those of his master Sylvius, treat for the most part of subjects of chiefly medical interest, but, as we have just seen, facts and opinions of importance to chemistry are also recorded in them.

If we wish to arrive at the main result which the iatro-chemical doctrines produced upon the development of chemistry, we must particularly bear in mind the point already touched upon, viz. that the study of chemistry by physicians who had had a thorough education helped materially to shape its course on scientific lines. Notwithstanding the numerous errors and fantastic conceptions in which the iatro-chemists were involved, we come across many very striking views,—views which exercised a marked influence upon the whole tendency of the succeeding epoch. Of these we would mention here: (1) the recognition of the more intimate components of salts, and the clearer comprehension of what was meant by the terms "chemical compound" and "chemical affinity," by a knowledge of which the chief aim of chemistry, *i.e.* the investigation of the true composition of bodies, was effectively advanced; and (2) the recognition of the analogy between the processes of combustion and the calcination of the metals on the one hand, and respiration on the other. These were doctrines of very great weight indeed. The phlogistic hypothesis, too, which predominated during the greater portion of the eighteenth century, was indicated by many of the iatro-chemists; *i.e.* many of the latter had ideas upon combustion which approximated to those of the phlogistonists. Lastly, van Helmont's work upon gases exercised the greatest influence on the development of pneumatic chemistry, from which the impulse to the great reform of our science at the end of last century sprang.

It is thus evident that many of the aims of the phlogistonists were intimately connected with the observations and opinions proper of the iatro-chemists. And while the medico-chemical opinions of the latter were rudely upset after the middle of the seventeenth century, their facts and theories appertaining to chemistry were the means of guiding the latter into scientific paths.

Agricola, Palissy, and the other Promoters of Applied Chemistry during the Iatro-chemical Age.[1]

Independently of the main iatro-chemical current, chemistry in its applications to industries was fostered by men who possessed, for their time, sound chemical knowledge.

[1] Cf. Kopp, *Gesch. d. Chem.*, vol. i. pp. 104, 128; and Höfer, *Histoire de la Chimie*, vol. ii. pp. 38, 67 *et seq.*

The chief of these were Georgius Agricola, who directed his attention specially to metallurgy; Bernard Palissy, who developed the ceramic art; and Joahnn Rudolf Glauber, who, without ceasing to be an iatro-chemist, devoted his powers for the most part to technical chemistry. The following paragraphs give a few details explanatory of the services rendered to the science by the knowledge and experiences of those men; but what we are chiefly concerned with here is their significance from a more general point of view.

Georgius Agricola [1] was born at Glauchau in 1494, and became a noted physician; he died while mayor of Chemnitz in 1555. He was thus a contemporary of Paracelsus. Although, like the latter, a medical man, he followed totally different lines. Without troubling himself about the storms which raged round medicine in his day, he devoted himself by choice to the study of mineralogy and metallurgy, being impelled thereto by the flourishing mining and smelting industry of Saxony, while at the same time he continued to practise as a doctor. His chemical knowledge and wide experiences are detailed by him in his principal work: *De Re Metallica, libri XII*, which remained for a long time the most important text-book of mineralogy. Through this, as well as through his other writings—of which *De Natura Fossilium, libri X*, and *De Ortu et Causis Subterraneorum* were also of especial mineralogical value,— there runs quite a different tone from what we find in Paracelsus. They are characterised by a clearness of expression, a temperate conception of the operations described, and a distinct description both of the apparatus employed and the processes followed,—qualities which stamp Agricola as a true investigator. It was through his writings, especially through the first of those named above, that the more important operations in the working up of ores for their metals first became generally known; and he was likewise the first to explain intelligibly the

[1] Cf. G. H. Jacobi's dissertation:—*Der Mineralog Georgius Agricola und sein Verhältniss zur Wissenschaft seiner Zeit* (Leipzig, 1889). ("The Mineralogist, Georgius Agricola, and his relation to the Science of his Time.")

manufacture of other products obtained by smelting, and of various preparations of technical importance. His works are indispensable to the history of metallurgy.

His quiet objective modes of thought and investigation did not, however, prevent him in his more mature age from attributing a certain degree of liklihood to the alchemistic problem, to which he had devoted himself warmly in his youth; at the same time he had no sympathy with the wild exaggerations which even then prevailed.

Working on lines similar to those of Agricola, and at about the same period, the Italian Biringuiccio of Siena busied himself with the processes of metallurgy, as detailed in his work *Pirotechnia*, which appeared in 1540. This too is marked by the clearness and exactitude with which various technical procedures are described. Biringuiccio held aloof from the iatro-chemical questions and the alchemistic doctrines of his day.

Bernard Palissy became distinguished as an investigator, and as a man who allowed himself to be guided solely by the results of experiment, at a time before the inductive method was commonly recognised as the means of attaining to the truth. It was in the domain of ceramic art that his principal work lay; and, although frequently disappointed in the results he obtained, his untiring efforts at improvment in it were ultimately followed by success. The simple and clearly written works of Palissy enable us to appreciate the labours and struggles of this remarkable and steadfast man, who, beginning as a common potter destitute of the higher education, became the great authority on his subject.[1] He took his first lessons from the book of nature, as he himself tells us; [2] putting observation and experiment in the foreground, he combated every speculation which was not based upon these, especially such doctrines as had merely the stamp of authority to back them. There could hardly have been any man of his

[1] Höfer, who was the first to recognise the services of Palissy as they deserved, speaks of him as "*un des plus grands hommes dont la France puisse s'enorgueillir*" (*Histoire de la Chimie*, vol. ii. p. 92).

[2] "*Je n'ai point eu d'autre livre, que le ciel et la terre, lequel est connu de tous et est donné à tous de connoistre et lire ce beau livre.*"

time more free from prejudice; his clear understanding and circumspect criticism enabled him to cast aside the doctrines of Paracelsus, and to make use of the weapons of ridicule against the mistaken beliefs of alchemy. His life extended over nearly the whole of the sixteenth century, and might be said to consist of a series of vicissitudes. Along with Agricola he may be looked upon as the chief exponent of experimental chemistry in his time. His acuteness was further evidenced in the domains of mineralogy and agricultural chemistry, to the founding of which branches of science he largely contributed.

Johann Rudolf Glauber, who was born at Karlstadt in Franken (Bavaria) in 1604, and who died in 1668 at Amsterdam, fostered applied chemistry ardently, and enriched it by valuable observations. It was in this direction that he chiefly worked, his iatro-chemical labours holding but a secondary place. His life was an extremely restless one, which may not improbably account for the unsettled and almost discontented tone which runs through many of his writings. Without a classical education, and imbued with the prejudices of his age, he has been well designated the Paracelsus of the seventeenth century. He was, in fact, addicted to fantastic and superstitious ideas, and therefore also to the extravagances of alchemy; on the other hand, he possessed exceptional talents of observation and invention, regarding which some details will be given in the next section of this book. In theoretical points of chemistry, too, he gave proof of his clear-sightedness, explaining, for example, many of the effects of chemical affinity in the decomposition of salts by acids or bases,—and so on. He was the first to explain a case of what we call double decomposition,—the mutual action of mecuric chloride and antimony trisulphide upon one another. Mention must also be made here of his perspicacity in questions of national economy, his writings upon which are to be found mixed up with his chemical papers, especially in the six-volume work *Teutschland's Wohlfarth* ("The Weal of Germany"). Time after time Glauber sought to demonstrate that his country should work up and improve its own products, and not leave this for other nations to do; instead of buying at a dear rate manufactured articles whose raw material was obtained from Germany, that country ought to make and export them herself.

With Glauber and Tachenius the iatro-chemical period closes. Both of them belonged in many of their chemical ideas and also in point of time (during the last years of their lives) to the succeeding era, between which and the previous one it is impossible to draw an absolutely sharp line. Both aided chemistry by observations of extreme value, and materially advanced the experimental method, which became from thenceforth the sure guiding star of chemical research.

EXTENSION OF PRACTICAL CHEMICAL KNOWLEDGE IN THE IATRO-CHEMICAL AGE.[1]

As was to be expected from the whole tendency of this period, during which chemistry became so intimately united to medicine, the gain of knowledge lay chiefly in respect to chemical preparations, which it was hoped to apply as medicines. The efforts to discover new remedies had the result of causing chemical compounds, whether novel or already known, to be investigated more carefully and scientifically than had ever been done before. The products of the animal body were zealously studied, and a small beginning was made in physiological chemistry by the examination of milk, blood, saliva, etc., which in its turn increased the interest felt in organic compounds. In technical chemistry less progress was made than in chemistry which was related to medicine. An advance in the knowledge of the composition of substances and in the observation of reactions, *i.e.* in qualitative analysis, first became noticeable towards the end of the iatro-chemical period.

Technical Chemistry.—The most eminent exponents in this direction, chief among whom were Agricola and Palissy,

[1] Cf. Kopp, *Gesch. d. Chemie*, vol. ii. pp. 111, 126; vols. iii. and iv. *passim*.

have been already referred to. In their works, as also in the writings of Biringuiccio, Cæsalpin and others which are devoted to technical chemistry, special weight is laid on the particular operations by which technical products are obtained, these operations being minutely described.

In Metallurgy Agricola and Libavius were the first to point out a method by means of which it was possible to estimate approximately the amount of metal in an ore; the science of testing thus gradually developed itself from such beginnings. The more scientific treatment of applied chemistry is further shown by the fact that by-products began to be used which had previously been neglected, *e.g.* the sulphur which escaped during the partial roasting of pyrites was condensed, the tutty from zinc ores was utilised for brass, and so on.

A knowledge of the individual metals, and of the methods by which they could be obtained and worked up, became extended in the sixteenth century by Agricola and other authors making into common property what had hitherto been only known to the few; *e.g.* the separation of gold from silver by means of nitric acid, which was first carried out on a large scale in Venice towards the end of the fifteenth century, and the amalgamation process, first applied in Mexico about the middle of the sixteenth century for extracting silver from its ores, but only introduced into Europe towards the end of the eighteenth. It is in the sixteenth century that we find the first reliable observations on the production of ruby glass by means of gold. Salts of the latter metal and also of silver were more carefully investigated, with reference particularly to their medical application; and some of their characteristic reactions—by which it became possible to distinguish them from other substances—were also noticed.

With respect to copper and its precipitation from a solution of copper vitriol by means of iron, we find even chemists of discernment like Libavius holding fast to the old idea that a transmutation had occurred; but others, *e.g.* van Helmont and Sala, recognised the pre-existence of the copper. The metallurgical operations necessary for obtaining iron became generally known through Agricola's writings,[1] thus the production of steel by the puddling process was first described by him. Steel was at that time regarded as a very pure iron. Of the other metals, a knowledge of zinc and bismuth was gradually acquired, although there was often uncertainty about them, and they were frequently confounded with antimony. Tin, lastly, was much used in the sixteenth century for tinning iron. But the iatro-chemical age interested itself less in the metals themselves than in the salts prepared from them, since there was always the chance of these proving useful in medicine. (See under *Preparations.*)

Pottery and Glass Mauufacture.—The ceramic industry in particular made considerable progress, thanks to the untiring efforts of Palissy; his only guide was the experience gained from innumerable trials, but he succeeded in affixing beautiful and durable enamels on earthenware vessels, especially on those of Fayence pottery. His observations on this point, and also on the application of different clays for ceramic purposes, and the burning-in of colours, are given in his work *L'Art de Terre,* which at the same time aims at showing the value of the experimental method as opposed to theory alone. Porta was also busy in Italy about the middle of the sixteenth century with work similar to Palissy's.

The manufacture of glass did not lag behind that of pottery. From the Venetian factories, whose sixteenth-century productions still astonish and delight the connoisseur, the art of making glass of the most various colours and of different degrees of refrangibility spread to other countries. The work of the Florentine Antonio Neri, entitled *De Arte Vitraria,* which appeared in 1640, not improbably contributed materially to spreading a knowledge of special operations, his large experience on the subject being detailed in this book. Great skill was also attained

[1] The significance of Agricola's work in this field is clearly seen in the account given by L. Beck in his *Geschichte des Eisens,* vol. ii. p. 22, etc.

even at that date in the imitation of precious stones, as Porta's recipes show. One of the most important discoveries of the time was that of cobalt blue by Christoph Schürer, a Saxon glass-blower, who obtained it on fusing the cobalteous residue from the manufacture of bismuth with glass; it soon became a much-prized article of commerce, being known under the names zaffre (from sapphire), and, later on, smalt.

Dyeing.—One of the results of the discovery of America and of the ocean route to the East Indies was seen in the increased importation of indigo and cochineal, which gave a fresh impetus to the dyeing industry. Many improved methods of fixing these and other colours upon cloth—*e.g.* the use of a solution of tin, the judicious mordanting of the stuffs with alum, iron solution, etc.—were found out in the sixteenth century. The dyer of that time might consult the first text-book on this subject, written by the Venetian Rosetti, which appeared in 1540. Glauber, too, made numerous observations on dyeing processes, and aided not a little in advancing a knowledge of these.

A new industry sprang up towards the end of the fifteenth century in the rapidly extending distillation of brandy; up to that time spirit of wine was looked upon as a medicine only, but now it began to be more and more used, sufficiently diluted, as a drink. The development of this branch of trade resulted in great improvements in distilling apparatus, which also came to be of service in laboratories. The interest which this industry excited is seen from the numerous works upon the art of the distiller which appeared at that time.

The applications of chemistry were in fact extended in the most various directions, among others to agriculture, if only in a modest degree; thus we find the gifted Palissy calling attention to the importance of soluble salts in manures, and recommending the addition of mineral substances, *e.g.* marl, to these. Here we have the earliest beginnings of a rational chemistry of agriculture.

Development of Pharmacy and of the Knowledge of Chemical Preparations.

Pharmaceutical chemistry is most distinctly a creation of the iatro-chemical age, during which it was taught that the chief aim of chemistry lay in the discovery of medicines that could be prepared artificially. In accordance with this dictum, not only were the preparations already known, but also those others which had been newly discovered after much seeking, tested for their action upon the organism. The circle of chemical facts was thus greatly widened by these iatro-chemical labours. The influence of the latter upon chemistry was made further apparent by the fact of the drug-shops in which artificial preparations were made, becoming the nurseriés of hard-working chemists, who, especially in the succeeding generation, played an important part in the building up of the scientific system.

Inorganic Compounds.—The preparation of mineral acids showed improvements, and their investigation was marked by advances which, however, only became of practical value later on, when the acids began to be employed technically. Glauber taught how to prepare hydrochloric acid from rock-salt and oil of vitriol, and also fuming nitric acid from saltpetre and white arsenic.

To Libavius belongs the merit of simplifying the mode of preparing sulphuric acid, and of proving that the acid obtained in various ways—from alum, vitriol (sulphate of iron), or sulphur and nitric acid—was one and the same substance. The behaviour of the acids just named to metals, salts and organic compounds led to a knowledge of a great number of bodies which had been either unknown hitherto, or at least had never been produced in this particular way; and thus, from their modes of preparation, deductions as to their composition often became possible. Among such substances were the chlorides formed by the action of hydrochloric acid upon many of the metals, which up to then had been prepared by heating the latter with sublimate, and hence the presence of mercury in the resulting products was assumed. Glauber, to whom we owe a knowledge of many of them,—*e.g.* zinc, stannic, arsenious and cuprous chlorides—disproved this erroneous assumption; he and his contemporaries regarded these salts as compounds of the metal and hydrochloric acid.

Salts were destined to play a very great part in medicine. Especial interest was taken in the alkaline salts, both from a theoretical point of view—their composition being a frequent theme of discussion—and also from a practical, on account of their technical and officinal applications.

Potash saltpetre, which was prepared on a large scale on account of its increasing use in the manufacture of gunpowder,[1] was also prized as a medicine when fused. The observation made by the pseudo-Geber—so important for a knowledge of its composition—that saltpetre results on saturating potashes with nitric acid, was first made use of technically in the iatro-chemical age. Sulphate and chloride of potash, which were prepared by many different methods and known under various names, were employed as medicines,—the former by Paracelsus, and the latter by Sylvius and Tachenius (as *sal febrifugum Sylvii*). Carbonate of potash, too, prepared from tartar and the ashes of plants, was another medicament. Even iatro-chemists of eminence like Tachenius believed in a chemical difference between various potashes, according to their modes of preparation,—an error which Boyle was the first to correct; still more frequently do we meet with a confounding of potash salts with those of soda, *e.g.* their carbonates and chlorides. Glauber's sulphate of soda, obtained from the residue left in the manufacture of hydrochloric acid, and known under the name of *sal mirabile*, was highly prized by physicians. Whether borax, which was used in soldering during the iatro-chemical period, was also employed as a medicine is doubtful.

[1] Agricola describes the preparation of saltpetre in his work *De Re Metallica*.

Salts of ammonia were largely used, both officinally and technically, especially sal ammoniac, whose manufacture was attempted in Europe so early as the seventeenth century; its artificial formation from volatile alkaline salt and hydrochloric acid was known to the iatro-chemists of that time (Sylvius, Tachenius, Glauber), but it was only at a much later date that its true composition was indicated. The near relation thus found to exist between carbonate of ammonia and salmiac led conversely to the preparation of the former from the latter by means of carbonate of potash; from the apparently different action of samples of volatile alkaline salt of diverse origin (from blood, urea and salmiac), it was supposed that they were different compounds, but this error was recognised by Tachenius. Of other salts of ammonia we may mention the sulphate, discovered by Libavius, the nitrate, by Glauber, and the acetate; the last, known as *spiritus Mindereri* (from its discoverer, the physician Raymund Minderer), was much valued as a medicine

But few of the salts of the earths were known, and there was uncertainty as to their composition. Lime and alum earth (alumina), for instance, were supposed to be pretty much the same. Of their salts, alum—prepared by adding putrefied urine to the crude alum lye (the aqueous extract from roasted aluminous shale)—was much prized for its technical value, and was manufactured in large quantity; the alum of that day was thus essentially ammonia alum. Agricola himself characterised gypsum as a compound of lime, while chloride and nitrate of calcium were known in the seventeenth century, and possibly even before then. Agricola and his contemporaries were also aware that silica (*i.e.* pure sand)—which was for long reckoned as one of the earths—fused with potashes to a glass which was soluble in water, and the clear-sighted Tachenius saw in this behaviour an indication of the acid nature of the substance.

The salts of the heavy and of the noble metals, and various preparations of the semi-metals (arsenic, antimony and bismuth) were of much importance for iatro-chemistry, and therefore also for the development of the chemical knowledge of that time. Basil Valentine, as we have already seen, had worked up a large number of antimonial preparations and had recommended them for internal use prior to the appearance of Paracelsus and his school. And although, in consequence of the abuses resulting from secret medicines containing antimony, sharp edicts were issued prohibiting their employment, preparations of antimony came notwithstanding more and more into favour, this being greatly due to the efforts of Sylvius. Metallic antimony itself was prescribed in pills, which were called "the everlasting," since it was believed that they acted merely by contact, and that therefore, after passing through the body, they could be used again and again.[1]

It was during this period that "Kermes" mineral, *sulphur auratum*, and powder of algaroth[2] were added to the medical treasury; antimoniate of potash—prepared by detonating antimony trisulphide with saltpetre—was also much used as a medicine. To Glauber more than any one else is due a clearer knowledge of the chemistry of this and other antimony compounds.

There was still great obscurity with regard to white arsenic and the metal prepared from it, and also with respect to other arsenic compounds; among the latter we may mention arseniate of potash, which was prepared by fusing the trioxide with saltpetre, and for which Paracelsus appears to have had a great predilection as a medicine (*arsenicum fixum*). Glauber was the first to prepare chloride of arsenic ($AsCl_3$). Preparations of bismuth were less used for medicinal purposes, although the similarity between bismuth and antimony, which often led to confusing the one with the other, did not escape the iatro-chemists. Basic nitrate of bismuth was much prized as a cosmetic, while the oxide, according to Agricola, was used as a paint.

[1] Lemery in his *Cours et Chimie* (1675) remarks upon the use of these pills as follows: "*Lorsqu'on avale la pillule perpetuelle, elle est entrainée par sa pésanteur, et elle purge par bas; on la lave et on la redonne comme devant, et ainsi perpetuellement.*"

[2] So called after the Veronese physician Victor Algarotus, who praised it as *pulvis angelicus*.

Of the compounds of zinc, the oxide, zinc vitriol (which Agricola terms *chalcanthum candidum*), and the chloride became better known; the last of these was prepared by Glauber by heating calamine strongly with hydrochloric acid, and it therefore contained basic salt. From tin Libavius obtained its tetrachloride, by distilling it with sublimate; assuming in this the presence of mercury, he termed it *spiritus argenti vivi sublimati*, but later on it was commonly known as *spiritus fumans Libavii*. The solution of this compound, obtained by treating tin with aqua regia, began to be applied by Drebbel in many dyeing operations about the year 1630.

The discovery and investigation of ferric and plumbic chlorides, the latter of which was used instead of white lead as a paint, is likewise due to Glauber. The methods of preparing many metallic salts already known were also much improved, as is seen, for instance, in the description given by Agricola of the preparation of iron and copper vitriols.

The iatro-chemists devoted much attention to the production and medical application of quicksilver compounds. It was given to Paracelsus to overcome the prejudices of many against mercurial medicines, although most of the physicians of the old school would have nothing to do with them. Paracelsus and his disciples had no hesitation in making use of metallic mercury—finely divided in pills,—sublimate, and the so-called turpeth mineral (*i.e.* basic mercuric carbonate or sulphate, both of which went under this name). In this way a much better knowledge of various mercury compounds was gradually arrived at, some of these compounds being already known and some newly discovered. Among the latter were calomel and white precipitate (from sublimate and ammonia), both of which were prized as medicines. It was during this period that chemists gradually learnt that cinnabar consisted of mercury and sulphur, and that mercury itself belonged to the true and not to the half-metals.

Of the compounds of silver, *lapis infernalis* (the nitrate) was found useful in medicine, principally through Sala's recommendation, and the sulphate and chloride of silver were also known. The production of the latter on precipitating a solution of silver with hydrochloric acid or common salt dissolved in water, was made use of analytically as a test both for silver and for chlorides.

Indeed the beginnings of qualitative analysis in the wet way are to be sought for in the iatro-chemical age, in so far that conclusions regarding the presence of one or another constituent were drawn from the appearance and behaviour of precipitates, and of salts which crystallised out from solution. Tachenius laid especial weight on distinguishing such precipitates by their colours, and he was himself able to detect several metals in solution together by means of certain reagents, such as tincture of galls, the carbonates of potash and ammonia, caustic potash, etc.

Organic compounds became known in rapidly augmenting numbers, in consequence of the increasing attention paid to the products of vegetable and animal assimilation; the actual knowledge of such bodies continued, however, very superficial and incomplete, since their composition remained quite obscure. Of the acids, acetic acid became better known, and several of its salts were used in medicine with good effect. It did not escape Glauber that the distillate from wood contained an acid which strikingly resembled that of vinegar. The iatro-chemists taught how to prepare concentrated acetic acid by the distillation of verdigris, whence it was known as copper spirit or radical vinegar; and this latter substance Tachenius was inclined to regard as van Helmont's alkahest. The two acetates, sugar of lead and the basic acetate, were also examined more accurately by Libavius, and employed as medicines.[1]

Salts of tartaric acid, of which tartar had been known for a long time, came to be valued as medicines in the sixteenth century; the discovery of the free acid itself belongs to a

[1] The liquid which distils over on heating sugar of lead and which we now know to contain acetone, was investigated repeatedly; from its designation of *quintessence*, a specially high value seems to have been put upon it.

much later date. The designation *tartarus*, applied to tartar, was likewise the generic name in the iatro-chemical age for other very different salts, *e.g.* for the salts of potash, in so far as they were prepared from tartar, and also for sediments from solutions, especially those from animal secretions. The part which *tartarus* played in the theoretical considerations of iatro-chemistry has already been spoken of. The salts of other vegetable acids were also frequently termed *tartarus*, *e.g.* salt of sorrel, which appears to have been often confused with tartar. Neutral tartrate of potash, known as *tartarus tartarisatus*, from its preparation from tartar and salt of tartar (K_2CO_3), and the double tartrate of potash and soda, called Seignette salt after the man who accidentally discovered it, likewise became known to chemists.

A compound of even greater importance to the medical treasury than the tartrates just mentioned was tartar emetic, the preparation of which from oxide of antimony and tartar was described by the Dutch physician Mynsicht, and afterwards more accurately by Glauber.[1] A tartar containing iron (*tartarus chalybeatus*) became known through Sala's *Tartarologia*. Paracelsus also made use of the distillate from tartar—which is now known to contain pyro-tartaric acid besides other substances—as a medicine (*spiritus tartari*).

Succinic acid, the near relation of which to tartaric has only become clear in our own time, is described by Libavius and Croll under the name of *Bernsteinsalz* (*flos succini*), what they referred to being the distillation product of amber; Lemery was the first to recognise its acid nature, about 1675. The acid juice of the apple and other fruits was employed for preparing various medicines (*e.g.* the *tinctura martis pomata*), before any attempt was made to isolate the acid itself. Free benzoic acid, however, obtained by subliming gum benzoïn, was discovered and minutely described by the French physician Blaise de Vigenère (1522–96) towards the end of the sixteenth century, while Turquet de Mayerne described

[1] It may just be mentioned here that the taking of small quantities of tartar emetic, prepared by allowing wine to stand in goblets made of antimony, had been a common practice long before this.

the improved method of preparing it in the dry way, which is still practised at the present time. The juice of gall apples, which contains tannic acid, and the extract of oak bark were used by many iatro-chemists from the time of Paracelsus to test for iron in solutions, especially in mineral waters; but no one succeeded in isolating either tannic or gallic acid itself.

Although the old observation—that the fats were altered chemically by the alkalies and metallic oxides—did not lead the iatro-chemists to a knowledge of the fatty acids, it guided many of them, the acute Tachenius in particular, to the correct assumption that "oil or fat contains a hidden acid." It was only one hundred and sixty years later that Chevreul's work upon fats laid the firm foundation for the present views upon their chemical constitution.

Spirit of wine—the *aqua vitæ* of the alchemists—continued to grow in importance during the iatro-chemical age, as it had done in the alchemistic. This applied to it not merely from a theoretical point of view, as being a product of various fermentation processes to which much attention was paid, but also from a practical, since Paracelsus and his disciples used it largely in the preparation of essences and tinctures.[1]

To the German physician Valerius Cordus is due the first exact knowledge of the ether produced from alcohol by acting upon it with sulphuric acid, although his instructions for preparing it were only published after his death, and the ether then accepted in the Pharmacopeias as *oleum vitrioli dulce verum* (about 1560). His work, however, became forgotten so soon, that we find even such an accomplished chemist as Stahl unaware of it. A mixture of alcohol and ether, which later on enjoyed a wide popularity under the name of Hoffman's drops, had probably been employed by Paracelsus as a medicament. The knowledge of compound

[1] The name alcohol (*alcool*) for spirit of wine, which has been in common use since the time of Libavius, had formerly quite another meaning, having been applied indifferently to antimony sulphide, vinegar and various other compounds.

ethers remained very fragmentary, scarcely any addition having been made to it since the observations of Basil Valentine (p. 57).

The work done upon other organic substances led to their practical application in medicine and in daily life, and also to improvements in the modes of preparing them, *e.g.* in the extraction of sugar from the sugar-cane, the juice being clarified by white of egg and lime; but scientific knowledge with regard to such bodies remained at the lowest level.

CHAPTER IV

HISTORY OF THE PERIOD OF THE PHLOGISTON THEORY, FROM BOYLE TO LAVOISIER

Introduction.—The reasons for naming this period of about one hundred and twenty years the period of the phlogiston theory, or of phlogistic chemistry, have been already stated shortly (p. 4). For the first portion of this era the designation is in truth not absolutely fitting, since Robert Boyle—the man who above all others gave a new direction to chemistry at the time—did not concur in the phlogistic views. The development proper of the phlogiston theory really took place after his death. Nevertheless the period from Boyle to Lavoisier may be so named, because the most important part of chemical research during that time had to do with the phenomena of combustion and—what was recognised as analogous—the calcination of the metals. All the eminent chemists of that day directed their attention to this problem both theoretically and experimentally. It formed, especially towards the end of this period, the centre around which the whole of chemistry circled; it became a stumbling-block to the adherents of the old doctrines, and led to a reform of the science so fundamental and far-reaching that the chemistry of to-day still lives under it.

The iatro-chemical theories strove after the impossible, and therefore quickly succumbed; the marked one-sidedness apparent in them, the gratuitous explanations of life-processes, and the total neglect of the anatomy and morphology of the organs, made their decline inevitable. An opportunity was thus given to chemistry to loosen and finally break the bands which medicine had wound around her, and to take up an independent position of her own. She still remained for a time under the protection of the healing art, to which she was indeed an indispensable aid; but from the time of Boyle onwards, the great aim of chemistry was recognised as being the discovery of new chemical facts, for the sake of arriving at the truth alone.

The spirit of true investigation which penetrated the natural sciences at the end of the sixteenth and beginning of the seventeenth centuries also began to extend itself to chemistry, the development of physics exerting an especially powerful influence upon the younger sister-science. The inductive method, too, acquired a continually growing and a lasting influence as a guide, the nature of which was indicated by Francis Bacon[1] substantially as follows:—

"The true kind of experience is not the mere groping of a man in the dark, who feels at random to find his way, instead of waiting for the dawn or striking a light. . . . It begins with an ordered—not chaotic—knowledge of facts, deduces axioms from these, and from the axioms again designs new experiments." Equipped with such axioms, chemistry might enrol itself among the exact sciences.

The learned societies which came into existence in the second half of the seventeenth and beginning of the eighteenth centuries, and whose periodicals spread abroad the results of chemical investigations, aided materially towards the healthy development of the science. The incitement they gave to researches, which could then be submitted to verification by other workers, was also of great value. Finally, they promoted the reciprocal action of chemistry and allied branches of science upon each other, an action so fruitful in its results, by bringing their respective exponents into closer connection.

[1] *Novum Organon*, Aphorism 82, paragraph 3. Bacon in the above paragraph gave expression to no new idea, but merely called special attention to the value of experience, a point already recognised by his predecessors Palissy, Leonardo da Vinci, Paracelsus and others. Liebig in a series of essays has proved conclusively how unjustifiable it is to designate Bacon as the originator of the inductive method, and how little he was permeated by the spirit of true research (see Liebig's *Reden und Abhandlungen*, 1874).

The Royal Society, which was formed about the middle of the seventeenth century by the amalgamation of the two smaller scientific societies of Oxford and London, and which began to publish the *Philosophical Transactions* in 1665, furnishes a good instance of what has just been said. The Italian academies, especially the *Academia del Cimento* of Florence (1657), devoted themselves mostly to physical and mathematical studies. In Vienna the *Academia Naturæ Curiosorum* was started in 1652, taking the name of *Cæsarea Leopoldina* in honour of its patron Leopold I. The *Académie Royale* originated in Paris in 1666 out of friendly meetings which were held at the house of the physicist Mersenne; the *Mémoires de l'Académie des Sciences* began to appear in 1699. The Berlin Academy was founded in 1700 by Frederick I., Leibniz being its first president; and during the earlier half of the eighteenth century the northern countries followed suit with similar learned societies, that of St. Petersburg being started in 1725, that of Stockholm in 1739, and that of Copenhagen in 1743.

That an extraordinary interest was felt at this time in scientific questions is readily seen from the literature of the day, which reflects the excitement—sometimes feverish in its intensity—raised by isolated discoveries, like that of phosphorus, or by disputed problems, such as the question of the cause of combustion.

The modes in which chemical questions were treated did indeed approximate to the methods followed in recent times, but in one respect there was a striking distinction between them. The chemical investigation of the phlogistic period took very little note (and then only incidentally) of the proportions by weight in which substances entered into reaction; it turned its attention almost alone to the qualitative side of the phenomena. The introduction and subsequent development of the phlogistic doctrines were only possible because of the utter neglect of quantitative relations. Even acute observers who noticed that metals increased in weight upon calcination, and thus came into direct conflict with the phlogistic view, evaded the only correct explanation of this,—

and, with it, of the phenomena of combustion—by far-fetched conceptions. This blinding of the understanding by an erroneous theory, consequent upon the refusal to look into all the conditions which might have helped to clear up the question, is peculiar to the period of phlogistic chemistry.

In spite, however, of the fundamental error which ran through it, the period was a highly fruitful one for chemistry; it forms the indispensable introduction to the most recent phase of development of the science. And although it was itself fettered by many erroneous ideas, still the phlogistic age contributed largely to the refutation of mischievous errors, *e.g.* those belonging to the iatro-chemical doctrines and the false beliefs of alchemy.

GENERAL HISTORY OF THE PHLOGISTIC PERIOD[1]

Robert Boyle and his Contemporaries.

Boyle has been rightly spoken of as the investigator who, by his creative genius, pointed out the new path to the period just then beginning; it would be even better to say that with him this new path originated. The spirit of pure investigation, free from the fetters of alchemistic and iatro-chemical conceptions, animated this remarkable man, whom chemistry has to thank for teaching her the real aims which she should pursue. The leading ideas of his scientific programme, which are laid down in the *Preliminary Discourse* (in Shaw's edition of Boyle's works, three vols., 1725), deserve to be quoted here:—

P. xxvi. "· · · I saw that several chymists had, by a laudable diligence, obtain'd various productions, and hit upon many more phenomena, considerable in their kind, than could well be expected from their narrow principles; but finding the generality of those addicted to chymistry, to have had scarce any view, but to the preparation of

[1] Cf. H. Kopp, *Gesch. d. Chemie*, vol. i. p. 146 *et seq.*; Höfer, *Hist. de la Chimie*, vol. ii. p. 146 *et seq.*

medicines, or to the improving of metals, I was tempted to consider the art, not as a physician or an alchymist, but a philosopher. And, with this view, I once drew up a scheme for a chymical philosophy; which I shou'd be glad that any experiments or observations of mine might any way contribute to complete."

P. xviii. ". . . And, truly, if men were willing to regard the advancement of philosophy, more than their own reputations, it were easy to make them sensible, that one of the most considerable services they could do the world is, to set themselves diligently to make experiments, and collect observations, without attempting to establish theories upon them, before they have taken notice of all the phenomena that are to be solved."

Experimental methods,[1] taken in conjunction with the careful observation of actual phenomena, form therefore, according to Boyle, the only sure foundation for speculations. To have made this the common property of chemistry, which from thenceforth strove to work out its fundamental principles by means of experiment, is the undying service rendered by Boyle.

His life [2] was devoted to fostering the natural sciences, especially chemistry. The seventh son and fourteenth child of the Earl of Cork, he was born on the 25th of January, 1626. After an exceptionally careful training at Eton, he became a student at Geneva, and continued his studies in the quiet of his estate of Stalbridge until 1654, when he settled at Oxford, carrying on there a constant intercourse with other eminent men of learning. While at Oxford, he belonged to a society called *The Invisible College*, the stimulating effect of which doubtless led to the formation of the Royal Society. From 1668 he lived in London, where he continued to work actively, as he had done at Oxford, for the Royal Society, which had been founded in 1663; he became its president

[1] Thus he says that from these alone can we look for progress in all useful knowledge.

[2] For a pleasant account of Boyle's life and works cf. Thorpe's *Essays in Historical Chemistry*, p. 1 *et seq.*

in 1680 and held that office until his death in 1691. His noble and unpretentious character, with its accompanying modesty, and his simple religious tone, called forth astonishment and admiration both from his contemporaries and his successors. What a contrast between this modesty and the rude assumption of Paracelsus, or the self-appreciation of van Helmont and many other *savants* of the iatro-chemical age!

The services which Boyle rendered in the development of chemistry stretch over the most various provinces of the science. Isolated observations of importance, by which he enriched—indeed, fundamentally extended—applied chemistry, the knowledge of chemical compounds and their analysis, the chemistry of gases, and pharmacy, will be treated of in the special part of this book. We have at present only to do with the general significance of his work and of his theoretical views for chemistry.

The term "element," which before Boyle's time was a very fluctuating and therefore uncertain one, received through him a more positive meaning. In his work, *Chemista Scepticus* (1661), he criticises the Aristotelian and the alchemistic elements, which were still accepted by many in the iatro-chemical age. He enunciated the axiom that only what can be demonstrated to be the undecomposable constituents of bodies are to be regarded as elements; and he considered it hazardous to advance opinions as to the properties of the elements in general, without having first obtained a firm foundation in their actual properties individually. With a far-seeing glance he looked forward to the discovery of a much greater number of elements than was at that time assumed, at the same time contending that many of the substances then held to be elementary were not really so.

Hand in hand with this wholesome simplification of views upon the elements, there went fruitful ideas upon the union of the elements to compounds, and also upon affinity as the cause of chemical combination. Boyle was the first to state with perfect clearness that a chemical compound results from the combination of two constituents, and that it possesses properties totally different from those of either of its constituents alone. This definite opinion enabled him to draw a sharp distinction between mixtures and chemical compounds.

In order to explain the formation or decomposition of compounds, Boyle advanced a corpuscular theory which gave evidence of his acuteness and showed how far he was ahead of his contemporaries. In his opinion all substances consisted of minute particles, and chemical combination took place when particles of different matter which mutually attracted each other came together. If another substance interacted with this new body, whose particles possessed a greater affinity for those of one of the components of the latter than these components had for each other, then decomposition ensued. In such simple manner did Boyle endeavour to explain the formation and decomposition of chemical compounds.

No one before him had grasped so clearly and treated so successfully the main problem of chemistry,—the investigation of the composition of substances. In doing this he had the solid ground of experience and experiment under his feet, and could always bring forward evidence for the probability of his views. His endeavours to get at the root of the composition of bodies gave a refreshing impetus to analytical chemistry, which indeed before his time could hardly be said to exist; and we are at the same time indebted to him for fixing the meaning of a "chemical reaction." Boyle appears to have been the first to make use of the term *analysis*, in the sense in which it has since been employed by chemists.

Boyle likewise devoted much attention to the question of the cause of combustion and other similar phenomena, and although his attempts at explaining these were not very successful, his remarkable experiments on the part played by air in combustion helped materially to the later solution of the problem. His work on air and gases led him in 1660 to the memorable discovery of the now well-known law that "the volume of a gas varies inversely with the pressure" (Mariotte found this out independently seventeen years later).

Boyle's writings, which were already widely read in his own lifetime, are characterised by simplicity of style and clearness of expression; they offer an agreeable contrast to the works of many of the other chemists of his time, who sought to hide their deficiencies in clear thought and accurate knowledge by metaphorical and mysterious language. In addition to other papers published in the *Philosophical Transactions*, the following works of his, which were brought out both in English and Latin, are to be especially mentioned:—*The Sceptical Chymist* (*Chemista Scepticus*) first published anonymously in 1661, and afterwards in numerous editions with Boyle's name as author; *Tentamina quædam Physiologica* (1661); and *Experimenta et Considerationes de Coloribus* (1663).

Among the contemporaries of Boyle who also advanced the natural sciences, especially chemistry, and of whom Willis, Hooke, Wren and Hawksbee must be mentioned here, there was one in particular who, although a practising physician himself, rendered good service to chemistry by his observations on combustion and calcination, viz. John Mayow (born 1645). His assumption—that atmospheric air contained a substance[1] (also present in saltpetre) which combined with metals when they were calcined, and which sustained respiration and converted the venous blood into arterial—was bound to result in the right interpretation of the phenomena of combustion, when the observations which had led to it were sufficiently extended. Mayow's early death in 1679 was perhaps the reason why this did not come about, the development of the new chemistry being greatly retarded in consequence.

Lemery and Homberg.—The *Académie Royale des Sciences* formed in France the centre of union for chemists in that country, the chief exponents of the science in Boyle's time, particularly during the last quarter of the seventeenth century, being Wilhelm Homberg and Nicolas Lemery. Both of them being good observers, their work

[1] Mayow termed this substance *spiritus igno-aëreus* or *nitro-aëreus*.

tended chiefly to the development of practical chemistry, which was especially indebted to Homberg for many valuable contributions. In the scientific explanation of technical processes they come a long way after Boyle; Homberg, in particular, was still trammelled by alchemistic views, and held fast to the idea that substances consisted of sulphur, mercury and salt.

Lemery, born in 1645, hardly did any independent work on the treatment of theoretical questions, but he well knew how to sift and put together the facts already known. This is shown in his *Cours de Chymie*[1] brought out in 1675, which was for long held to be the best text-book of chemistry, and was so widely used that the author himself lived to see thirteen editions of it published.

In addition to this literary work Lemery was exceedingly active as a teacher, the last thirty years of his life being taken up in that way; in his earlier years he was much involved in religious polemics, and hence was unable to turn his chemical knowledge to the best account during that period.

Lemery designated chemistry a "demonstrative science," and therefore sought to elucidate chemical operations by suitable experiments. In theoretical questions, *e.g.* in his views upon combustion and upon the composition of substances, he was for the most part an adherent of Boyle.

While Lemery was chiefly exercised, then, about the effective propagation of his science, Homberg—born in 1652 and permanently settled in Paris after a restless life and multifarious study—found particularly good opportunity, as body-physician and alchemist to the Duke of Orleans, of making numerous and sometimes important observations in practical chemistry. Some of his researches, *e.g.* that upon the saturation of acids by bases, contained fruitful germs

[1] Shortly before the publication of the *Cours de Chymie*, two other text-books appeared in Paris, both entitled *Traité de Chymie*, by Lefèvre (1660) and Chr. Glaser (1663), under the latter of whom Lemery had begun his studies. Glaser's book treats chiefly of pharmaceutical, and Lefèvre's of theoretical chemistry, which latter, however, was not much advanced by it.

which became developed later on in the hands of other workers. Most of the writings of these two men, both of whom died in the same year (1715), were published in the Memoirs of the French Academy.

Kunkel and Becher.—The most eminent German chemist in Boyle's time was Kunkel, in conjunction with whom Becher must also be named. Closely connected with the latter was Stahl, the originator of the phlogiston theory, of which the germs are to be seen in the views of both of the men first mentioned.

Johann Kunkel, born at Rendsburg in 1630, did excellent service to practical chemistry as an able experimenter and acute observer. Originally a pharmacist, he early showed the leaning towards alchemy which was decisive and fateful as regarded the whole course of his life; he was too honest not to see through many of the frauds of adepts, but at the same time was so firmly convinced of the possibility of the transmutation of metals that he gave his life-work to solving the problem. Employed as an alchemist by various princes (among whom were the Dukes of Lauenburg, the Elector John George of Saxony, and the great Elector of Brandenburg), whose desires he was unable to gratify, he led a restless life which came to a close at Stockholm in 1702, where, by the favour of Charles XI., he had found a more honourable position than any previously allotted to him. Kunkel's preconceived opinions caused his writings to be permeated by mischievous errors, and to contain work bearing upon alchemy. What a contrast between him and Boyle! While the latter was seeking to ascertain the real composition of substances, and to get at their demonstrable constituents, the former still held to the tenet that all metals contained mercury. Nevertheless, as a promoter of experimental chemistry, and therefore of practical chemical knowledge, Kunkel deservedly holds a high place.

Johann Joachim Becher, who was born at Speyer in 1635 and died in London in 1682, worked almost contemporaneously with Kunkel, but more for the theoretical explanation of already observed facts than for the practical side of the subject; in his unsettled life and his propensity towards new projects, he resembled the latter. He worked as an alchemist at various courts (in Mainz, Munich and Vienna), but he was too honourable to deceive his patrons, and too candid to allow of his remaining long in any one place. His bold technical projects almost always came to nothing; they show only too clearly their author's deficiency in practical chemical knowledge. In theoretical questions as to the composition of substances Becher attempted to revive the old ideas of Basil Valentine and Paracelsus in another form. In place of mercury, sulphur and salt, he set up three "earths," of which all inorganic ("sub-terrestrial") bodies should consist, viz. the mercurial, the vitreous and the combustible (*terra pinguis*). The nature of any material depended upon the proportions in which these three fundamental earths were contained in it. Of especial importance was Becher's assumption that, when substances were burnt or metals calcined, the *terra pinguis* escaped, and that in this escape lay the explanation of combustion; it was from this conception that Stahl's phlogiston theory originated. The opinions of Becher upon the production of salts and acids from these earths were also received with approbation by his disciples.

These theoretical views are to be found in Becher's first work, *Physica Subterranea* (1669), and in his last, *Theses Chymicæ* (1682). His doctrines acquired great celebrity through Stahl, whose work belongs for the most part to the eighteenth century, on which he conferred a character of its own by his development of the phlogiston theory.

Stahl and the Phlogiston Theory.

The theory of the phenomena of combustion and other analogous processes, which were to be explained by the assumption of the hypothetical *phlogiston*, was the point round which chemists in general gravitated during the eighteenth century; until the appearance of Lavoisier the phlogiston theory received the assent of most investigators.

Georg Ernst Stahl, born at Anspach in 1660, devoted himself to the study of medicine, and acquired, first at Jena and later on at Halle—to whose university he had been called as professor of medicine and chemistry in 1693—, the reputation of a distinguished physician and academic teacher. Appointed physician to the king in 1716, he removed to Berlin, where he laboured with success for the extension of chemical knowledge until his death in 1734. He worked at chemistry in the true scientific spirit; himself guided by the ardent desire to discover the truth, he was able to draw around him pupils animated by a similar aim. The most eminent among the Berlin chemists of the succeeding generation studied under him.

Even in his own lifetime the doctrines which he taught, together with a number of valuable detached observations, were widely disseminated by means of his writings, and especially by his lectures, the latter of which were published by several of his pupils.[1] Stahl, however, exercised his greatest influence both upon his contemporaries and upon the succeeding generation by his phlogiston theory, which eclipsed all his other chemical work.

Stahl himself freely recognised the close connection between his views upon combustion and calcination and the original ones of Becher; he went to work, however, quite differently from the latter, although his doctrine was grounded upon Becher's idea regarding the combustible constituent. This assumption of a constituent common to combustible bodies (a "fire material," a "sulphur," and so on) was indeed of older date than that of Becher's *terra pinguis*, which Stahl at once utilised, in order to build up his phlogiston theory thereon. This rests upon the hypothesis that combustible substances—among which the metals capable of calcination

[1] Among Stahl's writings we may name the *Zymotechnia Fundamentalis*, etc. (1697); *Specimen Becherianum*, etc. (1702); and *Zufällige Gedanken über den Streit von dem sogenannten Sulphure* ("Occasional Thoughts on the Dispute regarding the so-called Sulphur"), (1718). Of his pupils, Juncker was especially active in propagating the views of his master.

were reckoned—contain phlogiston as a common constituent, which escapes on combustion or calcination. Since, as was then held, every phenomenon bearing upon this could be readily explained by the aid of such an assumption, it was considered unnecessary to prove the actual existence of phlogiston itself directly. Stahl was able by means of it to group uniformly together and to explain a large number of chemical reactions. The more violently the combustion of any substance went on—so he taught,—the richer it was in phlogiston; coal, which can be almost entirely consumed, was therefore to be regarded as nearly pure phlogiston. In order to reproduce the original substance, its combustion-products had to be added to it again; in this manner the metals were "revived" from their calces, which, according to Stahl's notion, had resulted from the former through the escape of the phlogiston. When a metallic calx was heated along with coal, the phlogiston so abundantly contained in the latter combined with it, the metal being thus reproduced; consequently a metallic calx was a constituent of a metal. Upon a like sophism rested Stahl's assumption that sulphur consisted of sulphuric acid and phlogiston. He saw in the production of sulphur, on heating sulphuric acid or a sulphate with coal (phlogiston), a synthesis of the former, and therefore a proof that sulphur was a compound body. Upon the further logical conclusion,—that the products of combustion of any substance must be lighter than that substance itself, seeing that they are constituents of it, no importance was placed. And no attention was paid to the numerous observations which showed that this was not the case,—that, indeed, a calcination of the metals was accompanied by an increase in weight. It was facts like those just named which, after a prolonged struggle, brought about the overthrow of the phlogiston theory.

To Stahl belongs the merit of grouping together the phenomena of oxidation and reduction, as we now term these, albeit by the aid of a false hypothesis. The addition of phlogiston is equivalent to reduction, and its withdrawal or escape to oxidation. The analogy between respiration and the decomposition of animal matters on the one hand, and combustion on the other, did not escape Stahl, who likewise assigned the chief *rôle* in these processes to phlogiston.

The value of his theory lay therefore in the interpretation which it afforded of a variety of processes from one common point of view. The simplicity of this explanation blinded both himself and the generation which followed him to such a degree that they left unnoticed all the glaring contradictions between actual facts and the phlogistic doctrine. Notwithstanding this, however, the latter was not an obstacle to the development of chemistry, seeing that chemists like Black, Cavendish, Marggraf, Scheele, Bergman and Priestley, who so greatly extended the science by their wide-reaching discoveries, were phlogistonists in the full sense of the word.

Hoffmann and Boerhave.—Before speaking of the further destinies of the phlogiston theory, and, in connection with this, of the state of chemistry at that date, the work of two of Stahl's contemporaries who contributed materially to the advancement of the science must be considered, viz. Friedrich Hoffmann and Hermann Boerhave. Both of these men were eminent physicians and accomplished chemists, but they were not exactly adherents of Stahl's phlogiston doctrine, although they held similar views with regard to combustion.

Hoffmann, born at Halle in 1660, *i.e.* in the same year as Stahl, after acquiring a thorough knowledge of medicine, mathematics and the natural sciences, practised first as a physician and then became professor of the science of medicine in Halle, where he ultimately died in 1742, after an interregnum spent in Berlin. His most important work was done in medicine and in pharmaceutical and analytical chemistry. He combated with success the iatro-chemical doctrines of Sylvius and Tachenius, which still held their ground with many physicians, exposing their absurdities and showing to what nonsensical deductions such exaggerations led.

Many of his investigations and discoveries in pharmaceutical and analytical chemistry will be touched upon in the special history of this time. Hoffmann's views on combustion were very similar to those of Stahl. With respect to the calcination of the metals and the reduction of their oxides, however, he expressed opinions which approximate to those held at the present day, believing, as he did, that metallic calces contained a *sal acidum* in addition to a metal, the former of which escaped when the calces were reduced. This assumption did away with the similarity between combustion and calcination; these phenomena became indeed rather opposed to one another thereby, and with this the special use of the phlogiston theory vanished. Hoffmann was a very voluminous author, and his collected works, entitled *Opera Omnia Physico-medica*, show clearness of style and precision of expression.

Hermann Boerhave, born in 1668 at Voorhout near Leyden, was originally destined for the study of theology, but devoted himself to medicine, gaining at the same time an excellent knowledge of the natural sciences, and especially of chemistry. From the year 1709 onwards, he was able to utilise his catholic education to advantage as professor of medicine, botany and chemistry in Leyden, and attained to the highest distinction; he died there in 1738.

Boerhave's place in the history of chemistry is due not to any striking experimental researches, but to the exceptional acuteness which he showed in noting and collating chemical phenomena from one common point of view. His large text-book *Elementa Chemiæ* (1732)[1] was intended to contain all the most important work done in chemistry, and for a long time it remained by far the best guide to the study of the science. His estimate of the latter as an absolutely independent science, subordinate to no other,

[1] The admirable English editions of Boerhave's text-book were edited with great care and success by Dr. Peter Shaw. As a writer in the *Saturday Review* has pointed out, the third edition of Shaw's translation, which appeared in 1753 in two quarto volumes, contains a mass of original notes of great value and, especially, some detailed catalogues of early Greek writings upon alchemy.

and whose aim should be the investigation and perception of chemical facts, was at once a beneficial and an elevated one. In accordance with this view we find him condemning the abuses which the iatro-chemists had introduced into chemistry. The work of the alchemists he did not criticise sharply enough; in his endeavours to test the assertions which they made, he believed that he found here and there some corroboration of them, and was thus probably not disinclined to decide in favour of the adepts in cases where experience had not as yet spoken her last word. On the other hand, he refuted many statements, such as those which told of the fixation of mercury and of the production of the latter from lead salts, and thus contributed to clear up and rectify alchemistic opinions and assertions.

Boerhave appears to have concurred in the phlogiston theory in many points, at least he expressed no opinions contrary to Stahl's fundamental views, although he did not agree in regarding the calces of the metals as the earthy elements of these latter. Like many other investigators, Boerhave studied the processes involved in calcination, and to him is due the valuable experimental contradiction of the view put forth by Boyle and others,—that, during calcination, a ponderable fire-stuff is taken up, and thus the increase in weight of the metals explained.

The Development of Chemistry, and particularly of the Phlogiston Theory, after Stahl's time.

The influence of Stahl's doctrine manifested itself more immediately in Germany, where it received the almost unqualified support of chemists, Berlin remaining the centre point of this theory. Among the men who upheld and sought to propagate it, Marggraf was the most eminent. Kaspar Neumann (born 1683) and Johann Theodor Eller (born 1689), contemporaries of Stahl, were also active adherents of the doctrine in the capital city of Prussia. Both of them, as professors at the medico-chirurgical institute, were in a high degree active in maintaining and spreading a knowledge of chemistry. Their own observations were, however, of little importance; those of Eller were chiefly upon subjects of medical physiology, and are full of untenable speculations. Stahl's disciple and pupil Johann Heinrich Pott (born 1692) enriched chemistry by many valuable observations, but he was unfortunate in his explanation of these, regarding boracic acid, for instance,—a substance which he had himself investigated carefully—as consisting of copper vitriol and borax. The results which he achieved were not at all commensurate with his untiring perseverance, which he showed, among other ways, in his endeavours to prepare porcelain. Although an adherent of the phlogistic doctrine, Pott did not bring forward anything new in its favour; with regard to the nature of phlogiston itself, he could only express the opinion that it was "a variety of sulphur."

Andreas Sigismund Marggraf[1] (1709–1782) was the last and most eminent adherent of phlogistic views in Germany. Destined originally for an apothecary, he acquired a knowledge and practical experience of chemistry, pharmacy and metallurgy as assistant to Neumann at Berlin, and by sedulous study at the high schools of Frankfurt on the Oder, Strasburg and Halle, and finally at the Freiberg School of Mines; this knowledge, accompanied as it was by exceptional gifts of observation, put him in a position to carry out researches of the greatest value. One has only to think of the observations made by him in his work on phosphoric acid,—observations which, considering the highly defective state of chemical analysis at that date, fill us with admiration; of the proof which he furnished of the difference between alum and the so-called bitter earth (magnesia), substances which had hitherto been generally confounded; and, above all, of his investigation of the juice of the red beet, in which he discovered cane sugar (see special section of this book). It was during this research that Marggraf introduced

[1] For an account of his life and works see A. W. Hofmann's charming *Erinnerungen aus der Berliner Vergangenheit*, p. 10 *et seq.*

the microscope into chemistry, as a valuable aid in distinguishing between different substances.

With this great talent for observation he united the gift of drawing what were generally sound conclusions from his work. In one point, however, Marggraf, like all phlogistonists, was not in a position to do this; although he had himself proved that phosphorus increases in weight by conversion into phosphoric acid, he could not free himself from the idea that phlogiston escaped during this process of combustion. And he could never be brought to see that this conception was an erroneous one, although the antiphlogistic doctrine was brought out several years before his death. Marggraf's papers are almost all contained in the Memoirs of the Berlin Academy; most of them were published during his lifetime in two volumes, under the title *Chemische Schriften.*

The French Phlogistonists.—The chief exponents of chemistry in France during the eighteenth century until the downfall of the phlogistic system were Geoffroy, Duhamel, Rouelle and Macquer, who concurred essentially in Stahl's views. They enriched the science not only by important facts, but also now and again by useful working theories.

Stephan François Geoffroy (the elder, to distinguish him from his less celebrated younger brother, Claude Joseph, whose work was chiefly pharmaceutico-chemical) was born in Paris in 1672, and helped for some time in his father's drug shop; he gave himself up, however, to chemical and medical studies, and laboured with great success as professor of medicine in the *Jardin des Plantes* from the year 1712 until his death in 1731. Geoffroy made himself a name throughout the scientific world by his researches upon chemical affinity; his *Tables des Rapports* (tables of affinity), in which the results of his most important observations are collected, exercised a great influence upon the doctrine of affinity. His theoretical views were less happy, *e.g.* he looked upon the iron found in the ashes of plants as having been produced artificially during the process of ignition. In the questions of combustion and calcination he approximated very closely to Stahl's view; the metals, for example, he regarded as composed of earths and a species of sulphur. Geoffroy rendered a real service by the energy with which he attacked alchemistic frauds, subjecting these as he did to critical examination in the memoir *Des Supercheries concernant la Pierre Philosophale*, presented to the French Academy.

Geoffroy's treatises were published partly in the Memoirs of the French Academy, and partly in the *Philosophical Transactions*. His long-celebrated work, *Tractatus de Materia Medica*, shows what a high value he placed upon chemistry as a sister science and aid to medicine.

Duhamel de Monceau (born 1700, died 1781), of the school of Lemery and Geoffroy, spent his entire life in Paris, where his versatility gained for him a high reputation. His sterling work was not by any means in pure chemistry alone, but also in physics, meteorology, physiology, botany, and—particularly—in chemistry as applied to agriculture. We must make especial mention here of the fact that he furnished definite proof of the difference between potash and soda, by preparing the latter pure; he also showed that it was the base of rock-salt, borax and Glauber's salt. The first proposals to prepare soda artificially from rock-salt came from him, a fact which shows his far-sightedness.

Whilst Duhamel worked purely as an academician, Guillaume François Rouelle (born 1703, died 1770) was mainly occupied in teaching,[1] in which he greatly excelled; some of his pupils, particularly Lavoisier and Proust, arrived at

[1] The numerous records of Rouelle's activity as a teacher, which have come down to us, enable us to form a clear picture of the conditions of chemical teaching in those days, and at the same time to appreciate the remarkable personality of the man. The lectures on chemistry were delivered by two professors, one of them treating the theory of chemical processes, whilst the other, in conjunction with him, showed and explained how they were carried out practically. While the former (Bourdelin) fatigued his audience by abstract reasonings, Rouelle inspired the students of practical chemistry by the vivacity of his discourse, during which he frequently became so excited as to throw off his periwig and some of his articles of clothing (cf. Höfer, *Hist. de la Chimie*, vol. ii. p. 378).

the highest eminence. At the same time he was also busy as an investigator, as many admirable observations and conclusions drawn from these show. Rouelle fixed the meaning of the term "salt" (in the Memoirs of the Academy for 1745) from a far more general point of view than van Helmont or Tachenius had done. The composition of a substance alone was sufficient to tell him whether it belonged to the class of salts or not. Salts were produced by the combination of acids of every kind with the most various bases; and, in addition to neutral salts, he drew a distinction between acid and basic ones. With views so clear as these, Rouelle was far ahead of his contemporaries.

Among the latter was Pierre Josèphe Macquer (born 1718, died 1784), who was likewise an active and successful teacher at the *Jardin des Plantes*, and who also aided effectively in the spread of chemical knowledge by means of his text-books.[1] His own individual work lay less in theoretical than in applied chemistry, to which he made valuable contributions (especially in the manufacture of pottery and in dyeing).

From the beginning of his career to its end Macquer was a phlogistonist, and did all that he could to reconcile the continually augmenting discrepancies between theory and facts; he paid no heed to proportions by weight, for it was only in this way that he could maintain the phlogistic hypothesis. And even although it was proved to be erroneous and untenable several years before his death, he was still unable to renounce it.

The English, Scotch and Swedish Phlogistonists.—Joseph Black, professor of chemistry in the Universities of Glasgow and Edinburgh successively, who was born in 1728 and died in 1799, advanced chemistry in an exceptional degree by his splendid experimental researches, which were published in the *Philosophical Transactions*; in especial by his, for that time, masterly investigations on carbonic acid and its compounds with the alkalies and alkaline earths, which

[1] The principal of these were: *Élémens de Chymie Théorique* (1749); *Élémens de Chymie Pratique* (1751); and his *Dictionnaire de Chymie* (1778).

were planned and carried out with the utmost ingenuity. His observations led to a clear knowledge of processes which had formerly been explained quite wrongly, and they drew the attention of investigators in a special manner towards gases; the work done with these had the effect of causing chemistry to proceed on new lines, and was, in fact, the necessary forerunner of the latest epoch of the science. In addition to this Black threw open a new field to physics by the discovery of latent heat in 1762, in which his wonderful gift of experimenting came to his aid.[1]

In order to appreciate his labours at their true value, and to compare them with those of other chemists who busied themselves with similar questions, we have only to fix our attention on his researches upon the alkaline earths and the alkalies. The carbonates of these were before Black's time regarded as simple substances; and it was further assumed that when limestone was burnt fire-stuff was taken up, and that this went over into potashes or soda when these were causticised by means of lime. Black, on the contrary, proved by his researches that when limestone or *Magnesia alba* was calcined, something escaped which led to a loss of weight and which was identical with van Helmont's *gas sylvestre*. This gas—which he termed *fixed air*, because of its being held bound by caustic alkalies, lime, etc.—he proved to be also present in the mild alkalies; and these latter became caustic when deprived of their carbonic acid by lime or magnesia. In this truly classical research we meet with methods which bear the impress of an entirely new departure. That Black devoted great attention to the proportions by weight of the compounds which entered into the reaction is seen in all his investigations; and it is thus easy to understand how he gave up the phlogiston theory and concurred in the doctrine of Lavoisier when the correct explanation of combustion and similar processes became possible through the discovery of oxygen.

[1] The Swedish physicist, J. C. Wilcke, also discovered latent heat about the same time and independently of Black.

Black, by his fundamental labours, did away with many errors, and thereby prepared the way for the definite knowledge of the true composition of important chemical compounds. Notwithstanding this, the evident conclusions which followed from his researches on causticity were unfavourably criticised by many of the chemists of his time, and indeed their correctness disputed; it is strange to find that even Lavoisier could not bring himself candidly to recognise Black's services in this respect, and that he rather ranged himself on the side of the latter's antagonists, who were in reality unable to weaken one of his arguments.

In his countryman, Henry Cavendish, Black had a most distinguished co-worker, who, while investigating quite independently of him, did so upon similar lines, and to the great benefit of chemistry. Cavendish, born at Nice in 1731, devoted himself very quietly but not the less efficiently to the natural sciences, which he studied thoroughly, especially to physics and chemistry; he died in London in 1810.[1]

There is but little to be said about his life, for his unsociable and retiring nature led him to shun anything like publicity, and indeed it was only with reluctance that he was induced to publish the results of his remarkable work; for this reason many valuable observations of his remained unknown for some decades. Although Cavendish inherited a large patrimony, he adhered throughout to a severely simple style of living.

His masterly researches—so important both from a physical and from a chemical standpoint—upon hydrogen (*inflammable air*), which he was the first to distinguish as a peculiar gas differing from all others, and also those upon carbonic acid, constitute him one of the founders of pneumatic chemistry and one of the originators of the new era. To him we owe the proof, of what value need not be said, that water consists of hydrogen and oxygen; further,

[1] The details of Cavendish's life and a picture of his peculiar disposition are to be found in Wilson's *Life of the Honourable H. C. Cavendish* (1848). Compare also Thorpe's clever memoir in his *Essays in Historical Chemistry*, p. 70 *et seq.*

the proofs that atmospheric air is a mixture of nitrogen and oxygen in constant proportions, and that nitric acid can be produced by the chemical combination of these two latter gases in presence of water. All these were discoveries of the greatest moment. In them Cavendish himself forged the most powerful weapon for the overthrow of the phlogiston theory, notwithstanding which we find him still faithful to the latter. His opposition to the antiphlogistic doctrine, which he himself helped to found by his own investigations, can only be explained by the fact that he did not pay enough attention to the proportions by weight in the processes of combustion, but explained the latter in a way which appeared to him sufficiently convincing, viz. by regarding hydrogen as identical with phlogiston.

Besides this Cavendish showed an absolutely marvellous exactitude in his researches upon gases, whose specific gravities and volume-ratios in chemical reactions he established. With what ingenuity he thought out and carried through physical experiments is well exemplified in his work on the specific heats of metals, and in his attempt—the first one which was successful—to determine the specific gravity of the earth. Another instance will be fresh in the memory of most readers, viz. Cavendish's surmise, from the results of his own experiments on the combination of oxygen and nitrogen, that there was possibly still another gas present in the air in small quantity (argon). When one considers this wonderful versatility and remembers the thorough mathematical training that Cavendish had gone through, one can but wonder the more that he laid too little stress upon proportions by weight in chemical reactions.

The most zealous champion for the phlogistic idea at that time was Joseph Priestley, to whom the chemistry of gases owes an extraordinarily large number of new observations and important discoveries. In Priestley were united an eccentric mind, in which fantastic speculations found a place, and a simple and child-like disposition. He combated the antiphlogistic doctrines until his death (in 1804) as no other man did, although his own researches often went to strengthen, even to lay the foundations of, the latter. In contrast with the quiet existence of Black and Cavendish, wholly devoted to science, a wandering life full of vicissitudes and even of persecutions was destined for Priestley,[1] doubtless for the most part because of his relations to the English Church and his own intolerance. Theology was his own special subject, and he was already a minister when he first came more closely into contact with scientific questions. Born at Fieldheads near Leeds in 1733, and acquainted with poverty in his early years, he afterwards earned a modest living as a teacher of languages (he taught Latin, Greek, French, Italian and Hebrew), and then as a minister of the Gospel. His versatility was further shown by the fact that he also occasionally gave lectures in logic, history, law, anatomy, etc. The numerous philosophical and theological books which he wrote, some of them very comprehensive, are probably now altogether forgotten, although Priestley himself considered these his best work. A personal acquaintance with Benjamin Franklin led him to make scientific researches, an early result of which was his *History of Electricity*. Later on, in the comparative leisure of librarian to Lord Shelburne, he found time for chemical investigations, his most important work being done at this period (1772—9).

After some years spent as minister of a meeting-house in Birmingham, Priestley was obliged to leave the latter town for London in 1791, an attack on Burke's writings upon the *French Revolution* having raised popular opinion against him, and indeed resulted in open mob-riot. A few years later he emigrated to America, and settled at Northumberland near Philadelphia, where he died in 1804. Although there is much of dilettantism in the mode in which Priestley treats scientific problems, he rivets our attention by the charm of his intense originality and perspicacity.

Endowed with an unusual gift for experimenting and

[1] Thorpe's admirable paper (*Essays*, p. 28) gives a graphic account of Priestley's life and many-sided activity. Compare also Priestley's *Scientific Correspondence*, edited by H. C. Bolton (New York, 1892).

observing, he was able to treat the most difficult problems of pneumatic chemistry, although lacking a thorough scientific education. He prepared and investigated a large number of gases which, with the exception of carbonic acid and hydrogen, were practically unknown before his time. Of all his discoveries, that of oxygen (in 1774) was the most important; it will be treated of later on. It is true, as we now know, that Scheele had indeed preceded Priestley in many of these observations, but he had omitted to publish his results soon enough. Priestley's beautiful researches on this gas did not, however, lead him to the correct explanation of combustion; he remained, on the contrary, true to the doctrine of phlogiston. But his mistaken ideas respecting this and similar processes did not prevent him drawing from his own observations sagacious conclusions with regard to the series of recurrent changes which oxygen undergoes in animal and vegetable metabolism,—a far more complicated process than that of combustion, which, tied as he was by a false hypothesis, he was unable to explain.

Contemporaneously with the three last-named British chemists, two most distinguished investigators, Torbern Olof Bergman and Karl Wilhelm Scheele, were labouring in Sweden as upholders of the phlogistic theory, which their brilliant discoveries and observations only served so deeply to undermine, that its supersession was inevitable. Bergman had acquired such a wide knowledge of the natural sciences that he taught with eminent success as professor of physics, mineralogy and chemistry at Upsala. Born in the year 1735, he died at the early age of forty-nine, doubtless from the effect of overwork upon a weak constitution. His chief services to chemistry, to which from 1767 he principally devoted himself, were in the domain of analysis, which he treated systematically and enriched by valuable methods. He knew well how to make his chemical experiences useful for the definition and classification of minerals, and thereby laid the foundation of mineralogical chemistry and chemical geology. The current views upon chemical affinity thus gained through him precision and clearness; the scientific character of chemistry was materially raised by such observations, and a general survey of chemical processes rendered much easier. His papers appeared originally in the Memoirs of the Academies of Stockholm and Upsala; later on they were collected together, and published in five volumes in 1779-1788, under the title *Opuscula Physica et Chemica.*

Karl Wilhelm Scheele will remain for all time one of the most distinguished of chemists; and his fame is not lessened by the fact that he continued all his life through a zealous supporter of the phlogistic doctrine. In spite of this fact, of the unfavourable conditions under which he lived, and of the short span of his life, he contributed to chemistry a wealth of new observations—many of them discoveries of supreme value—which furnished a rich mine for the experimental work and theoretical discussions of future generations.

Much new light has been thrown on Scheele's life and scientific work by A. E. Nordenskiöld's recently published book: *Karl Wilhelm Scheele: Nachgelassene Briefe und Aufzeichnungen* ("Karl Wilhelm Scheele: His Letters and Journals") (Stockholm, 1892). This materially supplements the earlier biographies of Crell, Sjösten-Wilcke, etc., and gives us more especially a clear account of the genesis and of the dates of Scheele's magnificent discoveries, while at the same time we learn what a number of his observations, of great importance, have hitherto remained unknown.

Scheele, born on the 9th of December, 1742, at Stralsund, the capital of Pomerania, which at that time belonged to Sweden, began at fourteen years of age his apprenticeship in Gothenburg with Apothecary Bauch, who soon recognised and appreciated the boy's remarkable gifts. Restricted almost entirely to a few antiquated text-books, together with the fairly good chemical inventory of the apothecary's shop, Scheele, by his unwearied experimenting, acquired such a knowledge of the properties and reactions of many substances that, by the time he went to Malmö (in 1765) he had, although still only an apprentice, gained more experience than the majority of the chemists of the time. At Malmö, and also in the succeeding posts he held (Stockholm, 1768-

1770, and Upsala, 1770-1775), he increased his knowledge of the most important branches of chemistry without, however, becoming so well known at the time as he deserved. It was only when, through Gahn's good offices, he came into close relation with Bergman—a connection which began in a misunderstanding and coolness, but which developed into a friendship—that Scheele continued to gain steadily in reputation. After taking over the pharmacy at Köping in 1775, he was able to devote himself more closely to scientific work, and with still more brilliant results. The records of his researches followed one another rapidly in the *Transactions* of the Stockholm Academy, into which he had been received as *Studiosus Pharmaciæ* in 1775. In 1777 he published the results of his investigation on air, oxygen, combustion and respiration in a volume entitled *Chemische Abhandlung von der Luft und dem Feuer* ("A Chemical Essay on Air and Fire"). After his early death at barely forty-four years of age—a death undoubtedly hastened by a too close devotion to science—his collected works were published in two volumes in German by Hermbstädt (Berlin, 1793), under the title: *Sämmtliche Physische und Chemische Werke.*

It is not merely as an investigator and discoverer, but as a high-principled and unassuming man, that Scheele merits our warmest admiration. His aim and object was the discovery of the truth. The letters of the man reveal to us in the pleasantest way his high scientific ideal, his genuinely philosophic temper, and his simple mode of thought. "*It is the truth alone that we desire to know, and what joy there is in discovering it!*" With these words he himself characterises his own efforts.

It is not proposed to enter minutely at this point into his varied investigations; a general account only of his services to science will be given here, and the more important parts of his work will be referred to in short detail later on.

Endowed with a most wonderful gift of observation Scheele was able to bring to a successful conclusion researches carried on with but very limited means at command. A brilliant proof of this is given in his investigations upon black oxide of manganese (*De Magnesia Nigra*), which many competent workers before him had studied without succeeding in making its nature clear. During this research Scheele discovered in rapid succession four new substances —chlorine, oxygen, manganese and baryta—of which the two first especially were of the utmost importance for the proper understanding of chemical processes.

The way in which he isolated and noted the characteristics of oxygen and also, previous to this, of a long series of hitherto unknown gases, prove him to have been a magnificent experimenter. And similarly we see him as an incomparable observer in the discovery of analytical methods and in the opening out of entirely new fields of inorganic chemistry (see special section). Scheele was the first to note the fact that there are various stages in the oxidation of such metals as iron, copper and mercury, notwithstanding that he still adhered to the phlogistic hypothesis in explaining the composition of those products. With this knowledge he was far ahead of Lavoisier, Proust and others.

In a manner nothing short of marvellous Scheele brought his inventive genius to bear upon organic chemistry, which had till then been left almost untouched; working out in every direction new methods for isolating the products of vegetable and animal metabolism, he prepared a large number of acids and other organic compounds hitherto unknown. Scheele was a pioneer in nearly every branch of chemistry, being unique in power of observation and in the quick comprehension of facts, although, it is true, not always happy in his interpretation of these, fettered as he was by the phlogiston theory. Scheele's discoveries will be referred to separately in the various sections of the *Special History of Chemistry*.

In order to properly appreciate the condition of the phlogiston theory in the seventh and eighth decades of the eighteenth century—that is, shortly before its downfall,—the development up to that date of a special section of chemistry, viz. pneumatic, must be considered. The work done with gases, and, more especially, the knowledge acquired of their properties and behaviour, had led finally to the correct interpretation of combustion. The special history of the phlogistic period thus falls to be treated of now.

DEVELOPMENT OF PARTICULAR BRANCHES OF THEORETICAL AND PRACTICAL CHEMISTRY IN THE PHLOGISTIC PERIOD.

Pneumatic Chemistry and its Relations to the Doctrine of Phlogiston.—The influence which the investigation of gases, especially of oxygen, exercised in shaping chemistry is sufficiently well known. Oxygen forms to some extent the centre-point of chemical research during the last quarter of the eighteenth century, for the knowledge of the part which it played in combustion and similar processes led to the setting aside of a doctrine that had dominated all theoretical views for a hundred years; and, further, because results of the greatest importance were conjoined with its study, inasmuch as this contributed materially to the development of the atomic theory.

The services of the men whose observations did most towards building up the chemistry of gases have already been mentioned generally; it will suffice here to treat in more detail certain of these observations together with a few others. Boyle's researches show a marked advance over those of van Helmont in the mode in which he collected gases and worked with them; at the same time neither he nor his contemporaries felt quite sure whether carbonic acid and hydrogen, whose characteristic properties he knew, differed materially from atmospheric air. This uncertainty is also seen in the work of later investigators, *e.g.* Hales; the erroneous idea that gases were ordinary air with various admixtures, had fixed itself firmly in the minds of chemists. To Black is due the merit of proving the precise difference between carbonic acid and air, by showing the "fixation" of the former by caustic alkalies. Cavendish, who recognised in hydrogen a peculiar gas, likewise helped to do away with the misconception. That Scheele had already discovered numerous gases by the year 1770, and had proved them to be individual substances, is clearly shown in his letters and journals (cf. pp. 124—5). Finally, we would mention here the remarkable supplemental researches of Bergman on carbonic acid (1774).

The methods of collecting gases had improved considerably since Hales—and, before him, the little-known Moitrel d'Élément—had effected a separation of the generating vessel from the receiver. Air was found to be a fluid capable of measurement which possessed weight, and which, like all other fluids, could be transferred from one vessel to another. The apparatus which Black, Priestley, Scheele and others used, and those which we employ at the present day, gradually developed themselves from that of Hales. Priestley was the first to describe the collection of gases over mercury, and he succeeded by this device in discovering gaseous ammonia, hydrochloric acid, silicon fluoride and sulphurous acid,—all of which had been overlooked so long as water only was used for this purpose. Scheele had anticipated Priestley in the isolation of some of these, as well as of nitric oxide and sulphuretted hydrogen (about 1770), but had not published his observations.

The discovery of so many gaseous substances of such different character greatly excited the chemical world. The properties of each gas were carefully examined; and, after Mayow's researches, and especially after the more exact determinations by Cavendish, the density was taken as the criterion of one gas differing from another and from atmospheric air. Due regard was also paid to the greater or lesser absorption of gases by water, as a distinct test for some of them; Bergman, for instance, determined with fair accuracy the solubility of carbonic acid in water. But the true composition of gaseous bodies remained unknown during this epoch, great uncertainty prevailing even about the simplest of them until Lavoisier had pronounced his opinion as to the elementary nature of oxygen and hydrogen. How could this indeed be otherwise, so long as the presence of phlogiston was assumed in most gases? Hydrogen was considered identical with phlogiston by many chemists soon after the middle of the eighteenth century, Cavendish and Kirwan setting the precedent for this; others looked upon coal as being rich in phlogiston, if not as the latter itself. The most various and often confused opinions were expressed regarding the composition of carbonic acid, carbonic oxide, nitric oxide, sulphurous acid, sulphuretted hydrogen and other gases, these opinions being made to fit in with the views of the phlogistic doctrine prevalent at that time.

Of greater moment than these varying opinions upon the constitution of the gases just named were the long unsettled questions: "Is atmospheric air a simple or a compound body, and—if the latter—what are its constituents or ingredients?" These questions were solved experimentally by chemists belonging to the phlogistic era, more particularly by Scheele and Priestley; but it was left to Lavoisier to interpret their observations correctly. We must now speak of the most important of the facts then brought to light, which bore upon the composition of the air.[1]

The first observation which aided in overthrowing the old assumption of air being a simple substance, was the behaviour of an enclosed volume to a body burning and to metals heated in it. Boyle was iorced by his researches in this direction to the supposition that one ingredient of the air was necessary to respiration and combustion, and to the calcination of the metals; but he was unable to isolate this ingredient, as was also Mayow, who, with his assumption of a *spiritus ingo-aëreus*, which brought about combustion (cf. p. 107), came pretty near to the right interpretation. It was, however, only a hundred years later, after oxygen and nitrogen had been prepared successfully, that the question approached its solution. Nitrogen, which various investigators had already worked at, was first isolated by Scheele; but Rutherford, who discovered it independently in 1772, by the absorption of the carbonic acid produced by

[1] Cf. Ramsay's recent volume, *The Gases of the Atmosphere* (Macmillan and Co., 1896).

combustion or respiration in an enclosed volume of air, preceded Scheele in publication. It followed from their observations that this gas, which was incapable of sustaining either combustion or respiration, must be one of the ingredients of the atmosphere. The other was isolated and examined by Scheele and Priestley. The journals already alluded to make it clear that as early as 1771—1773, *i.e.* during the years of his sojourn at Upsala, Scheele prepared oxygen by heating black oxide of manganese with sulphuric or arsenic acid, and also from nitrates and from the oxides of mercury and silver, and noted its characteristics clearly. Priestley, who likewise observed the gas at about the same time, without, however, recognising its peculiar nature,[1] first isolated it for certain on August 1st, 1774, by heating red oxide of mercury; and, as he published his results earlier than Scheele, he has hitherto been regarded as the first discoverer of oxygen, whereas we now know the converse to be the case. Both observed that this gas was capable of supporting combustion and respiration in an intensified degree. Priestley named it "dephlogisticated air," and Scheele at first *aer vitriolicus*, later "fire air" and also "life air."

The momentous discovery of oxygen enabled both of them to recognise air as being a mixture of two kinds of gas;[2] Priestley calls nitrogen "phlogisticated air," and Scheele terms it "spent air." They both found substances which absorbed the one constituent of the air (oxygen). Here, again, Scheele showed the greater versatility, for while Priestley employed for this purpose saltpetre gas (nitric oxide), Scheele made use of phosphorus, hydrate of protoxide of iron, mixtures of iron and sulphur, and moist iron filings. They made the further important observation that, upon burning a candle in an enclosed volume of air, exactly as much "fixed air" (carbon dioxide) was generated as oxygen had vanished.

Notwithstanding all this they did not get at the right

[1] Hales and Bayen, too, had observed oxygen previous to this, but also without recognising its peculiar nature.

[2] Scheele, in his treatise *Von Luft und Feuer* ("On Air and Fire"), puts as the heading to a series of his investigations this sentence:—"The air must be made up of elastic fluids of two kinds."

explanation of combustion, respiration and calcination, whose analogy to one another they clearly saw: so prejudiced were they by the idea that phlogiston escaped during these processes, that the path distinctly marked out by their own observations was left for another to tread. Lavoisier was destined to do this, as he easily threw aside the trivial phlogistic prepossessions that he cherished at the beginning of his scientific career. The others, indeed, upheld a contradictory explanation of combustion and analogous processes, in order to remain loyal to the phlogistic doctrine. But that it was Priestley and Scheele who, by their exhaustive researches on oxygen and the part which it played in the processes just mentioned, furnished the experimental material for the correct understanding of these, and not Lavoisier, is beyond all question.

After the discovery of oxygen and of its chief properties the days of the phlogistic theory were numbered, although many of the most eminent chemists still held to it in spite of accumulating contrary evidence. The greatest difficulty in the way of the old doctrine was the fact, already known for a long time, that, in those cases where phlogiston was supposed to escape, the products became heavier instead of decreasing in weight. The exact researches on the calcination of the metals,[1] had their results been studied without any preconceived opinions, ought to have led to the correct explanation, viz. that one ingredient of the air combines with the metals to form calces; for not only was the increase in weight observed, but also the disappearance of a portion of the air. But instead of drawing from this the conclusion that the phlogistic hypothesis was untenable, chemists endeavoured to make the observed facts fit in with the latter by putting a strained interpretation on them. Even Boyle, acute as he was, tried to help himself by the false assumption

[1] The earliest of such investigations, which yielded extremely valuable observations on the increase in weight of the metals and the part played by the air in their calcination, were undertaken by Jean Rey, Hooke, Mayow and Boyle in the seventeenth century. Rey and Mayow came very near to explaining the results of their experiments correctly.

that the increase in weight was due to a ponderable fire-stuff.[1] It was sought to show by pure philosophy alone, without the faintest shadow of proof, that air was essential to calcination and similar processes, by assuming that it must be present in order to take up the escaping phlogiston. This expedient, first brought forward by Becher and Stahl, was made use of again and again by later phlogistonists.

While these latter imagined that they had thus correctly interpreted the part played by the air, they followed Stahl's example in paying no heed to the observed alteration in weight, either regarding this as accidental or making the most unhappy attempts at explaining it. Thus we find Juncker, a pupil of Stahl's, pointing out that the metallic calces were denser than the metals, and therefore heavier,—an utter confounding of the absolute weight with the specific gravity, and also a wrong assertion, since Boyle had already shown in certain instances that the calces were specifically lighter than their corresponding metals. Equally unscientific was the assumption that the phlogiston which escaped in these processes possessed a negative weight, and that, therefore, the residual product must be the heavier; even Guyton de Morveau and Macquer fell into this gross error. True, the most able chemists of the phlogistic period did not concur in these untenable views, but maintained that it was the business of physicists to investigate such points.[2] As a matter of fact, it remained for the physicist Lavoisier to give the right explanation of this, and, with it, that of combustion and similar processes.

[1] Boerhave showed the weakness of such an assumption by proving that the weight of certain metals, *e.g.* silver, remained the same, whether they were at the ordinary temperature or at a red heat. He, therefore, expressed the opinion that an increase in weight on calcination depended upon the addition of a "saline ingredient" (*salziges Theilchen*) from the air.

[2] Some chemists there were who did not regard the above observations on the increase in weight of metals when calcined as meaningless; Tillet, for example, who made a communication to the French Academy in 1762 upon the increase in weight of lead, calling special attention at the same time to the fact that a fit explanation of this had still to be given.

Development of some particular Theoretical Views in the Phlogistic Period.

It is necessary to make one's self acquainted with the growth of the more important chemical ideas of this time, in order to properly appreciate the advances which they show upon those of the preceding periods, and also in order to comprehend the connection existing between the theoretical views of the phlogistic era and of that new one which begins with Lavoisier. We have to deal here with the meanings attached to the terms "element" and "chemical compound," and also with the ideas of the phlogistonists upon chemical affinity.

Views regarding Elements and Chemical Compounds.—The position which Boyle took up with respect to the question of the elements has been already spoken of; he it was who established the scientific term "element," in that he regarded as elements those actual constituents of compound bodies which were capable of isolation and which could not themselves be broken up into simpler substances. With the increase of means for deciding the question whether any substance is in this sense an element or not, the boundary line between elements and chemical compounds became more and more altered in position, but at the same time sharper. Boyle further cherished the idea that the elements attainable by chemists were not the ultimate constituents of matter.

Notwithstanding the clearness with which Boyle set forth the conditions which an element, according to his view, must fulfil, we find among his contemporaries and their successors a tendency to go back to the alchemistic elements, and even to the Aristotelian. Willis, Lefèvre and Lemery associated earth and water with the three elements of Basil Valentine and Paracelsus; Becher also adhered to those three under other names, adding water to them; and even Stahl was unable to free himself from ideas of this kind.

The erroneous assumption of the phlogiston theory—that the products of combustion and calcination, *i.e.* acids and metallic oxides, were simple, and the original substances compound—had the most serious consequences in keeping back a knowledge of the true elements. While Boyle appeared inclined to reckon the metals among the latter, their compound nature was never questioned from the time of Stahl until the fall of the phlogistic doctrine; and, conversely, the metallic calces and compounds produced in an analogous manner (*e.g.* sulphuric acid, phosphoric acid and water) were regarded as elements. Sulphur and phosphorus belonged of course to the compounds. Phlogiston itself, the supposed existence of which was due to this inversion of actual relations, was regarded, on the other hand, as an element. Only after this purely hypothetical state of matters had been set aside by the proof that instead of the escape of phlogiston the absorption of oxygen must be allowed, and instead of the assimilation of phlogiston the withdrawal of oxygen, did Lavoisier bring light into the prevailing confusion—a confusion which was being continually increased by the addition of contradictory facts.

With respect to the term "chemical compound," and the formation of such, ideas were developed during this period which contained much that was sound, and which indicated an advance over previous ones; this is, of course, apart from the erroneous assumption that those bodies which were afterwards recognised as being simple (many metals and some non-metals) were compounds of their oxides with phlogiston. By the clearness of his views Boyle contributed materially to an insight into the nature of chemical compounds, and to a recognition of their dissimilarity to simple substances. Boyle, Mayow and especially Boerhave gave utterance to the weighty tenet that the characteristic properties of substances which combine together chemically do indeed disappear after such combination, but that nevertheless the latter are not lost, but are still present in the compound. At that time it was necessary to defend this truth, which became more distinctly formulated later on in the law of the Conservation of Matter, against the old delusion that the formation of a compound was synonymous with the creation of a new substance. How clearly the investigators just named had grasped the meaning of the term "chemical compound," is shown by the sharp distinction which they drew between it and a mixture of its components.

Analytical chemistry, which was meantime gradually developing, aided towards a better understanding of the composition of substances, for by its means certain constituents of salts and of other compounds could be distinguished from one another. So long, however, as analysis remained merely qualitative, and no account was taken of the proportions by weight in which substances combined, any considerable development of the meaning of the term "chemical compound" was impossible; this was reserved for the succeeding age.

The defective knowledge of the quantitative composition of substances forced chemists back upon conclusions drawn from analogy, when they wished to obtain a survey of the compounds known. It was to the endeavour to explain similar phenomena by the assumption of a common principle that the phlogistic theory owed its origin. Acids, salts and metallic calces were looked upon as being of analogous composition, both because of their behaviour and their modes of formation. The distinct recognition of the fact that salts were produced by the combination of acids with bases was one of the greatest achievements of the phlogistic period. Before the term "salt" assumed such a definite form, indistinct ideas on the subject were very prevalent; we have only to recall that even such a man as Stahl used the word for acids and alkalies as well as for salts proper. After Boerhave, Geoffroy and Duhamel had succeeded in giving greater precision to the conceptions regarding these classes of compounds, Rouelle was able (in 1745) to define salts once for all as the products of the union of acids with bases,—and he further drew a sharp distinction between neutral salts (*sels neutres parfaits*) on the one hand, and basic and acid salts on the other.

The characteristics of salts which formerly obtained—their solubility in water and their taste,—therefore fell to the ground, seeing that Rouelle included the insoluble silver and mercurous chlorides among them.

But while Rouelle's views regarding the alkaline salts were perfectly sound, he could not throw off the old idea that the vitriols and other metallic salts consisted of metal and acid; it fell to Bergman to show that this was erroneous, by the proof that it is the metallic calces and not the metals themselves which combine with acids to salts.[1] What an advance is shown by those definite conceptions on the composition of salts, as compared with the vague ideas that even Stahl not long before had given utterance to, viz. that salts were made up of an earth and water!

Views regarding Chemical Affinity and its Causes.—The old assumption that those bodies have an affinity for one another which have something in common, that affinity, in fact, is conditioned by this, according to the axiom *similia similibus*, held its ground in speculative minds even into the eighteenth century. The word *affinitas*, which expresses this idea, and which was already employed by Albertus Magnus, presupposes therefore the similarity of substances which interact with one another. Boerhave, on the contrary, stoutly maintained that it is unlike substances which show the greatest tendency to combine with each other; and, notwithstanding that the reason given for the combination of bodies is exactly the opposite of what was originally taught as such, viz. their dissimilarity, the name "chemical affinity" or "affinity" for this force has been generally retained.[2]

After the time of Glauber, and especially after that of Boyle, much attention was paid to the processes in which the forces of affinity manifest themselves. Cases of so-

[1] The following passage from the pseudo-Geber's *Testamentum* shows that even then people were on the way towards the true explanation of this. The passage is: *Ex metallis fiunt sales post ipsorum calcinationem.*

[2] These terms were temporarily replaced by others, *e.g. rapport* (Geoffroy), *attractio* (Bergman).

called simple elective affinity (*attractio electiva simplex*, a term which originated with Bergman) were interpreted correctly by both the chemists just named, and also by Mayow; for instance, the expulsion of ammonia from salmiac by fixed alkali, by the assumption that the attraction of the latter for hydrochloric acid was greater than that of this acid for the ammonia (*flüchtiges Laugensalz*). Observations of this kind on the expulsion or precipitation of bases or acids from salts, by substances endowed with stronger powers of affinity, soon induced chemists to work out the order in which analogous bodies were separated from their compounds by others. The observations on the precipitation of metals and on the expulsion of various acids from salts by means of sulphuric and nitric acids, among others, may have tended in an especial degree to make clear the different strengths of affinity in analogous bodies. The collation of the results of numerous investigations on the behaviour of acids and bases to salts, and of metals to metallic salts, yielded tables of affinity, *Tables des rapports* (first published by Geoffroy in 1718 in the Memoirs of the Paris Academy), in which similar substances were so arranged that their affinity to the dissimilar ones placed outside the table gradually decreased.

The following table will serve to elucidate Geoffroy's principle :—

Sulphuric Acid.	Fixed Alkali.
Fixed alkali	Sulphuric acid
Volatile alkali	Nitric acid
Absorptive earth	Hydrochloric acid
Iron	Vinegar
Copper	Sulphur.
Silver.	

These tables of affinity remained in use for a considerable period, although it was apparent that they stood in need of amendment, and were frequently modified and enlarged. Their deficiencies became especially obvious when chemists began to recognise more fully the influence of heat upon the progress of chemical reactions, and observed that some, whose course under ordinary conditions was perfectly well known, proceeded in an exactly opposite direction at a higher temperature; Stahl, for instance, had noted this correctly in the interaction of calomel and silver at a lower, and of chloride of silver and mercury at a higher temperature. Such reciprocal reactions led to the proposal to prepare tables of affinity for medium and high temperatures, both for wet and dry (*i.e.* fusion) reactions. Bergman made the attempt in 1775 to work out this proposal of Baumé's by investigating the mutual behaviour of a very large number of compounds, with the result that the doctrine of chemical affinity was materially advanced, in so far as this was possible by such empirical work.

The results of his extended researches were utilised by Bergman for setting up a theory of affinity, which will be most conveniently considered in conjunction with Berthollet's doctrine of affinity (see the history of the doctrine of affinity in recent times). But even prior to the efforts of both of these men, the cause of this affinity was a subject of frequent reflection and of far-reaching speculation. Boyle's lucid conception—that the small particles (of which, in his view, different bodies were made up) attract each other—has been already mentioned. The greater or lesser degree of this mutual attraction of heterogeneous substances depended upon the form and position of each small particle. He did not, however, specially work out this idea, which lay at the root of his corpuscular theory, doubtless because he was so sagacious as to see that he could not possibly arrive at any knowledge with regard to the shape of atoms. Lemery, on the other hand, gave a loose rein to his fancy upon this question. According to him, the combination of two substances—*e.g.* of an acid with a base—depended upon one of the small particles being sharp and the other porous; by the fitting of the points into the cavities, combination was effected. He further attempted to explain the throwing down of precipitates, the solution of metals in acids, etc., in a similar manner.

The force which the mutual attraction of the particles calls forth was regarded by many, *e.g.* by Buffon (who occasionally took part in the discussion of theoretical chemical questions), as identical with that of gravitation. But Bergman, who was also inclined to this assumption, justly pointed out that, since these particles act upon one another at the smallest possible distances, this force must be exerted differently from that of gravity; and Newton, who also turned his attention to the point, likewise assumed a difference between affinity and gravitation.

It was, however, impossible that this subject which dealt with the phenomena of affinity could develop greatly in the phlogistic period, since the proportions by weight in chemical processes were hardly thought of at all. But the purely qualitative investigation of a large number of reactions, from whose outcome conclusions were to be drawn regarding the interaction of individual components, had the effect of maturing much good fruit, so that the unresting efforts of chemists to enlighten themselves upon such questions turned out by no means useless.

This indeed applies generally to the attempts of that age in questions of theoretical chemistry—attempts which were on the whole unhappy. The chief gain was on the practical side, in the rich material accumulated by observation, the complete application of which was reserved for the new era.

The most important achievements in practical chemistry during this period will be touched upon briefly in the following section, in so far as they have not already been described in the general part.

History of Practical Chemical Knowledge in the Phlogistic Age.

The question of the composition of substances — that problem which had been recognised as fundamental from the time of Boyle—could only be solved by the experimental method; it was analytical chemistry, which had developed

acids, bases and neutral substances, and for this purpose he studied and made particular use of the colouring matters in the juices of litmus, violets and corn-flowers. Besides these general reagents, which served to distinguish important classes of compounds, Boyle introduced many other characteristic ones which allowed of the recognition of individual substances in the form of precipitates. For the detection of sulphuric and hydrochloric acids, respectively, he used solutions of calcium and silver salts, and *vice versa*. Ammonia he recognised by the production of a cloud when it came in contact with hydrochloric or nitric acid; copper salts by the blue solution which they gave with excess of volatile alkaline salt; solutions containing iron by the black colouration they yielded with infusions of tanning stuffs[1] (from gall apples, oak leaves, etc.). He was also sometimes happy in the way in which he applied careful observations on the precipitation of certain metals by others, as tests for these.

The salt solutions found in nature, mineral springs in particular, had before this time stimulated the iatro-chemists to search out the substances which they contained. Some advances in the analysis of mineral waters became noticeable at the end of the seventeenth and in the eighteenth centuries, and we find at the same time the chemists engaged on the subject inspired with the wish to prepare those natural products artificially; but the knowledge requisite for doing this, *i.e.* a knowledge of the true, and especially of the quantitative, composition of these waters, was wanting even at the end of last century. Hoffmann investigated a large number of mineral waters, and proved the presence in them of carbonic acid, iron, common salt, and salts of magnesia and lime, showing at the same time how to test for these; he also pointed out the characteristics of alkaline and sulphur waters. In addition to this, he demonstrated the incorrectness of previous statements as to the presence of gold, silver and arsenic in such waters, and explained the connection

[1] The prescription for preparing black iron ink from gall apples and iron vitriol is due to Boyle.

between the occurrence of such exceptional salts as alum and copper vitriol and the nature of the soil at those places. He frequently made use of crystalline form to distinguish different salts.

The observations made by Marggraf materially enlarged the acquaintance with reagents suitable for the detection of substances, and also the knowledge of the composition of many compounds. He used, for instance, a solution of prussiate of potash to test for iron, and applied the different colourations which the salts of potash and soda impart to a flame for their detection—a point that had also been observed independently by Scheele. The behaviour of many salts to caustic potash enabled Marggraf to arrive at their composition; thus he proved that gypsum consisted of lime and sulphuric acid, and that this acid was also present in heavy spar. As already mentioned, he made use of the microscope for getting at the crystalline forms of different substances.

That Scheele owed his mastery in the discovery of new substances to the gift of deducing their presence from certain reactions, and that he, therefore, greatly extended analytical chemistry by a multitude of observations, hardly requires to be stated. But, although in his knowledge of the chemical behaviour of bodies he was equalled by no one of his contemporaries, he unfortunately did not apply this knowledge systematically, as Bergman did, thereby laying the firm foundation for the methodical use of reagents, and, with it, of qualitative analysis. The reactions which the latter made use of as tests for baryta, lime, copper, sulphuretted hydrogen, and sulphuric, oxalic, arsenious and carbonic acids, etc., are those in vogue at the present day. Bergman also drew attention to the general application of the fixed alkalies for precipitating solutions of metals and earths; to many other reagents, such as sublimate, sugar of lead, and liver of sulphur; and also to modes for estimating precipitates and separating salts. The first methods, by which it was possible to test minerals and especially ores completely, were due to him, viz. their digestion with hydrochloric or nitric acid, or their fusion with carbonate of potash. There can be no doubt, however, that he was indebted to Scheele for many observations; the latter, for instance, fused up minerals with alkalies so early as 1772-3, perceived the difference between soluble and insoluble silicic acid, and carried through the separation of iron and manganese by means of acetic acid.

Qualitative analysis in the dry way made considerable advances in the eighteenth century by the increasing use of the blowpipe, the value of which in the examination of ores was recognised more especially in Sweden. Gahn and Bergman, together with the mineralogist Cronstedt, were chiefly instrumental in introducing it into chemistry;[1] in their tests they employed borax, soda, cobalt solution and other reagents, and also made use of the difference between the inner and outer flames, though Scheele was manifestly the first to recognise and explain correctly the reason of this difference. But it was through Berzelius that the blowpipe became universally employed and felt to be an almost indispensable aid in analysis.

Attempts not merely to test for substances qualitatively, but also to determine their quantity, were few in number up to the time of Lavoisier, and yet it is evident from many statements made by Boyle, Homberg, Marggraf, Scheele, Bergman and others, that they sometimes endeavoured to take the proportions by weight into account. How otherwise is it possible to explain Marggraf's accurate determination of the weight of the precipitate obtained by dissolving a given quantity of silver and precipitating the solution with common salt; or Black's estimation of the weight of the precipitate obtained by adding carbonate of soda to a solution of sulphate of magnesia which corresponded to a definite amount of *magnesia alba*, in order to prove the constant proportion of fixed air in the latter? Mention must also be made here of

[1] After investigating the point with great care, J. Landauer (*Ber.* xxvi, p. 898) has brought forward proof to show that it was Cronstedt who really rendered the chief service here, and not Anton Swab, as has recently been contended. G. v. Engeström was the author of the first manual on the use of the blow-pipe; this was published in 1770 as an appendix to a work on mineralogy.

the determination of the weights of metallic precipitates (*i.e.* the metals themselves) by Bergman and others. Bergman was probably the first to proceed on the principle that an element should not be itself isolated and estimated according to its own weight, but separated in the most convenient form as an insoluble precipitate, *e.g.* lime earth as oxalate of lime, and sulphuric acid as sulphate of baryta.

In pneumatic chemistry, too, the necessity became strongly felt of being able to detect different gases in presence of one another by means of reagents, and to estimate their relative volumes quantitatively. For this purpose special absorptives were used, by the action of which the differences in the gases had first been noticed. Thus caustic potash was found to be suitable for the absorption and measurement of carbonic acid, and saltpetre gas (nitric oxide), hydrate of protoxide of iron, moist sulphuret of iron, or phosphorus, for that of oxygen. Of course the results of such quantitative analysis were very inexact.[1] But Cavendish succeeded in making an extremely accurate determination of the oxygen in air by the method suggested by Volta, viz. by exploding with hydrogen. Unlike previous experimenters, he found the composition of the air constant, the oxygen amounting on the average to 20·85 per cent.; the mean, as determined at the present day, is 20·9 per cent.

As the foregoing short account shows, a great deal of preparatory work, which chiefly required perfecting in the quantitative direction, stood ready to hand at the period which began with Lavoisier. The most important features and principles of chemical analysis were contained in these preparatory researches, and only waited for development.

[1] As the result of very imperfect methods, Priestley and Scheele found that the proportion of oxygen in air varied between 18 and 25 per cent. The term "eudiometry" [εὔδιος, fine (applied to weather), and μέτρον, a measure] came into use then, because it was supposed that the purity of the air was arrived at by the determination of its oxygen; and it has continued to be employed in gas analysis in spite of its inaptness.

The State of Technical Chemistry in the Phlogistic Period.

Many chemists of the time, among whom we may mention Boyle, Kunkel, Marggraf, Macquer and Duhamel, frequently directed their efforts to applying their scientific experience of chemical processes to the advancement of particular branches of industry. Technical chemistry thus made good progress during this period. We come across the beginnings of great chemical industries, and are able to perceive the development of a knowledge of technically important chemical preparations, whose manufacture has increased during this century in an undreamt-of degree.

The distinction between applied and pure chemistry was universally recognised towards the middle of the eighteenth century. Serviceable text-books, treating of particular branches of technical chemistry, were not wanting, the conjunction of theory and practice so necessary for the w lfare of the latter being thus cared for. Analysis was also successfully brought into the service of chemistry, especially in the working-up of ores. Even so early as 1686 Charles XI. of Sweden had recognised the value of such investigations, and had caused a technical laboratory to be built. Here, under Hiärne's superintendence, all sorts of natural products (such as ores and other minerals, soils, etc.) were examined, and researches were instituted, with the object of rendering chemical products of practical use, and of applying in daily life the various results obtained.

In metallurgy the several modes of procedure underwent only slight changes, but, as a consequence of the clearer comprehension of chemical reactions, light was thrown upon many processes which had hitherto been wrongly explained. The results of the researches of Bergman, Gahn and Rinman came to be used in the manufacture of iron and steel, the difference between these being traced to its true reason only at the end of the phlogistic period. Marggraf taught an easier mode of preparing zinc from calamine in closed chambers, with exclusion of air as far as possible, and thus made this useful metal more available. The manufacture of brass was materially improved by Duhamel de Monceau, and that of cast-iron and steel by the versatile Réaumur. The production and working-up of particular metals, *e.g.* the engraving, tinning and gilding of iron, the silvering of copper, etc., were developed in many ways by Boyle and Kunkel.

A highly productive field was opened up for the ceramic industry by the accidental discovery of porcelain, the manufacture of which, although carried out on a large scale at Meissen, remained a secret until it was successfully solved at Sèvres in 1769 by the carefully planned experiments of Réaumur and other later chemists, notably Macquer. Improvements and novelties in the manufacture of glass were introduced by Kunkel and Boyle, *e.g.* in the preparation of ruby glass and in glass painting. Dyeing was likewise enriched by the experiences of various chemists. New colours, chief among which was Prussian blue (discovered quite accidentally by the dyer Diesbach in 1710), together with paints, such as mosaic gold and Scheele's green, were made available for industrial purposes. And chemists, among whom Stahl, Hellot and Macquer must be particularly mentioned, endeavoured not only to prepare and apply colours by practical recipes, but also to aid the manufacturer by speculations upon the modes in which dyeing processes are brought about. Dyes were divided by them into two classes, according as they were capable of being fixed upon cloth with or without mordants, and Bancroft (in 1794) distinguished these as *adjective* and *substantive* dyes. Scheele was the first to give a correct explanation of the formation of lead white, a substance much prized as a white paint.

Those technically important preparations, of which an intimate knowledge was first gained in the phlogistic age, constituted a valuable introduction to the chemical industries of to-day. At that time the tendency of chemists was to inquire whether this or that substance was technically useful, just as in the preceding period they had tested chemical compounds for their application to medicine. The manufacture of acids and alkalies, the chemical industry which constitutes the basis of nearly all others, was last century only in its infancy, although even then some of these products began to be made in considerable quantities. Thus Boyle tells us that nitric acid was manufactured from saltpetre in special "distilleries" (*Brennereien*) to more advantage than was the case before, by improved methods worked out by Stahl and others. Rouelle was the first to show how it could be concentrated by distilling it with oil of vitriol. Sulphuric acid was first manufactured on the large scale in England (by Ward of Richmond) about the middle of the eighteenth century, by burning sulphur with the addition of saltpetre. The perishable and at the same time costly glass balloons in which the process was carried out were soon replaced—at first in Birmingham—by leaden chambers, which are still indispensable for this manufacture; the continuous working of these chambers is an achievement of our own century. The preparation of fuming sulphuric acid from "weathered" iron vitriol had been known long before that of oil of vitriol itself, which last, moreover, received its name because of its production from this salt. The manufacture of the fuming acid, based upon the old observations of the pseudo-Geber and Basil Valentine, was first carried on at Nordhausen in the Harz (whence its name of Nordhausen sulphuric acid, still in vogue), being removed subsequently to Bohemia. The time for the technical application of hydrochloric acid and the chlorine generated from it was not yet come; hydrofluoric acid, however, was used for etching glass by Schwanhardt of Nürnberg so far back as the seventeenth century.

The alkalies and their carbonates were, as in ancient times, obtained from the ashes of plants, carbonised tartar and incrustations on the soil, to be used for the production of soap, glass, etc. The discovery of the practical preparation of soda from common salt, which revolutionised industrial chemistry, was reserved for the beginning of the present epoch; but even so early as the first half of the eighteenth century some remarkable observations were made which showed that it was possible to convert salt first into sodic sulphate, and then the latter into soda—reactions which, as he himself tells us, were turned to use by Leblanc, the gifted originator of the soda industry.[1]

Duhamel de Monceau, one of those who showed how to transform common salt into soda, deserves praise for introducing suitable processes for the preparation of various products of technical importance,—salmiac, starch, soap, etc. We find, in fact, the clearer knowledge of chemical reactions resulting in improvements in old processes generally, and many new manufactures created or at least prepared for, *e.g.* the now enormous beet sugar industry by Marggraf's discovery.

Knowledge of other important Compounds during the Phlogistic Period.

The increase in the knowledge of the elements and of chemical compounds—which, although of no special technical value then, were partly destined to become so—was quite remarkable in the phlogistic period, so that it is worth while to take a short survey of these here. To the elements known at that time (although they were not regarded as such) various new ones were added, of which we may mention phosphorus, chlorine, manganese (isolated by Gahn in 1774), cobalt (Brandt, 1742), nickel (Cronstedt, 1750), and platinum (Watson, 1750). The discovery of these was usually preceded by a thorough investigation of their compounds, although chance sometimes came into play, *e.g.* in the isolation of phosphorus. This last discovery excited chemists in an unwonted degree and produced an extraordinary sensation among educated circles in Germany, England and France, on account of the marvellous properties of the new body. Brand, a Hamburg alchemist, succeeded in 1669 in obtaining phosphorus by distilling the residue from evapor-

[1] A notable observation made by Scheele about the year 1770 deserves mention here, viz., that soda can be prepared by treating a solution of salt with oxide of lead, filtering, and passing carbonic acid through the filtrate. This process was patented by Turner in 1787.

ated urine, and gave it the same name as the Bologna stone or *phosphor* (which was sulphide of barium, prepared by heating the sulphate with carbon), already known. The two leading chemists of the day, Boyle and Kunkel, endeavoured for years to discover the secret of its preparation, and ultimately succeeded, contributing thereby at the same time to a better knowledge of the element.

Of the chemical compounds prepared artificially, it was the combustion- and calcination-products of the elements, *i.e.* acids and metallic oxides, which awakened the most interest, in accordance with the tendency of the age; and accompanying this, the salts formed from these bodies were carefully studied. A good deal has already been said with regard to the knowledge of these substances. Although the views as to their composition were quite erroneous, the correct interpretation which came later was materially aided by the accurate investigation of their behaviour.

Of acids as combustion products, phosphoric acid deserves the first mention. It was discovered by Boyle, and its nature elucidated by an admirable research of Marggraf's, who showed how it was produced by burning phosphorus, and also by treating the latter with nitric acid; he likewise explained its production from urine. Further, that the amount of phosphorus present in the latter depended upon the nutriment taken, was distinctly stated by him. Scheele and Gahn were the first to prove the presence of phosphoric acid in bones. It has already been mentioned that the earliest accurate knowledge of the combustion-products of sulphur, coal, and of gases containing oxygen generally, belongs to the second half of the eighteenth century. Cavendish proved the composition of nitric acid by its synthesis from nitrogen and oxygen (in presence of water), but the clear result of his researches was obscured by phlogistic accessories. The discovery and accurate examination of nitrous acid—"volatile nitric acid"—was due to Scheele in 1768; his able treatise on the subject was only published recently along with his letters (*loc. cit.*, p. 9).

The many investigations which were made on the products of calcination of the metals and semi-metals greatly advanced the knowledge of these. We may mention here the recognition by Scheele of white arsenic as the calx of the metallic arsenic and the oxidation of the former to arsenic acid in 1775, his discovery of molybdic and tungstic acids, and the investigation of the behaviour of quicksilver calx upon heating—so pregnant in its results.

The knowledge that a salt consisted of an acid and a base facilitated the survey of many compounds widely apart from one another. Marggraf, for instance, showed that sulphate of potash had an analogous composition to gypsum and heavy-spar, although it was so unlike these. The definite distinction of alum earth from lime earth, of the latter from magnesia[1] (Hoffmann and Black), and of potash from soda (Duhamel, Scheele, and others) belonged, with many other discoveries, to the phlogiston theory in its prime, and was of great service to the succeeding period. A large number of new salts became known, among others salts of manganese and bismuth (including the basic nitrate of bismuth, so much valued as a cosmetic), compounds of cobalt, nickel, platinum, etc. And the qualitative composition of many salts, whose nature had hitherto been quite misunderstood, was correctly explained, *e.g.* that of alum, borax, calamine and other compounds.

Organic Preparations.—The knowledge of organic compounds was likewise much advanced, especially by Scheele, who devised methods for discovering and isolating organic acids. While new fields were thus opened up at the close of the phlogistic period, those organic substances which were already known were also further investigated. It is true that the real composition (even qualitative) of all these carbon compounds remained unrecognised, and this complete ignorance hid itself behind meaningless expressions and periphrases; thus oil and water, or a combustible and

[1] Silicic acid, which had for long been reckoned among the earths as "vitrifiable earth," was first characterised by Scheele as a fire-proof acid in the year 1773 (*Letters*, p. 69).

a mercurial principle, were assumed as the constituents of alcohol. It was again Lavoisier who pointed out the right path here, by proving that carbon, hydrogen and oxygen were the constituents of this as of most other organic substances, and by indicating modes for determining the proportions by weight of the elements just named.

Spirit of wine and the ethers which could be obtained from it, together with common ether itself, were the subjects of frequent investigation, so that they came to be prepared fairly pure. Spirit of wine especially was employed in analysis for the separation of different salts, and attempts were made to deduce the amount of alcohol in aqueous solutions of it from its specific gravity; the beginnings of alcoholometry are to be found with Réaumur in 1733 and Brisson in 1768. With respect to its formation in spirituous fermentations opinions were very confused; many, indeed, disputed this formation, assuming its pre-existence in the wine must, etc.

Ether, which was termed *spiritus vini vitriolatus* or *æthereus*, became known through the labours of Frobenius (about 1730), Hoffmann, Pott, Baumé and others, and was used medicinally admixed with spirit of wine (Hoffmann's drops). The erroneous idea that it contained sulphur prevailed for a long time, until this was finally done away with by the investigation of Valentin Rose the younger (in 1800).[1] The name "sulphur ether" arose from this. At that time any pungent volatile liquid was termed an ether.

Nitrous[2] ether, muriatic ether, and acetous[3] ether, so named because of their respective origins, were likewise carefully investigated, and were valued as officinal preparations. Scheele's acuteness of observation is well shown by the fact that he recognised the necessity for having a mineral acid present during the formation of ethers of weak

[1] Prior to this date, Hoffmann and Macquer correctly assumed that ether was formed from alcohol by the elimination of water.

[2] Our present nitrous ether, admixed with a little nitric ether, aldehyde, etc.

[3] Ethyl acetate.

acids, such as acetic and benzoic, a point which had been overlooked before his time.

The knowledge of the organic acids was materially extended during the phlogistic period, especially towards its close. Acetic acid, which had been longest known of any, was now prepared in the concentrated pure state as the glacial acid, and its combustibility was observed by Lauraguais. Kunkel, Boyle and others believed in the identity of the acetic acids prepared by fermentation and by the distillation of wood, without, however, being able to adduce definite proof of this; the latter was furnished by Thénard in 1802. The resemblance between formic acid, discovered by Wray in 1760, and acetic acid was early noticed, and led to confounding the one with the other, until Marggraf definitely proved their dissimilarity.

Scheele showed how to prepare a large number of acids from plant juices, by first forming their lime or lead salts, and then decomposing these with suitable mineral acids, usually sulphuric. In this way he discovered tartaric acid, which had hitherto been overlooked in spite of the fact that tartar had been known for a long time; also citric, malic and oxalic acids, the last of which he prepared by acting upon sugar with nitric acid, and which he recognised as being identical with the *acetosellic acid* he had obtained from wood-sorrel. By treating milk sugar with nitric acid he was led to the discovery of mucic acid, and by investigating sour milk to that of lactic acid, while he found uric acid in (bladder) stones. For other acids, already known, he devised improved methods of preparation, *e.g.* for gallic and benzoic. Lastly, his discovery of prussic acid in 1782, by decomposing yellow prussiate of potash with sulphuric acid, is worthy of note. The masterly investigation of it which he made enabled him to give its qualitative composition with accuracy; one only requires to translate his phlogistic language into modern chemical terms.

The fatty oils and animal fats were frequent subjects of investigation, without their composition and chemical behaviour, especially towards the alkalies, becoming any clearer;

and this in spite of an important observation made by Scheele in the discovery of glycerine, or *Oelsüss*, as he termed it, by acting upon a fatty oil with litharge. The importance of this observation was only recognised at a much later date. Only the rudiments of preparatory researches are to be seen in the chemistry of the sugars and of other products of animal and vegetable metabolism, such as the ethereal oils, albumens, etc.

Condition of Pharmaceutical Chemistry.

The interests which chemistry and pharmacy had in common resulted in their exercising a beneficial action upon one another. A large number of famous investigators owed to the practice of pharmacy their stimulus to the study of purely chemical phenomena; of these we may mention Kunkel, the Lemerys (father and son), Geoffroy, Rouelle, Neumann, Marggraff and Scheele. While they themselves and others contributed a wealth of the most valuable observations, indeed of fundamental discoveries, to chemistry, pharmacy was at the same time materially advanced, not only by those discoveries, but also by special pharmaceutical researches. The chief gain for pharmacy lay in its intimate fusion with pure chemistry. On the other hand, the work required in apothecaries' shops proved itself the best preparatory training for future chemists. The scientific taste was nourished by excellent text-books on pharmaceutical chemistry, *e.g.* Baumé's *Éléments de Pharmacie Théorique et Pratique* (1762), Hagen's *Lehrbuch der Apothekerkunst* (Hagen's Text-Book of Pharmacy), and was firmly established by the founding of pharmaceutical laboratories; the growth of the latter belongs, however, more to the present epoch.

Many additions were made during this period to the medical treasury by pharmaceutical chemistry. Of the new medicines which then came into vogue, and whose nature was often involved in mystery until they ceased to be secret remedies, the following important ones may be mentioned:—Carbonate of ammonia, which was contained in the famous

"English drops"; sulphate of potash, valued under Glaser's designation of *sal polychrestum*, which was obtained by detonating sulphur with saltpetre; sulphate of magnesia, first prepared from the Epsom (spring) water by Grew in 1695, and termed *sal anglicum*, and, later on, bitter salt; and *magnesia alba*, obtained from the mother liquors in the preparation of saltpetre by means of carbonate of potash. Among the preparations of antimony, the *Kermes minerale*, whose composition was only arrived at correctly during the present century, came into repute. Ferric chloride in alcoholic solution was a favourite secret medicine in the first half of the eighteenth century under the name of "gold drops" or nerve tincture; its nature, however, soon became recognised. Hoffmann's drops and the compound ethers were likewise used officinally. Goulard introduced basic acetate of lead after the middle of the last century as a remedy for external use, and it is called by his name to this day.

Many observations were made with regard to substances of special antiseptic action, Kunkel pointing to the mineral acids for this. The antiseptic properties of iron vitriol and alum were made use of in the impregnation of wood with these salts, according to the proposal of the Swede, Faggot. In 1782, Scheele recommended the conservation of vinegar by boiling it in closed vessels; he was thus the discoverer of the sterilisation method, now of such supreme importance.

Concluding Remarks.—The period of phlogistic chemistry must be looked upon as the indispensable forerunner of the new era which began with Lavoisier. The erroneous conception which underlay the important phenomena of combustion and calcination, and which spread itself over many other processes, most assuredly did not prevent the young science of chemistry from developing in a healthy manner. Without doubt it was the experimental method which contributed most to this. Hand in hand with this development we find an increasing improvement in the means for observing chemical processes and for establishing the properties of substances. These advances were due partly to improved apparatus (for instance, the apparatus required for collecting and measuring gases), and partly to the use of physical methods of research; and here we may note the more frequent determinations of the specific gravity of bodies in different states of aggregation, and the use of the microscope. The time had not yet arrived when the balance was to be employed with such great advantage for the exact determination of proportions by weight in chemical reactions, although a number of noteworthy beginnings of quantitative analysis are to be found.

It is especially to be noted as characteristic of this period that chemistry now became fully awake to her own proper task, which was to investigate the composition of substances, and to find out the constituents from which they could be prepared. Analytical chemistry was to aid in solving this problem; but useful and important results were achieved by the synthetic method also.

The independent scientific character of chemistry showed itself in the forms which its relations to other sciences assumed. The previous dependence upon medicine and pharmacy ceased; instead of being their servant, chemistry became their helper and adviser. It also came into close contact with physics, mineralogy and botany, which resulted in mutual advantage to all of them, and made chemistry the indispensable helpmeet of the others. We have only to think of the services rendered to those sciences by chemists, *e.g.* to physics by Boyle, and to physics and mineralogy by Bergman. This coalition with the various other sciences had the effect of opening up new common ground both for these individually and for chemistry. We find the first scientific treatment of mineralogical and physical chemistry during the phlogistic period, and the advances made in organic prepared the ground for physiological chemistry.

Nothing is less justifiable, therefore, than to assert that chemistry was at that time no science, and that it was Lavoisier who created one out of what was, before his time, a science only in name. The record of the services of Boyle, Stahl, Black, Bergman, Scheele, Cavendish, Priestley, Marggraf and others, is sufficient to prove the error of such an assumption.[1] In spite of the false hypothesis which lay at the root of the phlogiston theory, it was the latter itself, together with the work which resulted from it, that formed the necessary foundation for the correct standpoints and the numerous researches of the succeeding period.

[1] Cf. Dumas's *Leçons sur la Philosophie Chimique* (1837), p. 137; and also the sentence with which Wurtz began his *Histoire des Doctrines Chimiques* (1868): "*La chimie est une science française; elle fut constituée par Lavoisier*," etc. Volhard investigated this statement and so completely overthrew it (*Journ. pr. Chem.*, N. F., vol. ii. p. 1 *et seq.*), that recent attempts to minimise the force of his criticism have not only missed their mark, but are unjustified in their form and style (see especially Grimaux' *Lavoisier* (1888), pp. 128 and 363). The sentence by Grimaux (p. 128): "*Toute la science moderne n'est que le développement de l'œuvre de Lavoisier*" can only be regarded as an extravagant exaggeration, exceeding even that of Wurtz, just quoted. The most eminent among the antiphlogistonists, moreover, never thought of calling in question the scientific tendency of the chemical views which they themselves combated.

CHAPTER V

HISTORY OF THE MOST RECENT PERIOD (FROM THE TIME OF LAVOISIER UP TO NOW)

THE beginning of the latest period of chemistry, to which the present generation of investigators still belongs, is rightly associated with Lavoisier's reforms, which turned the chemical science of his day into new paths; he demonstrated the supreme importance of the proportions by weight in chemical reactions, which were wrongly interpreted when these were disregarded. This applied in an especial degree to the processes of combustion and similar phenomena, which Lavoisier was the first to explain correctly. Of course this explanation only became possible after Scheele's and Priestley's discovery of oxygen. If we desire, therefore, to associate the commencement of the new era with any particular event, it must be with the latter important discovery, which has been already described in the history of the preceding period.

Lavoisier's combustion theory, with oxygen as its centre-point, now stepped into the place of the phlogistic doctrine, which had attained to the dignity of a dogma; the chemistry dominated by the latter was thus changed into the so-called antiphlogistic system. A complete transformation of all the ideas respecting combustion and calcination, and therefore respecting the composition of the most important substances, took place,—truly a reform in the fullest sense of the word. For, all the reactions in which the escape of phlogiston had hitherto been assumed depended, as Lavoisier taught, upon the taking up of oxygen; and, conversely, those processes which had been explained by assuming the absorption of phlogiston, depended upon the separation of oxygen.

Lavoisier showed that substances like sulphuric and phosphoric acids and the metallic calces, which according to the phlogistic doctrine were looked upon as elements, were really compounds: while those regarded as compounds, *e.g.* the metals, sulphur and phosphorus, he assumed to be elementary.

It will be appropriate here to enter shortly again into the chief points of dispute in which the phlogistic doctrine became involved at the time of the discovery of oxygen (about 1775), and by which its fall was accelerated. The facts to which the phlogiston theory was unable to accommodate itself were many in number. To chemists who regarded hydrogen as phlogiston—a frequent assumption—the great difficulty arose of proving whence the phlogiston came which escaped during the calcination of the metals, and the combustion of sulphur, phosphorus and coal in closed vessels. The reduction of the metallic oxides by hydrogen did indeed appear to allow of a perfect explanation from the phlogistic standpoint, if one paid no heed to the simultaneous formation of water and the diminution in weight of the oxides. But how could a reduction of the metallic calx take place without the presence of phlogiston (hydrogen)? This occurred in the case of those calces which were converted into metal when heated alone in closed vessels. For the production of quicksilver from red oxide of mercury and of silver and gold from their oxides by heat, the phlogistic doctrine was able to offer no explanation. It was, indeed, those reactions which led to the discovery of oxygen that brought about the collapse of the theory, and rendered possible the establishment of the antiphlogistic system. And a few years later the keystone was added to the latter by the proof that water, which had hitherto been looked upon as an element, was a compound of oxygen and hydrogen.

GENERAL HISTORY OF CHEMISTRY DURING THIS PERIOD.

Lavoisier and the Antiphlogistic Chemistry (from 1775 *to the end of the Eighteenth Century).*

Lavoisier's great achievement consisted in abolishing old prejudices, and in the masterly application of scientific principles to the explanation of chemical processes. A wealth of important facts was handed down to him by the phlogistonists; he himself did not add much to this in the way of new chemical observations, but he sifted and collated, from a point of view hitherto unattained, that which was ready to hand, giving at the same time the correct explanation of many processes. We shall not be wrong if we place such services to the credit of his highly-trained physical and mathematical mind, which early freed itself from the bonds of the phlogistic hypothesis. As a physicist Lavoisier was bound to take into account alterations in weight, *e.g.* in the calcination of metals; the properties of the products obtained interested him in a lesser degree. This explains why he himself made no independent chemical discoveries; but the unique service which he rendered in being the first to give a comprehensive and correct explanation of the observations of others remains incontestable.

Lavoisier lived to see his work appreciated in the highest degree; he saw the fruit of his labours, the antiphlogistic system, come out victorious in the fight with the phlogistic, and propagate itself beyond France.—Anton Laurent Lavoisier was born on August 26th, 1743, a year after Scheele, but how different were the outward circumstances of the two! While the latter was early thrown upon his own resources, and was in the fullest sense of the word a self-educated man, Lavoisier, the son of a distinguished barrister, had a splendid training given him, and enjoyed special opportunities for acquiring a thorough knowledge of mathematics and physics, which exercised a permanent influence upon the whole tendency of his thoughts and methods of investigation. In botany, too, in mineralogy and geology, meteorology and anatomy he was well versed. Of his teachers, the mathematician La Caille, the botanist B. de Jussieu, and the mineralogist Guéttard may be mentioned, while it was Rouelle who initiated him into chemistry. Even whilst still very young, Lavoisier gained great repute by his scientific investigations, so that we find him received (as Associate) into the French Academy in 1768, the immediate cause of this being a prize essay upon the most suitable method of street-lighting for large towns.

His earliest chemical work[1]—particularly the research upon the supposed transformation of water into earth, the results of which he published in 1770—afford clear evidence of his physical methods. In this he proved that the total weight of the closed glass vessel plus that of the water which had for a long time been boiling in it remained unaltered, but that the weight of the earth produced was exactly equivalent to the loss in weight of the vessel; the logical conclusion to be drawn from this was that the earth came from the glass and not from the water. What this earth was he did not investigate; on the other hand, Scheele was led to the same conclusions as Lavoisier by examining it qualitatively.

The latter here recognised and laid stress on the use of the balance as a reliable guide in chemical work. Soon after this he busied himself with investigating the reactions involved in the combustion of substances and in the calcination of the metals, making use here of some previous obser-

[1] With regard to Lavoisier's writings, the reader is referred to the *Œuvres de Lavoisier (publiées par les soins du Ministre de l'Instruction Publique)*, which were published in Paris in 1862; and to the analyses of his most important papers, given by H. Kopp in his *Chemie in der neueren Zeit* (1874), and by Höfer in his *Histoire de la Chimie*, vol. ii. p. 490 *et seq.* In addition to these, Grimaux' book, *Lavoisier, 1743—1794* (published in 1888), is a valuable authority on Lavoisier's life and work, even allowing for the fact that the laudation of the famous chemist is overdone in it (cf. note, p. 157). This circumstance, together with the criticism meted out to opponents, and exaggerations of various kinds, seriously detracts from the value of what is in itself a great historical treatise.

vations by others on the increase in weight during such calcination. With the aid of an exceedingly delicate balance he sought, in the first instance, to estimate exactly the alterations in weight which occurred during these processes, and to get at the reason for this. The results of these labours, materially amplified by Priestley's and Scheele's observations on oxygen and its chemical behaviour, formed the foundation of Lavoisier's theory of combustion.

His position had, in the meantime, become a brilliant one; as Farmer-general (he began in 1768 by being an assistant —*adjoint*), and, shortly after, as chief director of the saltpetre industry, of which the Government had a monopoly, he had plenty of leisure to devote to his own investigations, and to assist the State both by his advice and by the introduction of valuable improvements (*e.g.* in the manufacture of potash saltpetre, gunpowder, etc.). His numerous reports on technical questions are evidence of his industry, his versatility, and his wide-reaching influence. For such kind of work he had ample opportunity as member of various Commissions, *e.g.* of the *Société d'Agriculture*, the *Bureau de Consultation*, the *Commissions des Poids et Mésures*, and so on.

Closely related to his work upon combustion were the important researches which he carried out in conjunction with Laplace upon the latent heat of ice and the specific heats of various bodies. It was his clear physical conception of the nature of heat, as opposed to that of many phlogistonists (who were unable to get rid of the assumption of a ponderable caloric), which enabled Lavoisier to interpret correctly those chemical reactions in which heat was evolved,—the phenomena of combustion in particular.

Notwithstanding the extraordinary services which Lavoisier rendered to science, and, through the latter, to his country, by applying his knowledge and experience with never-flagging zeal for her benefit, he did not escape the fate which befell so many of his fellow-citizens. Impeached under the Reign of Terror, he was condemned to death, and executed together with twenty-eight other *Fermiers-généraux*, on the 8th of May, 1794.[1] Amongst all his numerous friends and admirers, only a few, including Hauy and Borde, and only one chemist, Loysel, had the courage to protest against this, but without effect. His more influential colleagues, like Guyton de Morveau, Monge, and especially Fourcroy,[2] who took part in politics, and who had assuredly been able during his five months' imprisonment to do something for his deliverance, did not dare to offer any opposition to this terrible crime.

Lavoisier published most of his works in the Memoirs of the French Academy, over sixty papers by him being contained in its volumes for the years 1768–87; some others are to be found in the *Journal de Physique* and in the *Annales de Chimie*.[3] His projected plan of publishing an edition of his collected works was only carried out long

[1] Much light has been thrown upon this sad event by documents published by Ed. Grimaux, which relate to the death of Lavoisier. It has been conjectured that Marat hastened the proceedings against him from a feeling of petty revenge, because of Lavoisier having unfavourably criticised a treatise of his, entitled *Recherches Physiques sur le Feu*, which appeared in 1780. For Marat, in his infamous *Ami du Peuple*, had repeatedly denounced Lavoisier and had brought about the impeachment, although he did not himself survive to see the arrest of Lavoisier and his colleagues. In the sentence, which was passed after an imprisonment and inquiry extending over five months, it was stated that he was condemned to death "as convicted of originating or participating in a plot against the French nation, the aim of which was to aid the enemies of France; especially in that he had practised every kind of extortion upon the people, and had caused tobacco to be admixed with water and pernicious substances, to the detriment of the health of the citizens who used it."—Cf. Grimaux' work, *Lavoisier, 1743-94, d'après sa Correspondance, ses Manuscripts, etc.* (Paris, 1888).

[2] Grimaux' publication, just cited, and also Berthelot's *Notice Historique sur Lavoisier* (*Mon. Scient.*, 1890, p. 125), reflect seriously upon the indifference to Lavoisier's fate shown by Fourcroy, de Morveau and others.

[3] The dates upon which Lavoisier's papers appeared are of importance for their criticism; we have especially to remember here that the yearly volumes of the *Mémoires de l'Académie* did not correspond with the dates of their publication, but that they were usually brought out several years afterwards (*e.g.* the *Mémoires* for 1772 in 1776, and those for 1782 in 1785). The effect of this disarrangement has been great confusion with regard to the actual time at which this and the other treatise was written by Lavoisier, because of subsequent alterations in the papers. But, so far as it has been found possible to verify them, those dates are given here.

after his death (1862–1892). His *Opuscules Physiques et Chymiques*, which appeared in 1774, contained his ideas upon the nature of gases and his views upon the processes of combustion. In his *Traité Élémentaire de Chimie (présenté dans un ordre nouveau et d'après les découvertes modernes)*, published in 1789, he gave a summary of the most important facts of chemistry, and explained them according to the antiphlogistic theory, which thus received its first text-book; by means of translations of this book the new doctrine was materially propagated.

The researches of Lavoisier which were of greatest moment for the development of chemistry were those which contributed to the founding of the antiphlogistic system, and which led to the overthrow of the phlogistic; those, namely, which treated of the phenomena of combustion, calcination and respiration. The chief work of his life consisted in his recognising and explaining the part played by oxygen in these processes, and in this lies his abiding service.

The previous observations of Rey, Mayow and others, who had attributed the increase in weight of the metals during their calcination to an absorption of air, contained only the first germs of the correct explanation of these processes. From the year 1772 Lavoisier busied himself with investigations bearing upon this subject, the first results of which he delivered in a sealed note to the French Academy on November 1st of that year. This note stated that by the combustion of sulphur and phosphorus, and by the calcination of the metals, the weight of these substances increased from the absorption of a large amount of air; and that, by the reduction of litharge with coal in an enclosed space, a considerable quantity of air—a thousandfold the volume of the litharge—was generated. Lavoisier was at this time in the same position as Mayow had been, that is, still quite uncertain as to which portion of the air caused this increase in weight, as to the air itself being a mixture of gases, and especially as to the nature of the process which went on in the reduction of the litharge; he was inclined to regard the generated gas (carbonic acid) as the fluid originally combined with the lead. This uncertainty was brought about by his paying too little heed to the qualitative side of the chemical reactions.

By repeating these and similar researches, however, Lavoisier soon arrived at a clearer perception of the matter, and he especially recognised his error with regard to the reduction of the oxide of lead. In 1774 he gave further details of these observations, in particular of the calcination of tin;[1] the investigation was in its main points a repetition of Boyle's, but Lavoisier was able to draw more correct conclusions from it than Boyle had done. A sealed retort, in which some tin had previously been placed, was weighed both before and after being heated, and found equally heavy each time, whence the conclusion was drawn that no fire-stuff had been absorbed; on the retort being opened after cooling, air rushed in, and the whole apparatus showed an increase in weight exactly equal to that which the tin had undergone by calcination. Lavoisier concluded from this that calcination depends upon the absorption of air, *i.e.* that the latter is the cause of the increase in weight.

But although we find in these results the beginnings of his combustion theory, there was still wanting the definite knowledge as to which portion of the air combined with the metals and the combustible substances. Oxygen was in the meantime discovered independently by Scheele and Priestley, and they recognised in it the constituent of the air which was necessary for combustion; but Lavoisier held the key to the explanation of his researches as soon as he received news of this discovery. How he turned this to advantage is shown in a paper written in 1775,[2] in which the *rôle* of oxygen for the general explanation of the reactions in question is fully appreciated; it was this gas which combined with the metals, sulphur, phosphorus, coal, and so on. The production of carbonic acid from saltpetre and coal led him to the conclusion that oxygen must likewise be present in this salt—a point that Mayow indeed recognised a hundred years before this,

[1] *Œuvres de Lavoisier*, vol. ii. p. 105.
[2] Cf. *Œuvres*, vol. ii. p. 125.

only that the latter terms it *spiritus nitro-aëreus* instead of oxygen. Strangely enough, no reference is made by Lavoisier to the influence which Priestley's discovery of oxygen (communicated to him by Priestley himself) exercised upon his researches with oxide of mercury and upon his explanation of previous experiments.[1]

Lavoisier in due course arrived at perfect clearness in his explanations, for instance, with regard to the composition of atmospheric air; it was in 1776 that he observed that the combustion-product of the diamond consisted of carbonic acid alone, and in the following year he showed that, by burning phosphorus in a closed vessel, one-fifth of the volume of air in the latter was used up, and non-respirable air remained behind. The results of these researches, together with the observations made by Scheele and Priestley, of which he had in the meantime obtained fuller knowledge, and the investigations which he made in 1777 on the combustion of organic substances, the products of which he proved to be carbonic acid and water, enabled Lavoisier to establish the main points of his Combustion or Oxidation Theory as follows[2]:—

[1] The attitude which Lavoisier sometimes took up with respect to the observations and discoveries of others awakens painful feelings; it is melancholy to see an investigator of such splendid gifts so unjust regarding the services of others. Thus Lavoisier makes no mention in his first chemical paper, on the composition of gypsum, of Marggraf's important researches, although these were among the best known of any, while more than their due recognition was awarded to the other chemists who had worked at the same subject. In a similar manner he ignored, in the account of his researches on the composition of water, those of Cavendish which proved the same point (*i.e.* its composition), and of whose results he had positive knowledge through Blagden's information. Black's splendid investigations upon fixed air, from which Lavoisier without doubt received the greatest assistance towards his conception of the fixation of gases, he treated in a cold and depreciatory manner, whilst the most trivial objections raised against Black were examined with the utmost minuteness and care. These are unfortunately blots upon Lavoisier's reputation, notwithstanding the lustre with which it has become surrounded through the idealistic historical writings of Dumas, Wurtz, Grimaux and others. Cf. also Thorpe's *Essays*, p. 87, and especially p. 110 *et seq.*, in which many of the disputed points in question are cleared up.

[2] *Œuvres*, vol. ii. p. 226, in the *Mémoire sur la Combustion en général.*

(1) *Substances burn only in pure air* (air éminemment pur).

(2) *This air is consumed in the combustion, and the increase in weight of the substance burnt is equivalent to the decrease in weight of the air.*

(3) *The combustible body is, as a rule, converted into an acid by its combination with the pure air, but the metals, on the other hand, into metallic calces.*

The last sentence contains an idea of great moment, which Lavoisier developed later into his theory of the composition of acids, according to which these latter contain oxygen as the oxygenating or acidifying principle (*principe oxygine ou acidifiant*). To establish this assumption, he both made investigations himself and referred to and utilised those of others; in this way he states that sulphuric acid consists of sulphur and oxygen, phosphoric acid of phosphorus and oxygen, and nitric acid of saltpetre gas (nitric oxide) and oxygen. The true composition of the last acid was first determined by Cavendish, through its synthesis from nitrogen and oxygen in presence of water. Hydrochloric acid being a powerful acid, likewise contained oxygen, according to Lavoisier's assumption, and this applied in still stronger degree to the chlorine produced by its oxidation. Lavoisier further occupied himself with the question—What kind of oxygen-compound does hydrogen yield? without, however, arriving at the correct explanation of this independently; for he expected to find an acid as the product of its combustion, and therefore looked for one. It is the undisputed merit of the phlogistonist Cavendish to have proved that water alone is produced by the combustion of hydrogen.[1]

This fundamental observation first proved itself fruitful,

[1] With regard to this point and also to Watt's share in recognising the composition of water, cf. H. Kopp's detailed memoir: *Ueber die Entdeckung der Zusammensetzung des Wassers* (Braunschweig, 1875). See also Berthelot's essay on Lavoisier (*Mon. Scient.*, 1890, p. 138), and Thorpe's *Essays*, p. 110. Berthollet's testimony (*Ibid.*, note, p. 139) leaves no doubt whatever that even Lavoisier's own friends admitted without any reservation Cavendish's priority in this discovery.

however, in the hands of Lavoisier, who was thus enabled to give at once the real composition of water (out of hydrogen and oxygen), while at the same time estimating the relative proportions of these approximately. He also correctly interpreted the decomposition of water by red-hot iron, and its formation from the reduction of metallic oxides by means of hydrogen. The generation of the latter gas on dissolving metals in acids was likewise satisfactorily explained. It was precisely this reaction which had strengthened the phlogistonists in their opinion that the metals contained phlogiston, which, being identical with hydrogen, escaped on dissolving these in acids. The composition of water having been arrived at, Lavoisier now saw that the hydrogen came from the water, and that the oxygen of the latter united with the metal to oxide, which then in its turn combined with the acid.[1]

With the knowledge of this, which came in the year 1783, the last obstacles with which the antiphlogistic system had to contend were overcome: the phlogistic theory could maintain itself no longer, but collapsed. Up to this date Lavoisier was almost alone in the fight against it, having only received material aid from eminent physicists and mathematicians, like Laplace, Monge, Cousin, etc. But now chemists of standing began to apply his ideas, at first in France (Berthollet 1785, de Morveau 1786, and the diplomatically cautious Fourcroy not until 1787), and very soon in other countries also (Kirwan, *e.g.*, in 1792). Lavoisier's critical treatises, which were directed to showing the untenability of the phlogistic theory, conjoined with his *Traité de Chimie*, gave the final blow to that doctrine.

The main features of Lavoisier's work, which was the means of leading chemistry into new paths, have now been described; but some of his observations and speculations, *e.g.* his researches on the composition of organic compounds, and his comprehensive ideas regarding metabolism in the organic world, will be treated of in the special history of this time. The systematic application of quantitative methods of re-

[1] Laplace and Meusnier took an active share in these investigations.

search, and the unbiased treatment of chemical processes from a rather physical point of view, led him to interpret correctly the most important phenomena of chemistry, the explanation of which had been sought for in vain by several generations of investigators, fettered as they were by the phlogiston theory. The material which these latter had collected together, especially the observations of Black, Scheele, Priestley and Cavendish, were indispensable to Lavoisier; we have only to recollect that the discoveries of most importance for his system—of oxygen, and of the true composition of water—were not made by himself. But his genius, far transcending that of any of his contemporaries, enabled him to get at the root of phenomena which they failed to comprehend. After recognising that phlogiston had no existence, and that oxygen was the gas necessary for combustion, calcination and respiration, he translated the obscure and wholly erroneous reactions in which phlogiston was assumed into simple antiphlogistic language.

Although the quantitative method of research was followed and duly valued by individual chemists both before and during the time of Lavoisier, *e.g.* by Boyle, Black, Marggraf, Cavendish, Scheele, and especially Bergman, still none of these investigators made use of the balance as an aid to chemical work with such a definite aim and perfect conviction of its significance as he. Lavoisier was penetrated by the truth that no matter is lost during chemical reactions, and he gave admirable expression to this conviction of the *conservation of matter* by indicating chemical reactions by equations, writing down as equal the substances before their interaction with each other and the products of this interaction.[1] What

[1] In his *Traité de Chimie* (1789) there is the following notable passage in connection with his researches on fermentation: "*Rien ne se crée, ni dans les opérations de l'art ni dans celles de la nature, et l'on peut poser en principe que, dans toute opération, il y a une égale quantité de matière avant et après l'opération, que la qualité et la quantité des principes est la même, et qu'il n'y a que des changements, des modifications. C'est sur ce principe qu'est fondé tout l'art de faire des expériences en chimie. On est obligé de supposer, dans toutes, une véritable égalité ou équation entre les principes des corps qu'on examine et ceux qu'on retire par l'analyse.*"

many others accepted as being correct, without emphasising it particularly, was for him a law upon which he based his speculations and researches. The weight of a compound body was equal to the aggregate weights of its constituents. Although this last sentence now sounds so simple and self-evident, it had to be proved to those who regarded heat as material; for the evolution of heat which took place during chemical combination was bound to be accompanied by a decrease in weight, if a caloric was assumed. Lavoisier was kept from falling into this grievous error by his conception of the nature of heat. His *matière de chaleur* had no weight; this he concluded from experiments in which he burnt substances in closed vessels, proving thereby that no diminution in weight occurred. Many of his expressions show that his views upon its nature approximate to the Mechanical Theory of Heat.[1] The phlogistonists, on the other hand, who saw in heat a ponderable substance, were bound to suffer shipwreck with such a false basis to start from.

The antiphlogistic system, the outcome of the proper interpretation of those processes which were designated combustion, calcination, reduction, etc., meant, in fact, a complete reform of chemistry. The more important of the changes which the latter underwent have been already detailed, but it will be convenient here to refer shortly to the most striking alterations thus effected in the views regarding elements and chemical compounds. Contemporaneously with the definite formation of these opinions went the attempts to introduce a scientific nomenclature, which likewise fall to be treated of now.

Boyle's view with respect to the term "element" was retained by Lavoisier; the latter, therefore, regarded as elements those substances which could not be decomposed into simpler ones. But then what immense alterations he made in details here! The metals and the most important non-metals were ranked among the elements; compound bodies like the alkalies, ammonia and the earths were indeed numbered among these also, but not without great doubt

[1] Cf. *Œuvres*, vol. ii. p. 285.

being expressed as to their elementary nature. Oxygen, also recognised as an element, became, on account of its part in combustion and its capacity for combining with so many other elements, the centre point of the antiphlogistic system, which indeed owed its inception to the knowledge of the behaviour of other elements towards oxygen. The importance which Lavoisier attached to this gas is clearly shown in his theory of acids, just mentioned, and in the statement that the bases which combine with acids likewise contain oxygen. The composition of a large number of compounds—oxides, acids and salts—was thus now rightly interpreted, the phlogistic hypothesis having regarded as simple the substances belonging to the first two of these classes.

The extent of Lavoisier's knowledge and that of his disciples, and especially their views with respect to elements and compounds, is to be seen in the work entitled *Méthode de Nomenclature Chimique*, which was published by the former in 1787 in conjunction with Guyton de Morveau, Berthollet and Fourcroy. The three last were the first French chemists of note to give up the phlogiston theory and to follow the "new chemistry." To Guyton de Morveau belongs the credit of making the first attempt towards a convenient chemical nomenclature, and thereby of inciting to the publication of the above book.

In this work all substances are divided into elements and compounds. To the former belonged—in addition to light and heat—oxygen, hydrogen and nitrogen; these formed the first class. The second group contained the acid-forming elements,—sulphur, phosphorus and carbon, to which were added the hypothetical radicals of hydrochloric, hydrofluoric and boracic acids. The third class comprised the metals, the fourth the earths, and the fifth the alkalies; but Lavoisier considered the elementary nature of the last of these as so improbable that in his *Traité de Chimie* (1789) he no longer included them among the elements. For the nomenclature of the latter, the old names of the metals and of some of the non-metals (*e.g. soufre, phosphore,* etc.) were retained, while

Lavoisier's new names for others of the non-metallic elements (*e.g. oxygène, hydrogène, azote*) were introduced.

Compounds were classified as binary and ternary, and these designations were to a great extent retained later on, although it was found necessary to extend their meaning as chemistry developed. To the binary compounds belonged, in the first instance, the acids, whose names were composed of two words, one of which (*acide*) was common to all, the other being special to each acid, *e.g. acide carbonique, sulphurique, azotique.* In the case of two acids of one and the same element, the name of that one which contained the less oxygen ended in *eux, e.g. acide sulphureux.* The second group of binary compounds embraced the oxygen compounds of the metals, which, as bases, were placed opposite the acids; they were given the generic name of *oxydes,* that of the particular metal in question being added (*e.g. oxyde de plomb,* etc.) The *sulphures* (*e.g.* sulphuretted hydrogen and the metallic sulphides), *phosphures* and *carbures* likewise belonged to the class of compounds of two elements, as did also the compounds of the metals with one another.

The principal ternary compounds were the salts, produced by the combination of bases with acids; their generic name was derived from the latter, with the addition in each case of that of the metal, alkali, or earth in question (*e.g. nitrate de plomb, sulfate de baryte,* etc.).

The advance which is shown by this classification of chemical compounds is very great. In place of false assumptions and designations devoid of any system, we find a correct idea of the qualitative composition of substances, and a rational nomenclature corresponding to this. The development of the latter, and the international form which was given to it by Berzelius, will be treated of below.

Guyton de Morveau, Berthollet and Fourcroy.

These three investigators, who, along with Lavoisier, laid the foundation of a scientific chemical nomenclature, exercised a further influence on the development of chemical doctrines by their other work, the most important of which falls to be considered here.—Guyton de Morveau, born at Dijon in 1737, began life as a lawyer (*avocat*), but gave up this career in order to devote himself wholly to chemistry. His first attempt at a chemical nomenclature brought him into close contact with the French Academy, and in particular with Lavoisier, the outcome of which was the book cited above. Elected a deputy in 1791, Guyton de Morveau did his best to render his chemical knowledge and its practical application of use to his country; we have only to recall here his efforts to employ the air-balloon for strategic purposes in the battle of Fleurus, his activity in helping to found the *École Polytechnique*, in which he subsequently became a professor, and his services as Director of the Mint, etc. The part which he played in politics was less beneficial—it was, in fact, pernicious; for, although an influential member of the National Assembly and of the Convention, he did nothing which could tend to lessen the excesses of the Revolution. He died in Paris in 1816.

To the main service which he rendered, viz. that of having been efficacious in introducing a rational system of nomenclature for chemical compounds, in place of the unmeaning names and confusing synonyms[1] hitherto in use, he added the further one of developing this system by experimental researches in analytical and technical chemistry. He also aided in spreading abroad a knowledge of the labours of Bergman, Scheele and Black, by making good translations of their works.

Claude Louis Berthollet, born at Talloire in Savoy in 1748, had his home in Paris from the year 1772, and showed a wonderful activity in the most various branches of chemistry, especially after the year 1780, when he was

[1] Thus sulphate of potash had five different names, most of which were unintelligible, viz. *sal polychrestum Glaseri*, *tartarus vitriolatus*, *vitriolum potassæ*, *sal de duobus*, and *arcanum duplicatum*. A large number of the names in common use at that time for gases, salts, acids and bases have been grouped together by Nordenskiöld in an appendix to Scheele's *Letters* (p. 467).

elected to the French Academy. He found vent for his great organising talents as a teacher in the Normal and Polytechnic Schools (after 1794), in Napoleon's historical expeditions to Italy and Egypt, in which he took part, and in undertakings for the public benefit. He attained to the highest honours both under the Empire and after the Restoration, and died at Arçeuil, near Paris, in 1822. During the last years of his life, regular meetings attended by eminent *savants* were held at his house, their proceedings being published in the *Mémoires de la Société d' Arçeuil* (1807-1817). At first a phlogistonist, Berthollet frankly declared for Lavoisier's doctrine in 1785.

His experimental researches were especially valuable and fruitful during this period, *i.e.* from 1785 until his death. Mention may be made here of those upon ammonia, prussic acid, sulphuretted hydrogen and chlorate of potash, and upon the practical application of chlorine; he worked out with substantial correctness the composition of the three hydrogen compounds just named. But his researches and speculations upon chemical affinity were of more general and far-reaching significance his *Essai de Statique Chimique* exercised at that time and st exercises a most powerful influence upon this question. The cardinal points of his doctrine of affinity will be given in detail in conjunction with the results obtained by Proust (whose work arose from Berthollet's), the latter of which led to the knowledge of definite chemical proportions, and therefore belong to the history of the development of the Atomic Theory.

Anton François Fourcroy (born 1755, died 1809) understood as a teacher how to inspire his pupils with enthusiasm, and worked in this way with quite remarkable vigour for the propagation of the antiphlogistic system, aiding the latter also by his writings. The chemical articles which he wrote (after 1797) for the *Encyclopédie Méthodique* contain panegyrics upon the antiphlogistic chemistry which, in his excess of patriotic zeal, and possibly not without an egotistical *arrière pensée*, he termed *chimie française*. Fourcroy expounded the antiphlogistic doctrine in larger works also, among others in his *Système des Connaissances Chimiques*, and his *Philosophie Chimique*, etc.

Born one of an impoverished family, he had to earn the money required for his studies under the most pressing circumstances. His work in medicine and natural history led to the honour of his inclusion in the French Academy in 1785, a year after he had succeeded Macquer as professor at the *Jardin des Plantes*. Later (especially after the Reign of Terror), when Fourcroy was on the Public Education Committee, he found an opportunity of utilising the experiences which he had gained as a teacher. Under Buonaparte (then Consul) he became himself Minister of Public Instruction, the education of the country being reorganised for the most part according to his views, and special regard paid to scientific studies. It was certainly due indirectly to him that chemistry bore such wonderful fruit in France during the succeeding decades. Lastly, he took the leading part in founding the Polytechnic and Medical Schools, the *École Centrale*, and the Natural History Museum.

Fourcroy's great merit lay in his activity as an organiser and teacher. And although his experimental investigations yielded no results of great general significance, they served as preparatory work in many branches, *e.g.* in those of physiological and pathological chemistry. His conjoint researches with Vauquelin, in which the latter undoubtedly had the principal share, were of special importance with regard to organic compounds, which had been but little worked with up to that time.

The results of most of these researches were published in the *Annales de Chimie*, which was founded at Lavoisier's instigation by Fourcroy, Berthollet and Guyton de Morveau. This journal, which started into life during the first year of the Revolution (1789), lived through the storms of the period and formed the point of union for French chemists; it was at the same time the organ of the new doctrine, as opposed to the older *Journal de Physique*, in which the last adherents of the phlogiston theory endeavoured to uphold the latter. The *Mémoires* of the French Academy appeared in 1789 for the last time; the Academy itself ceased to exist four years after that date, to be replaced in 1795 by the *Institut National*, out of which the present *Académie Française* originated in 1815, shortly after the Restoration.

After Lavoisier's death the chief representatives of chemistry in France were the three men just named, together with Vauquelin the younger. The latter had won by his researches the right of being numbered among those who gave effective aid in firmly establishing the antiphlogistic system. Vauquelin, born at Hébertot in 1763, was first brought into contact with chemistry as an apothecary's apprentice; a fortunate destiny led him to Fourcroy's laboratory, in which he found employment as assistant. He soon became Fourcroy's collaborateur, and attracted the attention of chemists in general by his brilliant work. From 1793 onwards he filled various posts of distinction, and laboured with success in many different directions, succeeding Fourcroy as Professor of Chemistry to the Medical Faculty after the death of the latter; he died in 1829. Vauquelin did not content himself with merely teaching chemistry by lectures, but gave systematic practical instruction in his laboratory to young men who were desirous of it, and thus trained many chemists who afterwards rose to fame.

Vauquelin's work, which is characterised by great carefulness and exactitude, extended over the most various branches of chemistry. His investigations of minerals promoted the development of mineralogical chemistry, and led him to the discovery of new bodies, *e.g.* chromium and beryllia. His splendid gifts of observation likewise showed themselves in organic chemistry, in the discovery of quinic acid, asparagine, camphoric acid and other substances. His papers are to be found for the most part in the *Annales de Chimie*, of which he was one of the editors after 1791, but some of them are contained in the *Annales des Mines* and other journals. An "Introduction to Chemical Analysis," which appeared in the *Annales de Chimie* in 1799, may be mentioned here; a German translation of this led to its becoming better known and appreciated than would otherwise have been the case. In 1812 Vauquelin published his *Manuel de l'Essayèur.*

Fourcroy's contemporary and Berthollet's celebrated opponent, Josephe Louis Proust, belongs—in virtue of his chief work, which helped materially to found the doctrine of chemical proportions—to the succeeding period, under which he will therefore be spoken of. Other French chemists, *e.g.* Pelletier, Gengembre, Bayen, Parmentier, etc., who gave in their adhesion to the doctrine of Lavoisier during the lifetime of the latter, were also active in chemical research, but they produced no work of general significance; some of the observations made by them will be referred to in the special history of the chemistry of the time.

The State of Chemistry in Germany at the End of the Eighteenth Century.

German chemists proved themselves much less accessible to the antiphlogistic doctrines than Lavoisier's own countrymen. The more eminent among them only began to slacken in their warfare against the new views, and to accommodate themselves to these, during the last decade of the eighteenth century. Of those who lived during that period, and who were active both as investigators and teachers, Klaproth deserves the first mention. Richter likewise participated in the working out of a most important question for general chemistry, in that he was the originator of "stöchiometry"; his investigations are to be looked upon as valuable preparatory work for the chemical atomic theory, and they will be referred to under this. None of the other German chemists of that time produced work of general importance, although they laboured with success in particular departments of the science. Some of the most noteworthy of these efforts will find their place in the special history of certain branches of chemistry; Buchholz, Trommsdorff, Wiegleb and Westrumb may be named here as having enriched pharmaceutical and technical chemistry by valuable observations. Hermbstädt and Girtanner were among the German chemists who first frankly recognised the antiphlogistic system, and they effectively aided in propagating it in their own country by means of their writings.

Martin Heinrich Klaproth, born at Wernigerode in 1743 (*i.e.* in the same year as Lavoisier), only began to teach chemistry—at the Berlin School of Artillery—when somewhat advanced in life, as he continued true to his apothecary's calling till 1787; but this did not prevent him from carrying out in his earlier years investigations of the utmost value, at first under the guidance of Valentin Rose, and later on, independently. It was to these latter researches that he owed his reception into the Berlin Academy. When the University was founded in the Prussian capital, he—although sixty-seven years of age—was elected its first Professor of Chemistry in 1810, and in this post he continued until the beginning of 1817—the year of his death.

Klaproth was distinguished by the care and thoroughness with which he carried out all his work; the quantitative method of research was materially developed and improved by him, and he thereby helped on the recognition of the cardinal principles advocated by Lavoisier. After Klaproth had convinced himself of the correctness of the antiphlogistic doctrine, by thoroughly testing the reactions which took place in combustion and calcination, he became one of its truest adherents; and his example led many other German chemists in the same direction. Other scientists, too, who were not precisely chemists, took a part in the contest to which these theories gave rise; thus we find Alexander von Humboldt publicly declaring for Lavoisier's doctrine in 1793.

Klaproth's researches in analytical chemistry were rightly looked upon at that time as patterns for the younger generation of chemists. Like Vauquelin's efforts, they aimed at establishing the composition of minerals by means of improved analytical methods, and thereby laying the foundation for a chemical classification of these. His observations were so exact as to result in the discovery of various elements and earths—*e.g.* uranium, titanium, cerium and zirconia—while, at the same time, he corrected and amplified results which had been arrived at by others upon many new substances—*e.g.* tellurium, chromium and beryllium. We shall frequently have occasion to refer to Klaproth's meritorious work in the history of analytical and mineralogical chemistry. His conscientiousness further showed itself in the way in which, contrary to the custom prevalent among chemists at that day, he published the results of his analysis; instead of merely stating the conclusions presumably arrived at from his experiments, he gave the actual figures of these, and so made it possible to subject them to a minute criticism or correction.

The above sketch of Klaproth's work may be fitly concluded by quoting the following sentence of A. W. Hofmann's[1]:—"Endued with a modesty totally alien to all presumption, recognising to their full extent the services of others, and tender of his fellow-men's weaknesses but unsparing in the criticism of his own work, Klaproth will remain to us for all time the model of a true investigator of science."

Klaproth's experimental researches were published in various journals, *e.g.* in the "Memoirs" of the Berlin Academy and in Crell's *Chemische Annalen* ; he himself collected these scattered papers together into a five-volume work, entitled *Beiträge zur chemischen Kenntnis der Mineralkörper* (1795-1810), to which a sixth volume, *Chemische Abhandlungen gemischten Inhalts,* was added in 1815. His literary activity was further shown in the publication of the *Chemisches Wörterbuch* (1807-1810), and in the revision of the works of others, *e.g.* B. Gren's *Handbuch der Chemie* (1806).

That chemistry in general was carefully fostered in Germany during the two last decades of the eighteenth century is also proved by the fact that various journals were

[1] *Chemische Erinnerungen.*

started during that period, whose main object was the publication of papers on chemistry. Among these were L. von Crell's *Chemische Annalen*—whose editor merits our praise,—which were a continuation of the *Chemisches Journal*, begun in 1778; Scherer's *Allgemeines Journal der Chemie*, which was incorporated with Crell's *Annalen* after 1803; and the *Annalen der Physik*, founded by Gren and Gilbert in 1798, and which since 1825 have appeared as *Poggendorff's Annalen der Physik und Chemie*.

The State of Chemistry in England, Scotland and Sweden towards the End of the Eighteenth Century.

The most distinguished chemists in England and Sweden at the time of Lavoisier's attack upon the phlogiston theory, viz. Black, Cavendish, Priestley, Scheele and Bergman, were avowed opponents of the new doctrine. Black alone among them, after considerable hesitation, frankly recognised its truth. Cavendish, whose own discoveries contributed in great degree to the downfall of the phlogistic view, could not bring himself fairly to renounce it. The others, whose brilliant work had likewise forged the best weapons for its overthrow, died without being convinced of its untenability. Other English chemists, for instance Henry, Kirwan, and Hatchet, also tried to hold fast by the phlogistic hypothesis so long as it appeared possible to say anything in its favour. Kirwan especially, who was one of those who believed phlogiston to be identical with hydrogen, continued the fight against the new doctrine till 1792, in which year he subscribed to it himself. Its first adherent in England was Lubbock, who concurred in Lavoisier's views so early as 1784. The four chemists just named, being representatives of their science at that day, merit this brief mention; they advanced particular branches of chemistry by their work, but did not influence its general tendency. Thus their countryman, John Dalton, who soon after this made such a wonderful step in advance, showed only the greater individuality in pointing out the new path, by following which chemical research has since made such enormous strides.

After the deaths of Bergman and Scheele, Sweden had at the close of the eighteenth century no chemist who enriched the science with facts of general importance, though Ekeberg and Gahn worked energetically at analytical and mineralogical chemistry. It was only at the dawn of this century that Berzelius' star arose, the light from which was to illumine nearly every branch of chemistry during its first four decades. A period singularly rich in scientific facts for chemistry thus began with him, while in his contemporaries, Davy and Gay-Lussac, the science possessed two other workers of the highest power. Dalton's Atomic Theory, founded as it was upon the doctrine of chemical proportions, formed the basis of all their efforts.

Development of the Doctrine of Chemical Proportions. Dalton's Atomic Theory.

The idea of atoms as forming the ultimate constituents of matter often arose of old in speculative minds, without, however, an exact chemical atomic theory being evolved from it. Boyle's corpuscular theory was and remained merely a product of ingenious speculation, which ended in the assumption of a primary material, and therefore bore no fruit. Only after a series of proven facts had led to the presupposition of atoms, and after this assumption had enabled those facts to be satisfactorily explained, could there be any talk of founding a chemical atomic theory. The merit of establishing this is without a shadow of doubt due to John Dalton. But before it could be brought to completion, the meaning of the term "chemical proportions," according to which simple substances united to form compound ones, had to be firmly fixed; and an important part of this problem was worked out by two chemists before Dalton, viz. Richter and Proust.

Richter, whose work was to all intents and purposes unknown to Dalton at the time when he conceived his atomic theory[1], founded the doctrine of chemical proportions without, perhaps, seeing its great importance himself, while Proust proved that the ratio in which two elements combine chemically with one another is constant, or, if there is more than one compound of these elements, the ratio alters by definite increments. If we but consider that the atomistic hypothesis, from which the chemical atomic theory sprang, originated with an observation by Dalton which followed from Proust's demonstrations, and which was comprised within the law of multiple proportions, we see how intimate was the connection between the latter and these preparatory labours (cf. note 1, p. 189).

Jeremias Benjamin Richter, born at Hirschberg in Schlesien in 1762, became a mining official (*Bergsekretär*) at Breslau, and then chemist (*Bergassessor and Arkanist*[2]) in the porcelain manufactory at Berlin, in which city he died in 1807. His researches—from which the doctrine of proportions by weight was mainly established, and which showed that acids combined with bases to form salts—together with the conclusions which he drew from them, were published by him in his *Anfangsgründen der Stöchiometrie oder Messkunst Chemischer Elemente* ("Rudiments of Stöchiometry, or the Art of Measuring Chemical Elements"), (1792-1794), and in his work entitled, *Ueber die neueren Gegenstände in der Chemie* (" Upon recent Discoveries in Chemistry "), which was published in eleven parts at irregular intervals between 1792 and 1802; this latter was in great part a continuation of the first-mentioned book.

Many chemists before him had busied themselves with the same task—the determination of the amounts of acid and base in salts; in addition to Kunkel, Lemery, Stahl and Homberg, special mention must be made here of Wenzel (who was born at Dresden in 1740, and died while director of the Freiberg foundries in 1793), who placed beyond a

[1] Angus Smith, *Memoir of John Dalton and History of the Atomic Theory*, p. 214.

[2] *Arkanist*, meaning literally "secret chemist," was the German title in use at that time.

doubt the fact that acids and bases combine in constant proportions, grounding this conclusion upon the results of numerous and, for the most part, thoroughly serviceable analyses. Richter was in a position to deduce the important "law of neutralisation" (*Neutralitätsgesetz*) from his own researches upon the quantities of bases and acids which combine to form neutral salts—researches carried out with great circumspection. Translated from his writings, obscured as these were by much phlogistic verbiage,[1] into the chemical language of to-day, this runs somewhat as follows: "When equal amounts of one and the same acid are rendered neutral by different amounts of two or more bases, the latter are equivalent to one another, and *vice versa.*" It follows quite clearly from his statements that he regarded those quantities of oxides which contain equal amounts of oxygen as equivalent to one another, *i.e.* as requiring like quantities of a given acid to neutralise them. Richter had come to the right conclusion as to the capacity of iron and quicksilver to unite with oxygen in two proportions, from the composition of the corresponding salts. With these weighty observations he thus anticipated the precisely similar ones of Proust. Scheele had previously attained to the same knowledge (cf. p. 127).

Notwithstanding that Richter's work contained such far-reaching discoveries, these remained almost unnoticed, their value being manifestly not recognised. This was partly due to the peculiar phlogistic language—obscure and clumsy—in which he clothed the results of his researches. A curious speculation in which he indulged may also have caused his whole work to be unfavourably criticised,—his assumption, namely, that a definite arithmetical relation existed between the combining weights of the bases and acids.[2] Judicious as he was in other points, he believed that he had found a proof that the combining weights of the bases

[1] Although he had ceased to be a phlogistonist, Richter still made frequent use of phlogistic expressions, which often obscured his writings.

[2] Even before his scientific career had begun, Richter was animated with the conviction that "chemistry was a branch of applied mathematics."

and acids form approximately regular series,—the former arithmetical, and the latter geometrical. The importance which he assigned to his "law of progression," and his continuous efforts to furnish proofs in support of it, manifestly prevented him from perceiving the significance and range of his law of neutralisation; indeed, he held this speculation as being the more important of the two.

The chemical world was to a certain extent made acquainted with the truths lying dormant in Richter's papers by G. E. Fischer, who put his countryman's observations into intelligible language; he collected together in a clear manner the scattered numerical values which Richter had arrived at as representing the amounts of bases and acids which combined with one another, and thus prepared the first table of equivalent weights.[1] Notwithstanding that the attention of chemists was in this way drawn to Richter's researches, it was a long time before they became thoroughly known and estimated at their true value. It was thus that facts proved by him were rediscovered by others much later, *e.g.* the combination of bases which contain equal amounts of oxygen with equal quantities of the same acid, by Gay-Lussac, who was without doubt unacquainted with this portion of Richter's work. As Kopp pertinently remarks in his *Entwickelung der Chemie in der neueren Zeit, S.* 152 ("Development of Chemistry in Recent Times," p. 152): "The history of our science affords few examples of important and well-proven facts being overlooked for so long a time and to such an extent; and, further, when the appreciation of these facts did finally come, of the merit of their discovery being minimised so far as the discoverer himself was concerned, and the credit given in great part to another."

It was only long after his death that Richter's services were recognised to their full extent.[2] Starting from the

[1] This table was published by Fischer in his translation of Berthollet's *Recherches sur les Lois de l'Affinité*. The fact that the latter adopted Fischer's grouping in his work, *Essai de Statique Chimique*, vol. i. p. 134, made Richter's labours known in France also.

[2] Cf. especially C. Löwig's memoir, *Jeremias Benjamin Richter, der*

observation that the neutrality is not disturbed by the mutual decomposition of two neutral salts, he created the doctrine of equivalents; he was the originator of "Stöchiometry,"[1]—"the art of chemical measurement, which has to deal with the laws according to which substances unite to form chemical compounds."

Josèphe Louis Proust.—The work of this investigator, who, independently of Richter, also partially proved the validity of the law of chemical proportions, fell later in point of time than the most important of Richter's researches. Born at Angers in 1755, Proust went through Rouelle's course of study, and then applied his knowledge of pharmacy and chemistry in the first instance as manager of the apothecary's shop attached to the Salpêtrière Hospital in Paris, and later as a teacher in different Spanish universities. It was in Madrid, where he settled after 1791, that he carried out his most celebrated investigations. The war deprived him both of his post and of his splendidly equipped laboratory in 1808, and it was only towards the end of his life that his necessities were relieved by a pension, while he was at the same time received into the Paris Academy; he died at his native town of Angers in 1826.

His most important work was the result of a series of questions which Berthollet had propounded. At the end of the eighteenth century (*i.e.* from 1798 onwards), the latter's *Recherches sur les Lois de l'Affinité*, which he collected together in 1803 in his *Essai d'une Statique Chimique*, created an extraordinary sensation. Grounding his objections upon

Entdecker der chemischen Proportionen (Breslau, 1874) ["Jeremias Benjamin Richter, the Discoverer of Chemical Proportions" (Breslau, 1874)]. According to Fischer, Richter's work was particularly emphasised by Gehlen, Schweigger and Berzelius. The discovery of the law of neutralisation was ascribed by Berzelius to Wenzel, in consequence of a misunderstanding on the part of the former; and it was left to H. Hess of St. Petersburg to point out this error, thirty-three years after Richter's death.

[1] Richter himself says that he was unable to devise a better name for this than the word "*Stöchiometrie*, from στοιχεῖον, signifying something which cannot be further divided, and μετρεῖν, which denotes the finding out of relative proportions."

speculations apparently well founded, this gifted writer disputed the fact that constant proportion was the rule with regard to the constituents of chemical compounds. His ideas upon chemical affinity, by which the combination of substances with one another is regulated, will be treated of in detail in the special history of this part of our science. Suffice it to say here that, starting from the axiom that chemical processes are dependent upon the relative masses of the reacting bodies, he arrived at the conclusion that, in a chemical compound which results from the union of two substances, so much the more of the one substance must enter into it, the more of that substance there is available, always supposing that no exceptional circumstances stand in the way of this mass-action. Berthollet's great reputation may have been the reason why none of the other leading chemists of the day raised any objections, although they certainly did not concur in this view. For, with respect to many compounds, salts especially, the constancy of the combining proportions of their constituents was a fact beyond all doubt to men like Richter, Wenzel, Klaproth, Vauquelin and others.

Proust took up the cudgels against Berthollet, and, by means of exact experiment, overthrew one by one the theoretical conclusions of his opponent. This memorable controversy, which, beginning in 1799, was continued for eight years, and which was conducted on both sides with consummate ingenuity and supplemented by laborious investigations, ended in the conclusive proof of constant combining proportions.

To what extent Dalton was influenced by Proust's labours in his researches in a similar direction, it is hard to say; but they were certainly not without some effect upon him, the dispute between Berthollet and Proust being followed with the keenest interest in scientific circles.

So early as the year 1799 Proust had proved the constant composition both of natural and of artificial carbonate of copper,[1] and had called special attention to the unvarying

[1] *Ann. de Chimie*, vol. xxxii. p. 30.

proportions by weight in true chemical compounds, as opposed to the varying ones in mixtures. Still more important than these were observations—to be supplemented later on by himself and others—upon the two stages of oxidation which tin shows,[1] and upon the two compounds which iron forms with sulphur;[2] for he particularly emphasised the point that not only were the proportions between the metals and oxygen or sulphur constant in the individual compounds, but also that the combining proportions increase by leaps, and not gradually, when two elements unite to form more than one compound. Berthollet thought that he had proved exactly the opposite in his researches on the formation of oxides and salts[3] (*e.g.* the nitrates of mercury), viz. that metals can form oxides with gradually increasing amounts of oxygen. But Proust[4] showed that his experiments were wrong, and that he had deduced his conclusions from the analysis of mixtures and not of definite compounds. The superiority of Proust in experimental points was clearly manifested, since he proved to Berthollet that many of the substances which the latter regarded as oxides contained chemically-combined water; it was Proust who first classed the hydrates among chemical compounds. In fact, he succeeded by generalisation and by firmly establishing his view—that combination between the other elements and oxygen or sulphur only takes place in one or, at most, in a few proportions—in completely routing the weak arguments of his opponent, many of which were advanced without any experimental proof to support them.[5]

Proust had repeatedly laid stress upon the validity of combining proportions, without however trying to get clearly at the reasons for this. How near he was to recognising the law of multiple proportions, which Dalton deduced from his own researches—researches similar to Proust's, and not excelling these in exactitude! One is led to the surmise

[1] *Journ. de Phys.*, vol. li. p. 174. [2] *Ibid.*, vol. liv. p. 89.
[3] Cf. *Essai de Statique Chimique*, vol. ii. p. 399 *et seq.*
[4] *Journ. de Phys.*, vol. lix. pp. 260, 321.
[5] *Ibid.*, vol. lxiii. pp. 364, 438.

that if Proust had calculated the results of his experiments on the composition of binary compounds otherwise than he did, he would have discovered that law. The happy idea occurred to Dalton of reckoning the amounts of one element, which combined in different proportions with another, in terms of a given chosen quantity of the latter; the result of this was that the multiple proportions became manifest, and these he explained by the aid of the atomic hypothesis.

DALTON'S ATOMIC THEORY.

John Dalton,[1] the eldest son of a poor weaver, was born at Eaglesfield in Cumberland on September 6th, 1766, and had to make his own living at an early age as an elementary teacher. Endowed with a strong bent towards mathematics and physics, he acquired a sound knowledge of these subjects, and was thus enabled to carry out independent investigations in them, and to take the post of mathematical and physical master in a college at Manchester in 1793. It was there, in 1794, that he made the important discovery of colour-blindness, which he noticed in the first instance in himself; as a consequence of this the phenomenon goes by the name of *Daltonism* to the present day. He soon included chemistry also in his studies, the most important problem of which he was destined to solve. In his modesty Dalton had no thought of acquiring for himself a brilliant position in life, and a wide sphere of action; after 1799, in fact, he supported himself by taking private pupils. The highest reward for his truly philosophic mind consisted in the elucidation of the truth. He died at Manchester in 1844.

Dalton's earlier researches on the physical behaviour of gases (their expansion by heat and absorption by liquids)

[1] For Dalton's life and work, compare the *Memoirs of the Life and Scientific Researches of John Dalton, by W. C. Henry, M.D.* (Cavendish Society, London, 1854), and Lonsdale's *Life of Dalton*; the latter author has preserved to us a number of traits which were characteristic of Dalton's simple and kindly nature.

were of great influence upon his later chemical labours. For it was through them that he acquired the experimental dexterity which stood him in such good stead when analysing those gases, whose composition led him to the law of multiple proportions.

The discovery of this law, and the conception of the atomic theory which arose from it,[1] date from about 1802-1803. After that time Dalton applied himself to the task of building up a firm foundation for these by amplifying his observations; he only published his discovery in 1808, when the first volume of his *New System of Chemical Philosophy* appeared. But the outlines of the atomic theory had, with Dalton's concurrence, been made public by Thomas Thomson—an enthusiastic admirer of Dalton—in his *System of Chemistry* a year before this, so that the first influence of this great scientific event upon the chemical world is to be dated from then. The second volume of Dalton's above-mentioned work, with material additions to the researches originally published, appeared in 1810, and the third volume so late as 1827, by which time its contents were mostly out of date.

The first of Dalton's observations which formed the starting point for the atomic theory consisted in the determination of the composition of oil-forming gas (ethylene),

[1] Interesting details respecting the steps by which Dalton was led to the formulation of his atomic theory have recently been given in H. Debus's *Die Genesis von Dalton's Atomtheorie*, Parts I and II [*Ztschr. phys. Chem.*, xx, 3 (1896); xxiv, 2 (1897)], and Roscoe and Harden's *A New View of the Origin of Dalton's Atomic Theory, a Contribution to Chemical History*, etc. (Macmillan and Co., 1896). Cf. Also H. Debus's pamphlet:— *Ueber einige Fundamentalsätze der Chemie, insbesondere das Dalton-Avogadro'sche Gesetz* (1894). According to these authors, Dalton arrived at the atomic hypothesis *deductively* about the year 1801, and not from the result of his researches on the composition of gases. The discovery of the law of multiple proportions was thus not the cause of the atomic theory being brought forward, but, on the contrary, succeeded the latter.—This is a point which is obviously of the first importance in the history of chemistry. But the author feels that before he can himself express a definite opinion on the subject, a more minute re-examination of the whole question is required than the time at his disposal permits, before the issue of the present edition. In the meantime, therefore, the different views are merely placed before the reader.

and light carburetted hydrogen (methane). From his analyses of these two gases he concluded that, for the same quantity of carbon, twice as much hydrogen was contained in the latter as in the former, *i.e.* that the proportions of hydrogen were as 2 : 1. This regularity induced him to investigate other compounds in the same direction; thus, in the case of carbonic oxide and carbonic acid, he found that, for the same amount of carbon, the ratios of oxygen present in these were again respectively as 1 : 2. His conviction that there must be a law underlying these so simple relations hardly required any further strengthening after he had met with similar simple numerical proportions in the results of his analysis of nitrous oxide, nitric oxide, nitrous acid and nitric acid (*i.e.* the anhydrides of the two last).[1] He had, therefore, proved that when different quantities of one element combined chemically with one and the same quantity of another, these amounts stood in a simple relation to one another—a relation which could be expressed by whole numbers. The law of multiple proportions was thus discovered; it had, indeed, been deduced from experiments which were of necessity not very exact, as was to be expected from the state of chemical analysis at that time.

Dalton, however, did not remain content with this important result, but sought an explanation of the numerical relations which he had discovered. This was afforded him by the atomistic hypothesis, in the assumption, not new in itself, that substances consist of ultimate particles not further divisible—of atoms. This hypothesis gave a satisfactory explanation of the facts comprised within the law of multiple proportions, for one now only required to substitute absolute numbers for the relative ones, *i.e.* to assume that in carbonic oxide (for instance) one atom of carbon was combined with one of oxygen, and in carbonic acid one atom of carbon with two of oxygen, and so on. Upon the firm basis of this assumption Dalton erected his Atomic Theory, the

[1] Dalton was, however, wrong in his analysis of nitric acid, which he made out to consist of nitrogen and oxygen in the proportions of 1 atom to 2.

essence of which is given in the two succeeding paragraphs:—

(1) *Every element is made up of homogeneous atoms whose weight is constant.*

(2) *Chemical compounds are formed by the union of the atoms of different elements in the simplest numerical proportions.*

His speculations upon the atoms themselves, which Dalton assumed for the sake of simplicity to be spherical in shape, and also the hypothesis that they do not come into direct contact with one another but are separated by a heat zone, have but a merely subordinate significance as compared with the above two sentences; they exercised no influence on the development of the chemical atomic theory.

Dalton now sought to deduce the relative atomic weights from the proportions by weight in which the elements unite to form compounds, proceeding to this task, which constituted the main feature [1] of his *New System*, with wonderful confidence. Since he had no certain means of arriving at these numeric proportions of the combining atoms, assumptions had to be made, and these were of the simplest kind. The following statements by Dalton refer solely to compounds of two elements.

When only one compound of two elements A and B is known, we must assume that it is made up of one atom of the one and one atom of the other: $A+B$ (binary compound, or atom of the second order. Dalton spoke of an elementary atom as an atom of the first order).

If two compounds of two elements A and C are known, their composition is expressed by the symbols $A+C$ and $A+2C$ (ternary compound, or atom of the third order).

When the composition of three compounds of two

[1] Dalton's own words are:—(to ascertain) *the relative weights of the ultimate particles, both of simple and compound bodies, the number of simple elementary particles which constitute one compound particle, and the number of less compound particles which enter into the formation of one more compound particle.*

elements A and D had to be decided, then, according to Dalton, the following combinations were the probable ones: $A+D$, $A+2D$, and $2A+D$. Atoms of the fourth order (*e.g.* $A+3E$), etc., were also allowed by Dalton, although he favoured the more simple proportions. Compounds whose atomic numbers were as 2 : 3 or 2 : 5, he explained as resulting from two atoms of a higher order than the elementary atom (*e.g.* nitrous acid from one atom of nitric oxide and one of nitric acid).[1]

Dalton's statement that the atomic weight of a compound is equal to the sum of the atomic weights of its constituent elements appears to us nowadays self-evident; but we must not forget that at that period, in spite of Lavoisier's energetic protest, the false idea of heat being material had by no means been discarded by all chemists, many of them still believing that a loss of matter occurred when heat was evolved from the combination of two elements.

Setting out then from the above premises, Dalton endeavoured to determine the relative atomic weights of the elements as follows:—Starting with water, as the only compound of hydrogen and oxygen (peroxide of hydrogen being at that time unknown), he estimated the proportions in which both of these were present, and then took hydrogen as the unit to which oxygen and other elements were to be referred.

[1] Dalton's precise words, as given in his *New System*, second edition, vol. i. p. 213, are as follows:—

"If there are two bodies, A and B, which are disposed to combine, the following is the order in which the combinations may take place, beginning with the most simple, namely:

"1 atom of A + 1 atom of B = 1 atom of C, binary,
"1 atom of A + 2 atoms of B = 1 atom of D, ternary," etc.

Again, at p. 214:—

"1st, When only one combination of two bodies can be obtained, it must be presumed to be a *binary* one, unless some cause appear to the contrary.

"2d, When two combinations are observed, they must be presumed to be a *binary* and a *ternary*.

"3d, When three combinations are obtained, we may expect one to be a *binary* and the other two *ternary*.

"4th, When four combinations are observed, we should expect one *binary*, two *ternary*, and one *quaternary*, etc."

The relative values of the latter, as deduced from the composition of their oxygen and hydrogen compounds, were according to his view their atomic weights. In this way he determined the relative atomic weight of nitrogen from the composition of ammonia, which, as the only compound of hydrogen and nitrogen, consisted of one atom of each of those elements; and that of carbon from the analyses of carbonic oxide and carbonic acid, using in this case the value he had obtained for oxygen in the analysis of water.

As the analytical methods which he employed were liable to many sources of error, it was impossible that his results could be accurate; but the great merit belongs to Dalton of having propounded the principle of the determination of the relative atomic weights, or, to speak more correctly, of the combining weights of the elements. How far his first "atomic-weight numbers," as published by Thomson in 1805, differ from the values current to-day, is seen from the following table:—

"Relative Atomic Weights."	According to Dalton.	Their current Values.
Hydrogen	1	1
Oxygen	6·5	7·98
Nitrogen	5	4·66
Carbon	5·4	6

Dalton published a greatly extended and, to some extent, improved table of "relative atomic weights" in the first volume of his work (1808), in which 7 is the value given for oxygen; the numbers which he obtained are too low throughout, and deviate from the true values by several units in the case of the elements of higher atomic weight.[1] His attempt to apply the atomic hypothesis to organic compounds must

[1] This table of atomic weights shows his endeavours to round off the numerical values, from his perception of the insufficiency of the methods employed, as is seen in the following instances; the figures appended below in brackets, after those of Dalton, give the correct combining weights: sulphur 13 (16), iron 38 (56), zinc 56 (64·9), copper 56 (63·3), silver 100 (108), mercury 167 (200).

also be mentioned here, although it turned out unsuccessful, the results of his organic analyses being far from exact.

Nor must we forget Dalton's efforts to build up a system of notation which should illustrate atomic composition. The atoms of the elements were represented by various circular symbols, *e.g.* oxygen by an empty circle ○, hydrogen by ⊙, nitrogen by ⦶, and sulphur by ⊕. These signs, placed conveniently near to each other, indicated the supposed constitution of chemical compounds; for water the symbol ⊙○ was used, for ammonia ⊙⦶, for sulphuric acid[1] ○⊕○ (with ○ above ⊕), and so on.

But the simpler and easily decipherable system of notation which Berzelius introduced some time after this, prevented Dalton's from ever coming into general use.

Further Development of the Atomic Theory.

The reception which Dalton's atomic theory found among chemists was almost wholly favourable, although there were not wanting a few to depreciate the new doctrine, and even to ascribe the merit of its origination to others. In Great Britain it found from the beginning an enthusiastic adherent in Thomas Thomson,[2] who, however, rather did it harm than good by his excess of zeal, a fatal tendency to speculation sometimes causing him to quit the sure ground of exact experiment. It was of particular importance, at the

[1] Dalton did not know the compound SO_3, but supposed that this formula gave the composition of sulphuric acid.

[2] Thomas Thomson (born 1773, died 1852) exercised no slight influence on the growth of theoretical chemical views, especially in England, both by his experimental researches in chemistry, and by his text-books. That it was he who first gave to the public the principles of Dalton's atomic theory has been mentioned already. As a historian of chemistry he was also active, his *History of Chemistry* appearing in 1830-31. Most of his papers were published in the *Annals of Philosophy*, which he himself edited. As professor in the University of Glasgow (1818-1841) he was eminently successful, founding there the first chemical laboratory for general instruction in Great Britain.

time a theory so far-reaching was set up, that the facts on which it rested (still few in number) should be amplified and deepened by reliable observations.

The estimations made by Thomson of the relative atomic weights of elements and compounds were still more defective than Dalton's, and became influenced subsequently in an inexcusable manner by Prout's erroneous hypothesis,—and that, too, after Berzelius had begun his long series of classical labours with the accurate determination of atomic weights. On the other hand, Thomson's investigation of the potash salts of oxalic acid helped to confirm the atomic doctrine, since they showed that the quantities of potash which reacted with a given amount of oxalic acid were to each other as 1 : 2 : 4 by weight. An analogous observation was made by Wollaston,[1] who found that in the neutral and acid carbonates of potash the proportions of carbonic acid relatively to the same weight of potash were as 1 : 2. The applicability of the law of multiple proportions was thus also proved for salts.

The position which from that time (about 1808) the most distinguished investigators of the day—Davy, Berzelius and Gay-Lussac—took up with regard to Dalton's atomic theory, renders an account of their most important work and their general services appropriate at this point. The researches of Gay-Lussac upon gases, and even more the unresting efforts of Berzelius to work out sure foundations for the determination of the true atomic weights, had the deepest influence on the development of the atomic doctrine, which is now the basis of chemistry.

Davy and Gay-Lussac; their life and work.—Davy was at first sceptical with regard to Dalton's rights as the origin-

[1] W. H. Wollaston was born in 1766 (the same year as Dalton), and died in 1828. Originally a physician, he soon gave himself up to the study of physics and chemistry, enriching the former especially by important observations. At the same time he became favourably known by his chemical researches, particularly by his work on the platinum metals. Most of his papers are to be found in the *Philosophical Transactions*, but a few of them in the *Annals of Philosophy*.

ator of the atomic theory, and indeed, in 1809, he claimed for Higgins the priority for this doctrine, the latter having made use of the atomic hypothesis to explain chemical facts so early as 1789 (in his work, *A Comparative View of the Phlogistic and Antiphlogistic Theories*). Higgins certainly expressed opinions which, on a superficial glance, appeared similar to those of Dalton, stating as he did that the smallest particles combine in simple numerical proportions to form chemical compounds. But these views were brought forward without any internal organic connection, and, moreover, they were not based upon experiment. It became clear to Davy later on that Higgins had no claim to be regarded as the originator of the atomic theory, and he then frankly recognised Dalton's service.

Humphry Davy, born at Penzance in Cornwall in 1778, was destined for a distinguished career, to be cut short by an early death, his creative genius being impaired during the last years of his life by prolonged illness. So early as 1813, when only thirty-five of age, he had to leave off work and seek renewed health on the Continent, in Italy for the most part. From that time he travelled a good deal. After 1820 he lived and worked again in England, but left it in 1827, never to return, for he died in 1829 at Geneva on his homeward journey.

While only a surgeon's assistant, Davy acquired by his own energy such a wide knowledge of chemistry and the natural sciences, that at twenty years of age he was able to take the post of chemist in the newly-founded Pneumatic Institution at Bristol. The aim which this institution had set before itself was to test the various artificially prepared gases for their physiological and medical action. It was here that Davy carried out his researches on nitrous oxide, whose intoxicating and stupefying action he discovered, and on the effect of other gases (admixed with nitrogen) on the organism, *e.g.* hydrogen and carbonic acid ; in this way he laid the foundation of his fame as a great experimenter. So early as 1801 we find him assistant professor at the Royal Institution of London (very soon to become professor), and shortly afterwards a member of the Royal Society, whose president he became in 1820.

His most memorable work, which effected a complete transformation in many branches of chemistry, was accomplished during the first thirteen years of this century. We need only mention here the isolation of the metals of the alkalies and alkaline earths by the galvanic current, through which a whole series of hitherto undecomposed substances were recognised as compound. An almost still more important result of these observations was the discovery of the elementary nature of chlorine, which up till then was held to be a compound; this opened out entirely new standpoints, which led to a transformation of the views upon the constitution of acids. When it was proved that there were acids which did not contain oxygen, a material alteration in Lavoisier's theory became for the first time necessary. Discoveries of such range as this characterise the period in which Davy developed his wonderful activity. His most important experimental researches will be described partly in the further course of the general history of this period, and partly in the synopsis of the progress of particular branches of chemistry.

Davy contributed greatly by his popular lectures, especially by those given for the Board of Agriculture, to heighten the public interest in chemistry during the first decade of this century. He it was, too, who showed in what high degree chemistry could and should meet the requirements of technical industries and of daily life ; we have only to think in this connection of the miner's safety lamp which he constructed.

Davy's genius in grasping chemical relations was especially apparent in his efforts to discover the connection between electricity and chemical affinity, both of which he regarded as resulting from a common cause. He was the first to set up an electro-chemical theory grounded upon experiments, which were devised and carried out in a masterly manner, and in this way he opened out the province in which

Berzelius was to work with such effect in the decade following.[1]

Wherever Davy, with his aptitude for experiment and acuteness of mind, treated chemical problems, he achieved great results. Within the narrower limits of special research also, *e.g.* in his invesigations on ammonium amalgam, phosgene, euchlorine, iodine, solid phosphuretted hydrogen, and the phenomena of combustion, the fruits of his labours were at once perceptible ; his work always left a deep mark. After the year 1801 Davy published his most important papers in the *Philosophical Transactions*, but some are to be found in the *Annales de Chimie* and in the *Journal de Physique*. Of his few larger works,[2] the *Elements of Chemical Philosophy* (1810-12) became best known, especially as it was soon translated into French and German. After his death all his works were collected together and published by his brother John Davy.

In addition to the interest which Davy's wonderful services to science call forth, there is to be added the purely human interest in his personality. The nobility and poetry of his nature are shown both in the journals which he kept during his extended journeyings in France, Germany and Italy, in his letters, and his *Memoirs*.[3] The inventions made by him for the public good raise still higher our regard for this remarkable investigator.

[1] Davy's electro-chemical theory of affinity will be described along with that of Berzelius in one of the succeeding paragraphs.

[2] The judgment which Berzelius passed upon Davy's literary activity, in a letter written to Wöhler in 1831, is of much interest (cf. *Ber.*, vol. xv. p. 3166). The latter had been deploring that he was overwhelmed with literary work, whereupon Berzelius replied as follows: "Had Davy been forced to occupy himself as much with writing as you have to do now, I am convinced that he would have advanced chemistry by a hundred years ; but he remained only a 'brilliant fragment' (*glänzendes Bruchstück*), because he was not compelled from the beginning to initiate himself thoroughly into every part of the science as into one organic whole."

[3] *Memoirs of the Life of Sir Humphry Davy*, by J. Davy (London, 1839).—A delightful monograph on Davy, based for the most part on Dr. John Davy's work, has recently been written by T. E. Thorpe (*Century Science Series*, Cassell and Co., 1896).

Davy's historico-critical attitude towards Dalton's atomic doctrine has been already spoken of. But although he subsequently gave the latter credit for originating this theory, he continued sceptical with regard to Dalton's conclusions.[1] He would not admit that Dalton's *atomic weights* were really such ; in his view these were merely the *proportion numbers* of the elements, for the determination of whose atomic weights there was no sure basis to go upon. Wollaston had before this given utterance to a similar circumspect criticism of Dalton's bold speculations, having published in 1808 his opinion that the numbers arrived at by Dalton gave, not the atomic weights, but the *chemical equivalents* of the elements. Gay-Lussac, too, whose labours began at that time to exercise such a powerful influence on the development of chemistry, rejected the assumption of atomic weights, and merely allowed that the ratio (*rapport*) of one element (*e.g.* hydrogen, nitrogen, or iodine) to another (*e.g.* oxygen) was established by analytical and synthetical determinations.

Gay-Lussac, whose critical attitude to Dalton's atomic theory has just been touched upon, helped on the latter in a quite exceptional degree by his wide-reaching discovery of the so-called "Law of volumes"—more, indeed, than he was willing to confess.

Josèphe Louis Gay-Lussac, born in 1778 at St. Léonard in the old province of Limousin, after acting as Fourcroy's demonstrator became in 1809 professor of chemistry at the *École Polytechnique* (at which he had been a pupil up to the year 1800), and at the same time held the chair of physics at the Sorbonne. In 1832 he resigned his chair at the Sorbonne to fill that of general chemistry at the *Jardin des Plantes*; he died in 1850. After his initiation into science by Berthollet, and while still very young, Gay-Lussac aroused the marked attention of his contemporaries by his physical investigations on the behaviour of gases—investigations which touched more or less on the province of chemistry. Brief mention may also be made here of his bold balloon

[1] Cf. particularly his *Elements of Chemical Philosophy*.

ascents in 1804, undertaken at first along with Biot and afterwards alone. His researches made after 1805, upon the laws deducible from the combining volumes of gases which unite chemically with one another, had most incisive results. What rich fruit this yielded for chemistry as a whole, and not merely for the chemistry of gases, will be shown later on. Gay-Lussac's name is further associated with the discovery of the definite relation which exists between the volume of a gas and its temperature; it was only after this law had been worked out, a law which supplemented that of Boyle and Mariotte, that reliable measurements of gases could be made.

In his work which bore upon special branches of chemistry Gay-Lussac likewise proved himself a masterly investigator; to exactitude in observing, and acuteness in explaining his observations, he added a wonderful lucidity in expounding his researches and the conclusions at which he arrived. His work on iodine and cyanogen and their compounds would alone suffice to ensure him a place among the most distinguished chemists. How stimulating and full of matter were his papers! The one upon cyanogen, especially, was the basis on which the radical theory was afterwards developed, for cyanogen was characterised by Gay-Lussac as the first compound radical. Even his minor work bears the classical stamp; of it we may mention here his researches on the compounds of sulphur, and on the various stages of oxidation of nitrogen, and his conjoint work with Thénard[1] upon the alkali metals. Together with Liebig he investigated the fulminates. Hidden in many of these pieces of work

[1] L. J. Thénard, born in 1777, a pupil of Vauquelin and Berthollet, became professor at the *École Polytechnique* and in the *Collége de France*, and worked energetically for the promotion of the study of natural sciences in France. His name is indissolubly united with that of Gay-Lussac, their conjoint work leading to a knowledge of many chemical processes, and contributing to the improvement of important methods. Thénard's *Traité de Chimie Élémentaire*, a text-book which was most widely used, thanks to the happy synoptical arrangement of its contents, was of great merit; the first French edition of it was published in 1813-16, and the first German edition (translated from the fifth French by Fechner) in 1825-33. Thénard died in 1857.

there lay germs which were to expand into important discoveries; for example, his observation on the action of chlorine upon wax laid the foundation for subsequent researches upon substitution reactions.

By his work on technical subjects, Gay-Lussac proved that he understood how to bring his results in analytical chemistry to bear upon these. He is to be regarded as the originator of volumetric analysis; and the improved analytical methods which he thus introduced, and which have since come into general use, have helped materially to advance chemical industries. We shall meet with his work in almost every important branch of chemical investigation,—in analytical, technical, physical and pure chemistry.

Gay-Lussac published most of his experimental results in the *Annales de Chimie*,[1] but a few of them are to be found in the *Mémoires de la Société d'Arçeuil* and in the *Comptes Rendus*. Of his papers which appeared separately, mention may be made here of a number upon methods of investigating and testing commercial products, silver ores, etc., which, as a member of various commissions, he worked out; also of the *Recherches Physiques et Chimiques* (1811), which he edited conjointly with Thénard.

Prout's Hypothesis and its Effects.

During the period in which Davy and Gay-Lussac were carrying on their brilliant work, and before the star of Berzelius had attained to its full lustre, a literary-chemical event occurred which made a profound impression upon nearly all the chemists of that day, viz. the advancement of Prout's hypothesis. This was one of those factors which materially depreciated the atomic doctrine in the eyes of many eminent investigators. On account of its influence upon the further development of the atomic theory, this hypothesis must be discussed here, although it has happened but seldom

[1] After the year 1816 this journal was edited by Arago and himself under the title *Annales de Chimie et de Physique*.

that an idea from which important theoretical conceptions sprang, originated in such a faulty manner as it did.

In the year 1815 a paper[1] appeared in which the relation between the atomic weights of elements and the specific gravities of their vapours was treated of; in this paper, and still more positively in a second,[2] published in the following year, the tenet was set up by their anonymous author that the atomic weights of the elements—taking that of hydrogen as unity—were expressible by whole numbers, *i.e.* that they were multiples of the atomic weight of the lightest element.[3] From this there followed the hypothesis proper of Prout (who had, in the meantime, become known as the author of the above two papers),—that hydrogen may be regarded as the primary matter from which all other elements are formed by various condensations.

This idea, so lightly thrown out, and which adapted itself so usefully to the incomplete investigations of others,[4] possessed both then and at various later periods a great charm for many chemists. Even before these papers had been published, Dalton's friend Thomson had alluded to the fact that, according to his own experiments and those of others, the atomic weights of several of the elements were multiples of those of oxygen. He endeavoured, indeed, to establish the same point several years after this, without considering that the numbers which Berzelius had found in the meantime differed widely from his own, which had therefore become of very doubtful value. Thomson was the victim of this preconceived opinion; he went so far as to see in Prout's assumption a fundamental law of chemistry.

[1] *Annals of Philosophy*, vol. vi. p. 321.

[2] *Ibid.*, vol. vii. p. 111.

[3] The author altered the numerical values of the atomic weights in a highly arbitrary manner, so that they should not merely be whole numbers, but should also show regular differences among each other, as is seen from the following examples:—

Calcium 20	Iron 28	Chlorine 36
Sodium 24	Zinc 32	Potassium 40.

[4] Prout himself was a physician, and his own investigations were few in number and anything but conclusive.

Although Berzelius and, later, Turner and others proved the untenability of Prout's hypothesis, many chemists still inclined towards it. In his text-book of 1827, L. Gmelin gave the "mixture weights" (*Mischungsgewichte*) as far as possible in whole numbers, which he was assuredly not justified in doing after Berzelius' classical researches. Later still, about the year 1840, Dumas and Stas, who had determined the atomic weights of carbon, oxygen, chlorine and calcium with great exactitude, and also Erdmann and Marchand in their numerous investigations in a like direction, betrayed a strong inclination to this hypothesis, the weakness of which was afterwards proved by Stas himself and by Marignac. The predilection shown by many chemists for this conception, which led to such far-reaching deductions, helped to discredit the whole atomic doctrine in the minds of thoughtful investigators.

Like Davy and Gay-Lussac, who, it is true, did not specially occupy themselves with the problem of determining the atomic weights of the elements, Berzelius kept himself entirely free from those prepossessions; and, since even at that time he devoted all his energies to the solution of questions allied to this, his opinions possessed the very greatest value. Firm, and not led away by the alluring simplicity of Prout's hypothesis, he held fast to his aim,—the accurate, purely experimental determination of the atomic weights, and by his masterly work he firmly established the then unsteady edifice of the atomic doctrine.

BERZELIUS—A SURVEY OF HIS WORK.

The life of this investigator, who developed and enriched chemistry in its most important branches as hardly any other man has done, was the quiet and uneventful one of a student. He was guided in his work by the great and comprehensive aims,—to investigate carefully the composition of chemical compounds, and to arrive at the laws according to which they are formed.

Jakob Berzelius was born at the little town of Westerlösa in Östergötland, Sweden, where his father was a schoolmaster, on the 29th of August, 1779. A love for chemistry appears to have developed itself in him at a very early date, but his desire to devote himself to its study at Upsala was only attained (in 1798) under many difficulties and disappointments. The lectures and instruction given by his teachers Afzelius and Ekeberg were uninspired by the spirit after which Berzelius strove. We therefore find him turning to the study of medicine, without, however, losing sight of chemistry as an important aid to the latter. His early work, especially that which he carried out along with Hisinger upon the action of the galvanic current on salts, made him known in his own country, so that in 1802 he was appointed assistant professor in medicine, botany and pharmacy at the University of Stockholm, and, five years later, professor of medicine and pharmacy. In 1815 he was called to the chair of chemistry in the newly-founded Chirurgico-Medical Institute there. His lectures, which were at first purely theoretical, according to the established custom, he began to enliven by judiciously chosen experiments; while a very imperfectly equipped laboratory enabled him to carry through the exact experiments which were to firmly establish the doctrine of chemical proportions. In those modest rooms were accomplished the famous researches, most of them by himself alone, but some in conjunction with specially gifted pupils. The names of those latter are sufficient in themselves to show the wonderful results which he achieved by his teaching; among them we may mention here Heinrich and Gustav Rose, Mitscherlich, Wöhler, Chr. Gmelin, Magnus and Mosander.

From the year 1818, when he was nominated permanent secretary to the Stockholm Academy, of which he had been a member since 1808, and still more after 1832, when Mosander succeeded him in his chair, Berzelius devoted himself to literary work with an effectiveness which has hardly been equalled by any chemist either before or after him. His energetic life came to a close on the 7th of

August 1848. In 1818 he was ennobled by King Charles XIV., and in 1835 made a baron by the same monarch.

To give a short and at the same time succinct account of the great scientific achievements of Berzelius is no easy task, for these did not merely touch upon the main points of chemistry, but penetrated deeply into them, and gave rise to weighty reforms. After occupying himself for the first seven years of his independent scientific work with researches in various branches of the science, especially physiological chemistry, and proving himself thereby to be an exceptional observer, his efforts rose—from 1807—to a higher level. For, from that date, his entire energy was devoted to one great aim; the minute investigation of chemical proportions and, with that, the development of the atomic doctrine he looked upon as his life-task. At the time when he began his work upon the combining proportions of the elements, the atomic doctrine was unknown to him. His first researches were inspired by J. B. Richter's papers and then by Davy's discoveries, before he was aware of the results of Dalton's labours which had led to the atomic theory. How Berzelius built up the doctrine of proportions by improving analytical methods and by the clear-sighted interpretation of his own researches and those of others, and how he created solid foundations for the determination of atomic weights, will be described in the following section.

But we must just mention here that he greatly enriched analytical chemistry by the discovery of new methods. These were, indeed, indispensable to him for the attainment of his great aim, for it was only by means of the most accurate possible analyses that the constancy of combining proportions could be definitely proved. This was, however, by no means the only branch of chemistry which was indebted to him, for analysis in his hands was made to open up other and larger domains. His first attempt to work out the composition of minerals on the basis of the atomic theory, *i.e.* with the aid of the law of multiple proportions, was made so early as the year 1812, and his establishment of a chemical mineral system created an extraordinary interest.

Of still more far-reaching effect were his successful endeavours to show that organic compounds were likewise subject to the law of multiple proportions. After materially improving the methods of analysis of organic bodies, he was able to demonstrate in 1814 that simple atomic relations prevail among the constituents of organic acids and of their salts. The atomic theory thus became the guiding star both for Berzelius and for the whole science.

Berzelius assumed that atoms were electrically polarised, and looked upon this as the cause of the combination of elements in definite proportions. His electro-chemical theory, developed from this assumption, and his dualistic system, which was the immediate result of this theory, will be described in detail along with other similar attempts at explaining the phenomena of affinity.

Experiment formed the basis of his speculations. By connected observations on the chemical behaviour of simple and compound bodies, he extended the most important branches of his science in a marvellous degree.

Of his numerous researches on inorganic substances, that upon selenium is a classical model, worthy to rank alongside of Gay-Lussac's upon iodine. We may also call to mind here his remarkable investigations upon ferro-cyanogen compounds, sulpho-salts and fluorine compounds, among many others. All his experimental work shows the originality of a master mind; and although his inventive genius was not so great as that of Davy, his strict methods of procedure and conscientious observations led him to discoveries of the first importance.

The work of Berzelius in organic chemistry is less imposing than that which has just been sketched, but we have only to recall his discovery of racemic acid, and his important investigations on its isomerism with tartaric acid, to see that here, too, he made a deep mark. As he was the first to apply the principles of the atomic theory to organic substances, so he sought to introduce his electro-chemical and dualistic views here also. These efforts of his to simplify complex relations were not however in this instance permanently successful, for, although his radical theory had a fruitful influence for a time, it was unable to hold its ground against the unitary conception. Much of his work in mineralogical and physiological chemistry was fundamental in its nature, and was even that of a pioneer, since it had as its immediate result (especially in mineralogical chemistry) the setting up of entirely new points of view and new aims.

The grand creative genius of Berzelius and the joy he had in his work are not only apparent in his experimental researches, but show themselves also in his activity as a teacher, whether as manifested in personal intercourse with his pupils or as finding expression in writing. In his little laboratory there assembled young men from far and near, most of them already well versed in chemical knowledge, to learn from his experience and then to further propagate his doctrines. From Germany especially, where at that time there was hardly any provision for practical chemical work, came aspiring students, who subsequently advocated the principles of his school and extended its influence.

Berzelius' literary activity is most strikingly shown in his *Lehrbuch der Chemie*,[1] of which five editions, each of them completely revised, appeared. Along with the absolute thoroughness which we also admire in his experimental work, clearness of description is united in this book with precision of expression. He did not merely confine himself to the simple exposition of known facts, but criticised with absolute impartiality the experiments from which these were deduced. His text-book remained a model, unapproached by any other, during succeeding decades. The many-sidedness of Berzelius and his power of work were also strikingly shown in the *Jahresbericht über die Fortschritte in der Physik und Chemie* ("Annual Report on the Progress of Physics and Chemistry"),

[1] This book came out for the first time in 1808-1818 in three volumes (Swedish); the second Swedish edition (four vols., 1825-31) was translated into German by Wöhler, while the subsequent editions were printed in German only. The third (four vols., 1833-35) and the fourth (four vols., 1835-41) were done into German by Wöhler "from the Swedish MSS. of the author," while the fifth "original edition" (five vols., 1843-48) was written by Berzelius himself with Wöhler's co-operation.

twenty-seven volumes in all, which were published by him in Swedish from the year 1821 until his death; these were also brought out in German by Gmelin for the first three years, and subsequently by Wöhler (in Tübingen). He had undertaken to report to the Stockholm Academy upon the work published on those subjects, a task which he performed with diligence and perspicacity. With regard to work which came at all within his own province, he knew to perfection how to fill the *rôle* of critic, although on some occasions he was led by the characteristics of particular experimental researches to express a judgment which betrays a certain prepossession. Notwithstanding this, however, his *Jahresberichte* are and will remain indispensable sources of information for any one who wishes to understand the currents and changes of opinion in the chemistry of that time.

The experimental researches of Berzelius were as a rule first published in Swedish in the *Transactions of the Stockholm Academy*, but most of them were afterwards given out in German, and a few in English and French (in *Gilbert's*, *Poggendorff's*, and *Liebig's Annalen*, the *Annales de Chimie*, *Annals of Philosophy*, etc.). They are characterised by the same excellences as his text-book.

The above sketch of his main achievements is sufficient to indicate the qualities which distinguished Berzelius as a classical investigator. Thoroughness and perseverance in everything which he undertook; exactness in all his observations, and the capacity for arranging these distinctly and explaining them clearly; inviolable adherence to the results of experience (which was his guide before everything else), and an equally firm adherence to results which, in his opinion, had been correctly arrived at from a number of data; these were the characteristics which distinguished this great man.

The desire to retain whatever of good the science possessed was developed in him in an exceptional degree; indeed, in sustaining this conservative attitude he went so far as to see a danger to the steady development of chemistry in every innovation which called in question views already proved and found useful. Hence his fervent opposition to many new hypotheses which he had in the end to recognise as correct His great services in furthering chemistry were, however, not lessened by this peculiarity, which had its real cause in a profound sense of justice; on the contrary, by a prudent adherence to approved opinions, Berzelius often prevented the confusion to which the views he combated might probably have given rise, had they been accepted without reservation. Not that he was averse to healthy reform. But against anything violent—to his mind revolutionary—he fought with all his energy; he did not shun even hot polemics[1] when anything that he regarded as sound was at stake.

His pupil Heinrich Rose gave a comprehensive review of his general character in the "Memorial Speech of Berzelius,"[2]—a speech of great beauty and with a pleasant warmth of tone running through it. At the close of it (p. 59) Rose says: "The irresistible captivation which Berzelius exercised over those who enjoyed the privilege of a lengthened intercourse with him was only partly due to the lofty genius, whose sparks flashed from all his work, and only partly to the clearness, the marvellous wealth of ideas, and the untiring care and great industry that gave everything with which he had to do the stamp of the highest perfection. It was also—and every one who knew him intimately will agree with me in this,—it was also those qualities which placed him so high as a man: it was his devotion to others, the noble friendship which he showed to all whom he deemed worthy of it, the great unselfishness and conscientiousness, the perfect and just recognition of the services of others,—in short, it was all those qualities which spring from an upright and honourable character."[3]

[1] His controversies with Dumas, Laurent, Liebig and others have often been harshly and unfairly criticised, in that a false light has been thereby thrown upon his whole work. The younger generation of chemists, in especial, quickly forgot after his death the debt which was due to him for the imperishable services which he had rendered in the building up of the science. Indeed, derision and cheap ridicule of the mistakes he made are still to be found in recent works which treat of the development of chemical theories.

[2] Delivered at a public meeting of the Berlin Academy, 3rd July, 1851.

[3] The recently published letters of Berzelius and Liebig to each other,

We may close this section with the following words, in which the same chemist portrays in a few lines the wonderful work of his master: "When a man who is endowed with exceptional talents as an investigator enriches every branch of his science with the most pregnant facts, distinguishes himself equally in empirical and speculative research, and grasps the whole subject in a philosophic sprit; when he arranges each detail systematically and clearly, and gives the whole to the world in a doctrinal system, critically sifted and put in as perfect a form as possible; lastly, when he proves himself a noble example of a practical and theoretical teacher to a circle of pupils eager for knowledge,—that man so fulfils the highest demands of his science, that he will continue to shine forth as a brilliant model for ages to come."

The Firm Establishment of the Doctrine of Chemical Proportions and the Development of the Atomic Theory by Berzelius; together with the share taken in these by Gay-Lussac, Dulong and Petit, and Mitscherlich.

It has been already stated in the preceding section that Berzelius regarded the investigation of chemical proportions, and of the laws which regulate these, as his life task. Compounds of oxygen formed the starting-point for his researches and for the deductions which he drew from them, this element being indeed, after the time of Lavoisier, the centre round which the whole of chemistry ranged itself. Even in the first investigations, which he began to publish in 1810 in Swedish, and in 1811 in German (in *Gilbert's Annalen*, vols. xxxvii., xxxviii., and xl.), Berzelius furnished powerful proofs

which embrace the years 1831-1845, confirm in the most absolute manner the above kindly critique. This book, edited by Liebig's grandson, J. Carrière, and published by Lehmann (Munich and Leipzig) in 1893, will be welcomed as one of the best contributions that has been made to the history of chemistry of late years. It will assuredly help towards a truer criticism not merely of Berzelius and Liebig themselves, but also of many other eminent men, and at the same time assist towards a clearer view of various important points.

of the existence of chemical and, more particularly, of multiple proportions in the oxygen compounds of the elements. If we consider that he carried out this great work and the subsequent investigations connected with it (for which entirely new methods had to be devised), almost altogether by himself, we shall gain some idea of the wonder which such achievements created anong his contemporaries.[1]

A true scientist, Berzelius knew how to advance from the particular to the general; he first collated a number of important facts which, taken together, rendered possible the gradual but firm establishment of the atomic theory. Among these were the proofs that the proportion of sulphur to metal in the metallic sulphides was the same as that in the corresponding sulphates; that the amounts of oxygen in the equivalents of bases were likewise the same; and that in salts of every kind the ratios between the quantities of base, acid and water were simple ones,—and so on.

In the years 1812 to 1816 Berzelius investigated the

[1] Many passages in the works of Berzelius proved that he looked upon the firm establishment of the doctrine of chemical proportions, and, in connection with this, the determination of the atomic weights of the elements and the constitution of chemical compounds, as his chief task. His own words may be quoted here to show how he, impressed as he was with the incompleteness of previous work on the subject, strove to improve upon it: "I soon convinced myself by new experiments that Dalton's numbers were wanting in that accuracy which was requisite for the practical application of his theory. I now perceived that if the light which had arisen upon the whole science was to be propagated, the atomic weights of as large a number of elements as possible, and, above all, of the most commonly occurring ones, must be determined with the greatest accuracy attainable; and, together with this, the proportions according to which compound atoms (*zusammengesetzte Atome*) combine among each other, as, for instance, in salts, with the analysis of which I had been occupied for some time. Without work of this kind no day could follow the morning dawn. This was, therefore, the most important point for chemical research at the time, and I devoted myself to it with unresting energy. Several of the more important atomic weights I subjected, after lengthened intervals, to a closer scrutiny, making use of improved experimental methods. After work extending over ten years, the results of which have been published in the scientific journals, I was able in 1818 to publish a table which contained the atomic weights, as calculated from my experiments, of about 2000 simple and compound substances."—*Lehrbuch der Chemie*, vol. iii. p. 1161, fifth edition.

stages of oxidation of most of the metals and metalloids then known (to use his own term for the non-metals), and, by determining the composition of these oxides, confirmed the law of multiple proportions. And, notwithstanding that he sometimes proceeded from erroneous premises, *e.g.* from the assumption that chlorine and ammonia contained oxygen, his grasp of the subject was so complete that he was able to keep the main conclusions drawn from his experiments free from error.

Of special significance for the sound development of the atomic doctrine were his efforts (intimately connected with the work just mentioned) to deduce the relative atomic weights of the elements, and also of compounds, from the composition of chemical compounds as determined by analysis. He went about this with great circumspection, showing wonderful tact in the selection of proper footholds from which to approach the difficult task. Already in one of his earlier papers[1] we meet with the first statement of the "oxygen law," according to which the amount of oxygen in the acid of a salt stands in a simple numerical proportion to that in the base,—a statement which was the result of experience, and which Berzelius followed in many atomic weight determinations.

The propositions which Dalton had brought forward with a view of arriving at the atomic numbers of the constituents of chemical compounds were rightly designated by Berzelius as arbitrary. Among them, for example, was the assumption that the atomic proportion of two elements to one another, when only one compound of these was known, must be 1 : 1. Berzelius, too, set out from simple premises, and had to exercise all his ingenuity in order to find further support for such assumptions. One of these latter (advanced at the beginning of his work on the subject) was—that 1 atom of one element A combines with 1, 2, 3, or 4 atoms of another element B. The less simple combining proportions $2A : 3B$ or $2A : 5B$ were first allowed by Berzelius about the year 1819, and without any reservation only in 1827.

[1] *Gilbert's Annalen*, vol. xxxviii. p. 161.

With such propositions as a basis, even when including the definitely expressed "oxygen law" (which had been worked out in the meanwhile), Berzelius would have been hardly more successful in solving the question of the number of elementary atoms in a compound than Dalton and his immediate successors, had he not known how to appreciate the value of Gay-Lussac's important discovery of the "law of volumes" for clearing up the points in question. By making use of this, the simplest combining proportions in which different elements unite became all at once apparent, and, by applying it further, Berzelius was able to bring his experimental work to its first conclusion. His *Versuch über die Theorie der chemischen Proportionen und über die chemischen Wirkungen der Elektrizität* ("Essay upon the Theory of Chemical Proportions and upon the Chemical Action of Electricity") appeared first in 1814 in Swedish, in 1819 in French, and in 1820 in German.[1] In this memorable work for the history of chemistry he developed his conception of the atomic doctrine, and his ideas upon the relations between chemical affinity and electric polarity. His dualistic views stood out clearly here, and at the same time he devised a new language and nomenclature for his system. Of special importance was the collection of the results of his arduous investigations in tables of the atomic weights of elements and compounds; he was able to give original figures for about 2000 substances. In order to become thoroughly acquainted with the grounds which influenced Berzelius in his choice of these values, we must take into account the law of volumes above all other things, because, as has already been mentioned, he not only drew important inferences from it, but used it almost from the beginning of his researches as the basis of his atomic weight system.

[1] Edited by K. A. Blöde.

Influence of the Law of Volumes upon the Atomic Theory.

Among the greatest of the services rendered by Gay-Lussac was the research which he published towards the end of 1808 in the *Mémoires de la Société d'Arçeuil*, vol. ii. p. 207[1]. Having three years previously, in conjunction with Alexander von Humboldt, observed that exactly two volumes of hydrogen unite with one volume of oxygen to form water, he showed by comprehensive investigations that similar simple volumetric relations exist between all gases which combine chemically with one another, and further, that the gaseous products formed also stand in a simple volumetric relation to their components. He proved this, for example, in the formation of two volumes of carbonic acid from two of carbonic oxide and one of oxygen, and in the combination of hydrogen and chlorine and of ammonia and hydrochloric acid in equal volumes; he likewise showed that two volumes of ammonia were composed of three volumes of hydrogen and one of nitrogen, and two volumes of (anhydrous) sulphuric acid of two volumes of sulphurous acid and one of oxygen. Several of these proportions he was able to deduce from the results of other workers, *e.g.* Dalton, Davy and Vauquelin, who had determined the volumes with fair accuracy in their experiments on gaseous compounds, without, however, recognising the underlying law.

Having concluded from their similar behaviour with regard to changes of pressure and temperature that all gases possess a like molecular constitution, Gay-Lussac deduced from his researches, just quoted, the following important law:—The weights of equal volumes of both simple and compound gases, and therefore their densities, are proportional to their empirically found combin-

[1] In his *Classiker der Exakten Wissenschaften*, W. Ostwald has put within every one's reach these papers of Gay-Lussac and of A. von Humboldt, as well as the fundamental researches of Dalton and Davy already referred to. A like service as regards chemical classics has been rendered to English-speaking chemists in the *Alembic Club Reprints*, edited by Dr. Leonard Dobbin, and published by W. F. Clay, Edinburgh.

ing weights, or to rational multiples of the latter. In this sentence the old idea—that certain definite relations exist in nature between the weight and mass (*pondere et mensura*) of compounds—first found distinct expression.

Gay-Lussac was himself inclined to connect his law of volumes with the atomic theory,—indeed, he recognised in it a support for the latter. But he was unable to set aside certain difficulties which, in spite of the simplicity of the known volume-relations, came in the way, and he therefore adhered to his empirical standpoint.

The assumption obviously so closely related to the above, viz. that equal volumes of different gases contain equal numbers of smallest particles, and that, in the case of the simple gases, these are not undecomposable but consist of several atoms, was made so early as 1811 by Avogadro.[1] From such an assumption it followed that the masses of these smallest particles, *i.e.* the molecular weights of the gases, were proportional to the vapour densities. The particles were termed by him *molécules intégrantes*, and their constituents (*i.e.* our atoms), *molécules élémentaires*. Notwithstanding the fruitfulness of those conceptions, and the ease with which by their aid the mutual relations between the volumes of gases and the atoms could be explained, they remained almost unnoticed. The reason for this may to some extent have been that Avogadro generalised too boldly, extended his hypothesis to non-volatile substances, and brought forward no new facts in support of it.

But although the conclusions drawn from the law of volumes by the scientist just named remained unheeded at the time, the law itself bore rich fruit for the atomic doctrine. Dalton himself showed a disinclination to agree with the results of Gay-Lussac's researches, indeed, he doubted

[1] *Journ. de Phys.*, vol. lxxiii. p. 58. (This paper forms No. 8 of Ostwald's *Classiker*; cf. also No. 4 of the Alembic Club Reprints.) Amadeo Avogadro, born 1776, died while still professor of physics at Turin in 1856. It is through the treatise just mentioned that his name will always remain famous.

their correctness. Thomson and Davy too did not perceive that the law of volumes had any special significance from the atomic point of view, as, although they frequently made use of the volume-relations of gases to arrive at their composition, at other times they interpreted these relations wrongly; thus they assumed that a volume of hydrogen contained only half as many atoms as an equal one of oxygen.

Berzelius, however, recognised in the law of volumes a welcome corroboration of the atomic theory, and allowed himself to be guided by it in his views upon the number of atoms in chemical compounds, and, consequently, upon the numerical values of the atomic weights. His "volume theory" (*Volumtheorie*) contained the attempt to combine Gay-Lussac's law with the atomic theory. He set forth the atomistic view, which he had himself put into shape under the influence of the law of volumes, definitely and conclusively in two papers.[1] He started with the assumption that in the case of every simple substance, when it was in the gaseous form, one volume corresponded with one atom, and therefore made use of the designation "volume atoms" (*Volumatome*) for those smallest particles. Wherever it was practicable, he endeavoured to measure the volumes of the combining substances, and from these deduced the atomic numbers. The analysis of the compound, in which the volumes of the elementary constituents were known, led him to the true determination of the atomic weights of the latter. Thus, from the fact that water consists of two volumes of hydrogen and one of oxygen, he deduced the atomic composition of water which holds at the present day, together with the relative atomic weights of oxygen and hydrogen; and from the (volumetric) mode of formation of carbonic oxide and carbonic acid he arrived at the true composition of these compounds, and at the atomic weight of carbon, etc.

But, however much Berzelius was convinced at that date (1813) of the superiority of this conception over the "cor-

[1] *Ann. of Philos.*, vol. ii. pp. 359, 443 (1813).

puscular theory," which took no account of volume-relations, he did not fail to recognise the limits of application of his volume theory. To extend to non-volatile bodies the conceptions which he had gained from gases seemed to him hazardous; in fact, his doubts as to the possibility of regarding all elements and chemical compounds from the standpoint of the volume theory grew rapidly, as is easily seen in his *Essay upon the Theory of Chemical Proportions, etc.* (cf. p. 230), which was published a few years after this. But he had, at any rate, found in the law of volumes a valuable aid towards the determination of the atomic composition of numerous substances, and the deduction from this of the atomic weights of many of the elements.

A glance at the table of atomic weights which he published in 1818 shows how reliable the values found by him are, comparing favourably as they do with those of other observers. A later table given out by him in 1827 contained marked improvements on the former one, and brought his atomic weights still nearer to those current at the present day. But great uncertainty still prevailed with regard to the proportional numbers of many of the atomic weights, as compared with that of hydrogen or oxygen. Berzelius took oxygen (as the most important element, the "pole of chemistry") for the basis of his other atomic weights, making that of oxygen = 100. His ground for this preference[1] was that oxygen was capable of combining chemically with every other element; in fact, oxygen compounds were almost the only ones made use of at that time for the derivation of the atomic weights.

If we calculate his values upon that of hydrogen, which is now the customary unit, we obtain numbers that can be compared with those in use to-day. The following selection of such atomic weights from the year 1818 will serve to

[1] In his text-book (first edition, vol. iii. p. 99) he expresses himself as follows: "To refer the other atomic weights to that of hydrogen offers not only no advantages, but has, in fact, many inconveniences, seeing that hydrogen is very light and is seldom a constituent of inorganic compounds. Oxygen, on the other hand, unites all the advantages in itself. It is, so to speak, the centre-point round which the whole of chemistry revolves."

corroborate what has just been said (the current values are those in brackets):—

Carbon	12·12	(12)	Lead	416	(207)	Sodium	93.5	(23)
Oxygen	16	(15·96)	Mercury	406	(200)	Potassium	157.6	(39)
Sulphur	32·3	(32)	Copper	129	(63·3)	Silver	433·7	(108).
			Iron	109·1	(56)			

The question now forces itself upon us—What were the grounds which led Berzelius to assume twice as high a value for many metals (*e.g.* iron, lead, mercury, copper, chromium, tin, etc.), and four times as high a value for potassium, sodium and silver, as are now assigned to them? The reason lay in his presupposition of the simplest possible combining proportions, for at that time such proportions as 2:3, 2:5, 3:4, etc., appeared to him too complex; only one atom of an element was, in his then view, present in (a molecule of) a compound. The compounds formed by the oxidation of iron, for example, in which the proportions of oxygen were as 2:3, and which we now express by the formulæ FeO and Fe_2O_3, had for him the composition expressed by the formulæ FeO_2 and FeO_3, whence the atomic weight of iron came out double what we now have it. An analogous composition was attributed to other metallic oxides corresponding to the protoxide and sesquioxide of iron, so that the atomic weights of their metals were doubled. In like manner Berzelius was led, by the assumption that the ratio of oxygen in potassic peroxide and oxide was as 3:2, to the erroneous conclusion that the latter contained one atom of potassium combined with two of oxygen, and the peroxide one of potassium combined with three of oxygen; hence for potassium and the analogous monovalent elements sodium, lithium and silver, whose oxides have in reality the general formula Me_2O, atomic weights four times higher than the true ones were deduced.

Thus, in spite of Berzelius' gigantic labours, many points attaching to his system of atomic weights still remained uncertain; there were as yet too few reliable and comprehensive data to allow of the true relations of the values found to that of hydrogen or oxygen being firmly established.

Berzelius himself was convinced of the insufficiency of the methods by which he had determined the atomic composition of compounds, and, from this, the atomic weights of the elements. Apart from his somewhat arbitrary suppositions, he had merely found in the physical behaviour of gases—in the relation of their specific gravities to the combining weights—a good basis upon which to work out the question of the magnitudes of the relative atomic weights.

The year 1819 brought with it two important discoveries in physical chemistry which helped to clear up the above uncertainties; attention was called almost simultaneously by Dulong and Petit[1] to the relations between the atomic weights of the elements and their specific heats, and by Mitscherlich to the connection between similarity of crystalline form and analogous constitution. The latter discovery and the doctrine of isomorphism which grew out of it were largely made use of by Berzelius for determining relative atomic weights; but to Dulong and Petit's statement he paid much less heed, as it still required extension and corroboration. Both of these discoveries have exercised such a profound influence on the development of the atomic weight system that they must be discussed shortly here, in so far as they refer to the latter (cf. section devoted to the history of Physical Chemistry).

[1] P. L. Dulong, who was born in 1785 at Rouen, and died in 1838 while Director of the Polytechnic School at Paris, rendered imperishable service, more especially by his physico-chemical investigations. But, apart from these, his purely chemical labours—*e.g.* that upon chloride of nitrogen, in discovering which compound he lost an eye and several fingers (in 1811), that upon the oxygen compounds of phosphorus and nitrogen, and his fruitful speculations upon the constitution of acids—ensure him an honourable place in the history of the natural sciences.

T. A. Petit was born in 1791, and died while Professor of Physics at the Polytechnic School so early as 1820. To chemists he is known by his conjoint work with Dulong on the atomic heats of the elements (see above), his other researches being purely physical.

Dulong and Petit's Law.

From researches[1] carried out in part with substances not quite pure, and in part by methods upon which not much reliance could be placed, those two investigators drew the important conclusion that the specific heats of a number of the solid elements, the metals in particular, were nearly inversely proportional to their atomic weights. But, however bold these deductions were, deductions which they expressed in the sentence: "The atoms of simple substances have equal capacities for heat," or, "The atomic heats of the elements are equal," their confidence in them was on the whole justified by later and more accurate experiments; at any rate most of the metallic elements fulfilled the Dulong-Petit law approximately. The exceptions to it, shown by many of the non-metals in a greater or lesser diminution of the atomic heats, have only in some measure been explained in recent years by the proof that the specific heats of such elements vary greatly with the temperature. In the case of simple chemical compounds, too, a relation was soon found between their specific heats and atomic weights (by Neumann, in 1831).

When once its validity had been proved, the significance of the Dulong-Petit law for the determination of the relative atomic weights of the elements became immediately apparent. One had merely to determine the specific heat of an element in order to arrive at its atomic weight from this, taken in conjunction with the atomic heat (which was assumed to be a constant), *i.e.* the product of the specific heat into the atomic weight. Dulong and Petit at once went on to apply their law to this problem, and came to the conclusion—a conclusion which was later recognised as correct—that the atomic weights ascribed by Berzelius to several of the metals must be halved.

There was, however, as yet no pressing reason why the latter, on a dispassionate review of Dulong and Petit's

[1] *Ann. Chim. Phys.*, vol. x. p. 395 (1819).

results, should at once agree to this demand. That those results were of great importance for theoretical chemistry he willingly admitted, but he maintained that they had not yet been proved to be of such general application that a law could be formulated from them. He especially opposed any fundamental alterations of his own atomic weights, as he held that, if this were done, improbable atomic proportions would have to be assumed for the compounds of some of the elements. This attitude towards the Dulong-Petit law was only gradually abandoned by Berzelius, after further proofs bearing on the point had been adduced.

Influence of the Doctrine of Isomorphism upon Atomic Weight Determinations.

After the founding and development of crystallography by Romé de l'Isle and Hauy, various experimenters had observed that substances of different chemical composition crystallise together in one and the same crystalline form. As instances of this may be mentioned Gay-Lussac's observation that crystals of potash alum grow in a solution of ammonia alum, while still retaining their original crystalline form, and Beudant's, that copper vitriol is obtained in the same form as iron vitriol when a small quantity of the latter is added to a solution of the former, and so on. But neither this observation nor the definite statement by Fuchs upon the replacement of certain substances in minerals by others [his doctrine of "vicariating constituents" (*Vikariierenden Bestandtheilen*[1])] led to the recognition of the relation between crystalline form and chemical constitution.

This important discovery[2] was reserved for E. Mitscher-

[1] This means substitution without any accompanying change of crystalline form; thus, to give one or two examples, Fe″ can replace Ca″, and Al‴ can replace Fe‴ or Cr‴ in this way.

[2] *Berl. Akad. Abhandlungen der phys. Klasse*, 1818-19, p. 426; also *Ann. Chim. Phys.*, vol. xiv. p. 172; xix. p. 350.

lich,[1] who explained the occurrence of isomorphous crystals in substances of different nature by proving that they possessed a similar chemical composition. Thus he found, on examining the salts of phosphoric and arsenic acids, that only those of analogous composition and containing equal amounts of water of crystallisation were isomorphous. His subsequent investigations of selenates and sulphates, of the isomorphism of magnesium and zinc oxides, and of iron, chromium and aluminium salts, confirmed the intimate connection existing between crystalline form and chemical composition. At first, after making those observations, Mitscherlich was of opinion that isomorphism depended chiefly on the number of the elementary particles (in the molecule), but he soon convinced himself that the chemical nature of these had also to do with it.

Berzelius, who regarded the discovery of isomorphism as "the most important since the establishment of the doctrine of chemical proportions," endeavoured to arrive at the atomic weights of the elements by the aid of isomorphous compounds. For, according to him, isomorphism meant similarity in atomic constitution; chemists only required to know the composition of one compound in order to deduce

[1] Eilhard Mitscherlich was born in 1794 in Oldenburg, and died in 1863 at Berlin, where he worked as Klaproth's successor in the University from the year 1821; he enriched chemistry by beautiful discoveries, and especially advanced it very greatly in the physical direction. At the beginning of his career he devoted himself to oriental and linguistic studies, only taking up the natural sciences incidentally; but, after circumstances had compelled him to turn wholly to medicine and its allied subjects, his intercourse with Berzelius, to whom he went in Stockholm in 1819, was decisive as to his future course. His work will frequently be referred to in the special section of this book; but mention may be made here of his important investigation of manganic and permanganic acids, his work upon selenic acid, and that upon benzene and its derivatives. His successful attempts to prepare minerals artificially and his varied studies in geological chemistry give further proof of the many-sidedness of the man, his greatest achievement of all being the discovery of isomorphism, mentioned above. His *Lehrbuch der Chemie* is marked by originality both of form and contents. For an account of Mitscherlich's life and work, see Hofmann's *Chemische Erinnerungen, etc.*, p. 30, and the *Erinnerung an Eilhard Mitscherlich* ("Memorial of Eilhard Mitscherlich," by Alexander Mitscherlich, Berlin, 1894).

that of the remaining isomorphous ones from it. The quantities of the isomorphous elements which replaced one another, referred to a definite unit,—say oxygen or hydrogen—were regarded by Berzelius as the relative atomic weights. He made extensive use of this new aid to confirm the correctness of his atomic weight determinations.

The Atomic Weight System of Berzelius, 1821—1826.

At first, in 1821, Berzelius did not consider that any change in the atomic weights was called for, as the new facts could be made to accord with his determinations and deductions. But five years later he resolved, after minute consideration, upon certain modifications, which chiefly consisted in halving the atomic weights of many of the elements. The grounds which weighed with him in this he set forth in a conclusive manner.[1] What mainly necessitated the abandonment of his former assumptions was the composition of chromic oxide and chromic acid. The amount of oxygen in the latter (so he writes) was to that of the base as 3 : 1 in neutral salts, whence the composition CrO_3 followed for chromic acid; while in chromic oxide the proportion was as $Cr_2 : O_3$. But, in order to concede this last, he had to give to ferric and aluminic oxides (oxygen compounds isomorphous with and capable of replacing chromic oxide) the analogous compositions F_2O_3 and Al_2O_3, and to their metals, as a consequence, only half as large atomic weights as he had previously done. Iron protoxide received the simplified formula FeO, and the oxides of magnesium, zinc, nickel, cobalt, etc., which were isomorphous with it, were regarded as similarly constituted. The necessary result of all this was, as already stated, the halving of the atomic weights hitherto in use, so that these now conformed to Dulong and Petit's law. With the atomic weights of sodium, potassium and silver, which Berzelius likewise halved, the circumstances were peculiar. He had arrived at the conclusion, with respect to basic oxides, that the strong bases

[1] *Pogg. Ann.*, vol. vii. p. 397; vol. viii. pp. 1, 177.

(such as oxide of potassium) contained metal and oxygen in the proportion 1 : 1, and therefore gave potassium, sodium and silver double their proper atomic weights; for, according to our present ideas, two atoms of the metal are combined in these bases with one of oxygen. The following list by him of the atomic weights of some of the more important elements, with hydrogen as the unit, shows the approximation of the numbers to those in use to-day, and also the amendment[1] which some of them had undergone during the years 1818–26 (cf. table, p. 218)—

Carbon .	12·25	(12)	Lead .	207·4	(207)
Oxygen .	16	(15·96)	Mercury .	202·8	(200)
Sulphur .	32·24	(32)	Copper .	63·4	(63·3)
Nitrogen .	14·18	(14)	Iron .	54·4	(56)
Chlorine .	35·47	(35·4)	Sodium .	46·6	(23)
Phosphorus	31·4	(31)	Potassium .	78·5	(39)
Arsenic .	75·3	(75)	Silver .	216·6	(108)

The figures in brackets indicate the current values.

In this table of the year 1826 we find for the first time the atomic weights of nitrogen and chlorine as simple substances. Berzelius held longer than any other chemist to his assumption that they contained oxygen; the grounds which necessitated his giving up this hypothesis are entered into further on.

If we review these efforts of Berzelius at determining the atomic weights of the elements, we see that he was mainly guided, in the case of non-volatile bodies, by the composition of the oxygen compounds, *i.e.* by the determination of the proportion of element to oxygen, and

[1] Berzelius, who had devoted his whole energies to perfecting analytical methods and amending the atomic weight numbers, had afterwards to suffer harsh criticism from others who, by reason of improvements in such methods, attained to still more exact results; this applied in an especial degree to Dumas (cf. *Ann. Chem.*, vol. xxxviii. p. 141 *et seq.*), who determined the equivalent of carbon "with every imaginable precaution," and found its value to be 6. The difference between this number and that which Berzelius had found, viz. 6·12, caused Dumas to utter the most severe reproaches against the great master of analysis (cf. Berzelius' mild reply, *Lehrb. d. Chem.*, vol. iii. p. 1165, and Liebig's admirable protest against Dumas' procedure, *Ann. Chem.*, vol. xxxviii. p. 214 *et seq.*).

then, secondly, by the doctrine of isomorphism, while to the Dulong-Petit law he allowed only a slight influence. In those cases where the elements or simple compounds of the elements were known in the gaseous state, his volume theory came in as a help towards deducing the desired values. Berzelius still held fast to the idea that the amounts of the elements contained in equal gaseous volumes were proportioned to their atomic weights. But this assumption was soon overthrown by the remarkable results of an investigation which exercised such a profound influence on the views of many chemists that it must be described at this point.

Dumas' Attempt to alter the Atomic Weights.

In the year 1827 a young chemist, J. B. A. Dumas (cf. p. 272 *et seq.*), who had already made himself favourably known by other work, published a research,[1] the great merit of which lay in the working-out of an admirable method for the determination of vapour densities. By this method he succeeded in estimating the specific gravity of the vapours of several elements; and the relation existing between these comparable values was, according to Dumas (who took up here the same standpoint as Berzelius in his volume theory), that of the relative atomic weights. The elements which he adduced were iodine and mercury, and to these he added phosphorus and sulphur a little later.[2] The result of this was that he obtained different numerical values from those assumed by Berzelius for the atomic weights of the above elements, which had been held for a year past. Taking the atomic weight of hydrogen as 1, and that of oxygen as 16 (Berzelius' numbers), the above vapour densities gave the values 123 for iodine, 101 for mercury, 62·8 for phosphorus, and 96 for sulphur. Further, Mitscherlich determined the vapour density of arsenic in 1833, and calculated from this the atomic weight 150. True, these numbers bore a simple

[1] *Ann. Chim. Phys.*, vol. xxxiii. p. 337.
[2] *Ibid.*, vol. xlix. p. 210; vol. l. p. 170.

relation to the atomic weights of Berzelius, that of the latter for mercury (200) being double, those for phosphorus and arsenic (31 and 75) half, and that for sulphur (32) one-third as great as the values deduced by Dumas from his vapour-density determinations, and held by him to be the correct ones. The result of this alteration of the atomic weights by the latter was great confusion. While Berzelius remained true to his own numbers, holding mercuric oxide, for example, to be composed of mercury and oxygen in atomic proportions, Dumas assumed in it two atoms of mercury to one of oxygen, and gave it the composition and formula which Berzelius ascribed to mercurous oxide, viz. Hg_2O. Again, to phosphuretted hydrogen, in which Berzelius quite rightly assumed the proportions of three atoms of hydrogen to one of phosphorus, on account of its analogy to ammonia, Dumas gave twice as many atoms of hydrogen, and therefore the formula PH_6.

In making the above alterations Dumas' procedure was quite without method, and only served to complicate matters further. He drew a theoretical distinction between smallest physical and chemical particles, bearing Avogadro's speculations in mind; but this attempt at separating molecule from atom remained not only unfruitful, but resulted in confusion. The manner in which Dumas spoke of half an atom of oxygen, and of hydrochloric acid as composed of half atoms of hydrogen and chlorine, must have been unintelligible at that time,[1] and was sharply criticised by Berzelius.

A comparison of the atomic weights of Berzelius and Dumas with those of to-day shows us how fully justified the former was in adhering to his own, which he had arrived at after the most mature consideration; Berzelius' values have proved to be the right ones. In view of recent experience, however, he became more cautious in the use of his volume theory, and from henceforth only applied the law—that the atomic weights of the elements are propor-

[1] If Dumas had been fully acquainted with Avogadro's ideas, he would have expressed himself more distinctly, and have cleared up the opposing points which remained unsolved.

tional to the densities of their vapours—to the permanent gases.

The mighty reform which Dumas aimed at in this section of theoretical chemistry remained without result; and there is justification for the reproach brought against him by many, and more especially by Berzelius, of having introduced obscurity and disorder into the atomic weight system of the latter. For the sake of an unproven hypothesis Dumas neglected the most striking chemical analogies (*e.g.* that between ammonia and phosphuretted hydrogen), and frequently confused things which were perfectly clear. In consequence of the objections which he raised to Berzelius' atomic weights of the elements, the distrust of these latter by contemporary chemists grew in extent, so that we find even the most distinguished investigators like Gay-Lussac and Liebig doubting whether it was possible to determine the relative weights of the atoms with certainty. They would have satisfied themselves with establishing the equivalents, and leaving the atomic weights quite out of account. The opposition to the atomic weight system of Berzelius was at its height towards the end of the third and beginning of the fourth decade of the century. In Germany, especially, L. Gmelin advocated the establishment of the simplest "combining weights"; but the certainty of being able to determine the true equivalents of the elements was not in itself sufficient, although Faraday's discovery of the electrolytic law in 1834 appeared to guarantee a solid basis for this (see second paragraph below).

Michael Faraday, who was born in London in 1794, was endowed with such exceptional inclination for the study of the natural sciences and such experimental aptitude that he worked his way up from humble circumstances, although he had received no systematic training previous to his connection with Davy. Davy immediately recognised the extraordinary talents of the youth, and got him to assist him in his work. Faraday's most important discoveries belong to the domain of physics (his investigations on induction currents, electro-magnetism and diamagnetism). His electrolytic law, which was of such supreme importance for the electrochemical theory, is touched upon below. He made himself known to the chemical world more particularly by his beautiful investigations on the liquefaction of gases, by his work on the hydrocarbons from oil-gas (when he proved the isomerism of butylene with ethylene), and by that on the chlorides of carbon. He was one of the earliest to promote the study of physical chemistry, which owed to him its first great advance since the investigations of Dulong and Petit on specific heat, and those of Mitscherlich on isomorphism. The results of most of his experimental work were published in the *Philosophical Transactions*, but some in *Poggendorff's Annalen* and other journals. During the greater part of his life (he died in 1867) he worked at the Royal Institution, in which he became professor in 1828. In addition to his wonderful gifts as an investigator, Faraday possessed in an exceptional degree the power of clear and pleasant exposition; the memory of his "Lectures to Children" at the Royal Institution still survives (see his delightful little book, *The Chemical History of a Candle*). In private life the simplicity and amiability of his character made him greatly beloved.[1]

Faraday made the memorable observation (see above) that the same galvanic current decomposed electrolytes, *e.g.* water, hydrochloric acid and metallic chlorides, in such a manner that equivalent amounts of hydrogen or metal were separated at the negative pole, and the corresponding quantities of oxygen or chlorine at the positive.[2] He grouped those facts together under the title of "The Law of definite Electrolytic Action." In the determination of electro-chemical equivalents he saw a sure auxiliary means for fixing chemical atomic weights in doubtful cases. Berzelius, however, did not recognise any necessity in this case either for departing

[1] A pleasant account of his life is given by Thorpe in his *Essays*, p. 142 *et seq.*, as a critique upon Bence Jones's *Life and Letters of Faraday*.

[2] *Phil. Trans.* for 1834, or *Pogg. Ann.*, vol. xxxiii. p. 301.

from his own atomic weights, but—obviously because of a misconception—disputed the correctness of the numbers obtained by the electrolytic method.

The time for a clear grasp of the terms *equivalent*, *atom* and *molecule*, and for drawing a sharp distinction between these, was not yet come. Berzelius was therefore perfectly justified in adhering to his relative atomic weights, the best proof for which was to be furnished later. But, as already remarked, he now only made use of his volume-theory in a greatly modified degree, in consequence of the results obtained by Dumas and Mitscherlich. With regard to vapours, he foresaw (in 1835) the possibility of the relation between volume and atomic weight being a variable one (he drew a distinction between gases and vapours, and only strictly applied the law of volumes to the latter).

How, in the course of the succeeding decades, Gmelin's combining weights became gradually replaced by the atomic weights now in use (most of which had been brought forward by Berzelius), will be detailed later on. The reader's attention will be chiefly directed in the following sections to Berzelius' energy in a speculative direction, as shown in the setting up of his dualistic system; this last was the fruit of an electro-chemical theory which, along with Davy's, now falls to be briefly considered.

The Electro-Chemical Theories of Davy and Berzelius.

The perception that a close relation existed between electrical force and chemical reaction spread rapidly at the beginning of the century, after the decomposition of water into its constituents by the galvanic current had been proved by Nicholson and Carlisle (in 1800), and that of salts into their bases and acids by Berzelius and Hisinger (in 1803). The first fruit of the many and varied observations on the action of the current on chemical compounds, and on the accompanying electromotive force in chemical reactions, was

Davy's Electro-Chemical Theory,[1] which he thought that he had founded on a firm basis by his ingeniously devised researches, begun in the year 1800. He took as his starting-point the proved experimental fact that different substances, capable of combining chemically with one another, *e.g.* copper and sulphur, became oppositely electrified upon contact when insulated. Heating intensified the resulting difference of potential, until it vanished in consequence of the chemical combination of the substances. This latter, Davy then reasoned, is simultaneous with the equalisation of the potentials. The greater the difference between these before combination, the greater must be the chemical affinity of the different substances for one another. By the addition of electricity to the compounds, their constituents receive the same electric polarities which they possessed before combination; the positive constituents go to the negative pole, and the negative ones to the positive.

Davy inclined to the assumption that electrical processes and the phenomena of chemical affinity arose from a common cause. His electro-chemical theory was characterised by the axiom that the small particles of substances which have an affinity for one another only become oppositely electrified upon contact. But later researches, especially those of Berzelius, led to the abandonment of this principle, while, otherwise, many of Davy's original ideas were retained.

Berzelius brought forward the main outlines of his electro-chemical theory in 1812,[2] after having already at various times expressed his views upon the indissolubility of chemical and electrical processes, upon combustion as an electro-chemical phenomenon, and on the probability of the small particles being polarised. But the theory as a whole, with its far-reaching conclusions, was first published in his *Versuch über die Theorie der Chemischen Proportionen*, etc.,

[1] *Phil. Trans.*, 1807, p. 1; cf. also his *Elements of Chemical Philosophy. Die Electrochemischen Untersuchungem Davys, with Annotations,* constitutes No. 45 of W. Ostwald's *Klassiker*.

[2] *Schweigger's Journ.*, vol. vi. p. 119.

already mentioned at p. 213. In this we see clearly how he deduced his theory from facts, and then how, from the standpoint so obtained, he succeeded in penetrating and dominating with it the whole domain of chemistry. His doctrine, developed in this way from the electro-chemical point of view, continued the prevailing one for the next twenty years, until it had to yield to the pressure of facts with which it could not be reconciled.

Berzelius started with the primary assumption that the atoms of elements were in themselves electric; electric polarity, therefore, was an essential property of these smallest particles, which further possessed at least two poles, whose quantities of electricity were in most cases different, so that either positive or negative electricity predominated in the particle as a whole. Thus elements were divided into positive and negative, according to whichever of these electricities prevailed; and this last point was easily solved by noting whether the element in question was separated at the negative or positive pole of the galvanic battery upon electrolysis.[1] In like manner Berzelius assumed a polarity for compounds as well as for elements, although, in consequence of the neutralisation of the opposite electricities by one another in the formation of compounds, this polarity was thereby weakened. The *intensity of the polarity* was, according to him, a measure of the excess of one or the other kind of electricity. The dissimilar polar intensity of the small particles was regarded as the cause of their various affinities (*der verschiedenen Affinitätswirkungen*). And, as the forces of affinity were found to be dependent on the temperature, so polarity was also to be regarded as a function of heat.

Chemical combination of the elements or compounds consisted, according to Berzelius, in the attraction of the dissimilar poles of the small particles, and in the consequent neutralisation of the different electricities. If positive

[1] At first Berzelius designated the elements after the poles at which they were separated, *i.e.* he called the metals negative, and the metalloids positive.

electricity predominated in the original substance, then an electro-positive compound resulted, and *vice versa*. If the electricities neutralised one another, then an electrically indifferent product was the result. Oxygen, as the most electro-negative element, served Berzelius here (as it had done in his atomic weight estimations) as the standard by which to determine the kind of polarity of the various elements. Those elements which yielded basic compounds with oxygen, even although only their lowest oxides were basic, were classed as electro-positive, and those whose oxides were acids as electro-negative. Following this principle he arranged the simple substances in a series, in which oxygen as the first member was followed by the other metalloids, while hydrogen formed the bridge between the latter and the metals, the whole ending with sodium and potassium. In referring to this, Berzelius frequently stated that many elements which were positively polar with regard to some were negatively polar with regard to others, *e.g.* sulphur was positive to oxygen, but negative to the metals and hydrogen,—and so on. Oxygen alone he held to an absolutely negative element, because in no case did it behave as a positive one with respect to any other.

By the aid of such conceptions, which formed the substance of his electro-chemical theory, Berzelius was enabled to give a satisfactory interpretation of the facts which were at that time considered of greatest moment. The electrolytic processes, *i.e.* the separation of the positive and negative constituents of compounds at the negative and positive poles respectively, were explained in a simple manner by the assumption that the galvanic current reinvested the small particles of compound bodies with their original polarity. The many and various manifestations of affinity could in this way be referred back to a common cause.

Proceeding from this one hypothesis,—that electric polarity was a property of the atoms of substances,—Berzelius was able to bring light and order into the province of inorganic chemistry, which was at that time (1819) almost the only branch of the science to be considered. His electro-chemical theory led him, in the first instance, to a perfectly definite conception of the "constitution or rational composition of chemical compounds," and then to a nomenclature and corresponding system of formulæ developed from this. His efforts in this direction were crowned with the greatest success. Even at the present day we cannot do without the chemical language which he introduced, although, on the other hand, his dualistic views on the composition of chemical compounds have not survived so long. He was, however, the first to draw a precise distinction between the *empirical* and *rational* composition of chemical compounds. The *constitution* of the latter was, according to him, arrived at by investigating their proximate constituents (such being, for instance, Cu_2O, CuO, and $(C_2H_5)_2O$ in copper salts, ethers, etc.), and this task he regarded as one of the most important which falls to the lot of the chemist. He himself devoted his whole energies to its solution, the electro-chemical theory serving as a means whereby he might attain to this great end.

The Dualistic System of Berzelius.

The necessary consequence of the electro-chemical view was the assumption that every compound body consisted of two parts, which were electrically different; without such difference a chemical compound could not be formed. Further, the constitution of the latter was known when its positive and negative constituents were demonstrated. It was again compounds of oxygen,—acids, bases and salts,—by means of which Berzelius developed this, his dualistic doctrine. The elements which were combined with oxygen were the positive constituents, *e.g.* the metals in oxides, and the metalloids in acids. The electro-chemical antithesis was illustrated by the following formulæ :—

$$\underbrace{\overset{+}{Fe}\overset{-}{O}}_{\text{Iron protoxide}} \qquad \underbrace{\overset{+}{Ba}\overset{-}{O}}_{\text{Barium oxide}} \qquad \underbrace{\overset{+}{S}\overset{-}{O_3}}_{\text{Sulphuric acid}} \qquad \underbrace{\overset{+}{C}\overset{-}{O_2}}_{\text{Carbonic acid.}}$$

The anhydrous bases are the positive constituents of salts, and the acids—in which negative polarity predominates—the negative ones, as is shown by the formulæ—

$$\overset{+}{BaO}\cdot\overset{-}{SO_3} \qquad \overset{+}{ZnO}\cdot CO_2.$$

Berzelius considered that the strongest proof of the correctness of this theory lay in the electrolytic decomposition of compounds, especially of salts, into the above-mentioned two portions, which were separated at the poles of opposite electricity to their own. He further sought to explain the composition of double salts according to the dualistic hypothesis, giving, for example, sulphate of potash as the positive, and sulphate of alumina as the negative constituent of alum.

In the year 1819, when Berzelius published a detailed exposition of his electro-chemical theory, he was convinced that all acids contained oxygen. In his view water played in hydrated acids the part of a weak electro-positive constituent, and in metallic hydroxides that of a weak electro-negative one; the hydrates of sulphuric acid and of cupric oxide therefore received the formulæ—

$$\overset{+}{H_2O}\cdot\overset{-}{SO_3} \qquad \overset{+}{CuO}\cdot\overset{-}{H_2O}.$$

The binary conception, which had already been applied by Lavoisier to acids and bases, and even by Rouelle to salts, thus received the strongest support from the electro-chemical theory, and was materially developed in consequence. It will be shown in the next section how Berzelius was obliged to give up Lavoisier's one-sided theory of the oxygen acids.

The efforts of Berzelius to introduce a rational and generally applicable nomenclature go back to the year 1811.[1] His nomenclature is a continuation of that of Lavoisier, de Morveau and Berthollet, which however he greatly extended and amplified, his first efforts in this direction having been published in the *Versuch über die*

[1] *Journ. de Phys.*, vol. lxxiii. p. 257.

Theorie der Chemischen Proportionen, etc., already frequently mentioned. The division of the elements into metalloids and metals, according to their electro-chemical character; that of the positive oxygen compounds into suboxides, oxides and peroxides; and the corresponding division of the acids (which were designated according to their degree of oxidation), have been found to be so convenient that only very trifling alterations have had to be made in them. In like manner he designated the chlorine compounds corresponding to the oxides by adding different final syllables or prefixes, *e.g.* sub-chloride (*Chlorür*), chloride, perchloride, etc. In the nomenclature of the oxygen salts the name of the acid constituent preceded that of the basic, *e.g.* sulphate of oxide of copper.

He also endeavoured to apply similar principles in naming organic compounds, whose constitution had been determined on his own lines. But the time had not yet come when it was possible to devise a rational nomenclature for these.

Berzelius next established a system of chemical notation,[1] connected in the most intimate possible manner with his chemical nomenclature, which had given expression in clear language to the electro-chemical views on the composition of substances; this notation was to attain the same end in a more concise manner. In doing this he rendered an immense service, for it thus became possible, by the aid of simple symbols, not merely to express the composition of chemical compounds, but to picture even complicated reactions in an easily intelligible manner. He gave to each element a symbol, which was usually the first or the first two letters of its Latin name, less often of the Greek one; thus the symbol *H* stands for hydrogen (*hydrogenium*), *S* for sulphur (*sulphur*), *O* for oxygen (*oxygenium*), *C* for carbon (*carbo*), *Ag* for silver (*argentum*), *Hg* for mercury (*hydrargyrum*), and so on. These symbols denote at the same time the atomic weights of the elements in question, referred to a definite unit.

[1] Cf. particularly the *Versuch über die Theorie der Chemischen Proportionen*, p. 116, *et seq.*

By placing the symbols alongside of one another, and adding a figure to indicate the number[1] of atoms when the latter amounted to more than one, the formulæ of chemical compounds was obtained: *e.g.* H_2O for water, SO_2 for sulphurous acid, CO_2 for carbonic acid, $Na_2O{\cdot}CO_2$ for carbonate of soda, etc.

What an advance upon Dalton's attempts towards the same end, his figures only serving to illustrate the simplest of compound substances! Dalton's notation was soon forgotten, never having indeed met with general approval, while that of Berzelius became indispensable to chemists, and still remains so.

Berzelius attached a special meaning to the symbols with a bar drawn across them, these being employed by him to indicate that the elements in question were in the state of double atoms, or, as he put it,[2] that "they remain connected together;"[3] this applied, for example, to the hydrogen in water, ~~H~~O, to the chlorine in anhydrous perchloric acid, ~~Cl~~O_7, and to the iron in the sesquioxide, ~~Fe~~O_3. This mode of notation, which had exceedingly bad results, arose from Berzelius taking oxygen as his unit, and using it as the standard for the saturation-capacities of other elements.[4] He was thus led to the assumption of the double atom constituting a chemical unit, and the above symbols with bars served him to give expression to this; at a later period, however, he gave up using them, and reverted to the true atomic weights. There were nevertheless many chemists who would not concur in this view, cherished by Berzelius for a time, of the atoms of certain elements being only present as pairs in compounds; these chemists assumed

[1] Berzelius at first denoted the number of oxygen atoms by dots, and that of sulphur atoms by commas, *e.g.* calcium oxide, $\dot{Ca}$; iron bisulphide, $\overset{,,}{Fe}$; this system remained longest in use among mineralogists.

[2] *Lehrb. d. Chemie*, fifth edition, vol. i. p. 121.

[3] " . . . *dass sie zusammenhängend bleiben.*"

[4] Berzelius designated oxygen as "the measure of the relative weight according to which an element entered into combination" (*dass Mass der relativen Gewichtsmenge, nach welcher ein Grundstoff vorzugsweise Verbindungen eingeht*).

simple instead of double atoms and, with this, equivalents instead of atoms. Blomstrand, who has shown in his admirable work, *Die Chemie der Jetztzeit* ("The Chemistry of the Present Time"), the close connection which exists between the views of Berzelius and those held to-day, describes the results of the system of notation and of the views just mentioned in the following eloquent words: "This erroneous conception was without doubt the almost sole reason why Berzelius' atomic theory found so little acceptance; it acted like a restraining curb in preventing the free development of the latter, and led little by little to a peculiar confusion with regard to the fundamental principles of chemistry, the distinction between atomic weight and equivalent becoming by degrees nearly effaced, until at last the volume-atomic weights and the whole atomic theory of Berzelius were almost forgotten by the great majority of the chemists of his school."

Like every innovation, the admirable system of notation which Berzelius recommended met with most violent opposition from many chemists, especially in England. People spoke of "abominable symbols" which were more calculated to introduce confusion than clearness.

In 1820, then, the dualistic system, with the electrochemical theory for its basis, stood fully equipped, and was soon utilised by the vast majority of chemists as a guide in the confusion which resulted from the daily accumulation of new facts. Berzelius further attempted to apply the dualistic hypothesis in organic chemistry, which, from the third decade of the century, was more and more attracting the attention of chemists. How it came into collision here with the unitary theory, and had finally to succumb to the latter, will be described further on.

Manifestations against Dualism—Theory of the Hydrogen and of the Polybasic Acids.

The tenet which was set up by Lavoisier, and which Berzelius defended with all his power,—that the character of acids depends upon their containing oxygen, and that consequently this element is an unfailing constituent of their salts,—this *theory of the oxygen acids* was already greatly shaken towards the end of the first decade of the present century, and was abandoned by most chemists during the second, as a knowledge of facts opposed to it increased. Finally Berzelius, who remained longest true to the older idea, convinced himself of the existence of acids free from oxygen. The gradual transformation of chemistry which resulted from the setting aside of this dogma (that all acids contained oxygen) was a thorough one, for the unadaptable dualistic system was thereby battled with, and its fall prepared for.

In order to thoroughly understand this change of views, it is necessary that a clear light should be thrown upon the facts which brought it about. The discovery of the alkali metals by Davy and the allied researches which he made on the nature of chlorine must be regarded as the starting-points from which the light of the new knowledge radiated. Before Davy, who had recognised in the galvanic current a powerful means for decomposing chemical compounds, isolated potassium and sodium from the alkalies by its aid,[1] the latter were regarded as undecomposable; and this even although, from the time of Lavoisier, it was considered probable that they were constituted analogously to the metallic oxides, and were therefore oxygen compounds. This view was also held at an even earlier date by Scheele, as his recently published journals show. The many fruitless experiments which Davy had made with the alkalies in solution were finally crowned with success when he exposed these substances, only slightly moistened, to the action of a strong current. His correct assumption, that the metals separated at the negative pole were true elements, did not indeed find immediate acceptance;—in fact he himself was temporarily in doubt as to whether they did not contain hydrogen, especially after the presence of the latter element in the alkalies had been proved by Gay-

[1] *Phil. Trans.* for 1808, p. 1.

Lussac and Thénard, both of whom from this point took an active part, by their researches,[1] in the solution of the problems in question. The idea that the alkali metals might be hydrogen compounds had crept in from an analogy drawn between them and ammonia; at that time the latter was supposed to contain oxygen, which was withdrawn from it in the formation of ammonium amalgam. The erroneous conclusion that the above metals contained hydrogen, which resulted from this false interpretation, was however put right by Gay-Lussac and Thénard, who explained the point correctly. (Cf. *below*. It was mainly upon the three reactions specified towards the end of the next paragraph that Gay-Lussac and Thénard relied here; from these the elementary nature of the alkali metals, as well as of chlorine, followed.) Consequently, from the year 1811, potassium and sodium were regarded as metals and therefore as elements.

With the elucidation of the above points, the question as to whether chlorine was really a compound substance, and not rather a simple one, rapidly approached its solution. According to the assumption of Berthollet and Lavoisier, hydrochloric acid contained oxygen combined with a *radical muriatique*, and the chlorine which was liberated by its oxidation was looked upon as oxidised hydrochloric acid, and was therefore named so (*oxy-muriatic acid*). At the time when Davy[2] and Gay-Lussac and Thénard[3] began their memorable investigations, hydrochloric acid gas was generally held to contain chemically combined water. But even with the most powerful reducing agents these chemists were unable to prove the presence of oxygen either in perfectly dry hydrochloric acid or chlorine, and this of itself made them incline to the belief that chlorine was an element and hydrochloric acid its hydrogen compound. The idea, however, of oxygen being a necessary constituent of all acids had taken such firm root that numerous fresh investigations

[1] *Ann. de Chimie*, vol. lvi. p. 205; vol. lxv. p. 325.
[2] *Phil. Trans. for* 1810, p. 231.
[3] *Mémoires de la Société d'Arçeuil*, vol. ii. p. 339.

were required before it could be got rid of. The most important of the observations which led to this were the following :—Hydrogen and chlorine unite to form anhydrous hydrochloric acid, which is decomposed by sodium with the liberation of half its volume of hydrogen and the formation of sodium chloride, while the latter also results directly from the combination of sodium and chlorine.

Upon the ground of those facts Davy was the first to express the distinct opinion that chlorine was an element, suggesting for it the name[1] by which it has since been known. At first Gay-Lussac and Thénard had misgivings about agreeing to this, fearing to disturb the uniformity of the chemical system. But, after the former had completed his famous investigation upon iodine, both he and Thénard, as well as other French chemists, were obliged to concur in Davy's view. Iodine and fluorine now received a place among the elements, next to their analogue chlorine.

Berzelius did not allow himself to be convinced all at once of the necessity for this thorough innovation, which entailed the abandonment of the theory of oxygen acids. The unity of chemical theory went with him before everything else; he saw in the projected reform an overthrow of the principles which had governed the older chemical system. After having given eloquent expression to his ideas on the subject in letters to Marcet, Gilbert, Thomson and others, he collected together the arguments in favour of the older view in a treatise[2] entitled: *Versuch einer Vergleichung der älteren und der neueren Meinungen über die Natur der oxydierten Salzäure, zur Beurtheilung des Vorzuges der einen vor der anderen* ("An attempt to compare the Old and New Opinions with regard to the Nature of the Oxidised Muriatic Acid, and to estimate the Advantages of the One over the Other"). His standpoint is clearly set forth in the following words: "I decline to give in my adhesion to the new doctrine until it has been made perfectly consistent and uniform with the new theoretical science which its

[1] *Phil. Trans.* for 1811, p. 1.
[2] *Gilbert's Annalen,* vol. l. p. 356.

authors claim to have built upon the ruins of the chemical theory that they have demolished. For I demand uncompromisingly from any chemical theorem that it shall agree with the rest of chemical theory and be capable of incorporation in it; if this be not the case, then I must reject it, unless, indeed, the evidence in its favour is of such an incontrovertible nature as to necessitate a revolution in the chemical theory with which it is at variance."

In one point, however, Berzelius soon gave up the opinion that every acid must contain oxygen, by recognising sulphuretted and telluretted hydrogens as hydrogen acids; this latter nomenclature (*hydracides*) was first made use of by Gay-Lussac. At that time Berzelius still held that oxygen was present in chlorine, iodine and fluorine, even after Gay-Lussac's famous research upon the salts of hydrocyanic acid had proved that these last were free from it. It was only after he had been able to make the results of his own investigations on ferro-cyanogen and sulpho-cyanogen compounds agree with the theory of non-oxygenated acids that he resolved to include chlorine and iodine among the elements. About the same time (1820) he gave up the idea that nitrogen and ammonia contained oxygen; but it was not until 1825 that he abandoned what remained of his old view, by including fluorine with chlorine and iodine among the salt-forming elements or halogens;[1] he drew a sharp distinction between the *haloid salts, i.e.* the salts produced by the combination of the above elements with the metals, and the *amphid salts*, or those containing oxygen.

Theory of the Hydrogen Acids.

Several years before Berzelius had given up the oxygen-acid theory, Davy,[2] and almost at the same moment Dulong,[3] made the attempt to bridge over the gap between the oxygen and hydrogen acids by a uniform interpretation of

[1] *Jahresber.*, vol. vi. p. 185; also in his *Lehrb. d. Chemie.*
[2] *Phil. Trans.* for 1815, p. 203.
[3] *Schweigger's Journal*, vol. xvii. p. 229.

their constitution. In these efforts we see the beginnings of the hydrogen-acid theory, which was to become of such great importance a few decades later on. From his observation that iodic anhydride was devoid of acid properties, but acquired them after combination with water, Davy drew the conclusion that hydrogen and not oxygen was the acidifying principle in the latter compound; hydrogen, in his opinion, was an essential constituent of all acids. The assumption, that hydrated acids and salts contained water or metallic oxides together with acid anhydrides, he held to be unproven and unnecessary. Dulong expressed himself in a similar sense after an investigation of oxalic acid and its salts; the former he regarded as a compound of hydrogen with carbonic acid, while in the latter he assumed an analogous combination of the metals with the elements of carbonic acid. In these discussions a dualistic conception of acids and salts was still apparent, hydrogen and the metals being placed opposite salt-forming radicals; but the way was now opened for a unitary theory of acids and salts.

Berzelius' criticism of those attempts to explain the constitution of important classes of compounds was unusually mild; but at the same time he adhered to his dualistic view, since he laid special weight upon the possibility of preparing the immediate constituents (of the acids), the radicals of the hydrogen-acid theory being but seldom capable of isolation.

As his electro-chemical theory became better known, and was received with approbation, the opposing views of Davy and Dulong lost ground; it was only in the thirties that they reappeared, with fresh arguments to back them up, after which they were gradually accepted. The following observation by Daniell (in the year 1840) upon the electrolysis of salts was brought forward as an argument in their favour: "When galvanic currents are passed through different electrolytes, *e.g.* acidified water, fused chloride of lead, or a solution of sulphate of potash, amounts of hydrogen, lead and potash are set free at the negative pole, which stand to one another in the ratios of their chemical equivalent-numbers." This is in accordance with Faraday's "Electrolytic Law," excepting that in the case of the sulphate of potash an equivalent of hydrogen is liberated in addition to an equivalent of the base. The current therefore appears to do double work here, in spite of the law just mentioned; for, if it be assumed that the immediate constituents of one equivalent of the salt are potash and sulphuric acid, then only one equivalent of potash—as the electro-positive portion—should result, and not one of potash plus one of hydrogen. But this apparent contradiction is done away with by adopting the view of Davy and Dulong, *i.e.* by assuming potassium as the positive, and the radical $\dot{S}O_4$ (*oxy-sulphion*) as the negative constituent. The two equivalents of potash and hydrogen are then seen to be secondary products of the decomposition of one equivalent of water by the potassium originally separated at the negative pole. The conclusion drawn from this observation on the constitution of salts was then of course extended to that of acids, in which hydrogen was assumed as the one constituent, and a radical—either containing oxygen, or free from it—as the other.

The theory of the hydrogen-acids became still more clearly defined after Liebig had brought forward his:—

Doctrine of the Polybasic Acids.[1]

This we shall consider here, although it only dates from 1834, because of its close connection with the above views of Davy and Dulong. Many chemists at that time, Gay-Lussac and Gmelin in especial, inclined to the assumption that the atoms of the various metallic oxides contained one atom of oxygen to one atom of metal, and combined with one atom of acid to form neutral salts; Berzelius too, after 1826, was of opinion that this combining proportion was the rule. But a view of such simplicity as this, according to which almost every acid was regarded as monobasic, could

[1] *Ann. Chem.*, vol. xxvi. p. 113.

no longer hold its ground after Graham's[1] famous investigation of the phosphoric acids.[2] For this chemist showed that ordinary, pyro-, and meta-phosphoric acids contained different amounts of "basic water" to 1 atom of P_2O_5, viz. 3, 2 and 1 atoms of water, these latter being replaceable by metallic oxides. The different saturation-capacities of those acids were in this way demonstrated, being held to depend upon the amounts of basic water which entered into their constitution.

Liebig built upon the ground which Graham had prepared, and with such success that, by the aid of his own admirable and comprehensive researches upon a large number of acids, he was able to firmly establish his *theory of polybasic acids.* By his investigations on citric, tartaric, cyanuric, comenic and meconic acids, he convinced most chemists that these resembled phosphoric acid in basicity (*i.e.* were polybasic). He distinctly and definitely resisted the application to them of the arbitrary tenet that the atoms of all acids are equivalent to one another, and he gave as the criterion of a polybasic acid its capability of forming compound salts with different metallic oxides (*e.g.* such a salt as $PO_4\left\{\begin{matrix}Na_2\\K\end{matrix}\right)$. Liebig was the first to distinguish between mono-, di-, and tri-basic acids.

In order to express the facts, he still made use of the definition of acids in the dualistic sense, according to which

[1] Thomas Graham, born in Glasgow in 1805, became in 1830 Professor of Chemistry at Anderson's College of that city, and then in 1837 at University College, London. In 1855 he resigned this post on being appointed Master of the Mint; he died in 1869. His admirable text-book, *Elements of Chemistry*, was used not only in England, but was recast and translated into German by J. Otto and H. Kolbe. Graham's originality was shown by his valuable physico-chemical investigations on the diffusion of gases, osmose, etc., which opened out new paths in the science, while at the same time he enriched general chemistry, especially inorganic, by his purely chemical work. Thanks to the generosity of Graham's old friend, the late James Young of Kelly, his collected researches have been published in one large volume, entitled *Chemical and Physical Researches* (Edinburgh, 1876). A full account of Graham's life, and of the great services which he rendered to chemistry, is given by Thorpe in his *Essays*, p. 160, *et seq.*

[2] *Phil. Trans.* for 1833, p. 253; or *Ann. Chem.*, vol. xii. p. 1.

these were regarded as compounds of one atom of acid anhydride with one, two or three atoms of water. But this he felt to be unsatisfactory, since it did not permit acids and salts to be regarded from a uniform standpoint. He pointed out with great acuteness the contradictions which were involved in the retention of this view, summing up his criticism as follows: "In order to explain one and the same phenomenon we make use of two different methods. We are obliged to ascribe to water the most various properties, calling it basic water, water of hydration and water of crystallisation, while at the same time we see it enter into compounds in which it assumes no one of these forms. And all because we have chosen to draw a sharp line of demarcation between haloid and oxygen salts—a line not observable in the compounds themselves, seeing that in all their relations they show similar properties."

Liebig was led to the theory of hydrogen acids from grounds of probability, and still more from grounds of convenience. The sentences in which he enunciates this doctrine explain his standpoint so clearly and tersely that they must be quoted here.

"Acids are particular compounds of hydrogen, in which the latter can be replaced by metals."

"Neutral salts are those compounds of the same class in which the hydrogen is replaced by its equivalent in metal. The substances which we at present term anhydrous acids only become, for the most part, capable of forming salts with metallic oxides after the addition of water, or they are compounds which decompose these oxides at somewhat high temperatures."[1]

Those sentences distinctly show us the influence which the accumulating observations on the substitution of hydrogen by other elements had exercised upon Liebig. This inclination of the latter to a unitary hypothesis was keenly felt

[1] Liebig here formulates sulphates as $SO_4 + \ddot{M}e$. The decomposition of the metallic oxides to which he refers is their reduction, thus—

$$BaO + SO_3 = Ba + SO_4.$$

by Berzelius,[1] who to the end of his life described Liebig's theory of the polybasic acids as one which "has led to the confusion of ideas, and has stood in the way of a more perfect knowledge." But in thus criticising views of such great importance, and which served in quite an exceptional degree to clear up the uncertain notions with respect to the term "equivalent," Berzelius stood almost alone.

Development of the Dualistic Doctrine in the domain of Organic Chemistry—The Older Radical Theory.

During the second, and still more during the third decade of our century, organic chemistry emerged from its modest beginnings, to play an important part even so early as in the forties. It was destined to be the medium for the development of important views and of doctrines evolved from these, thereby reacting beneficially upon its elder sister inorganic chemistry. At first it continued on pretty much the same lines as the latter, the dualistic hypothesis, which had kept its place so well with inorganic, being applied to organic compounds also. Here again Berzelius struck in as a reformer with all his accustomed energy, and guided for a time the fortunes of organic chemistry. A glance at the earlier history of the latter will show us how imperfect was the knowledge of this branch of our science before the second decade of the nineteenth century.

The Growth of Organic Chemistry previous to 1811.

So early as at the close of the seventeenth century mineral substances were classed apart from vegetable and animal, the three being treated separately in text-books of

[1] The letters between Berzelius and Liebig, already referred to, give instructive and at the same time interesting details upon this point, and upon the genesis and critical examination of Liebig's view; they also show us how the estrangement with Berzelius came about (cf. especially pp. 154, 159 *et seq.*, and 166).

chemistry, in that of Lemery, for instance; this division was in accordance with the classification of natural substances according to the three "kingdoms of nature," which was even then in vogue. It was from this empirical standpoint that the chemistry of organic compounds developed itself, after Lavoisier had proved qualitatively that the main constituents of these were carbon, hydrogen, oxygen and sometimes nitrogen, occasionally together with sulphur and phosphorus. How he sought to utilise this quantitatively also, by working out a method of organic analysis, will be described under the history of analytical chemistry. He it was at all events who laid the foundation for a thorough knowledge of the subject; for, before scientific investigation in this branch could become possible, the composition of organic compounds had to be established. Notwithstanding that but very little was known at that time about the chemical constitution of these, Lavoisier tried to form an opinion on the subject in particular cases. A point worthy of special mention was his view—a view which for long exercised great influence—that the organic acids were oxides of compound radicals, while he supposed that most of the mineral acids contained oxygen united with an element; this had indeed a distinct resemblance to the conceptions of the radical theory adopted at a later period.

While Lavoisier and other chemists after him remained true to the old classification of substances, Bergman began about the year 1780 to distinguish organic from inorganic bodies. But, in spite of the simplicity which this proposal had to recommend it, the line which remained drawn between vegetable and animal substances was only gradually removed as the knowledge increased that the same chemical compounds occurred both in vegetables and animals, as proved, *e.g.* in the case of several fats, formic acid, benzoic acid, etc. Still it was generally felt to be necessary to strictly separate organic from inorganic bodies, it being represented as an infallible distinction that the former could not be prepared directly from their elements. But even this barrier was destined to fall before very long, and both classes of compounds to be regarded henceforth from the same standpoints.

The Position of Berzelius with regard to Organic Chemistry.

At the beginning of this century chemists of such eminence as Dalton, de Saussure, Proust, and especially Gay-Lussac and Thénard, exercised all their ingenuity in trying to work out a reliable method for determining the quantitative composition of organic compounds, but the results of their experiments only partly approximated to the truth. Before Berzelius (1811), no one had attempted to give a definite answer to the question whether the composition of organic substances was, like that of inorganic, subject to the law of multiple proportions; whether, therefore, the former were to be looked upon as chemical compounds in the sense of the atomic theory. He himself had so far elaborated a method of analysing the salts of organic acids that he was able to deduce with tolerable certainty from his results the existence of simple chemical proportions between the elementary constituents of an acid and the oxygen of the base.[1] This first successful attempt to bring organic compounds under the atomic theory, in the same way as inorganic, was followed in 1813 and 1814 by investigations[2] carried on with improved processes, which strengthened his conviction that the law of multiple proportions applied in the fullest degree to organic compounds also. In determining these atomic weights, he recommended, as a principle to be followed wherever possible, that the substances in question should be analysed in the form of their compounds with inorganic bodies (*e.g.* acids as metallic salts).

But even although these researches—the first made in this direction—led to the recognition of an analogy between the two classes of substances, still Berzelius did not immediately make up his mind to regard organic compounds as constituted exactly like inorganic (*i.e.* with

[1] *Gilbert's Annalen*, vol. xl. p. 247.

[2] See especially *Annals of Philosophy*, vols. iv. and v.

respect to the arrangement of their constituent elements). On the contrary, he considered it necessary to draw a sharp distinction between the latter as binary, and organic compounds as ternary and quaternary; for these, as he stated in 1813, contain more than two elements. As a consequence of this, compounds like marsh gas, cyanogen and the hypothetical oxalic anhydride were classified as inorganic, an arrangement which was long retained (and still is, to some extent) on grounds of convenience, Gmelin being especially strong in his recommendation of it. But this empirical separation of the two series of substances soon proved to be quite inadequate, particularly after various oils had been recognised as binary compounds of carbon and hydrogen of complex composition.

Berzelius himself made the attempt, in his treatise[1] referred to above, to bridge over the gap between inorganic and organic bodies by assuming that the latter, like the former, are constituted binarily, but contain compound radicals in place of elements.

Gay-Lussac's beautiful researches on cyanogen had without doubt a powerful effect in reviving this idea, which had already been advanced by Lavoisier, for they proved the important fact that cyanogen, as a compound radical, can play the part of an element perfectly. This in its turn gave rise to further efforts to search for similar atomic complexes (*Atomkomplexe*) in other organic compounds. Gay-Lussac himself expressed the opinion that alcohol consisted of ethylene and water, and, as its vapour density proved, of equal volumes of these; while he assumed carbon and water as the immediate constituents of sugar. Hydrochloric ether was regarded by Robiquet as a compound of ethylene with hydrochloric acid, and anhydrous oxalic acid by Döbereiner as one of carbonic acid with carbonic oxide.

These efforts to look upon compound radicals as the immediate constituents of organic substances may be regarded as the beginnings of the radical theory. The

[1] *Versuch über die Theorie der chemischen Proportionen*, etc. (Dresden, 1820).

above attempts at a solution were, however, disapproved of by Berzelius, who raised a warning voice and declared them incompatible with the electro-chemical views. In accordance with the latter, the electro-negative oxygen was placed opposite to a compound radical as the positive constituent of a compound, thus showing that at that date Berzelius did not believe in radicals containing oxygen. At that time also he conceded the variability (*Veränderlichkeit durch Substitution*) of radicals, but went back from this later on, thereby putting an obstacle in the way of the healthy development of the radical theory.

The time for the completion of this doctrine was not yet come; but the theorising upon the proximate constituents of organic compounds was of much benefit, in that it gave a stimulus to the study of the latter. To the first task of determining their empirical composition was added the far higher one of investigating their chemical constitution by getting at the proximate constituents, as these were understood by Berzelius. The discovery of the first case of isomerism in the third decade of the century gave a powerful impetus to this, and caused the great importance of the task to be better appreciated, and a more correct idea of it to be formed. If we try to picture to ourselves the standpoint of the chemists of that day, we see how such startling observations of compounds having the same chemical composition, but differing totally in their properties, forced them of necessity to the conclusion that the cause of this phenomenon (termed isomerism) was to be sought for in a dissimilarity of the proximate constituents of the compounds in question. What a powerful and continually renewed charm was thereby given to the search for those different radicals of organic compounds!

Isomerism and its Influence on the Development of Organic Chemistry.

Up to about the year 1820 it was considered an axiom in chemistry that substances of the same qualitative and quantitative composition must possess the same properties. Even then, it is true, cases were known which appeared to contradict this natural assumption, viz. the different modifications of chromic oxide and of silicic acid, and, in especial, the proof given by Berzelius of the two varieties of tin dioxide. But little weight, however, was placed upon these observations; they were simply looked upon as exceptions to the general rule, and considered merely as indicating physical differences, as in cases of dimorphism, of which a number were known.

So little were chemists prepared for the existence of substances of the same composition, but of different chemical and physical properties, that most of them considered the first observed case of isomerism in organic chemistry as due to an error. In 1823 Liebig had found, on comparing his analysis of silver fulminate with that of silver cyanate, which Wöhler had investigated a year before, that the results of the analyses of both salts were alike.[1] Satisfied of the correctness of his own work, he thought that Wöhler had probably made some mistake, but became convinced that this was not the case upon repeating the investigation himself. From that date, accordingly, two compounds, which differed as widely as possible from one another chemically, were recognized as having the same composition.

While Berzelius attached full significance to the above observation, he did not immediately give in his adhesion to it,[2] but rather waited for further confirmation of the point; Gay-Lussac, on the other hand, felt no doubt whatever as to the correctness of the discovery, and explained the differences in the above salts by assuming a difference in the manner in which their constituent elements were combined. After Faraday's discovery,[3] in 1825, of a hydrocarbon in oil gas which had the same composition as ethylene, but which

[1] *Ann. Chim. Phys.*, vol. xxiv. p. 264.

[2] At first Berzelius was of opinion that an error had probably been made on one side or the other (cf. *Jahresbericht*, vol. iv. p. 110; vol. v. p. 85).

[3] *Annals of Philosophy*, vol. xi. pp. 44 and 95.

showed a totally different behaviour, and after Wöhler in 1828 had obtained urea from the transformation of the similarly composed cyanate of ammonium, chemists became more conversant with the existence of *isomeric compounds.* Berzelius only accepted those facts after hesitation, but ultimately convinced himself of their absolute correctness by experiments of his own. He proved that racemic acid had the same composition as tartaric,[1] and thereupon proposed the term *isomeric* for those substances which, with the same chemical composition, possess different properties. The general designation *isomerism* has since then been retained. Berzelius soon saw himself necessitated to define more strictly the meaning to be attached to this word;[2] he distinguished between *polymerism* and *metamerism,* as special cases of isomerism, in essentially the same manner as we still do to-day.[3] His power of generalising, even with but a scanty number of facts to go upon, was shown here in a very high degree.

The ideas of Berzelius with regard to the probable cause of isomerism in organic compounds are clearly shown in many of his utterances ; in his view isomeric compounds are those in which the atoms of the elementary constituents have grouped themselves differently into compound radicals. "The isomerism of compounds in itself presupposes that the positions of the atoms in them must be different." To conclude from this sentence that Berzelius looked upon the problem of elucidating the relative positions of the atoms in space as one which was soluble, is certainly not justifiable; what he no doubt had in his mind was the determining of the mutual relations of atoms in their compounds, and, especially, the establishment of the mode in which atoms are combined to form the proximate constituents or compound radicals of compounds. The accumulating observations of

[1] *Berzelius' Jahresber.*, vol. xi. p 44 (1832).

[2] *Ibid.*, vol. xii. p. 63 (1833).

[3] Berzelius regarded the different modifications of elements as a particular case of isomerism ; the designation *allotropy*, now employed for this, only dates from 1841.

cases of isomerism quickly brought the question of chemical constitution in this sense to the stage at which an experimental solution of it was deemed possible, and this was attempted by grouping together a number of organic compounds on the basis of the hypothesis of definite common radicals. The outcome of this attempt was the Radical Theory, in the shaping of which Berzelius and Liebig had the greatest share. To distinguish it from the more recently revived form of views of a similar character, it is known as the *older* Radical Theory.

The older Radical Theory.

Prior to 1830, as has been already stated, efforts were not wanting to explain the constitution of particular compounds by the assumption of compound radicals. The chief incitement to those efforts lay in the proof that cyanogen acted like an element in its numerous compounds, besides being known in the free state itself. The observation that alcohol is easily transformed into ether and ethylene may have given rise to the supposition that ethylene was a constituent of both of these.

This idea, which was held by Gay-Lussac, had new life imparted to it for the time being by Dumas and Boullay's attempt [1] to generalise it by extending it to derivatives of alcohol and ether. The radical "etherin," [2] C_2H_4, was assumed by them to be present in what afterwards became known as ethyl compounds, and was compared with an inorganic compound, ammonia. Like the latter, etherin was regarded as a base, capable of forming a hydrate with water, and ethers (analogous to salts) with acids. The following table will help to explain the endeavours to establish an analogy between organic and inorganic compounds

[1] *Ann. Chim. Phys.*, vol. xxxvii. p. 15 (1838).

[2] The radical C_2H_4 had, at Berzelius' suggestion, received the name *Ætherin.*

(some of the latter not having been isolated, as a matter of fact) :—

Etherin, C_2H_4	Ammonia, H_3N
Alcohol, $C_2H_4+H_2O$	
Ether, $2C_2H_4+H_2O$	
Hydrochloric ether, C_2H_4+HCl . .	Chloride of ammonia, H_3N+HCl.
Acetic ether, $2C_2H_4+C_8H_6O_3+H_2O$.	Acetate of ammonia, $2H_3N+C_8H_6O_3+H_2O$.[1]

This attempt, which is known under the name of the *etherin theory*, was so far the precursor of the true radical theory in that it had the comparison of organic with inorganic substances in common with the latter. In criticising it Berzelius was thoroughly justified in emphasising the point that it was quite admissible to group the above compounds in tabular form alongside of one another, while at the same time he expressed the opinion that their presumed constitution was highly doubtful.

But the real development of the existing idea that organic compounds owe their characteristics to the radicals which they contain, was mainly brought about by Liebig and Wöhler's memorable research, entitled *Ueber das Radikal der Benzoësäure* ("Upon the Radical of Benzoic Acid").[2] In this they proved incontestably that in numerous transformations of oil of bitter almonds, and of chlorine and bromine compounds prepared from it, a radical of the composition $C_{14}H_{10}O_2$,[3] which they termed *Benzoyl*, remained unaltered. They showed by convincing experiments that this radical may be assumed as present in benzoic acid, benzoyl chloride and bromide, benzamide, benzoic ether and benzoyl sulphide, and that it comports itself in these compounds like an element. This piece of work was not only of profound significance for the radical theory, but it has also exercised a most

[1] Dumas' atomic weights, taking H=1, were C=6, and O=16.

[2] *Ann. Chem.*, vol. iii. p. 249 (1832). The correspondence between Liebig and Wöhler (edited in 1888 by A. W. v. Hofmann and E. Wöhler) gives a welcome insight into the origin of this pioneering piece of work, while at the same time it constitutes the best memorial of the close friendship existing between the two men.

[3] Berzelius' atomic weights were: H=1, C=12, O=16.

powerful influence on the development of organic chemistry generally, the new methods given in it for the preparation of particular compounds having proved applicable to whole classes The authors laid greatest stress upon the proof of a "compound element, benzoyl, in a series of organic compounds."

Berzelius was so convinced by these astonishingly clear results of the correctness of their interpretation, that he concurred enthusiastically in the assumption of the radical benzoyl;[1] the facts were so strongly in its favour that he felt himself compelled to give up his axiom,—that oxygen cannot be a constituent of a radical. But unfortunately this was only for a short time, as he soon reverted to the opinion that the existence of oxygenated radicals was absolutely incompatible with his electro-chemical theory.

Most chemists of that day held that the radicals which were proved to be present in several compounds were to be regarded as atomic groups capable of existing separately, and that their isolation should therefore be striven after. Although benzoyl itself had not been isolated, as little doubt was felt with respect to its separate existence as with respect to that of calcium or of nitric anhydride, neither of which had yet been obtained. The natural result of Liebig and Wöhler's investigation was a strong incitement to chemists to search for the atomic groups peculiar to different series of compounds, whose modes of formation and behaviour pointed to a probable connection between them.

The radical theory proper, in the establishment of which Berzelius and Liebig took part during the ensuing years, arose out of such endeavours. A series of organic compounds, closely related to alcohol, furnished the most suitable object for such a view, these compounds being even at that date among the most carefully investigated of organic substances. In 1833 Berzelius[2] emphasised the necessity of assuming a

[1] In his letter to Liebig and Wöhler (*Ann. Chem.*, vol. iii. p. 282), Berzelius proposed the name *Proïn* or *Orthrin* (from πρωΐ and ὄρθρος respectively, meaning "morning blush"), because with this research a new day had dawned for organic chemistry.

[2] *Jahresber.*, vol. xiii. p. 190 *et seq.* The Berzelius-Liebig *Letters* (pp. 55 *et seq.* and 67) give many details as to the origin of this view.

binary structure for all organic as for all inorganic compounds renouncing at the same time the idea of oxygenated radicals. Benzoyl he explained as being the oxide of the complex C_4H_{10}, the peroxide of this being anhydrous benzoic acid. Ether he regarded as the sub-oxide of ethyl, and he gave to it the formula $(C_2H_5)_2O$; this last corresponded to the inorganic bases, and was combined with acids in ethers exactly as the metallic oxides were in salts. Alcohol, on the other hand, which is so nearly related to ether, was looked upon by him as the oxide of a radical C_2H_6, a view which entirely effaced the connection between the two compounds.[1]

Liebig,[2] noting this error, published in the following year his opinion that alcohol, as well as ether and its derivatives, were compounds of one and the same radical *ethyl*, to which, however, he gave the formula C_4H_{10} (in place of C_2H_5 by Berzelius). His view is apparent from the following table :—

Ether, $C_4H_{10}O$	Ethyl iodide, $C_4H_{10}I_2$
Alcohol, $C_4H_{10}O.H_2O$	Nitrous ether (*Saltpeteräther*), $C_4H_{10}O.N_2O_3$
Ethyl chloride, $C_4H_{10}Cl_2$ . . .	Benzoic ether, $C_4H_{10}O.C_{14}H_{10}O_3$.

He accordingly designated ether as ethyl oxide, and alcohol as hydrate of ethyl oxide, comparing the former with potassic oxide, and the latter with potassic hydroxide. Notwithstanding, however, his recognition of the fact that the same radical is common to both, he fell into an error which Berzelius had avoided, viz. he attributed to alcohol and the corresponding compounds twice the atomic weight that they really possess. But apart altogether from these mistakes of Liebig and Berzelius, the advantages of their *ethyl theory* were at once apparent. A broad pathway was opened out

[1] Berzelius conceived himself obliged to take this view of the atomic composition of alcohol and ether on account of their vapour densities; from these he deduced the correct molecular formulæ, without, however, being able to arrive at the true constitution of alcohol, as he did at that of ether.

[2] *Ann. Chem.*, vol. ix. p. 1, *Ueber die Konstitution des Aethers und seiner Verbindungen* ("On the Constitution of Ether and its Compounds").

for the conception that organic compounds were constituted analogously to inorganic. Ethyl played in a large number of compounds the same part as potassium or ammonium [1] did in others. Liebig finally extended this comparison to mercaptan and ethyl sulphide, then just discovered. It was due in a high degree to his eloquent advocacy of the assumption of "compound elements" that the radical theory found such wide recognition.[2]

The leading chemists of that day held firm to their expressed opinions regarding radicals:—Dumas to the assumption that etherin was the radical of alcohol, etc.; Berzelius to the view that alcohol and ether had different constitutions, although he did not absolutely deny the admissibility of the extended ethyl theory; while Liebig remained true to the latter. He differed most from Berzelius upon the question of oxygenated radicals, which were in his opinion indispensable; thus he had no doubt that carbonic oxide was a constituent of carbonic and also of oxalic acid. But in one point those chemists were all agreed, viz. that compound radicals existed as distinct constituents in their compounds.

Liebig by degrees took up another and broader view of the nature of radicals than Berzelius, who inclined more and more to the opinion that they were unalterable. In Liebig, on the other hand, we get frequent glimpses of the idea that the grouping of the elements to radicals must prove of essential service to a better understanding of the modes of decomposition and formation of compounds. This conception appears to have forced itself upon him from the result of an

[1] In the place of the assumption that ammonia itself is combined with acids in its salts, the view—originally held by Ampère (in 1816) and which had now the authority of Berzelius to back it—gradually spread, that in those salts ammonium, NH_4, acts analogously to the metals.

[2] We must not omit to state here that Kane, independently of Berzelius and Liebig, pointed out the analogy between a radical *Äthereum*, *i.e.* ethyl, which was to be assumed in ether, alcohol, etc., and the hypothetical ammonium; the paper, however, in which he expressed this view (which was published in 1833 in *The Dublin Journal of Medical and Chemical Science*, vol. ii. p. 348) remained quite unnoticed.

investigation[1] which Regnault[2] had undertaken at his suggestion. The latter had obtained a substance of the composition $C_4H_6Cl_2$, which he termed chloro-aldehyde, by decomposing ethylene chloride with alcoholic potash. Liebig[3] thereupon expressed his opinion that the radical C_4H_6 was a constituent of this chloride and of numerous other compounds; this radical he named *acetyl*, and he placed it parallel to the hypothetical *amidogen* (*Amid*), and its hydrogen compounds, ethylene and ethyl, to ammonia and ammonium, thus:—

C_4H_6, acetyl,	corresponds to	N_2H_4,	amidogen
C_4H_8, ethylene,	,,	,, N_2H_6,	ammonia
C_4H_{10}, ethyl,	,,	,, N_2H_8,	ammonium.

Liebig laid especial weight upon finding an expression for the constitution of aldehyde and acetic acid; these he looked upon as the protoxide and hydrated oxide of the acetyl radical, and he gave them the formulæ $C_4H_6O.H_2O$ and $C_4H_6O_3.H_2O$. This conception paved the way for the explanation of the conversion of alcohol into aldehyde and acetic acid, while at the same time it raised up doubt as to the rigid unchangeability of a radical.

The year 1837 may be looked upon as that in which the older radical theory attained to its zenith and stood out most securely, in spite of the many attacks which it had to undergo. Liebig and Dumas, who were convinced of the untenability of the etherin theory, joined together to make a thorough investigation of organic compounds with

[1] *Ann. Chem.*, vol. xv. p. 60.

[2] H. V. Regnault, who was born at Aix-la-Chapelle in 1810 and died at Auteuil near Paris in 1878, was a pupil of Liebig. Up to 1840 he gave his attention to organic chemistry, which he enriched by valuable work, but after that devoted himself to physico-chemical researches which will ensure him a distinguished place in the history of the science. His many-sidedness is shown in his admirable investigations on the respiration of animals, undertaken conjointly with Reiset. By means of translations, his *Cours Élémentaire de Chimie* (1847-49) became well known and appreciated in other countries besides France.

[3] *Ann. Chem.* vol. xxx. p. 229.

respect to the radical theory. In a paper[1] given out jointly in his own name and Liebig's, Dumas set forth his altered opinions and described the problems to be solved. Organic chemistry was regarded by both as the *Chemistry of Compound Radicals*, and was defined accordingly.[2] These radicals were compared with the elements, *e.g.* *ethyl*, *methyl* (whose existence in wood spirit was deduced from Dumas and Péligot's memorable research), and *amyl*[3] with the metals, *acetyl* with sulphur, and so on; and their compounds with the corresponding compounds of the elements, *e.g.* ethyl sulphide $(C_2H_5)_2S$, with sulphide of potassium, K_2S, etc.[4]

The chemists of that day did not, however, remain content with simply contrasting organic with inorganic compounds as an aid to getting at their formulæ; on the contrary, they applied in the happiest manner to the investigation of organic compounds the principles which they knew to hold good in inorganic chemistry, faithful to the axiom enunciated by Berzelius in 1817: "The application of what is known regarding the combination of the elements in inorganic nature, to the critical examination of their compounds in organic, is the key by which we may hope to

[1] *Comptes Rendus*, vol. v. p. 567. That this union of the two investigators was of short duration is easily intelligible when one considers the different modes of thought and dispositions of the two men. The criticism with respect to Dumas, which we find in the correspondence between Berzelius and Liebig, shows such a separation to have been inevitable. Gay-Lussac throws a clear light on the occurrence in a letter to Liebig which begins with the words :—*Maintenant, mon cher Liebig, je vous félicite d'être sorti de la galère où vous étiez entré. Je ne concevais pas votre mariage. . . ., etc.* (The *Berzelius-Liebig Letters*, p. 171).

[2] Cf. Liebig's *Handb. d. organ. Chemie*, p. 1.

[3] Cf. Cahours' investigation of fusel oil, *Ann. Chem.*, vol. xxx. p. 228.

[4] The following quotation from the paper cited above (note 1) shows the then standpoint of Dumas and Liebig: "Organic chemistry possesses its own elements, which sometimes play the part of chlorine or oxygen, sometimes that of a metal. Cyanogen, amidogen, benzoyl and the radicals of ammonia, of the fats, and of alcohol and its derivatives, constitute the true elements of organic nature, while the simplest constituents, such as carbon, hydrogen, oxygen and nitrogen, only appear when the organic substance is destroyed."

arrive at true ideas with respect to the composition of organic substances."

As the presence of such atomic complexes in organic compounds came to be assumed with more confidence, the term *radical* became more sharply defined. Liebig himself enunciated in 1838 three characteristics by which a compound radical was distinguished. In bringing forward his view he made use of cyanogen as an instance, and spoke as follows:[1] "We term cyanogen a radical because (1) it is the unchanging constituent of a series of compounds; (2) because it is capable of replacement in these by simple substances; and (3) because, in those cases where it is combined with one element, this latter can be exchanged for its equivalent of another element." At least two of the conditions here adduced had to be fulfilled in order that an atomic complex might be stamped as a radical. The existence of these conditions, moreover, could only be established by the most minute investigation of the chemical behaviour of organic bodies. That is to say, the nature of the radicals assumed in the latter could only be arrived at from the study of their reaction- and decomposition-products.

The radical theory gave such a powerful impulse to the science that its influence, even when it fell into error, cannot be too greatly prized. Chemists of the highest eminence were attracted to the task of investigating the constituents of compounds which were related to one another. Among the most fruitful of those efforts were a series of admirable researches upon the cacodyl compounds[2] by Robert Bunsen, begun in the year 1839 (*see below*).

Robert Wilhelm Bunsen, born at Göttingen on March 31st, 1811, became assistant-professor at the University there, then succeeded Wöhler at Cassel, and was appointed professor in the University of Marburg in 1838. His next post (only occupied for a short time) was at Breslau, after

[1] *Ann. Chem.*, vol. xxv. p. 3.

[2] *Ann. Chem.*, vol. xxxi. p. 175; vol. xxxvii. p. 1.; vol. xlii. p. 14; vol. xlvi. p. 1.

which he was called (in 1851) to Heidelberg, of whose University he remained a bright ornament until his resignation in 1889. Chemistry is indebted to him for a vast number of the most important researches in every branch of the science; his name will therefore be very often referred to in the special history of its various sections. Beginning with work in inorganic chemistry, he soon turned his attention to the organic compounds of arsenic, by investigating which he raised up a powerful support for the radical theory. His work upon gases led him to devise new methods, by sifting and combining which he created the gas analysis of to-day. The discovery of spectrum analysis by him and Kirchhoff—one of the grandest and most fruitful of the last fifty years—is fresh in every one's recollection. His labours in other branches of physical, analytical, inorganic and mineralogical chemistry will be referred to in the detailed description of these. Throughout he has shown himself an investigator of the most marked originality and a pioneer in the science; while his career as a teacher, extending over more than half a century, has been singularly successful in its results.

Bunsen's researches on the cacodyl compounds resulted in the proof that the so-called *alkarsin*, the product of the distillation of acetate of potash with arsenious acid, contained the oxide of an arseniuretted radical $As_2C_4H_{12}$ ($H=1$, $C=12$, $As=75$), this radical remaining unchanged in a long series of reactions of that oxide, and being even itself capable of isolation. This "compound element" containing arsenic (an unusual constituent of organic bodies) was thus shown to be a true radical.

The investigations of Gay-Lussac upon cyanogen, of Liebig and Wöhler upon benzoyl compounds, and of Bunsen upon the compounds of cacodyl, have been justly termed the three pillars of the radical theory. The assumption of radicals gained so immensely in probability from the results of these researches, that the hypothesis which lay at the root of the theory might now be regarded as well established. In any case the older radical theory formed an indispensable link in the chain of theoretical views, and marked an extraordinary advance upon the previous unconnected opinions. And even although this theory (as it then stood) exercised no very permanent effect directly, being soon overthrown by opposing currents, it showed itself in a high degree capable of further development. For, shortly after the catastrophe which came upon it, it was able to throw off a few restraining fetters and to start again into fresh life.

Before proceeding to describe the development of the hypotheses directed against the older radical theory, it will be convenient to give a short account here of the lives and chief labours of the three chemists who were mainly instrumental in changing the direction of organic chemistry during the third and fourth decades of the nineteenth century, and who furthermore exercised a powerful influence upon our science up to a very much more recent date.

Liebig, Wöhler and Dumas—A Survey of their more important Work.

Liebig and Wöhler, who were guided by similar scientific aims, and were at the same time close personal friends, must be spoken of together in the history of the science; the portrait of the one is incomplete unless supplemented by the characteristic features of the other. The fruit of their common labour is among the richest in the whole of chemistry. The selection of their letters, extending from 1829 to 1873, which was edited[1] by the late A. W. v. Hofmann, with E. Wöhler's co-operation, is a memorial to the steadfast friendship that existed between the two men, and at the same time a most important contribution to the history of chemistry.

Justus Liebig,[2] whose influence in shaping the radical

[1] Published by Vieweg, Brunswick, 1888.

[2] Cf. the Memoirs by H. Kolbe, *Journ. pr. Chem.* (2), vol. viii. p. 428; by A. W. v. Hofmann, *Ber.*, vol. vi. p. 465; and especially the various memorial papers (partly by A. W. v. Hofmann) on Liebig and Wöhler, *Ber.*, xxiii. Ref. p. 785 *et seq.* These last include, in an appendix, a fragment

theory and upon organic chemistry in general has just been touched upon, earned by his scientific work the right to be regarded as one of the most distinguished investigators of the century. Born at Darmstadt on 12th May 1803, his early years did not seem to give any special promise of the fiery spirit which he later developed, although it was not long before he felt himself drawn towards chemistry with irresistible power. He has himself given us a graphic description, in the autobiographical sketch already mentioned, of the way in which he gained a knowledge of chemical facts and phenomena, having determined at an early age to make chemistry a study, to the utter astonishment of his teachers and fellow pupils. He relates in pleasant manner how "that disposition developed in myself, which is found in chemists more than in students of other sciences, viz., to think in phenomena" (*in Erscheinungen zu denken.*) It was this capacity which caused "all that I saw, whether intentionally or unintentionally, to remain fixed in my memory with photographic accuracy."

He soon forsook the calling of apothecary, through which alone it was possible at that time to gain a practical knowledge of chemistry, in order to devote himself to academic studies. Relying on himself alone, he continued his early-begun investigations upon fulminate of silver, which he hoped would give him a certain definite position in science. But however independent the youth thus showed himself in this direction, he was unable to resist the influence of the natural philosophy (or, as it might be better expressed in English, physio-philosophy) current at that day. At a later period we find him speaking with bitterness of the two years that he had lost by it, during which time he studied under

of an autobiography of Liebig's. Compare also the *Letters* between Liebig and Wöhler, and Liebig and Berzelius. In 1895, W. A. Shenstone wrote a short *Life* of Liebig for the Century Science Series (Cassell and Co.), which gives in brief compass an excellent picture of the man and the chemist, though—from want of space—too little is said of his purely scientific work.

Schelling at Erlangen.[1] But he rescued himself from this by going in search of his science to where, at that time, it flourished most brilliantly,—to Paris, where Gay-Lussac, Thénard, Dulong, Chevreul, Vauquelin and others were hard at work. With recommendations from Alexander von Humboldt to Gay-Lussac and other influential chemists, he recovered himself in those surroundings (as he has himself so delightfully described), and soon became closely associated with Gay-Lussac, the result of which was their important investigation of the fulminates. This piece of work paved a way for him; in 1824 he was called as professor to Giessen, where he remained for twenty-eight years, but where at first he had to fight hard and continuously in order to maintain his position, his youth being a source of offence to the older professors.[2] In 1852 he accepted a call to the University of Munich, being led to this by the desire to throw off the fatigues of laboratory teaching and to live all the more ardently for research. His magnificent labours were brought to a close there by death, on 18th April, 1873, but the genius which inspired them, and which had acted with such powerful effect upon his contemporaries, continued to influence mankind. How powerful an influence he exercised—as shown in his greatness as a teacher, in the transformation of whole branches of the science, and in the setting aside of firmly rooted views which in his opinion were erroneous—we shall now attempt shortly to describe.

[1] In an essay entitled *Ueber das Studium der Naturwissenschaften* ("On the Study of the Natural Sciences"), published in 1840, Liebig expressed himself as follows: "I myself spent a portion of my student days at a university where the greatest philosopher and metaphysician of the century charmed the thoughtful youth around him into admiration and imitation; who could at that time resist the contagion? I too have lived through this period—a period so rich in words and ideas and so poor in true knowledge and genuine studies; it cost me two precious years of my life."

[2] G. Weihrich, in his pamphlet, *Beiträge zur Geschichte des chemischen Unterrichts an der Universität Giessen* (1891), has given a full and careful account of Liebig's academic work and of his relations to the University.

As a teacher Liebig stands almost alone. Berzelius, the great master, only drew around himself pupils who had already a considerable knowledge of the subject, and worked (directly) in a comparatively narrow circle. Liebig, on the other hand, founded a real school of chemistry, by sparing no pains in instructing his pupils individually from the commencement of their course of study. He was the first to give systematic teaching in chemistry, for up to that time there was no laboratory in existence which was devoted solely to that purpose. And he was also the first to recognise the necessity for having chemical institutes which should further not merely the science itself, but also the many other branches dependent upon it. His laboratory in Giessen served as a pattern upon which numerous others were in the course of years modelled, at first slowly but afterwards in more rapid succession. By the charm of his own personality Liebig stimulated his pupils and inspired them with enthusiasm, especially when the solution of a scientific question came up. Kolbe has described for us his unique character as a teacher in the following striking sentences:—"Liebig was not a teacher in the ordinary sense of the word. Scientifically productive himself in an unusual degree, and rich in chemical ideas, he imparted the latter to his more advanced pupils, to be put by them to experimental proof; he thus brought his pupils gradually to think for themselves, besides showing and explaining to them the methods by which chemical problems might be solved experimentally."

In addition to this Liebig gave a new form and meaning to his experimental lectures, so that here also he set up a standard. His pupils were legion; many of them afterwards spread abroad the doctrines of their master in universities, polytechnic institutes, technical schools, etc. Out of a long list of them which might be given here, the following may be mentioned:—A. W. v. Hofmann, Strecker, Fresenius, Will, H. Buff, Fehling, Henneberg, Schlossberger, Rochleder, Schlieper, Scherer, Redtenbacher, v. Bibra, Varrentrapp, Th. Poleck, Playfair, Muspratt, Stenhouse, Brodie, Gerhardt, Williamson, Wurtz, Frankland and Volhard.

The mental vigour which was shown in the results of Liebig's teaching is also seen in his literary activity, which awakens a feeling of astonishment by its many-sidedness, embracing as it does the most various branches of the science. Throughout it all we see the capacity of the true investigator to state points correctly and clearly, to grasp the connection between different processes distinctly, and to draw able and ingenious conclusions. These merits impart to Liebig's writings a great and ever-renewed charm. His numerous experimental researches, together with the joint ones with Wöhler, were mostly published in the *Annalen*,[1] which he began to give out in 1832. His extended investigations in physiological chemistry, which were begun in 1837, led him on to the grand achievement of setting forth the applications of chemistry to agriculture, physiology and pathology in three separate works.[2] In these he combated the current doctrines which were held with regard to the nutrition of plants and animals, basing his arguments upon exact experiments. Notwithstanding the great excitement which those publications produced, Liebig found leisure to write his *Chemische Briefe* ("Chemical Letters," 1844), by which he proved that chemistry might be treated popularly, and yet at the same time scientifically. It is almost inconceivable how he still found time remaining to devote to

[1] Till 1840 this journal was termed *Annalen der Pharmacie*, and after that date (with Wöhler as joint editor) *Annalen der Chemie und Pharmacie.*

[2] *Die Chemie in ihrer Anwendung auf Agrikultur und Physiologie*, 1840 ("Chemistry in its Application to Agriculture and Physiology," 1840); *Die Thierchemie oder organische Chemie in ihrer Anwendung auf Physiologie und Pathologie*, 1842 ("Animal or Organic Chemistry in its Application to Physiology and Pathology," 1842); *Der chemische Prozess der Ernährung der Vegetabilien und die Naturgesetze des Feldbaues*, 1862 ("The Chemical Processes in the Nutrition of Vegetables, and the Natural Laws of Tillage," 1862). In one of his letters to Berzelius (*Letters*, p. 210) Liebig tells us how and why he was led to take up this last branch of applied chemistry. An "insurmountable distaste and repugnance to this disputation in chemistry had taken hold of him; he was tired out (*auf die Spitze gestellt*) by the controversy about the substitution theory," etc. Whereupon he developed in broad lines the programme of his agricultural chemical work.

the *Handwörterbuch der reinen und angewandten Chemie* ("Dictionary of Pure and Applied Chemistry"), founded by Wöhler, Poggendorff and himself, and, after the death of Berzelius in 1848, to the *Jahresbericht über die Fortschritte der Chemie*. In addition to all these there are still to be mentioned his occasional papers,[1] some of which exercised a powerful effect; this applied in an especial degree to the two essays upon the state of chemistry in Austria and Prussia. In these, as in other papers devoted to questions of theoretical chemistry (*e.g.* in his writings directed against the views of Dumas, and of Laurent and Gerhardt), is shown the sparkling critical vein of this gifted man, who, from his rectitude and love of truth, never palliated what he felt to be erroneous or insincere. Occasionally Liebig may have gone too far in his critical utterances upon particular men; but the mainspring of his decided attitude with respect to them was always the boundless love of science and of truth, and an inflexible sense of justice.

As an investigator Liebig shows all his individuality. To organic chemistry he had devoted the full powers of his mind from the very beginning, without however neglecting any important part of inorganic. His very first work—that upon the fulminates—led to valuable results; for, through it the isomerism of cyanic and fulminic acids became recognised, a new field for investigation being thereby opened up. Another result of this laborious research upon these easily decomposable substances was the perfecting of organic analysis, to which Liebig gave its present form. By means of methods improved by himself, he established the composition of numerous organic compounds, especially of various acids. His work upon these last led him to a distinct conception of the term *basicity*; from this he developed his doctrine of polybasic acids (already touched upon), doing more to clear up the points involved here than any other chemist before him.

[1] These were published by M. Carriere under the title *Reden und Abhandlungen* ("Speeches and Essays"), by Justus von Liebig. (In 1845 he was made a baron by the Grand Duke of Hesse.)

His previous admirable researches upon compounds closely related to alcohol and acetic acid, *e.g.* ethyl-sulphuric acid, aldehyde, acetal, chloral, etc., rendered him specially capable of developing the radical theory and infusing fresh life into it. The work which he did upon sulphocyanogen compounds and upon the decomposition products of ammonium sulphocyanide showed him as a brilliant experimenter in all his many-sidedness.

But his most remarkable achievements were the researches carried out conjointly with Wöhler, which bring them both before us in their full freshness and power, and which will long continue to call forth the admiration of chemists. Wöhler's work upon cyanic acid and Liebig's upon the fulminates drew them together; their friendship is beautifully shown by the investigations which they undertook in common, during which each animated the other, while striving at the same time to do his best himself.[1] And how strikingly was the one man the complement of the other! Liebig—fiery, restless, and always advancing, able to utilise his rich experiences gained in the preparation and analysis of organic compounds for overcoming the hardest difficulties. Wöhler, on the other hand, quiet, almost prosaic, but not less conscious of his aim than Liebig himself, exercising patience in clearing up obscure points to which too little attention had been paid. The memorable research upon the radical of benzoic acid has been already detailed. The investigations upon amygdalin cleared up the difficult point as to how bitter almond oil was formed, and those upon uric acid, published in the same year (1837), enriched organic chemistry to an undreamt-of extent with a wealth of the most remarkable compounds,—compounds which have quite recently proved objects of the greatest interest to chemists. We are indeed not wrong in asserting that the organic chemistry of to-day is grounded mainly upon the pioneering labours of Liebig, and of Liebig and Wöhler together.

[1] Cf. the letters of both quoted in A. W. v. Hofmann's Memoir of Wöhler, *Ber.*, vol. xv. p. 3127 *et seq.*, and also the *Correspondence* already frequently referred to.

In addition to all this, inorganic chemistry was anything but neglected by Liebig, who enriched it by valuable observations on the most various subjects; we have only to recall his work upon the compounds of alumina, antimony and silicic acid, and many analytical methods which he worked out, *e.g.* the separation of nickel from cobalt. The results obtained by him in the laboratory were often of great service for technical chemistry; for instance, the improved preparation of cyanide of potash for the galvano-plastic process, and the reduction of a solution of silver by aldehyde for the production of mirrors.

Liebig's share in the development of organic chemistry, especially with regard to the views which had come to be accepted in it, became less marked towards the end of the thirties, as from that time he gave all his energies to the solution of a great question which had only an indirect bearing upon chemistry. The nutrition of plants and animals, the transformations of matter in animated nature —these were the grand problems which he strove to solve by experimental researches in an entirely new direction. The influences which emanated from him, the setting right of erroneous views, the ingenious interpretation of natural processes investigated by himself and his pupils, and the stimulus which invariably accompanied his labours and the deductions drawn from them,—all these can but be referred to here. The most important results of those researches will be spoken of under the history of physiological chemistry. Liebig's experiments on the nutrition of animals led him to distinguish clearly between nutrient substances among themselves, and between these and other substances which, though not directly nutrient, bring about metabolic changes in the organism.[1] By getting at the relative nutritive values of these materials he was enabled to introduce improved systems of feeding, and so to further the laws of health; we have only to recall here his extract of meat and his "children's food."

[1] ". . . *Unterscheidung der Nährstoffe unter sich und von den Genussmitteln.*"

He was thus in this respect a general benefactor of mankind.

We may close this attempt at depicting within narrow limits the scientific achievements of Liebig with the following eloquent words of A. W. v. Hofmann:—"If we sum up in our minds all that Liebig did for the good of mankind—in industries, in agriculture, and in the laws of health, we may confidently assert that no other man of learning, in his course through the world, has ever left a more valuable legacy behind him."

Friedrich Wöhler,[1] whose work blended so happily with that of Liebig, also proved himself by his own individual researches a master in his science. By far the greater portion of his work lay in the domain of inorganic chemistry which he furthered in a remarkable degree.

Wöhler's life may be sketched in a few sentences. Born in the village of Eschersheim, near Frankfort on the Main, on July 31st, 1800, he received in the latter city a splendid education at the hands of such eminent teachers as Karl Ritter, Grotefend, and F. C. Schlosser. There, too, he first made acquaintance with chemistry, to which he remained faithful, thanks to the influence of L. Gmelin, notwithstanding that he went through the medical curriculum at Marburg and Heidelberg. It was Gmelin, too, who recommended the young doctor of medicine to Berzelius, the latter receiving him with open arms. After barely a year's stay in Stockholm,—a year, however, rich in experiences and uneffaceable impressions, and of which he himself has given us such a clear picture[2]—Wöhler returned to Germany in the autumn of 1824, to become shortly afterwards a teacher in the Technical School (*Gewerbeschule*) at Berlin. In 1831 he had to leave the pleasant and stimulating society of his friends there (among whom we may mention Mitscherlich, the brothers Rose, Poggendorff and Magnus) to fill the post of professor in the newly-founded Higher Technical School at Cassel; while in 1836 he accepted a

[1] Cf. A. W. v. Hofmann's Memoir of Wöhler, *Ber.*, vol. xv. p. 3127 *et seq.*, and *Ber.*, vol. 23, *Ref.* p. 833.

[2] *Ber.*, vol. viii. p. 838 *et seq.*

call to Göttingen as successor to Stromeyer, where, till his death on 23rd September 1882, he remained a bright ornament of the *Georgia-Augusta* (the university of that town).

Wöhler's influence as a teacher, especially after his removal to Göttingen, may be described as enormous. Like his friend Liebig, he laid the greatest weight upon a thorough grounding in the rudiments of chemistry. The advantages which he had gained from his analytical work under Berzelius he now imparted to his pupils. Out of a long list of these, a few may be named who themselves subsequently continued to teach in the spirit of their master:—Th. Scherer, H. Kolbe, Henneberg, Knop, Städeler, Geuther, Limpricht, Fittig, Beilstein, Hübner and Zöller.

Wöhler was especially active in a literary sense during the earlier portion of his life, as is shown by his co-operation in the *Dictionary of Chemistry*, already mentioned, and his translations of the *Text-Book* and *Annual Reports* (*Jahresberichte*) of Berzelius. The first edition of his *Grundriss der anorganischen Chemie* ("Outlines of Inorganic Chemistry") occupying about 150 pages, appeared in 1831, the *Organic* following in 1840; both of these went through numerous editions.[1] His results in the investigation of minerals he collected together in 1853 in the valuable work, *Praktische Übungen in der chemischen Analyse* ("Practical Exercises in Chemical Analysis").[2] His experimental researches—most of which he published in the *Annalen der Chemie*, but some of the earlier ones in *Poggendorff's* and in *Gilbert's Annalen*—embrace almost every branch of inorganic chemistry. Some of them also led to the opening up of important branches of organic, *e.g.* his splendid work upon cyanic acid and its salts, the discovery of urea, and also the investigations carried on along with Liebig. In all of them, as

[1] From its sixth edition the *Organic Chemistry* has been admirably edited by Rudolf Fittig; the fourteenth and fifteenth (the last) editions of the *Inorganic* were given out by H. Kopp.

[2] The second edition appeared in 1861 under the title *Die Mineralanalyse in Beispielen* ("The Analysis of Minerals, illustrated by Examples").

also in his later labours, his remarkable gifts as an observer are apparent.

We cannot enter into details at this point either with regard to his work in analytical chemistry, which he enriched by admirable methods, or to that in inorganic. But a few investigations in the latter branch must just be mentioned, viz. those upon aluminium, boron, silicon and titanium, and their remarkable compounds, by which the resemblance between the two last-named elements and carbon was clearly brought to light.

The papers in which Wöhler describes the results of his experiments are written in a clear, forcible and simple manner, and attract our attention not merely by those characteristics—now-a-days somewhat rare,—but above all by the depth of their contents. That he had plenty of humour at command is proved by his letters to Liebig, and by the delicious satires [1] which he wrote when Dumas allowed himself to be carried too far by the deductions that he drew from the doctrine of substitution. Wöhler never rushed of his own accord into discussions upon important questions of theoretical chemistry,—a trait characteristic of his quiet disposition, and one which distinguishes him from Liebig, the born reformer, who looked upon this as a matter of duty.

As has been already said, the two investigators will remain inseparable in the history of chemistry. Liebig himself gives expression to this in one of his last letters to Wöhler, dated December 31st, 1871, in the following beautiful terms :—" Even after we are dead and our bodies long returned to dust, will the ties which united us in life keep our memory green, as an instance—not very frequent—of two men who wrought and strove in the same field without envy or ill-feeling, but who continued in the closest friendship throughout."

Jean Baptiste André Dumas,[2] who was born at Alais in 1800, and died at Cannes in 1884, rendered to his science

[1] *Ann. Chem.*, vol. xxxiii. p. 309 (see also *below*, p. 282); also the Liebig Berzelius *Letters*, p. 211, note.

[2] Cf. A. W. v. Hofmann's Memoir, *Ber.*, vol. xvii. *Ref.* p. 629 *et seq.*

extraordinary services, to which we shall frequently have occasion to refer. Beginning life as apprentice to an apothecary in his native town, he found this calling uncongenial, and set out on foot for Geneva in the autumn of 1816. Coming in contact there with such distinguished men as Pictet, Décandolle, de la Rive and others, he was stimulated to scientific researches which quickly attracted the attention of the *savants* just named. He made himself known particularly by the active part which he took in the physiologico-chemical investigations of Prévost. With the versatility which distinguished him, he soon began to take up problems in organic as well as in physical chemistry. In 1823, acting on A. v. Humboldt's advice, Dumas betook himself to Paris, finding there the most friendly reception at the hands of the eminent chemists of that city. At Paris he spent the rest of his life, filling various posts as a teacher and also other offices; he lectured with striking effect at the *Athenæum*, the *École Centrale des Arts et Manufactures*, the *Sorbonne*, and the polytechnic and medical schools.

No laboratory having been placed at his disposal, he established one at his own expense in 1832. After the year 1848 Dumas was frequently called into the public service, being for a long time minister of state, besides having to fill other offices, so that his work as a teacher was often interrupted. The keen interest which he felt in public affairs was shown in many cases by his active co-operation, *e.g.* in furnishing Paris with a water supply and in devising means to remedy the diseases of the silkworm and vine, etc. In 1868 he was further nominated permanent secretary of the Academy, of which he had long been a member.

We have still to make mention of the more important of Dumas' literary labours. The first of the larger works by which he became known was his *Traité de Chimie apliquée aux Arts* (1828); in its treatment of the matter, and especially its arrangement, this remained a model for many subsequent text-books on technology. The whole individuality of the man comes out in his *Leçons sur la Philosophie Chimique* (published in 1837 by Bineau from Dumas' lectures), in which he treats the development of chemical theories with great clearness and with a rare charm of style; this work, however, cannot be regarded as a strictly historical one. The numerous panegyrics which Dumas delivered are in their form, down to the minutest detail, carefully elaborated works of art; among them may be mentioned those upon Pélouze, Balard, Regnault and Faraday.

The *Essai de Statique Chimique des Êtres Organisés, par MM. Dumas et Boussingault* (1841), became especially well known; in this the life of plants and animals and, more particularly, the processes of metabolism, were treated from the chemical point of view. The opinions expressed here were in part instigated by the pioneering work of Liebig, whose influence however was not sufficiently recognised by the authors, so that he felt himself called upon to draw attention to his perfectly justifiable claims in very distinct language.[1] A debt of gratitude is due to Dumas for the pious service which he rendered in editing the reissue of Lavoisier's works.[2]

Most of the numerous experimental researches which we owe to Dumas were published by him in the *Annales de Chimie et de Physique,* of which he was one of the editors after 1840. In recalling his most important and productive labours, emphasis must be laid upon the great service which he rendered in working out various methods of general

[1] *Ann. Chem.*, vol. xli. p. 351. In this as well as in other instances Dumas unfortunately did not show in a favourable light. The historian is bound to notice such facts, since they cannot be erased from the scientific character of so eminent an investigator. Liebig criticised these peculiarities of Dumas with great severity (cf. *Ann. Chem.*, vol. ix. pp. 47, 129; also Kolbe's claim of priority, *Journ. pr. Chem.* (2), vol. xvi. p. 30; and the Berzelius-Liebig *Letters*, pp. 6, 7, 11, 34, 43, 45, 171, 238, etc.). Such occurrences are, to quote Liebig, "black leaves in the book of chemical history,—black, because they absorb the rays of light without thereby becoming luminous themselves." Dumas was unable to disprove or even to minimise the heavy charges which Liebig brought against him.

[2] Cf. p. 161, note.

application. His mode of determining vapour densities and that of estimating nitrogen have found universal appreciation. His admirable investigations in organic chemistry shed a brilliant light over wide branches of it, and guided many chemists for a time as to the direction in which they should work. Mention must be made too of his conjoint researches with Péligot[1] upon wood spirit and upon æthal (from spermaceti),—compounds whose analogy to alcohol he proved; and of his discovery and investigation of trichloracetic acid, which crowned the edifice of the substitution theory. The general character of his work naturally led Dumas to take an active share in the discussion of problems in theoretical chemistry. His rather unhappy participation in the question of the values of the atomic weights has been already noticed. The determinations which (partly in conjunction with Stas) he made of the atomic weights of carbon, oxygen and other elements deserve to be recorded as experimental work carried out with the utmost care and circumspection.

Apart from the shadow thrown upon Dumas' achievements by some of the incidents in his scientific life, his services will long continue to excite the highest admiration as evidences of a powerful and comprehensive mind. The immense influence which he exercised upon the form assumed by organic chemistry, and, in particular, upon the development of general views opposed to dualism, will be detailed in the following section.

The Development of Unitary Views in Organic Chemistry.—Substitution Theories.

At the time when Dumas brought forward his own as well as previous observations upon the substitution of hydrogen by chlorine and other elements as a basis for

[1] E. M. Péligot, born in 1811, was for a long time Professor of Chemistry at the *Conservatoire des Arts et Métiers*, and distinguished himself by much admirable work in inorganic, organic and technical chemistry (beet sugar industry); he died in April 1890. Cf. *Mon. Scient.* 1890, p. 885.

theoretical statements, the electro-chemical doctrine of Berzelius, and the radical theory which fitted in with it, were in high repute. The idea (deduced as it was from numerous facts) that electro-positive elements like hydrogen could be replaced by electro-negative ones like chlorine, oxygen and others, was bound to become a stumbling-block for the dualistic hypothesis, which could no longer after this be maintained in its integrity. The various attempts to explain the phenomena of substitution from general standpoints, which now fall to be detailed, were at the same time the significant utterances of a struggling unitarism against the binary view.

In connection with this, one has to recall to mind that, according to the position of Berzelius' dualistic doctrine at that time, the radicals were looked upon as unalterable atomic complexes. The consequence of the electro-chemical view was the assumption that negative elements like chlorine, bromine and oxygen could not enter into the composition of a radical. That the observations on the substitution of hydrogen atoms in organic compounds by those other elements was in direct contradiction to this assumption, appears to us now self-evident.

Dumas' Laws of Substitution.

Some isolated facts, which proved that a substitution of this kind could go on among the elements, were already known when Dumas turned his whole attention to the subject. Thus Gay-Lussac had established the formation of cyanogen chloride from hydrocyanic acid, Faraday that of sesquichloride of carbon (C_2Cl_6) from ethylene chloride, and Liebig and Wöhler the conversion of bitter almond oil into benzoyl chloride. It had not escaped these chemists that when the above compounds were subjected to the action of chlorine, an amount of hydrogen, equivalent to the chlorine which entered into them, was separated; indeed, the opinion was expressed (by some, if not all, of them) that the one element had replaced the other.

In the year 1834[1] Dumas, *à propos* of an investigation on the mutual action between chlorine and oil of turpentine, but more especially of his work upon the production of chloral from alcohol, condensed into two empirical rules the facts with regard to substitution, for which he proposed the designation *metalepsy* (*i.e.* exchange, μετάληψις). These were not intended to comprise a theory of substitution, as his first utterances on the subject show, but only to give expression to the facts. They were as follows:—

"When a compound containing hydrogen is exposed to the dehydrogenising action of chlorine, bromine, or iodine, it takes up an equal volume of chlorine, bromine, etc., for each atom of hydrogen that it loses.

"If the compound contains water, it loses the hydrogen of this without replacement."

The second of these rules was deduced from the transformation of alcohol into chloral, and was thus intended to explain the mode of formation of the latter, and at the same time to support Dumas' view of the constitution of alcohol, the latter being regarded by him as a compound of ethylene and water.

Dumas soon extended his statement to one of great significance, viz. that in chemical reactions generally an exchange of equivalents of one element for equivalents of others takes place. It was from this standpoint that he regarded the oxidation of alcohol to acetic acid, and that of bitter almond oil to benzoic acid, etc. etc., and he emphasised the point that each atom of hydrogen was here replaced by half an atom of oxygen. Those views, which gave evidence of great clearness of vision, were however obscured by certain additions which could not fail to create confusion with regard to the constitution of the compounds in question; thus, to give one instance only, formic acid was looked upon as a "metaleptic product" of alcohol, although such a relation could not be proved in this case.

[1] Cf. *Ann. Chim. Phys.* (2), vol. lvi. pp. 113, 140.

Laurent's Substitution or Nucleus Theory.

Dumas limited himself at that time (1835) to condensing the known facts into the two above-mentioned laws. But his countryman Laurent went further, in that he took into consideration the nature of the compounds produced by substitution, and compared them with the original ones. He was thus led to the proposition[1] that the structure and chemical character of organic compounds are not materially altered by the entrance of chlorine and the separation of hydrogen. This law, when taken in conjunction with the view that chlorine assumes the *rôle* of the substituted hydrogen, is the kernel of the Substitution Theory proper, of which Laurent must be regarded as the author; for Dumas denied at that time the analogy between substitution derivatives and the original compounds, and in reply to Berzelius, who attacked him for this assumption, threw the responsibility for it upon Laurent.[2]

The latter then strove to erect a system by developing the above doctrine, the result of his efforts being the so-called Nucleus Theory,[3] which was published in the year 1836; a short account of this must be given here, even although it never met with very hearty approval.[4] According to Laurent, organic compounds contained nuclei (*radicaux*), and he distinguished between original nuclei (*radicaux fondamentaux*), composed of carbon and hydrogen in simple atomic proportions, and derived nuclei (*radicaux dérivés*), which were produced from the first-named either through the substitution of hydrogen by other elements or by the taking up of additional atoms. He further stated that compound

[1] Laurent frequently enunciated this (cf. *Ann. Chim. Phys.* (2), vol. lx. p. 223; vol. lxi. p. 125; vol. lxvi. p. 326).

[2] *Comptes Rendus*, vol. vi. pp. 647, 695. Laurent stood up for his own view (*Ann. Chim. Phys.* (2), vol. lxvii. p. 303).

[3] Cf. *Ann. Chim. Phys.* (2), vol. lxi. p. 125.

[4] L. Gmelin did, it is true, make use of the subdivision of organic compounds, according to different nuclei, as a basis in his well-known textbook, and helped in this way to spread Laurent's views.

radicals like amidogen or nitroxyl might substitute in place of elements. This attempted classification of organic compounds, under the name of the neucleus theory, shows a distinct connection with the radical theory; but the one-sided view of the latter—that the radicals were unalterable—has here disappeared. While this change in principle marks an advance, the abandonment of the relation between organic and inorganic compounds was undoubtedly a great defect, since it involved the loss of a support indispensable for a natural classification of organic substances.

It was not difficult for the chief exponents of the radical doctrine to prove the insufficient basis of the nucleus theory, the more so that Laurent laid himself open to criticism not merely as a theoriser but also as an experimenter. His work was severely handled by Liebig, who came to the conclusion that Laurent's theory was unscientific and therefore pernicious. Berzelius likewise raised his voice energetically against it, and indeed went so far as to say that he considered a detailed criticism of it superfluous. But, as a matter of fact, Laurent was too much depreciated from this side; for, however much we may dissent from many of his untenable speculations, his effort to classify organic compounds on uniform principles, and to show their connection with one another, was not without merit. In addition to this he had effectively aided in overthrowing the dogma of the unchangeability of radicals. And, finally, we are indebted to him for the proof that Dumas' empirical rules of substitution are by no means always applicable.

Before Laurent, in conjunction with Gerhardt, had again brought forward his ideas in a more perfect form, Dumas[1] entered the lists to do battle against the radical theory, and, with this, against the dualistic idea in general. His beautiful discovery of "chloracetic acid" afforded him the immediate occasion for this, and he now gave in his adhesion to Laurent's opinions, which formerly he would have nothing to do with. The substituting atoms, *e.g.* the halogens, take up the *rôle* of the expelled hydrogen atoms, and the resulting

[1] *Ann. Chim. Phys.* (2), vol. lxxiii. p. 73 *et seq.*

halogen compounds must therefore show an analogy to the original ones,—this was for Dumas the clear result of his work upon trichloracetic acid; and he drew the same conclusion from the similar relations existing between aldehyde and chloral. To put his ideas into a more permanent form, he referred such related compounds to definite types, from which they were derivable.

Dumas' Type Theory (1839).

This effort, which reminds us strongly of Laurent's nucleus theory (since in this case also whole series of compounds were referred to fixed atomic complexes), bears in the history of chemistry the name of the Older Type Theory, to distinguish it from the newer one of Laurent and Gerhardt. Dumas was led to establish his theory of types[1] from the behaviour of trichloracetic acid, as observed by himself; he laid stress upon the fact that, in spite of the entrance of six atoms of chlorine in place of six atoms of hydrogen,[2] the character of this derivative remained essentially the same as that of acetic acid itself. Both compounds are monobasic acids, and both yield products of analogous composition with alkalies. From all this he concluded that "there are in organic chemistry certain types which remain unchanged, even when their hydrogen is replaced by an equal volume of chlorine, bromine, or iodine." Acetic and trichloracetic acids, aldehyde and chloral, marsh gas and chloroform, belong severally to the same chemical types. According to Dumas, one such type embraced compounds which contained the same number of equivalents combined in a like manner, and whose properties were in the main similar. We see here that the mutual relations of compounds belonging to one chemical type are the same as

[1] *Ann. Chim.*, vol. xxxiii. pp. 179 and 259; cf. also M. Berthelot's recent work, *Introduction à l'Étude de la Chimie des Anciens et du Moyen Age* (1889).

[2] Dumas assigned to acetic acid the formula $C_4H_8O_4$, and to (tri)chloracetic acid that of $C_4H_2Cl_6O_4$.

those which Laurent assumed between his original and derived nuclei.

But the term "chemical type" did not satisfy Dumas; he allowed it to merge into that of "mechanical type,"[1] this latter comprising all compounds which might be supposed to be formed from one another by substitution, even if they differed totally in properties. Acting on this idea, Dumas quite rightly classified alcohol and acetic acid under the same mechanical type; but, on the other hand, he brought together compounds which had no sort of connection with one another, *e.g.* formic acid and methyl ether. The ultimate result was that an empty scheme of formulation carried the day over what was really good in this doctrine—a doctrine developed from Laurent's nucleus theory. The endeavour to arrange organic compounds upon certain types outweighed and pushed aside the higher problems which Berzelius had sketched out for chemical science. The idea of definite atomic complexes or radicals, which was meant to pave the way for a knowledge of the chemical constitution of compounds, was superseded by the setting up of mechanical types, and thus the link intended to connect organic with inorganic compounds was completely snapped.

This total abandonment of the principles put forward by Berzelius, and found by him to be so serviceable, could not fail to arouse his liveliest opposition. Dumas had characterised Berzelius' electro-chemical doctrine as erroneous; the unitary conception was to step into the place of the dualistic which the latter theory involved. *Every chemical compound forms a complete whole, and cannot therefore consist of two parts. Its chemical character is dependent primarily upon the arrangement and number of the atoms, and in a lesser degree upon their chemical nature.* These propositions of Dumas stood in the sharpest opposition to the doctrine of Berzelius; they proclaimed a one-sided unitarism, which was therefore combated by Berzelius with every force at his command.

[1] Regnault had already (in 1838) spoken in a similar sense of *molecular types*, which remain unchanged in chemical reactions.

The Overthrow of Berzelius' Dualistic Doctrine.

Dumas did not scruple to say plainly that the dualistic doctrine was harmful and retarded the development of organic chemistry, and he made every effort to set it aside and to supplant it by the unitary theory. His attack upon Berzelius' doctrine (at that time held in high repute by most chemists) was vigorously answered both by the latter and by Liebig. Liebig[1] indeed admitted many points which were disputed by Berzelius, *e.g.* the fact of substitution, but he protested against Dumas' wide extension of this principle (of substitution). The assertion of the latter that every element of a compound might be replaced by another, and yet the type be retained, was characterised by Liebig as entirely unproven, and met with an ironical rejoinder.[2] Berzelius, who saw his whole system based upon the electrochemical theory threatened, directed his criticism in the *Jahresberichten* for 1838 and the next five years or so against the theory of types. In opposition to Dumas' unitary view he set up, as sharply as it was possible to do, the electrochemical and therefore dualistic theory as the fundamental principle; he adhered indeed essentially to his former standpoint, according to which electro-negative elements could in no case enter into the composition of radicals.

Berzelius sought to get over the difficulties which the substitution of hydrogen by chlorine and other elements involved, by arguing that compounds formed in this manner must have a constitution different from that of the original ones. But here he entered upon dangerous ground, and was thereby led, prudent investigator as he was, into the most utter contradictions of the principles which he had formerly held to be inviolable.

Berzelius first expressed himself upon the composition of acetic and trichloracetic acids. While the former (*i.e.* the

[1] *Ann. Chem.*, vol. xxxiii. p. 301.

[2] Cf. *Ann. Chem.*, vol. xxxiii. p. 308. The satirical letter given here was composed by Wöhler and published by Liebig.

anhydrous acid)[1] was regarded by him as the oxide of the radical acetyl, and given the formula $C_4\not{H}_3+O_3$, he looked upon trichloracetic acid as a so-called "copulated compound" or "conjugate compound" (*gepaarte Verbindung*[2]) of quite different constitution, viz. as a chloride of carbon copulated with oxalic acid, of the formula $C_2\not{Cl}_3+C_2O_3$.[3] But he could not at that time make up his mind to follow this to its logical conclusion, and to ascribe to acetic acid an analogous composition (*i.e.* to write it down as methyl copulated with oxalic acid), manifestly from the apprehension that he would in so doing surrender a principle of his electro-chemical doctrine. He attempted similarly to explain the constitution of other chlorine organic derivatives, by assuming copulæ (*Paarlinge*) containing chlorine, with the result that a different rational formula was assigned to the mother substance from that given to its derivatives.

These unfortunate attempts to explain by the speculative method the constitution of chemical compounds, that problem which, in his own opinion, was the most important in the science, led Berzelius completely astray. In order to carry through his doctrine of copulæ, he had to assume arbitrary radicals in organic compounds, without being able to adduce the least evidence in favour of such assumptions. Above all, he did not see what these really led to, for he overlooked the fact that his chlorinated copulæ could only be formed by the substitution of the hydrogen atoms of the radical by chlorine.

Melsens'[4] important observation, made in the year 1842,

[1] Berzelius formulated acetic acid as hydrate, $C_4\not{H}_3 \cdot O_3+\not{H}O$, *i.e.* as a compound of the anhydride (at that time unknown) with water.

[2] The idea that certain organic compounds are copulated or conjugated (*gepaart*) was definitely expressed for the first time in one of the earliest of Gerhardt's papers (*Ann. Chim. Phys.* (2), vol. lxxii. p. 184). In this paper he used the word copulation (*accouplement*) to signify the combination of organic substances with inorganic. The one portion of such compounds he termed the copula (*copule*), *e.g.* the organic substance which is copulated with an inorganic acid.

[3] For an explanation of these "crossed" symbols, see p. 236.

[4] *Ann. Chim. Phys.* (3), vol. x. p. 223.

that chloracetic acid is reconverted into acetic by the action of potassium amalgam, convinced Berzelius[1] that his view of the two acids having different constitutions was no longer tenable. He therefore decided to regard acetic acid in the same way as its chlorine derivative, *i.e.* as a copulated oxalic acid with the copula C_2H_3, formulating the two compounds thus—

$C_2H_3 + C_2O_3 . HO$ . . . Acetic acid.
$C_2Cl_3 + C_2O_3 . HO$ . . . Chloracetic acid.

But in doing this he made the important admission of the substitution of hydrogen by chlorine in the copula. And even although he did emphasise the point that the latter exercised no particular effect upon the compound to which it belonged, he none the less recognised hereby a fundamental principle of the doctrine of substitution.

But, notwithstanding this admission, Berzelius remained to the end of his life an opponent of the theory of types, and endeavoured to uphold the dualistic view by every means in his power. He had to undergo the pain, however, of finding his hitherto faithful adherents no longer able to follow him in this, and indeed of hearing them dissent publicly from his treatment of the question as to how the constitution of organic compounds was to be explained. Liebig, who had already before this taken the facts of substitution into account,[2] declared openly against Berzelius' far-fetched attempts at explanation,[3] the more so since the chlorine and bromine derivatives of aniline had been investigated in the Giessen laboratory by A. W. v. Hofmann, and had been accepted as evidence that the chemical character of a compound depends to a not inconsiderable extent upon the arrangement of its atoms. Liebig therefore turned himself to the unitary theory. The following words[4] show us the

[1] *Lehrb. d. Chemie* (fifth edition), vol. i. p. 709.

[2] *Ann. Chem.*, vol. xxxi. p. 119; vol. xxxii. p. 72.

[3] *Ibid.*, vol. l. p. 295 ("*Berzelius und die Probabilitätstheorien*"). The correspondence between Berzelius and Liebig, which has been so often referred to already, shows us in a truly dramatic way the gradual estrangement of the two men.

[4] *Ibid.*, vol. l. p. 297.

attitude taken up by Liebig, and we may be sure by others also, towards Berzelius at that time: "During the last years (of his life) Berzelius ceased to take an experimental share in the solution of the problems of the time, and turned the whole force of his mind to theoretical speculations; but these, not being the result of his own observations or supported by them, found no echo or approval in the science."

This much is certain, that, by carrying his speculations too far, Berzelius had not only shaken the edifice of his own doctrine, but had also greatly injured the radical theory, more particularly by heaping up one unproven hypothesis upon another. His opponents went so far as to assert that he had by his arbitrary assumptions "made a theory regarding substances which had no existence" in organic chemistry. It almost seemed as if his whole system was doomed to fall. One result of all this was that many chemists became visibly discouraged, and, holding all speculation as dangerous, either applied themselves to the empirical side of the science, or turned to other subjects. And yet, in spite of the slight regard in which the radical theory was held in many quarters, it soon became evident that, for the investigation of chemical constitution, the assumption of radicals, which had displaced the theory of types, was indispensable. In the course of the forties a fusion of the radical theory with the older doctrine of types took place on the unitary side; from the joint work of Laurent and Gerhardt there resulted the new theory of types. Upon the other side, at the same time, the much-derided copulæ were brought back to fresh life by H. Kolbe; with Frankland's aid a clearer notion of the meaning of copulated compounds was arrived at, and thus the way was smoothed for the establishment of the new radical theory and the doctrine of valency.

Fusion of the older Theory of Types with the Radical Theory by Laurent and Gerhardt.

Of the two investigators whose joint work effected a transformation of the old into the new theory of types, Laurent—as mentioned above—had been already active as the originator of the substitution theory proper. Although both of them were resolute opponents of the dualistic view, they had, nevertheless, no objection to make use of the conception of radicals, though to these latter they attached a meaning of their own. Besides Laurent and Gerhardt other chemists contributed materially to the establishment of the new theory of types, both by the ideas to which they gave expression and by the observations that they made. The stimulus thus given by Wurtz, Hofmann and Williamson thus falls to be recorded here also.

Laurent and Gerhardt exercised a strong mutual influence upon, and undoubtedly supplemented one another. Gerhardt was endowed with the special gift of bringing together isolated facts under one common point of view, and of drawing general conclusions therefrom. Laurent too was happy in perceiving the great importance involved in particular ideas, and he kept himself freer from prepossessions upon many points than his colleague.

A few sentences may be added here with regard to the lives of these two men. Auguste Laurent, born at La Folie near Langres in 1807, was initiated into chemistry by Dumas, thus acquiring a special knowledge of the organic part of it, to which with a certain one-sidedness he subsequently remained faithful. His work upon naphthalene and carbolic acid, together with their derivatives, is evidence of this. After filling various posts, the last of which was a chemical professorship at Bordeaux, Laurent became Warden of the Mint at Paris, where he remained in intimate connection with Gerhardt until his early death in 1853.

Charles Gerhardt was born at Strasburg in 1816, and began his scientific career well equipped with a wide general education; he studied chemistry at various places in Germany, finally under the stimulating guidance of Liebig, to whom he, like so many others, was so greatly indebted. After working for several years in Paris, he became Professor of Chemistry at Montpellier from 1844 to 1848, and after another prolonged residence in the first-named city (where he opened a school for chemistry, which however was not commercially a success), was called in 1855 to fill the chemical chair in the Faculty of Sciences at Strasburg, where he died in the following year. His important services in the development of organic chemistry, together with the joint theoretical views of Laurent and himself, are detailed below.

Gerhardt's Theory of Residues.

At the time that Gerhardt brought out his first scientific work, the fight between the radical and substitution theories was at its height. The latter found pronounced expression in Dumas' theory of types, and was opposed not merely to the dualistic views upon which the older radical theory was based, but to radicals in general. Gerhardt doubtless felt the disadvantages which the abandonment of the proximate constituents of organic compounds involved. Without forsaking the strict unitary standpoint of Dumas, he attempted to reintroduce the disdained radicals into chemistry under another name and with an altered meaning,—he set up the theory of residues (*théorie des résidus*).[1]

According to him, residues are atomic complexes which remain over from the interaction of two compounds, as the result of the stronger affinity of particular elements for one another, and which combine together because they are incapable of existing separately. Thus Gerhardt explained the formation of nitro-benzene from benzene and nitric acid, and, generally, the production of those bodies which he termed "copulated compounds" (*gepaarte Verbindungen*) in the following simple manner:—" When two substances react with

[1] *Ann. Chim. Phys.* (2) vol. lxxii. p. 184.

one another, an element (*e.g.* hydrogen) present in the one combines with another element (oxygen) present in the other to produce a stable compound (water), while the residues unite together." The latter he did not look upon as being actual atomic groups present in the compound in question, but as imaginary quantities; they were in his view absolutely distinct from the compounds of the same composition which were known in the free state, *e.g.* sulphurous acid (SO_2) or nitrogen peroxide (NO_2). Gerhardt gave expression to this difference by assuming the residues as being present in the "substitution-form." Further, the supposition of different residues in one and the same compound, according either to its mode of formation or decomposition, was also brought forward by him at that time.[1]

If we examine this conception of Gerhardt's more closely, we see that his views upon substitution are expressed in the same breath with those upon radicals as variable atomic complexes. He endeavoured, in fact, to explain the processes of substitution by the aid of this idea, in teaching that an eliminated element is replaced by an equivalent of another element or residue of the reacting substance.

Dumas and Laurent too had already said the same thing in a different way. But Gerhardt knew how to draw important conclusions from his theory with regard to the chemical nature of "copulated compounds"; it did not escape him that the saturation-capacities of the latter with respect to bases were quite different from those of the original acids before these had been "copulated" with an alcohol or a hydrocarbon. Thus nitro-benzene, an indifferent substance, was produced from nitric acid and benzene, and the monobasic ether-sulphuric acids from sulphuric acid and the alcohols. Gerhardt concluded from these and similar observations that "the basicity of a copulated compound is

[1] It must be mentioned here that the founders of the radical theory, Berzelius and Liebig, had expressed at one time (the former in 1834, and the latter in 1838) perfectly similar views as to the possibility of assuming different radicals in the same compound (cf. *Berzelius' Jahresbericht*, vol. xiv. p. 348; *Ann. Chem.*, vol. xxvi. p. 176).

equal to the sum of the basicities of the copulating substances minus 1." By means of this, his "Law of Basicity" (*Basizitätsgesetz*),[1] he was able to determine the chemical nature of acids about whose saturation-capacities doubt still prevailed at that time. With absolute definiteness he stated acetic acid to be monobasic, although it forms an acid sodium salt, and the same with regard to hydrochloric and nitric acids, because all these yield only neutral ethers; while sulphuric and oxalic acids were dibasic because, on copulation with an alcohol, they yield in the first instance monobasic ether-acids.

Gerhardt's first Classification of Organic Compounds.

Even before Gerhardt had attained to such clearness in this important question, he had directed his endeavours to the classification of organic compounds. His first attempt at this is contained in the *Précis de Chimie Organique* (1842). Here we find him strongly influenced by Dumas and his type theory; like the latter, he avoided the use of any formulæ which might appear to indicate the proximate or rational composition of chemical compounds. These he arranged in an ascending series according to their empirical formulæ, in such a manner that substances containing equal amounts of carbon constituted one group. Inclined to express himself in figurative language, he compared this classification of organic compounds to a ladder, whose lowest steps were formed of the substances of simplest, and whose highest of those of most complex composition. And since, from the oxidation of compounds rich in carbon, others which contain fewer atoms of that element are produced, he gave his arrangement the name of "combustion ladder" (*échelle de combustion*).

There was nothing of an unconstrained and natural classification here; on the contrary, the most diverse

[1] Cf. *Comptes Rendus*, vol. xvii. p. 312; *Comptes rendus des Travaux Chimiques par Laurent et Gerhardt* (1845), p. 161.

substances were collected into one class, provided only they fulfilled the necessary condition of containing the same number of carbon atoms. Not the slightest heed was paid to their chemical nature; acetic ether was placed alongside of butyric acid, and ethyl-oxalic acid alongside of succinic, solely for the reason given above. We note distinctly here the influence of Laurent, who not long before had made a mechanical classification of organic substances in a precisely similar manner (this, however, had made no impression).

Indeed, it is hardly conceivable to imagine how the older radical theory could have sustained a more severe blow than it did by the undue exaggeration of Dumas' theory of types. Gerhardt himself quickly felt this; his attempt at classification, which found its final and most definite expression in the new theory of types, showed distinctly that he had found a point of connection with the views of the radical theory, and that he strove to amalgamate the latter with the doctrine of substitution.

Before setting forth in detail these labours of Gerhardt, the efforts which he made—partly in conjunction with Laurent—to bring about uniformity of view with regard to the atomic weights of elements and compounds must be touched upon. The great and lasting service which those two men rendered in clearly defining what is meant by the term "molecule," and therewith reviving Avogadro's hypothesis, especially deserves our fullest recognition.

Gerhardt's "Equivalents."

At the beginning of the forties the uncertainty as to what atomic weights should be ascribed to the elements, and what atomic (*i.e.* molecular) weights to chemical compounds, had become one of great moment. The doubt which Gay-Lussac, Davy and others had previously urged against the assumption of definite atomic weights was again brought forward by Gmelin and his school. The atomic weight system of Berzelius, that work which he had accomplished after such immense labour, came very near to being given up, or at least greatly altered. In place of his atomic weights, based as they were upon solid foundations, "combining weights" were to be introduced, *i.e.* those values which were expressed by the simplest proportions of the substances entering into combination. All speculations upon relative atomic values were to be banished, and only the most sober possible formulation of chemical compounds attempted. The immediate result of this reaction was the halving of a large number of the atomic weights which Berzelius had introduced into the science. In place of the values assumed by him for carbon, oxygen, sulphur and most of the metals, other values only half as great were taken; these *equivalents* were: $C=6$, $O=8$, $S=16$, $Ca=20$, $Mg=12$, and so on.

Gerhardt began to oppose these equivalents in the year 1842, and was able to prove by cogent arguments that their assumption was inadmissible.[1] He showed, namely, that the amounts of water, carbonic acid, carbonic oxide and sulphuric acid, which were separated during the reactions of organic compounds, were never expressible by what was known as one equivalent, but by two or some multiple of two. The smallest equivalent formulæ for those compounds, according to Gmelin's view, were H_2O_2, C_2O_4, C_2O_2 and S_2O_4. But, argued Gerhardt, there must be an error underlying this: "the symbols H_2O_2 and C_2O_4 either express one equivalent, or they express two." If we assume the former of these, then the formulæ of the inorganic compounds must be doubled; if the latter, then the "organic formulæ" must be halved. Gerhardt did away with the contradiction which existed in the formulation of organic and inorganic compounds by reinstating Berzelius' atomic weights for the elements carbon, oxygen and sulphur, which were the ones of greatest moment here; *i.e.* taking $H=1$, then $C=12$, $O=16$, and $S=32$.[2] But he carried

[1] Cf. *Journ. pr. Chem.*, vol. xxvii. p. 439; also his *Précis de Chimie Organique*, vol. i. p. 49.

[2] Cf. *Journ. pr. Chem.*, vol. xxx. p. 1. It is very extraordinary that Gerhardt should have made no reference here to the identity of the atomic weights which he proposed with those of Berzelius.

this reform only half way; for, while he gave the proper values for the elements just named, he was led by special considerations to assume values for most of the metals which were only half as great as those proposed by Berzelius. Unlike the latter, who began by assuming that most of the metallic oxides had the composition indicated by the general formula MeO, Gerhardt compared these oxides with water, giving them therefore the formula Me_2O. He thus arrived at the correct atomic weights of the monovalent metals, but at incorrect ones for the divalent: *e.g.* for calcium the value 20 instead of 40, for lead that of 103·5 instead of 207, and so on.

Apart from this incompleteness there was an obscurity in Gerhardt's views with respect to atomic weights which could not fail to produce confusion; he both called the atomic weights just mentioned *equivalents*, and at the same time made use of this term for those amounts of chemical compounds which corresponded to their molecular weights, *i.e.* speaking generally, for quantities which are by no means necessarily equivalent chemically. Thus the quantities of hydrochloric, sulphuric and acetic acids represented by the formulæ HCl, H_2SO_4 and $C_2H_4O_2$, were in his mind equivalent to one another. We must here emphasise the point that Gerhardt attached another meaning to this word to what we now do; *equivalents* of chemical compounds were for him merely the comparable quantities of these.

Absolute clearness in the above points was only arrived at through Laurent's assistance. The latter definitely grasped the distinctions between molecular, atomic and equivalent weights, the correct determination of whose values constitutes the basis of our present views upon molecule and atom; it was he who brought Avogadro's hypothesis back to life again, and prepared the way for its development, so vitally important for chemical science.

The distinguishing between the terms Molecule, Atom, and Equivalent by Laurent and Gerhardt.

Gerhardt's most memorable efforts had for their aim the expression of the composition of all chemical compounds by means of formulæ based upon one common standard, *i.e.* formulæ comparable with one another. The formulæ of volatile compounds ought, according to him, invariably to indicate those quantities which occupy two volumes when in the gaseous state, taking the volume of one atom of hydrogen as equal to 1. This sound principle has ever since been fully recognised.

Acting upon this, he altered the *four-volume* formulæ of many organic compounds into *two-volume* ones by halving them. The false conception, much current at that time, according to which acetic acid (for example) received the formula $C_4H_8O_4$, alcohol that of $C_4H_{12}O_2$, and ethylene that of C_4H_8, had grown up as the result of the dualistic views upon the composition of organic compounds, and also of the use of several incorrect atomic weights.[1] It was precisely to organic compounds—most of them volatile without decomposition—that Gerhardt's law could be most extensively applied, the law, namely, that their formulæ depend upon the amounts by weight which are contained in equal volumes.

Much of Gerhardt's indistinctness, *e.g.* that produced by his using the word "equivalent" in a totally mistaken sense, was put right by Laurent. The latter pointed out with emphasis and clearness[2] that Gerhardt's equivalents were not even comparable with those of compounds, let alone of equal value with them; he showed that Gerhardt's equivalents of the elements must be regarded as their atomic weights, and the equivalents of compounds as their molecular

[1] For deducing the atomic composition of organic acids, the silver salts of the latter were chiefly made use of; for acetate of silver Berzelius had arrived at the formula $C_4H_6O_3.AgO$ ($Ag=216$), from which the composition of the acid, as given above, followed. Alcohol was regarded by Liebig as the hydrate of ethyl ether, and consequently formulated as $C_4H_{10}O \cdot H_2O$, whence the composition C_4H_8 was ascribed to ethylene, and so on.

[2] *Ann. Chim. Phys.* (3), vol. xviii. p. 266.

weights. Laurent's merit consisted in clearly grasping the meanings to be attached to these terms.

Laurent understood the molecular weight of an element or chemical compound as meaning that quantity which, under like conditions, occupies the same volume as two atoms of hydrogen; the quantity represented by the latter he looked upon as a molecule of hydrogen. For him, therefore, the molecular weights of chlorine, oxygen, nitrogen and cyanogen were expressed by the formulæ Cl_2, O_2, N_2 and $(CN)_2$, and the molecular weights of hydrochloric and acetic acids by the formulæ HCl and $C_2H_4O_2$, because the quantities indicated by those symbols filled, when in the state of vapour, the same space as two parts by weight of hydrogen. The agreement between his ideas and those of Avogadro is plainly evidenced here; but to Laurent belongs the further merit of developing these in a high degree. He defined the molecule as "the smallest quantity which can be employed in order to produce a compound." And he saw a proof of the correctness of this view in the fact that the atoms of chlorine, bromine, hydrogen, etc., always act chemically in pairs.

The atom, according to Laurent, is the smallest quantity of an element which can be present in a compound; for atomic weights he adopted the values proposed by Gerhardt, which agreed to a great extent with those of Berzelius. Equivalents, lastly, signified for him the "equivalent amounts of analogous substances" (*die gleichwerthigen Mengen analoger Körper*). This last definition led logically to the assumption that one and the same element has more than one equivalent, if it reacts with others in varying combining proportions.[1]

The joint work of Laurent and Gerhardt upon this

[1] "The idea of an equivalent includes in itself the view of a similar function; we know that one and the same element can fill the *rôle* of two or of several others, whence it must follow that different weights also correspond to those different functions. On the other hand, we find different weights of the same metal, *e.g.* iron, copper, mercury, etc., replacing the hydrogen of acids, and thus forming salts which contain the same metal but possess different properties. These metals have therefore various equivalents" (cf. *Comptes rendus des Travaux Chimiques par Laurent et Gerhardt* (1849), p. 1. *et seq.*).

question—so excessively important for theoretical chemistry—found very little acceptance amongst chemists; indeed, many of them actively opposed such a conception as that of variable equivalent values. The sound but not yet sufficiently grounded views of Laurent upon the magnitude of the molecules (*i.e.* molecular weights) of elements and compounds did not, however, succeed in making their way at that time—towards the end of the forties. Gmelin's combining weights were still for the most part adhered to, and at the date of the appearance of Gerhardt's *Lehrbuch der Chemie* (1853) were in such general use that the author, against his better judgment, used Gmelin's numbers for the chemical symbols in his first three volumes, *i.e.* he employed equivalent formulæ.[1] Stronger proof than that given by Laurent and Gerhardt had to be adduced in order to convince people that the atomic and molecular weights which they employed were the correct ones. It was the researches of Williamson, published at the beginning of the fifties, which were especially instrumental in leading to this. The true perception was again arrived at here from experience collected in the field of organic chemistry.

Influence of the Researches of Wurtz, Hofmann and Williamson upon the Development of the Theory of Types (1848–51).

The discovery by Wurtz[2] and Hofmann of organic derivatives of ammonia was of great importance for the firm

[1] Gerhardt gave his reasons for using this notation in the preface to his book (vol. i. pp. 1, 2) as follows: "*J'y ai même fait le sacrifice de ma notation, pour m'en tenir aux formules anciennes, afin de mieux démontrer par l'exemple, combien l'usage de ces dernières est irrational, et de laisser au temps le soin de consacrer une réforme que les chimistes n'ont pas encore généralment adoptée.*"

[2] C. A. Wurtz, who was born at Strassburg in 1817 and died at Paris in 1884, was a pupil of Liebig, Balard and Dumas; his life and works have been described very fully by A. W. v. Hofmann (*Ber.*, vol. xx. p. 815 *et seq.*), and by Friedel (*Notice sur la Vie et les Travaux de Wurtz*). From the year 1845 onwards, Wurtz filled the post of professor at various teaching institutions in Paris (including the *École de Médicine* and the *Sorbonne*), his influence becoming very great as time went on. The lucidity and general

establishment of the views finally comprised in Gerhardt's theory of types. In 1849 Wurtz observed the remarkable decomposition of cyanic ether by means of caustic potash, whereby he discovered methylamine and ethylamine, compounds so closely resembling ammonia.[1] Before this Berzelius had already expressed the view that the organic nitrogenous bases in general might be looked upon as substances which were copulated with ammonia. Liebig, on the other hand, regarded these as amido-compounds analogous to the ethers. Wurtz fluctuated between these two opinions, besides also suggesting the possibility of the organic bases being substitution products of ammonia, *e.g.* of "methyliak" (our methylamine) being ammonia in which one hydrogen atom was replaced by methyl. At first, however, he appears to have given the preference to the older view of Berzelius, according to which ethylamine, for example, was "ammonia copulated with etherin."

The "typical" view of these bases was first arrived at through A. W. v. Hofmann's brilliant researches upon amine bases; the production of these from ammonia and haloid compounds of the alkyls furnished a splendid proof of the correctness of the view that those compounds were formed from ammonia by the exchange of one or more hydrogen atoms for alcohol radicals.[2] The constitution of the *imide* and *nitrile* bases, like that of di- and tri-ethylamine, could

form of his lectures were such as to make it a pleasure to listen to them. From 1866 to 1875 he was Dean of the Medical Faculty, and in this position materially aided in raising the standard of instruction in practical chemistry and physiology for medical students. Among his writings were the *Léçons de Philosophie Chimique* (1864) and *La Théorie Atomique* (1879), works which treated of questions in theoretical chemistry and which found much acceptance because of their clearness and the charming style in which they were written; also his *Traité Élémentaire de Chimie Médicale* (1864), and the *Dictionnaire de Chimie Pure et Appliquée* (edited by him). His admirable experimental researches, by which he acted as a pioneer in opening up particular branches of organic chemistry, will be spoken of under the special history of the subject. Most of his work was published in the *Annales de Chimie et de Physique*, of which he became one of the editors in 1852, and in the *Comptes Rendus*.

[1] *Comptes Rendus*, vol. xxviii. p. 223 *et seq.*

[2] *Ann. Chem.*, vol. lxxiv. p. 174.

scarcely be explained in any other way than by their derivation from ammonia, through the substitution of hydrogen atoms by alkyl radicals. Before continuing this subject, a short account must be given here of Hofmann's life and work.

August Wilhelm von Hofmann, born at Giessen on 8th April 1818, after several years of philosophical and juristical studies devoted himself to chemistry under the guidance of Liebig, whose assistant he soon became. After filling for a short time the post of assistant-professor (*Privatdocent*) at Bonn, he accepted in 1845 a call (made at Prince Albert's instigation) to the newly founded College of Chemistry in London, which became a government institution in 1853; in 1855 he was also made a non-resident Assayer of the Mint (these appointments, which were held by eminent chemists, otherwise unconnected with the Mint, were abolished in 1870). He likewise taught at the School of Mines. In 1864 he removed back again to Bonn, and in 1865 was called to Berlin, as successor to Mitscherlich, where he continued to work with quite exceptional brilliancy until his death on May 5th, 1892.

Hofmann's work as a teacher was in every respect extraordinarily fruitful, the most striking proof of this being shown by the large number of his pupils who have since attained to eminence. His organising talent found scope in the building and fitting up of two admirable laboratories for general instruction at Bonn and Berlin. To success as a teacher there was also added, in a marked degree in his case, success as an author; here he showed the power of representing facts, and chemical doctrines founded upon them, in a delightfully clear and perspicuous manner. As an instance of this we may mention his *Einleitung in die moderne Chemie* ("Introduction to Modern Chemistry"). The Obituary Memoirs (upon Liebig, Wöhler, Dumas, Sella and Wurtz) by him are characterised by the loving care with which he enters into the life and works of the men whom he extols, besides being written in a most fascinating style.

As an investigator in experimental chemistry Hofmann meets us at every step; organic chemistry, more especially the field of nitrogen and phosphorus compounds, was thoroughly studied by him, and in part exhausted. Finally, reference must be made here to the wonderful influence which he exerted upon the coal tar colour industry, an industry which to a great extent arose out of his scientific studies. Most of Hofmann's papers were published in the *Annalen der Chemie* and in the *Berichte* of the German Chemical Society (at Berlin), which was founded by him in 1868, and of which he was for a long time President. In November, 1892, F. Tiemann delivered before the German Chemical Society a sympathetic and appreciative address upon his life and work (*Ber.*, vol. xxv. p. 3377).

It was only after these important investigations by Hofmann on the constitution of the organic ammonia bases that Wurtz clearly perceived that this relationship to ammonia was the only conclusive explanation of those compounds. He (Wurtz) condensed the result of the above researches into the words: "It was thus that the ammonia type was created."

To this Williamson,[1] by his distinguished researches,[2] added the water type, thereby with Wurtz and Hofmann providing the foundation for Gerhardt's theory of types. In his experiments Williamson started with the idea of replacing hydrogen in known alcohols by hydrocarbons, so as to obtain homologues of the former. The action of ethyl iodide upon potassium ethylate yielded him, however, ethyl ether, and not the expected ethylated alcohol. This result induced him to try whether, by the action of potassium ethylate upon methyl iodide, a mixture of ethyl and methyl ethers or only one homogeneous compound would be produced. The latter was found to be the case; methyl-ethyl oxide, a "mixed

[1] A. W. Williamson, born in 1824, was a pupil of Liebig and afterwards filled for a long time the chair of chemistry in University College, London, retiring from this post in 1887. Especially in the years between 1850-60 did he enrich organic chemistry with valuable observations, which led to deductions of general application. His work upon the formation and constitution of ethers, more particularly, was of the first importance.

[2] Cf. especially *Ann. Chem.*, vol. lxxvii. p. 37; vol. lxxxi. p. 73. Or *Journ. Chem. Soc.*, vol. iv. pp. 106 and 229.

ether," was obtained, and with this the much-discussed and at that time burning question of the molecular weights of ether and ethyl alcohol, and also that of the atomic weight of oxygen, were solved.[1] Liebig's idea that alcohol was the hydrate of ether had to be given up; on the other hand, Williamson's researches proved that the molecular formulæ of both compounds which had been assumed by Berzelius were the correct ones. The formation of ether by the interaction of alcohol and sulphuric acid, a process which had so greatly exercised the minds of the most eminent chemists, was thus now made perfectly clear by Williamson. Alcohol and ether he regarded as analogous to and built up on the type of water, as his definitions and formulæ show:—

$$\begin{matrix}H\\H\end{matrix}O\text{, Water;}\qquad \begin{matrix}C_2H_5\\H\end{matrix}O\text{, Alcohol;}\qquad \begin{matrix}C_2H_5\\C_2H_5\end{matrix}O\text{, Ether.}$$

This view (a view of which Laurent and other chemists had previously spoken favourably as being an admissible one) Williamson then proceeded to extend to many other substances, organic and inorganic, endeavouring at the same time to make its advantages evident. Thus he compared the acids, ketones (of whose true composition he had furnished beautiful experimental proof by a process similar to that mentioned above), and other compounds with water, *i.e.* he derived from water the compounds just named, by the substitution of one or both atoms of hydrogen by compound radicals or elements. The following examples will serve to illustrate his "typical" theory:—

$$\begin{matrix}C_2H_3O\\H\end{matrix}O\text{, Acetic acid;}\qquad \begin{matrix}K\\H\end{matrix}O\text{, Potassic hydrate;}\qquad \begin{matrix}NO_2\\H\end{matrix}O\text{, Nitric acid;}$$

$$\begin{matrix}C_2H_3O\\C_2H_3O\end{matrix}O\text{, Acetic anhydride (at that time unknown);}\qquad \begin{matrix}K\\K\end{matrix}O\text{, Potassic oxide;}\qquad \begin{matrix}NO_2\\K\end{matrix}O\text{, Nitrate of Potash.}$$

Williamson expressed himself as follows with regard to the applicability of the typical view:[2] "The method here

[1] Chancel arrived in a similar manner at the same result, independently of Williamson (cf. *Comptes Rendus*, vol. xxxi. p. 521).

[2] *Journ. Chem. Soc.*, vol. iv. p. 239.

employed of stating the rational constitution of bodies by comparison with water, seems to me to be susceptible of great extension; and I have no hesitation in saying that its introduction will be of service in simplifying our ideas, by establishing a uniform standard of comparison by which bodies may be judged of."

His confidence in the possibility of extending the "typical" idea came out still more strongly upon another occasion,[1] when he expressed the opinion that reference to the one type of water sufficed for all inorganic and for the best-known organic compounds; only that in the case of many substances, *e.g.* dibasic acids, the formula of water must be taken doubled. The views expressed here are also to be found for the most part in Gerhardt's theory of types. The most important result of Williamson's researches consisted, however, not in the one-sided typical mode of explaining the constitution of chemical compounds, but rather in the determination of the true molecular values of organic substances. The methods which he made use of in order to attain this end very soon proved themselves exceptionally productive; they led Gerhardt to the discovery of the acid anhydrides, and Wurtz to that of mixed hydrocarbon radicals, the investigation of both of which has finally settled the controversy as to the molecular formulæ of whole series of organic compounds.

Gerhardt's new Theory of Types.[2]

What has just been said is sufficient to show how effectively the "typical" view of organic compounds was furthered by the experimental researches of Wurtz, Hofmann and Williamson. Numerous nitrogenous compounds were referred to the ammonia type, and a still larger number of oxygenated ones to the water type. Gerhardt consummated

[1] *Journ. Chem. Soc.*, vol. iv., p. 350 (1851).

[2] Cf. *Ann. Chim. Phys.* (3), vol. xxxvii. p. 331; also *Traité de Chimie*, vol. iv. (1856).

his work by adding to these the hydrogen and hydrochloric acid types, and then he made the attempt to include all organic compounds under those few forms.

The endeavour to compare organic with inorganic bodies, which was already so strongly marked in the radical theory, was again distinctly apparent here; and here again it was ethyl compounds which mainly gave rise to the setting up of inorganic types as models for organic compounds. So early as 1846 Laurent[1] had thrown out the suggestion which was established in full detail by Williamson later on,—that alcohol and ether might be looked upon as derivatives of water, thus—

$$H_2O\text{, Water ;} \qquad \begin{matrix}Et\\H\end{matrix}O\text{, Alcohol ;} \qquad \begin{matrix}Et\\Et\end{matrix}O\text{, Ether.}^2$$

The inorganic acids and oxides too might be viewed (according to Laurent) as substitution-products of water. These compounds, so various in their natures, were regarded as built up after the same pattern.

In and after 1848 the American chemist Sterry Hunt published several papers,[3] in which he gave a wide extension to the typical view by showing how large numbers of oxygenated compounds, inorganic as well as organic, might be pictured as derived from water, and how hydrocarbons belong to the hydrogen type. But his work, being unknown in Europe, did not in any way quicken the growth of the similar ideas then running through many other minds. On the other hand, the above definite utterances of Williamson upon the reference of many organic compounds to water (as the form of compound of most general application) undoubtedly brought about a more rapid development of the doctrine of types. Not merely oxygenated bodies, but also non-oxygenated ones like amines, were without any hesitation taken as derived from water. Although Williamson thus lost his firm standing ground in consequence of the all too great elasticity of his formulæ, he gained, on the other

[1] *Ann. Chim. Phys.* (3), vol. xviii. p. 266 *et seq.*

[2] Cf. also Berzelius' view with regard to Ether, p. 256.

[3] *Amer. Journ. of Science* (2), vols. v., vi., vii. and viii.

hand, marked advantages from the extension of the type idea. He referred many compounds to the double or triple water type, and thereby introduced the notion of polyatomic radicals into chemistry. Sulphuric acid, for example, he referred to two molecules of water in which two atoms of hydrogen are replaced by sulphuryl (SO_2)—

$$\left.\begin{matrix} H \\ H \end{matrix}\right\} O \quad \left.\begin{matrix} H \\ H \end{matrix}\right\} O \text{ 2 Mol. water ;} \qquad \left.\begin{matrix} H \\ SO_2 \\ H \end{matrix}\right\} O_2 \text{ Sulphuric acid ;}$$

while phosphoric acid was derived in a like manner from three molecules of water, and so on.

Stimulated especially to it by his own important discovery of the anhydrides of monobasic organic acids,[1] Gerhardt collected the accumulated mass of "typical" ideas and condensed them into uniformity. Before everything else he desired to classify the large number of organic compounds in a synoptical manner, and for this the water, ammonia, hydrogen and hydrochloric acid types were to serve as models. In addition to this he made use of a principle for the co-ordination of organic substances, which had indeed been already applied by other chemists, but never in such a general manner, viz., he arranged them in different series, the members of each series belonging to the same type. His first classification of organic compounds (cf. p. 289) did not possess the advantages which such a grouping in series offered. Since then Schiel[2] had established the conception of homology by directing attention to the equal differences in the composition of analogous bodies, more particularly, of the alcohols, while Dumas had proved the same thing for the acids. And the researches of Kopp had further shown, with the utmost clearness, not only the chemical but also the physical resemblance of homologous compounds.

[1] *Ann. Chem.*, vol. lxxxii. p. 128. Those bodies, whose existence had been predicted by Williamson, were formerly supposed by Gerhardt to be incapable of preparation.

[2] *Ann. Chem.*, vol. xliii. p. 107 (1842).

Gerhardt now collated the results of those preparatory labours with great ingenuity, and associated with the series of homologous bodies, which differed in composition by the increment $(CH_2)_n$, other series of *isologous* and *heterologous* compounds. The former of these were, according to him, chemically analogous substances which show another composition-difference from homologous ones, *e.g.* ethyl alcohol, C_2H_6O, and phenol, C_6H_6O; propionic acid, $C_3H_6O_2$, and benzoic acid, $C_7H_6O_2$,—compounds which differ from one another by the increment C_4. Heterologous series contain such substances as are chemically dissimilar, but show a close connection with one another in their modes of formation. To such a series belong, for instance, ethyl alcohol, C_2H_6O, and acetic acid, $C_2H_4O_2$; amyl alcohol, $C_5H_{12}O$, and valeric acid, $C_5H_{10}O_2$.

As already mentioned, Gerhardt looked upon the members of such series as derivatives of one of his four types, resulting from these by the partial or complete substitution of their hydrogen atoms by *residues*. From the water type were derived (as Williamson had already taught) most of the organic compounds, including the alcohols, acids, simple and compound ethers, acid anhydrides, ketones, aldehydes and salts. Alongside of water the analogously constituted sulphuretted hydrogen was placed as an auxiliary type, and from it the sulphur compounds corresponding to the oxygen compounds just mentioned were derived, *e.g.* the sulphides, mercaptans, thio-acids, etc. The following examples will serve to illustrate what has just been said:—

H_2O	$\begin{matrix}CH_3\\H\end{matrix}O$	$\begin{matrix}C_2H_3O\\C_2H_3O\end{matrix}O$	$\begin{matrix}C_2H_3O\\C_2H_5\end{matrix}O$	$\begin{matrix}C_2H_3\\H\end{matrix}O$
Water	Methyl alcohol	Acetic anhydride	Acetic ether	Aldehyde.

Under the ammonia type were classified the amines, amides and imides, phosphines, arsines, etc., thus—

H_3N	$\begin{matrix}CH_3\\H_2\end{matrix}N$	$\begin{matrix}C_2H_3O\\H_2\end{matrix}N$	$\begin{matrix}C_4H_4O_2\\H\end{matrix}N$	$(C_2H_5)_3P$
Ammonia	Methylamine	Acetamide	Succinimide	Triethyl-phosphine.

The hydrogen type included the hydrocarbons, together

with the organo-metals; and the analogous hydrochloric acid type the chlorides, iodides, cyanides, etc., thus—

$$\begin{matrix} H \\ H \end{matrix} \quad \begin{matrix} CH_3 \\ H \end{matrix} \quad \begin{matrix} CH_3 \\ C_2H_5 \end{matrix} \quad \begin{matrix} C_2H_5 \\ Zn \end{matrix}; \qquad \begin{matrix} H \\ Cl \end{matrix} \quad \begin{matrix} CH_3 \\ Cl \end{matrix} \quad \begin{matrix} C_2H_5 \\ CN \end{matrix}.$$

Gerhardt was quite justified in terming this classification of organic compounds according to types a *système unitaire*, for all assumption of an opposite within the chemical compounds themselves, or of a binary structure, was here entirely eliminated. Each compound was looked upon as a complete whole; even in those cases where the dualistic conception appeared to be indicated (especially in that of salts), derivatives of water alone were seen.

The question now arises,—did Gerhardt himself believe that he would get nearer to the solution of that problem, which Berzelius had designated as being of supreme importance to chemistry, by setting up those types and referring organic compounds to them? Did he consider that he had thereby materially advanced the solution of the chemical constitution of organic bodies? The answer to this must be in the negative, if we mean "constitution" in Berzelius' sense. Gerhardt repeatedly expressed the opinion that it was impossible to arrive at the true constitution of these compounds, meaning by this the arrangement of their atoms (*l'arrangement des atomes*). In his view no strictly rational formulæ for organic compounds could be brought forward which would satisfy this demand, since several formulæ showing different proximate constituents or residues might be looked upon as equally correct, according to the modes of formation or decomposition of the compounds. Grounds of expediency alone must decide whether one formula was to be preferred to another; that formula which explained the larger number of methods of formation and decomposition of the particular compound in question was to be chosen. This elastic view was brought prominently forward by Gerhardt at every opportunity, especially in the fourth volume of his text-book, and he emphasised the point that the constitution of compounds, according to the type theory, was not the same thing as their rational composition in Berzelius' sense.

Formulæ were for Gerhardt merely pictures of the changes which chemical compounds underwent; they simply illustrated the modes of formation and decomposition of the latter. Types, on their part, even when their composition is exceedingly simple, "do not in any respect show how the atoms are grouped, but only the analogies of their metamorphoses. The type is the unit with which are compared all those compounds which show analogous decompositions, or which are the products of analogous decompositions."

After this exposition of Gerhardt's system in its main points, it will be intelligible why it has been spoken of as resulting from the fusion of the type theory of Dumas with the older radical theory. Gerhardt had made use of particular parts in both of these, and had recast them slightly for incorporation into his *système unitaire.* The idea that organic compounds are constructed on certain models, to which they can be referred, originated essentially in the older type doctrine, but, although hidden, it was also contained in the radical theory; in the latter, groups of organic substances had been directly compared with analogously constituted inorganic ones. Now it was of fundamental importance for the success of the new type theory that it borrowed from the radical theory the conception of atomic groups which behaved like simple substances; these groups could not, however, exist in the free state, as had formerly been supposed, but could only act in place of elements in compounds. This conception, coupled with that of the alterability (by substitution) of these atomic complexes, has proved to be absolutely correct, and at the same time of the greatest value. The question of the proximate composition of the above groups was left unanswered, and indeed untouched, by Gerhardt, the key to its solution being supplied from quite another quarter, *i.e.* by Kolbe and Frankland.

While the older type theory of Dumas ascribed no appreciable influence to the chemical nature of the constituents of a compound upon the character of the latter, Gerhardt showed his greater insight in this point also by recognising certain principles of Berzelius' school, even when he appeared mainly intent on opposing their spirit. He pointed out that the elements or atomic groups, which take the place of hydrogen in his types, determine according to their electro-chemical nature the nature of the resulting compounds. Thus he represented potash, $\begin{matrix}K\\H\end{matrix}O$, as a basic, and nitric acid, $\begin{matrix}NO_2\\H\end{matrix}O$, as an acid body, because the hydrogen of the neutral water was replaced respectively in these by an electro-positive and an electro-negative radical; but alcohol, $\begin{matrix}C_2H_5\\H\end{matrix}O$, as an almost neutral compound, ethyl being of pretty much the same nature as hydrogen itself. This return to views, which had formerly been combated so vigorously by that side, deserves to be especially noted.

The criticisms passed upon Gerhardt's type theory at that time varied very much. Many chemists, especially the younger ones, greeted it as an important conquest on the part of research. But, as a matter of fact, the favourable reception given to the typical view was due to grounds of a practical nature; men gave it as their opinion quite frankly, that the chief advantage which the reference of organic compounds to a few inorganic types brought with it, consisted in its thereby simplifying the study of organic chemistry. Liebig, who had criticised Gerhardt's earlier efforts at classifying organic compounds most severely,[1] acknowledged later on[2] the "utility of the so-called type theory"; but at the same time he laid stress upon the point that it left the weighty question of the formation of organic compounds untouched. Kolbe took up a more drastic attitude than this; he designated the grouping of organic compounds into the above four types a mere playing with formulæ. His own efforts he directed to replacing

[1] *Ann. Chem.*, vol. lvii. p. 93, *Herr Gerhardt und die organische Chemie.*
[2] *Ann. Chem.*, vol. cxxi. p. 163.

these purely formal types by real ones, which should stand in a natural connection to the compounds derived from them. Indeed, there was a serious danger that a door would be opened for empty formulation. We have only to recall that Odling and also Wurtz[1] endeavoured to simplify Gerhardt's types by referring those of water and ammonia to the double and triple hydrogen ones. With this the momentous question of the chemical constitution of organic compounds was diverted appreciably from the direction which had been given to it by the school of Berzelius and Liebig. The term "constitution," already very elastic in Gerhardt's theory, threatened to lose all meaning by formulation so exaggerated.

Extension of the Type Theory by Kekulé.

Gerhardt did not live to enjoy the cordial reception which was given by many chemists to the opinions laid down by him in the fourth volume of his text-book. His type theory underwent a not inconsiderable extension the year after his death (in 1857), by the assumption of the so-called *mixed types*, which aimed at making clear the relations of many organic compounds to two or more types. The more general application of this by Kekulé[2] was preceded by Williamson's idea that certain organic compounds might be derived from *multiplied* or *condensed types*. Just as chemical compounds proceeded from these through the substitution of several hydrogen atoms by polybasic radicals, so different types like water and ammonia, or water and hydrogen, etc., were conjoined in order to derive from them those substances which had previously been known as copulated compounds (*gepaarte Verbindungen*), to distinguish them from others which were readily classified under one type.[3] Kekulé recognised in the removal of this

[1] Cf. *Ann. Chim. Phys.* (3), vol. xliv. p. 305.

[2] *Ann. Chem.*, vol. civ., p. 129.

[3] The same idea which Kekulé generalised later on had indeed occurred to Gerhardt, in so far that he had referred the aminic acids (for example) to the mixed ammonia-water type.

barrier the main advantage which was to be derived from the assumption of mixed types, as is apparent from the following extract: "The so-called copulated compounds are not constituted differently from other chemical compounds; they can in like manner with these be referred to types in which hydrogen is replaced by radicals; and, in respect to formation and saturation-capacity, they follow the same laws which hold good for all chemical compounds."[1]

Before continuing the subject, a short account must be given of the career and labours of the famous chemist just named.—August Kekulé, born at Darmstadt on 7th September, 1829, became assistant professor of chemistry at Heidelberg in 1856, and then professor at Ghent from 1858 to 1865; in the latter year he was called to the University of Bonn, where he continued to labour until his death on the 13th of July, 1896. There could be no stronger testimony of his profound and wide-reaching influence as a teacher than in the large number of his pupils who have attained to eminence in their science. By his *Lehrbuch der organischen Chemie* ("Text-Book of Organic Chemistry," Erlangen, begun to be published in 1859), in which he endeavoured to work out the typical view and subsequently the structural doctrine to their logical conclusions, he exercised an immense influence upon the chemists of his time. More especially, by his happy conception of benzene (the basis of the "aromatic" hydrocarbons) as a hexamethine, he furnished the direction for one of the most important and wide-spreading branches of chemical research; and this still holds with undiminished force at the present moment. His researches on fulminate of mercury, unsaturated dibasic acids, and the condensation of aldehyde (to name only a few) proved him to be an admirable investigator. Mention may also be made here of his share in the editing of a former journal, the *Kritische Zeitschrift für Chemie*, etc., and of the present *Annalen der Chemie*, in the latter of which most of his experimental work was published. A warm tribute to his

[1] *Ann. Chem.*, vol. civ. p. 139.

memory, from the pen of H. Landolt, is to be found in the *Berichte*, vol. xxix. p. 1971; and the Kekulé memorial lecture by F. R. Japp, delivered before the Chemical Society on December 15th, 1897, is reprinted in the *Journ. Chem. Soc.* for February, 1898.

A few examples of formulæ will serve to make the use of the mixed types intelligible:—

$$\left.\begin{matrix} C_6H_5 \\ SO_2 \\ H \end{matrix}\right\} O, \text{ Benzene-sulphonic acid, referred to } \begin{matrix} H \\ \underline{H} \\ H \\ H \end{matrix}\Big\} O;$$

$$\left.\begin{matrix} H_2N \\ CO \\ H \end{matrix}\right\} O, \text{ Carbamic acid, referred to } \begin{matrix} \underline{H_3N} \\ H \\ H \end{matrix}\Big\} O.$$

Almost simultaneously with the above extension of the type theory, a suggestion was made by Kekulé which, thanks to special circumstances, was destined to give this theory a far more extended application. *À propos* of his researches upon fulminate of mercury,[1] he had expressed the opinion that the methyl compounds and the numerous bodies derived from them might be referred to the type of marsh gas, to which he gave the equivalent formula C_2H_4. He illustrated the connection of several compounds to the new type by the following examples:—

$$\underbrace{C_2H_4}_{\text{Methyl hydride}} \quad \underbrace{C_2H_3Cl}_{\text{Methyl chloride}} \quad \underbrace{C_2HCl_3}_{\text{Chloroform}} \quad \underbrace{C_2H_3CN}_{\text{Aceto-nitrile}} \quad \underbrace{C_2Cl_3(NO_4)}_{\text{Chloro-picrin.}}$$

Kekulé's formulation here is noteworthy, in that he uses atomic weights which he had formerly regarded as incorrect, *i.e.* H = 1, C = 6, and O = 8. And a remark that he makes strikes one as strange—viz. that the new type was not to be taken in the sense of Gerhardt's unitary theory, but in that of Dumas' types. From this one might infer that marsh gas was not intended to be placed alongside of Gerhardt's four types; but, notwithstanding this, to give it a place by itself does not seem to have been meant by Kekulé, since he adds, quite in the spirit of the newer type

[1] *Ann. Chem.*, vol. ci. p. 200.

theory, that what he mainly wishes to indicate by his formulæ are the relations in which the compounds enumerated stand to one another.

In the following year (1858), however, the meaning which he attached to methane as the mother substance of a large number of compounds became more clear. But a detailed account of his views upon this must be reserved for a later section of the book, when the transition of the type theory into the structure theory will come to be discussed.

Before, however, this development of chemical hypotheses could be consummated, much work had to be done in order to get nearer to a knowledge of the chemical constitution of organic compounds. The types themselves could not aid in the solution of this problem without their own nature being first elucidated. The key to the explanation of these relations was forged by the labours and speculations of Frankland and Kolbe. To these two investigators is primarily due the more profound insight into the constitution of organic substances as opposed to the typical and therefore superficial view (*der typisch schematischen*). Their researches contributed more than any others to bring about the change in direction taken by the type theory; they were, in fact, the indispensable preliminary to that transformation of theoretical opinions which completed itself towards the end of the fifties. The correctness of this statement will be seen from what follows in the succeeding sections.

It is true that the typists place quite another estimate upon the services of Frankland and Kolbe. The influence exercised by these two men on the remodelling of the type theory has not only been greatly minimised, but even the exact contrary has been asserted, viz. that "typical" hypotheses influenced them.[1]

[1] Such erroneous conceptions are always long of being dispelled. Thus, in the description of "the theories of to-day" in Wurtz's *Histoire des Doctrines Chimiques*, the influence of the above two scientists is very much neglected. It seems hardly credible that Frankland, the real originator of the doctrine of valency, should scarcely be mentioned in this publication. The same applies to the general section of Kekulé's *Lehrbuch der organischen*

Development of the Newer Radical Theory by Kolbe—
A Survey of his Principal Work.

Before speaking of Kolbe's scientific labours, which produced a deep and lasting effect on the development of theoretical chemistry, a short sketch of his life may be fitly appended here.[1]

Hermann Kolbe, son of the Pastor of Elliehausen near Göttingen, was born in 1818, and applied himself to the study of chemistry under Wöhler's stimulating guidance in 1838. The results of his first research were published in 1842, and for the next forty-two years he continued to enrich his science with a long succession of the most valuable experimental and theoretical work. His outward life, if we except perhaps the first few years immediately following his university curriculum, was that of a German scientist. From 1842–47 he was assistant to Bunsen at Marburg and then to Playfair in London, during which time he occupied himself mainly with practical chemical work; after this came the years of his literary apprenticeship (1847–51) in Brunswick, where he had gone at the request of the well-known publishers, Fr. Vieweg and Son to take up the editorship of the *Dictionary of Chemistry* started by Liebig. This work not being of such a nature as to satisfy him permanently, he willingly accepted in 1851 a call to Marburg, where, as Bunsen's successor, he developed exceptional powers as a teacher, especially in the years following 1858. In 1865 he was called to the University of Leipzig, and worked there with marked success until his death on 25th November 1884.

Chemie; there the debt due to Frankland is absolutely ignored, while the share in the development of organic chemistry taken by Dumas, Gerhardt, Laurent and Kekulé himself is minutely detailed. At a later date (cf. *Ber.* for 1880, p. 7) Wurtz unreservedly acknowledged Frankland's service by stating that he was the first to put forward the idea of the saturation-capacity of elementary atoms.

[1] Cf. the memoirs which appeared shortly after Kolbe's death by E. v. Meyer, *Journ. pr. Chem.* (2), vol. xxx. p. 417; Voit, *Bayer. Acad.*, 1885; and A. W. v. Hofmann, *Ber.*, vol. xvii. p. 2809.

The great influence which Kolbe exercised upon chemical science depended to an unusual degree upon his experimental work, which will be discussed later on, but at the same time also upon his eminence as a teacher, in which respect he may be spoken of along with Liebig. His method of teaching was very like that of the latter and had the best results; the student of practical chemistry was taught to observe and think for himself. Kolbe's gifts as a teacher were greatly enhanced by his sound common-sense and organising talent, which showed themselves in a marked degree in the building and fitting up of the new Leipzig laboratory (in 1868).

In addition to his work as a teacher, based as this was upon oral instruction, Kolbe was also extremely active in a literary sense. Apart from his numerous scientific papers, valuable articles for the *Handwörterbuch der Chemie* ("Dictionary of Chemistry"), and occasional pamphlets, he published a large *Lehrbuch der organischen Chemie* ("Text-Book of Organic Chemistry," Braunschweig, 1854–65), and smaller text-books both of inorganic and organic chemistry (1877–83). These books are distinguished by clearness in arrangement, precision of expression, a delightful style, and perspicacity and acuteness in discussion.

In his writings upon questions of theoretical chemistry, published for the last fourteen years of his life in the *Journal für praktische Chemie* (of which he succeeded Erdmann as editor in 1870), Kolbe gave play to a keen criticism, which became intensified as time went on, upon the defects and extravagances which he considered were due to the direction taken by modern chemistry. If those critiques were often strongly polemical and did not altogether avoid the danger of introducing the personality of many a brother chemist, still his only aim in them was the welfare of his beloved science, which he believed to be in serious danger. His efforts at exposing error were often wrongly interpreted by many of his contemporaries, just as Liebig's polemical writings were often perversely criticised.

The Re-animation of the Radical Theory by Kolbe—Frankland's Co-operation.

At the time when Kolbe published the first of his more important researches,[1] the doctrine advocated by Berzelius, that organic compounds contain definite radicals which act similarly to elements in inorganic compounds, had been driven into the background by the attack of unitarism. Many chemists were of opinion that the partly arbitrary supposition of hypothetical radicals could not advance the science any further. The assumption of copulæ (*Paarlinge*) in the so-called copulated compounds satisfied very few. In short, the old radical theory in its original form was held to be no longer capable of existence. The preference given by the school of Gmelin to the simplest views which were possible is sufficient evidence of this sense of discouragement. Facts alone were to decide; any intelligent grouping of these facts together was deemed useless.

Kolbe now united the conclusions deduced from his first researches with the declining theory of Berzelius; he endued the latter with new life by casting aside whatever of it was dead and replacing this by vigorous principles. From his own and other investigations he came to the conclusion that the unalterability of radicals, as taught by Berzelius, could no longer be maintained, since the facts of substitution had to be taken into account. He did, indeed, adopt Berzelius' hypothesis of copulæ, but attached another meaning to these, since he allowed that they exercised a not inconsiderable influence upon the compounds with which they were copulated.[2]

If we desire to sum up the main results of his labours just cited, and of his synthesis of trichloracetic acid, so immediately connected with them, we may do so as follows: Trichloro-methyl-hyposulphuric acid (our present trichloro-methyl-sulphonic acid), discovered by him, and trichlor-

[1] *Ann. Chem.*, vol. xlv. p. 41; vol. liv. p. 145.
[2] Cf. *Ann. Chem.*, vol. liv. p. 156.

acetic acid, together with the compounds free from chlorine obtained from these by reduction, were analogously constituted acids, copulated respectively with trichloro-methyl and methyl, thus—

$$C_2Cl_3 + S_2O_5 + HO \qquad C_2H_3 + S_2O_5 + HO$$
$$C_2Cl_3 + C_2O_3 + HO \qquad C_2H_3 + C_2O_3 + HO$$

True, the mode in which these two radicals were combined with the acids was not yet known, but the germ of the correct explanation with regard to the constitution of carboxylic and sulphonic acids, which was given by Kolbe at a later date, was already present here.

This germ was soon to undergo further development by investigations carried out at first by Kolbe alone, and afterwards together with Frankland in London. From their beautiful researches on the transformation of the alkyl cyanides into fatty acids,[1] they concluded with perfect precision that methyl, ethyl, and similar radicals were immediate constituents of acetic acid and its homologues. Kolbe himself was led to the same conclusion by his important work upon the electrolysis of salts of the fatty acids;[2] he saw in the methyl and butyl, separated at the positive pole from acetic and valerianic acids respectively, the proof of the correctness of this assumption. He believed, indeed, that he had isolated the radicals themselves; and even although he was wrong in so thinking (the hydrocarbons obtained by him having double the molecular weight that the radicals would possess), this affected but little the question of the constitution of the carboxylic acids. The chief goal of his endeavours, *i.e.* the discovery of the true composition of the above and similar acids, was still kept in view by him, notwithstanding this mistake.

The outcome of this work of his was that the view previously held with regard to these organic acids no longer satisfied him. He did not, however, abandon this all at once, but rather developed from it a theory which approxi-

[1] *Ann. Chem.*, vol. lxv. p. 288.
[2] *Ibid.*, vol. lxix. p. 258.

mated to the truth, and which soon showed itself capable of further improvement. Even so early as when writing the articles upon *Formulæ* and *Copulated Compounds* for his Dictionary (in 1848), he expressed and gave reasons for the view that the fatty acids were oxygen compounds of the radicals hydrogen, methyl, ethyl, &c., combined with the double carbon equivalent C_2.[1]

Acetic acid contained as its immediate constituent an atomic complex constituted similarly to that of the cacodyl compounds. *Cacodyl* itself, which was here for the first time interpreted as being arsenic copulated with two methyl radicals, corresponded to the so-called *acetyl* of acetic acid, *i.e.* $C_2H_3C_2$ (not to be confounded with the radical acetyl of to-day, which at that time was known as *acetoxyl*).

Even at this early date Kolbe expressed the significant opinion that in the acetyl ($C_2H_3C_2$) of acetic acid, "the last C_2 alone forms the connecting-link for the oxygen, the methyl being in some sort only an appendage." This idea, which recalls Berzelius' doctrine of copulæ, was based upon the point that it was unessential for the nature of the acids whether hydrogen or methyl, ethyl, etc., was copulated with the C_2.

He entered into these important ideas in detail in a treatise entitled, *Ueber die chemische Konstitution und Natur der organischen Radikale* ("Upon the Chemical Constitution and Nature of the Organic Radicals").[2] Taking his stand upon the basis of the older radical doctrine, he built this up into a living theory by eliminating from it all those principles which stood in contradiction to the facts. But at the same time he did not remain stationary upon the point of vantage he had thus gained.

Under the influence of the admirable researches of

[1] Kolbe, like many others, made use at this time of Gmelin's equivalent weights, in which H=1, C=6, O=8, S=16, etc. His formulæ were, notwithstanding this, molecular formulæ; thus he gave carbonic acid, acetic acid, alcohol, aldehyde and acetone the same atomic (*i.e.* molecular) weights as we employ for these substances to-day.

[2] *Ann. Chem.*, vol. lxxv. p. 211; vol. lxxvi. p. 1.

Frankland [1] upon the alcohol radicals and the organo-metallic compounds, which were begun at that time, Kolbe advanced step by step. With regard to this period, he stated definitely himself,[2] that "the want of clearness in (my) conception of the mode in which the so-called copulæ were combined, was a great weakness in the hypothesis of copulated radicals. . . . It is Frankland's merit to have been the first to throw light upon this, and therewith to have thereby completely done away with the idea of copulation, by recognising the fact that the various elements possess definite saturation-capacities."

Kolbe readily embraced his friend's views, and copulæ thus received a totally different meaning from what they had formerly done; henceforth they were to be regarded as integral parts of organic compounds and not as mere appendages.

This change in his opinions was not long of bearing fruit. And here it was again the fatty acids whose constitution he undertook to work out. In 1855 [3] he first gave definite expression to the view that the acids, considered as anhydrous, were derivatives of carbonic acid;

[1] Sir Edward Frankland, born at Churchtown, near Lancaster, on January 18th, 1825, studied chemistry at the Museum of Practical Geology in London, then with Liebig and Bunsen, and also under Kolbe's stimulus in Germany. He afterwards filled successively the chairs of chemistry in the Owens College, Manchester (1851–7), St. Bartholomew's Hospital, London (1857), Royal Institution (1863), Royal School of Mines (1865), and Normal School of Science, South Kensington (1881). This last chair he resigned in 1885, retiring then from professorial work. Frankland attracted the attention of chemists even by his earliest work, which led him to the discovery of the organo-metals, and also by his joint researches with Kolbe. The chief share which he took in the development of our present views upon the valency of the elements will be discussed in detail later on, while his other memorable investigations in organic chemistry will often have to be referred to under the special history of this branch. Frankland's papers have mostly been published in the English journals and the *Annalen der Chemie*; in 1877 they were collected into one volume, entitled *Researches in Pure, Applied and Physical Chemistry*. He is also the author of the text-book, *Lecture Notes for Chemical Students*.

[2] Cf. *Das chem. Laboratorium der Universität Marburg, etc.* (Braunschweig, 1865), p. 32.

[3] *Handwörterbuch der Chemie*, vol. vi. p. 802.

for instance, acetic was methyl-carbonic acid, *i.e.* C_2O_4, in which one oxygen-equivalent was replaced by methyl, C_2H_3. The hydrated acids he still regarded dualistically as compounds of the anhydrides with water.

The assumption that those acids were substitution-products of carbonic acid had developed itself from the views held regarding the organo-metallic compounds. Just as Frankland explained cacodylic acid as arsenic acid with two methyls in the place of two equivalents of oxygen, and stanno-ethyl oxide as the corresponding tin derivative, so did Kolbe happily interpret the constitution of other organic compounds. He soon advanced beyond the field of the organic acids, and developed the idea, similar to that mentioned above, that many organic substances are to be regarded as derivatives of carbonic acid, and many others as derivatives of sulphuric. How this idea expanded into a perfect whole is seen from his writings in the years 1857–58,[1] and also from those portions of his text-book which were written both at that time and shortly before it. These theoretical considerations and, with them, the revived radical theory attained to their completed form in a treatise published in 1859, entitled, *Ueber den natürlichen Zusammenhang der organischen mit den unorganischen Verbindungen, die wissenschaftliche Grundlage zu einer naturgemässen Klassifikation der organischen chemischen Körper* ("Upon the Natural Connection existing between Organic and Inorganic Compounds, being the Scientific Basis of a Rational Classification of Organic Chemical Substances").[2]

The main outcome of Kolbe's speculations is given in the following sentence: "Organic compounds are all derivatives of inorganic, and result from the latter—in some cases directly—by wonderfully simple substitution-processes." This idea runs through the whole treatise, and is illustrated

[1] *Ann. Chem.*, vol. ci. p. 257; this paper is a joint one with Frankland, *i.e.* Kolbe lays emphasis on the point that he is here giving utterance both to his own and Frankland's views. Cf. also Kolbe's pamphlet (1858), *Ueber die chemische Konstitution organischer Verbindungen* ("On the Chemical Constitution of Organic Compounds").

[2] *Ann. Chem.*, vol. cxiii. p. 293.

with the most convincing clearness by numerous examples taken from the wide field of organic chemistry.

The alcohols, carboxylic acids, ketones and aldehydes were derived, according to Kolbe, from carbonic acid, $(C_2O_2)O_2$, and its hydrate, $C_2O_2{}^{OHO}_{OHO}$, respectively. The polybasic carboxylic acids proceeded in the same way from two or three molecules of the hydrated carbonic acid, through the entrance of polyatomic radicals, just as the monobasic did from one molecule. Similar definite views were expressed by Kolbe with regard to other classes of organic compounds, *e.g.* the phosphinic and arsenic acids, amines and amides, and the organo-metals, which he derived in the simplest manner from inorganic compounds. He laid the utmost emphasis upon his formulæ being the unambiguous expression of precise opinions; with Gerhardt's assumption that various constitutional formulæ might, with equal justice, be set up for one and the same compound, he had absolutely nothing in common.

Kolbe himself gave a striking proof in the treatise above mentioned of the capacity for development of his views respecting the constitution of organic compounds. He comprised in his survey not merely those classes of compounds which were known, but advanced beyond them to others at that time unknown. From the relations so clearly recognised by him as existing between the alcohols and the carboxylic acids, he deduced the possibility of preparing new varieties of alcohols; he predicted the existence both of secondary and of tertiary alcohols,[1] and even went so far as to indicate a probable method for preparing and decomposing the first of these. No such brilliant deductive treatment of chemical questions had as yet been seen in organic chemistry. And the discovery of those classes of compounds which he had prognosticated had not to be waited for long; Friedel isolated secondary propyl alcohol in 1862, and Butlerow tertiary butyl alcohol in 1864.

The comprehensive speculations of Kolbe upon the

[1] Cf. *Ann. Chem.*, vol. cxiii. p. 307.

constitution of organic compounds could not have attained to the firm hold and the wide significance which they did, had they not been conjoined throughout with admirable experimental work. We shall frequently have occasion, in the special history of organic chemistry, to refer to those labours, through which the rational composition of important classes of compounds was first arrived at with certainty. Thus it was his researches upon lactic acid which showed it to be oxy-propionic, and the corresponding alanin to be amido-propionic acid. Glycollic acid and glycocoll were likewise shown by Kolbe to belong to the same class, the one being proved to be oxy-, and the other amido-acetic acid; he also recognised salicylic acid as oxybenzoic, and the so-called benzamic acid (*Benzaminsäure*) as amido-benzoic. He was thus in a position to clear up the constitution of compounds upon whose investigation chemists of such eminence as Kekulé and Wurtz had laboured in vain. Numerous substances, the names (*Trivialbezeichnungen*) given to which showed how little was known with respect to their constitution, received from Kolbe their proper place among other compounds. The conversion of malic and tartaric acids into succinic, which was carried out by Schmitt[1] at his suggestion, revealed at one stroke the hitherto unknown relations existing between the two first of these acids and the last. By his researches upon taurine, which he taught how to prepare artificially, he proved how both it and the isethionic acid produced from it were constituted analogously to alanin and lactic acid. And the same clearness shed itself over the rational composition of asparagine and aspartic acid, which he was the first to interpret correctly.

The above are merely the results of work performed within a short period of time, but they are amply sufficient to prove what undying service he rendered in investigating the chemical constitution of organic compounds. And no

[1] Rudolf Schmitt, born in 1830, filled the chair of chemistry at the Dresden *Technische Hochschule* from 1871 to 1893, having previously occupied other chemical posts at Marburg, Cassel and Nürnberg. His admirable experimental researches extend over many branches of organic chemistry, but deal more especially with the aromatic compounds.

mention has been made here of a large number of other researches carried out at his suggestion and with his co-operation; among these were the work of Griess upon the class of diazo-compounds, Oefele's discovery of the sulphines, and Volhard's synthesis of sarcosine.

In order to round off in some degree this short record of Kolbe's achievements, we ought further to recall several investigations made in the years following, *i.e.* after 1863, in which he was guided throughout by the aspiration to gain the furthest possible insight into the constitution of organic compounds. Among these we may refer to his proof of malonic acid resulting from cyan-acetic, and being therefore carboxyl-acetic acid, the discovery of nitro-methane, the series of memorable researches upon salicylic and para-oxybenzoic acids, and lastly, that upon isatoic acid, which was cut short by his death.

Kolbe's Attitude towards the older and the newer Chemistry.

In all Kolbe's investigations, whether speculative or experimental, we feel the salutary historic method by which they are characterised. He built upon the edifice already existing, and remained in his scientific efforts in continuity with the chiefs of the classical school. He was always glad to acknowledge that his success as a chemist was due primarily to Berzelius, and, after him, "to the great exemplars Liebig, Wöhler and Bunsen, who, to use a phrase of Berzelius, were true workers in chemistry" (*wahre Bearbeiter der Chemie gewesen sind*).

The criticisms passed upon Kolbe by his contemporaries, in so far as regarded his attitude to organic chemistry, differed very greatly. The exponents of the earlier period appreciated his services better than the disciples of the type theory,—a theory which he himself did not value at its true worth. A few remarks upon the relation between Kolbe's views and those of the typists will be in place here. As already stated, he spoke of the type theory as being unscientific; he saw in it not a real theory but merely a play upon formulæ. In spite of his definite utterances upon this point, however, it has been frequently asserted that he took Gerhardt's doctrine of types as his basis, and that therefore his derivation of organic compounds from carbonic acid, carbonic oxide, sulphuric acid, sulphurous acid, etc., coincided with that from the three types of hydrogen, water and ammonia. Kolbe did indeed connect organic with inorganic compounds, but he repeatedly emphasised the point[1] that these latter were real types, as opposed to the formal ones of the type theorists. His most ardent wish was to fathom the chemical constitution of organic compounds; but to merely classify the latter upon certain models or to go so far as to force them into arbitrary types, was in the highest degree distasteful to him. Kolbe attached special weight to the relations actually existing between organic and inorganic bodies, whence the emphasis laid in the title of his treatise, spoken of above, upon the "natural connection between these as forming a scientific basis for a rational classification of organic substances." Hence, also, his attempts, begun at an early date, to prepare organic compounds artificially from simple inorganic ones, with the object of thus gaining an insight into their chemical constitution.

We thus see Kolbe pursuing his own way, not led aside by the criticisms of his contemporaries, but working with wonderful effect, more particularly in advancing a knowledge of the rational composition of organic compounds. The older radical theory acquired through him new life, and the radicals themselves received a more profound meaning. While in the type theory the latter were looked upon as residues whose nature could be no further investigated, Kolbe devoted his whole energies to breaking up the radicals into their immediate constituents. To give but a few examples,—he showed cacodyl to be arsene-dimethyl, acetyl to be a compound of methyl and carbonyl, and the alkyls to be derivatives of methyl. These and other results of his investigations, together with the rich fruits of Frank-

[1] Cf. (*e.g.*) *Journ. pr. Chem.* (2), vol. xxviii. p. 440.

land's labours, were undoubtedly of the first importance, indeed indispensable, for the development of the new type doctrine into the structure theory.

These two men, the workers of greatest originality in the field of organic chemistry during the storm-and-stress period of the fifties, thus contributed most materially by their labours to the recognition of the fact that the peculiarity of Gerhardt's types rested upon the different saturation-capacities of the elements which they contained. The chief merit of having worked as a pioneer in this direction belongs to Frankland.

THE FOUNDING OF THE DOCTRINE OF THE SATURATION-CAPACITY OF THE ELEMENTS BY FRANKLAND.

In the foregoing section the influence exercised by Frankland on the views developed by Kolbe with regard to the constitution of organic compounds has been already distinctly emphasised. It was Frankland who, in his memorable paper,—*On a New Series of Organic Compounds containing Metals*[1]—furnished the proof that the copulation of radicals with elements (*e.g.* carbon, arsenic and sulphur), as taught by Kolbe, depended upon a property inherent in the elementary atoms of the compounds just named. The notion of copulation was recognised by Frankland as being one-sided, and the misconception which had crept in from its use was done away with by him,—the idea, namely, that the radicals present as so-called copulæ in organic substances exercised no appreciable influence upon those compounds with which they were supposed to be copulated.

From his experiences gained from the organo-metallic compounds, Frankland developed the doctrine of the valency of the elements. If, freeing our minds from all prepossession, we turn our glance backward, we recognise the germ of this doctrine as being already present in the law of multiple

[1] *Phil. Trans.*, vol. cxlii. p. 417; *Ann. Chem.*, vol. lxxxv. p. 329. This paper was read before the London Chemical Society in 1852.

proportions, which stated that the elements show different, but at the same time perfectly definite stages in their combinations. Among the facts known at a very early period was, for instance, that of one atom of phosphorus combining with three and five atoms of chlorine to definite compounds; but the expression for this and other similar observations, viz. that phosphorus and many other elements were possessed of more than one valency, *i.e.* could manifest varying saturation-capacities, had yet to be found. Further, no one had any clear conception of a limit to the saturation-capacities of elements, and, what was of the first importance, a sharp distinction between the terms "atom" and "equivalent" was still wanting. With regard to this latter point, the experiences gained respecting the substitution of the hydrogen of organic compounds by chlorine, oxygen, etc., and the deductions drawn from these had tended to elucidate matters. So early as 1834 Dumas had pointed out that 1 atom of hydrogen was replaced by 1 atom of chlorine, but only by ½ an atom of oxygen; those quantities were therefore equivalent to 1 atom of hydrogen. The idea of the "replaceable value" of certain metals also came more distinctly into prominence through the doctrine of polybasic acids, already spoken of; this was exemplified, for instance, in Liebig's statement that 1 atom of antimony was equivalent to 3 atoms of hydrogen, but one of potassium only to 1 atom of hydrogen. Notwithstanding this, however, a precise expression for such facts as these had not yet been found. In the course of the forties the conception of a chemical equivalent as distinguished from an atom, a conception which had been arrived at after so much labour, completely died out; the growing influence at that time of the Gmelin school affords us eloquent testimony of this backward step.

It is a remarkable fact that, for establishing the doctrine of valency, it was not the simple compounds of inorganic chemistry but the more complicated ones of organic that were called into service. The relations which in the former found clear expression, and were easily read in the law of multiple proportions, had to be first laboriously deciphered here from organic compounds.

As stated already, it was the organo-metals from which Frankland deduced the results which constitute the kernel of our present theory of valency. He acted as pioneer in this branch more than any other man, and distinguished himself by his admirable investigations. Before him (more particularly) Bunsen had accomplished his memorable work on the cacodyl compounds, and cacodyl itself had been designated by Kolbe as arsene-dimethyl. Relying upon his own observations on the stanno-ethyl compounds, and on the behaviour of the cacodyl derivatives and other bodies, Frankland proved with convincing clearness that the theory of copulæ was untenable. Frankland's train of reasoning was somewhat as follows:—If we start with the latter theory, we must assume that the power of the metals to combine with oxygen is not altered by their being copulated with radicals. But facts tell against such an assumption, as is seen at a glance from the following examples:—Tin-ethyl (SnC_4H_5; $C=6$) ought, according to that theory, to unite with oxygen in two proportions, but in reality it is only capable of taking up one equivalent of this element, and not two, like tin itself. Cacodyl, which is arsenic copulated with two methyls, does indeed form two oxides, from which it might be argued that the one with one equivalent of oxygen corresponded to arsenic sub-oxide, and the other with three equivalents to arsenious acid; but this hypothesis affords no explanation whatever of the fact that the latter compound is very readily oxidisable, whereas its supposed analogue cacodylic acid cannot be oxidised by any means.

These and similar contradictions were done away with by Frankland in the simplest manner, by the assumption that the so-called copulated compounds were derivatives of inorganic bodies in which oxygen had been replaced by its equivalent of hydrocarbon radicals. Stanno-ethyl oxide was explained as tin dioxide, SnO_2, in which one equivalent of oxygen was replaced by ethyl, and cacodyl oxide as arsenious acid, in which two equivalents of oxygen had been substituted by two methyls. Frankland then proceeded to extend this conception to other compounds in the most felicitous manner, and—what was especially important—thus brought the laws which are shown in the composition of organic and inorganic substances into relation with the fundamental properties of the elements which these contain.

He expressed his views upon this point in the following sentences,[1] which, from their great importance, have a claim to a special place in a history of chemistry: "When the formulæ of inorganic chemical compounds are considered, even a superficial observer is impressed with the general symmetry of their construction. The compounds of nitrogen, phosphorus, antimony, and arsenic, especially, exhibit the tendency of these elements to form compounds containing 3 or 5 atoms of other elements; and it is in these proportions that their affinities are best satisfied: thus in the ternal group we have NO_3 NH_3, NI_3, NS_3, PO_3, PH_3, PCl_3, SbO_3, SbH_3, $SbCl_3$, AsO_3. AsH_3, $AsCl_3$, etc.; and in the five-atom group, NO_5, NH_4O, NH_4I, PO_5, PH_4I, etc. Without offering any hypothesis regarding the cause of this symmetrical grouping of atoms, *it is sufficiently evident, from the examples just given, that such a tendency or law prevails, and that, no matter what the character of the uniting atoms may be, the combining power of the attracting element*, if I may be allowed the term, *is always satisfied by the same number of these atoms.*"

In this way was established the doctrine that a varying, but at the same time, within certain limits, definite saturation-capacity appertains to the atoms of the elements. For the ones which have just been named this was expressed by the numbers 3 and 5; Frankland did not assume any higher stage of saturation for them. By this treatise of his, so rich in ideas and facts, he opened up a new field in theoretical chemistry, which, assiduously cultivated as it has been ever since, has served both as the centre- and the starting-point for all chemical investigations. Under the influence of the theory of valency all theoretical chemical views thenceforth developed themselves, as will be clearly

[1] *Phil. Trans.*, vol. cxlii. p. 417; *Ann. Chem.*, vol. lxxxv. p. 368.

seen from the following sections. The happy interpretation of the constitution of the so-called copulated compounds was the immediate cause of this great advance, in so far that Frankland proved copulation to be a consequence of saturation-capacity.

After the definite valency of particular elements had been established by Frankland, it might have been imagined that every chemist could have deduced for himself the saturation-capacities of other elements from their behaviour. Frankland's pioneering work did not, however, produce fruit with such rapidity. How slowly his views found acceptance among chemists is proved by a paper of Odling's, published in 1854, and entitled *On the Constitution of Acids and Salts.*[1] The latter chemist still adhered firmly to the type theory. He argued that salts and acids, especially those containing oxygen, can be referred to the simple or multiple water type in such a way that the hydrogen of the latter is partially or completely substituted by elementary or compound radicals of definite *replaceable value.* This latter term was used by Odling to express what Frankland had done by the word *atomic.* Iron and tin had, according to Odling, two replaceable values, whose magnitudes he indicated by the dashes which have since then been so largely employed, thus: Fe'' and Fe''', Sn' and Sn''. Thus far he followed Frankland's conception of the saturation-capacity of the elements. For the polybasic acids he accepted Williamson's views, in that he assumed in them oxygenated radicals of definite replaceable value, which were introduced into the type $(H_2O)_n$. Just as sulphuric acid was built up on the double water type by the entrance of the diatomic radical SO_2, so he derived phosphoric and arsenic acids from the triple water type $(3H_2O)$ by introducing the atomic groups $(PO)'''$ and $(AsO)'''$; while in the carbonates the radical CO, with a replaceable value of 2, was assumed, and so on. But mischievous obscurations now began to be mixed up with this. As a result of his one-sided typical conception, Odling did not hesitate to assume that the diatomic radical SO_2 acted

[1] *Journ. Chem. Soc.*, vol. vii. p. 1.

as monatomic in dithionic acid,[1] and the diatomic radical CO as monatomic in oxalic acid; and this last (for example) he referred to the double water type, thus: $\left.\begin{matrix}(CO)'(CO)' \\ H_2\end{matrix}\right\} 2O''$. But, with all this, Odling deserves credit for being instrumental in causing a constant replaceable value to be ascribed to particular elements, to hydrogen and oxygen in especial, whereby the atomic weights of these two latter served as standards for fixing the replaceable values of other elements and compound radicals. Williamson afterwards helped most materially to clear up the meaning of Odling's formulæ, and to bring about a more intelligent conception of the constitution of chemical compounds.[2]

The utterances of Wurtz[3] and of Gerhardt[4] upon the saturation-capacity of the nitrogen atom also showed that Frankland's ideas acted but slowly; for the last-named had expressed himself on this point in almost exactly the same sense three years previously. In many cases chemists were content with merely the notion of compound radicals, without investigating the influence of the contained elements upon the saturation-capacities of these complexes; this applied in an especial degree to the radicals composed of carbon and hydrogen, with whose replaceable value (that of the radicals) various eminent investigators occupied themselves.

The Recognition of the Valency of Carbon.

A considerable time elapsed before any definite utterance was made with regard to the valency of the carbon of alcohol radicals—the *organic element* in the true sense of the term. Instead of deducing this fundamental property from its oxygen compounds, carbon monoxide and dioxide, a more

[1] This he formulated :—$\left.\begin{matrix}(SO_2)'(SO_2)' \\ H_2\end{matrix}\right\} 2O$.

[2] Cf. *Journ. Chem. Soc.*, vol. vii. p. 137; or *Ann. Chem.*, vol. xci. p. 226.

[3] *Ann. Chim. Phys.* (3), vol. xliii. p. 492 (1855).

[4] *Traité de Chimie*, vol. iv. pp. 595 and 602 (1856).

tedious method was adopted; it was the investigation of carbon-containing radicals which led to the final solution of the question. Among the researches which were of effective service here, we must first mention that by Kay,[1] made at Williamson's suggestion, upon "tribasic formic ether"; this compound, which resulted from chloroform and sodium ethylate, was regarded as a derivative of three atoms of ethyl alcohol, in which the three atoms of basic hydrogen had been replaced by the "tribasic radical of chloroform, CH." Ranking alongside of this important piece of work came that of Berthelot upon glycerine.[2] Aided materially by Wurtz's expositions, Berthelot characterised this compound as a triatomic alcohol, since he assumed in it a *tribasic radical*, C_6H_5 ($C=6$), replacing three atoms of hydrogen in the triple water type. To the alkyls which took the place of three atoms of hydrogen, diatomic ones were soon added, ethylene being so designated by H. L. Buff.[3] The brilliant discovery by Wurtz of the first known diatomic alcohol, glycol,[4] served as a corroboration of this view.

Chemists were, it is true, upon the track of the cause of the different replacing values of those radicals $(CH)'''$, $(C_6H_5)'''$, and $(C_2H_4)''$, for we find utterances by Gerhardt and Wurtz to the effect that ethylene was dibasic, because one atom of hydrogen had been withdrawn from the monobasic ethyl, and glyceryl tribasic, because it contained two atoms of hydrogen less than the corresponding propyl. But no one had attained to a complete explanation of these radicals; their saturation-capacities had never been distinctly referred back to that of carbon.

In a paper entitled, *Ueber die Konstitution und die Metamorphosen der chemischen Verbindungen und über die chemische Natur des Kohlenstoffs* ("On the Constitution and Metamorphoses of Chemical Compounds, and on the Chemi-

[1] *Journ. Chem. Soc.*, vol. vii. p. 224.
[2] *Ann. Chim. Phys.* (3), vol. xli. p. 319.
[3] *Ann. Chem.*, vol. xcvi. p. 302.
[4] *Comptes Rendus*, vol. xliii. p. 199.

cal Nature of Carbon"),[1] which was published in 1858, Kekulé drew the following nearly allied conclusion. He applied to carbon what had already for a long time been recognised with regard to other elements,—to nitrogen and its chemical analogues in the first instance. The reasons given by him for carbon being tetravalent are contained in the following sentences:—" If we look at the simplest compounds of this element, CH_4, CH_3Cl, CCl_4, $CHCl_3$, $COCl_2$, CO_2, CS_2 and CHN, we are struck by the fact that the quantity of carbon which is considered by chemists as the smallest amount capable of existence—the atom—always binds four atoms of a monatomic or two of a diatomic element, so that the sum of the chemical units of the elements combined with one atom of carbon is always equal to four. We are thus led to the opinion that carbon is tetratomic." This train of thought is almost the same as that which led Frankland to deduce the tri- and penta-valence of nitrogen, phosphorus, arsenic and antimony,[2] the latter chemist having also arrived at the saturation-capacities of these elements from a study of their simplest compounds. It follows from this that the above utterance of Kekulé cannot be regarded as a thoroughly original achievement, all the more since the tetravalence of carbon had already been recognised both by Kolbe and Frankland, and especially as it formed the basis of the latter's statements upon the constitution of organic compounds.[3] In curious contrast

[1] Couper, too, independently of Kekulé, and shortly after the appearance of the paper just cited, expressed the view that the atom of carbon was tetravalent (cf. *Comptes Rendus*, vol. xlvi. p. 1157).

[2] Cf. p 325.

[3] Cf. Kolbe's publication entitled *Zur Entwickelungsgeschichte der theoretischen Chemie* ("Contribution to the History of the Development of Theoretical Chemistry"), Leipzig, 1881, p. 26 *et seq.*, especially p. 33. Others, too, have claimed for Kolbe the merit of being the first to perceive the tetravalence of carbon, *e.g*, Blomstrand, who thus expresses himself in his *Chemie der Jetztzeit* ("Chemistry of the Present Time"), p. 110: "No other chemist can lay the same claim as Kolbe to be regarded as the originator of the doctrine of the saturation-capacity of carbon. Alongside of him must be placed Frankland, whose uninterrupted researches, conceived and carried out with equal felicity, continually furnished new supports in aid of the doctrine mentioned above,—a doctrine which comprises in itself every-

with the high value which most chemists have placed upon this service of Kekulé's is the depreciatory way in which he talks of it himself.[1]

Kekulé's service in this point must be sought for in the fact that he endeavoured to get at the root of the problem as to how two or more carbon atoms combine with one another, and how their mutual affinities are satisfied. The immediate result of these speculations was the doctrine of the "linking of atoms" (*Verkettung der Atome*) in chemical compounds. Indirectly, Kolbe's and Frankland's views had a most material share in developing this crowning edifice of the structure theory.

thing that relates to saturation, and which has found in Kolbe's carbonic acid theory by far its most important application." A. Claus (*Journ. pr. Chem.* (2), vol. iii. p. 267) has written in a similar sense. Kekulé is therefore not justified in claiming for himself the merit "of having introduced the idea of the atomicity of the elements into chemistry" ("*den Begriff der Atomigkeit der Elemente in die Chemie eingeführt zu haben*"), (cf. Kekulé, *Ztschr. Chem.* for 1864, p. 689). This idea was without doubt primarily due to Frankland, who expresses himself clearly and unequivocally on the point in his *Experimental Researches* (1877), p. 145, as follows: "This hypothesis which was communicated to the Royal Society in the second of the following papers" (cf. p. 325 of this book), "on 10th May 1852, constitutes the basis of what has since been called the doctrine of atomicity or equivalence of elements; and it was, so far as I am aware, the first announcement of that doctrine."

[1] Thus Kekulé says, at the close of his above-mentioned treatise, p. 109: "Lastly, I feel bound to emphasise the point that I myself attach but a subordinate value to considerations of this kind. But since in chemistry, when there is a total lack of exact scientific principles to go upon, we have to content ourselves for the time being with conceptions of probability and expediency, it appears appropriate that those views should be published, because they seem to me to furnish a simple and tolerably general expression precisely for the latest discoveries, and because therefore their application may perhaps conduce to the finding out of new facts."

DEVELOPMENT OF CHEMISTRY UNDER THE INFLUENCE OF THE DOCTRINE OF VALENCY DURING THE LAST FORTY YEARS.

The chemical atomic theory had been in existence for nearly fifty years before the natural inference was drawn with sufficient exactitude from it that each elementary atom possesses a definite saturation-capacity, and that this is expressible in some cases by a constant factor, but in most cases by a varying one. In recognising this a great advance was made,—an advance which showed itself particularly in the fact that, after the establishment of the valency theory by Frankland, people attained to a more definite conception of the chemical constitution of inorganic, and more especially of organic compounds. From thenceforth continuous efforts were made to solve this problem, first recognised in its fullest signification by Berzelius, by the aid of the ideas which Frankland had either himself expressed or had induced in others. Chemists endeavoured, by breaking up compound bodies (in part actually and in part on paper only) and distributing the elementary atoms according to their supposed saturation-capacities, to work out the mutual relations of these ultimate constituents. In this way there shone forth from valency a light which now illumines the whole field of chemistry.

The theory of the linking of atoms was considered by most chemists as the necessary result of the idea that a saturation-capacity (with respect to other elements), expressible by figures, belonged to the atoms of each individual element. With the development of this view, in organic as well as in inorganic chemistry, many brains have been busily engaged for the last forty years. The idea of a definite saturation-capacity for each element has formed a necessary aid in the solution of numerous important points which have come up during this period, *e.g.* the question of the nature of valency, the reasons for many cases of isomerism hitherto unexplained, etc., and it still remains an indispensable guide in all scientific chemical investigations.

Beginnings of the Structure Theory—Kekulé and Couper.

The theory of types, according to which all organic compounds were referred to a few simply constituted bodies, had been rendered objectless by Frankland's conception of that property of elements which we now term valency. The types now presented themselves as hydrogen compounds of mono-, di-, tri-, and tetra-valent elements. Had Frankland's ideas at once received the attention which they merited, the detailed development of the theory of types, as given by Gerhardt in the fourth volume of his text-book, could have been entirely dispensed with.

Out of Frankland's idea of saturation-capacity there grew the further notion that the elementary atoms could be combined among themselves by one or more affinities, according to their nature, and that a disappearance of individual affinities took place as the result of this. This idea was first advanced by Kekulé, and shortly after by Couper, in the treatises already referred to (in 1858). These therefore contain the beginnings of the structure theory.[1]

After having deduced the "tetratomicity" of carbon from the composition of a number of simple compounds of that element, Kekulé expressed himself upon the constitution of compounds which contain more than one atom of carbon as follows:[2] "In the case of substances containing several carbon atoms we must assume that at least some of the atoms (of the other elements present) are held bound by the affinities of the carbon atoms, and that the latter are themselves linked together, whereby a part of the affinity of the one (carbon atom) is necessarily tied by an equally large part of the affinity of the other."

"The simplest and therefore the most probable case of

[1] The term "structure" (*Struktur*) was first introduced by Butlerow (*Ztschr. Chem.* for 1861, p. 553); through it he quite unintentionally awakened the erroneous idea that the actual spacial arrangement of the atoms could be arrived at by the aid of the above hypothesis.

[2] *Ann. Chem.*, vol. cvi. p. 154.

such a combination (*Aneinanderlagerung*) of two carbon atoms is that in which one affinity of the one atom is tied by one affinity of the other. Of the four affinity units of each of the two carbon atoms, two are thus taken up in keeping both atoms together ; six consequently remain over, to be available for atoms of other elements."

Here, therefore, there was set up the hypothesis that the carbon atoms join together,[1] and lose in consequence a portion of their affinities. Starting with the assumption that more than two atoms of carbon can coalesce in the same manner, Kekulé generalised this particular case by establishing the value $2n + 2$ for the saturation-capacity of the complex C_n. He did not, however, remain stationary at this point, but represented further that "a more compact combination of the carbon atoms" might be assumed in other organic compounds, *e.g.* benzene and naphthalene. As the "next most simple coalition of carbon atoms" he conceived the case of the mutual interchange of two affinity-units. The relations, too, of other polyvalent elements to the carbon atoms were taken into account by him, and he gave illustrations to show that these were bound either by all their affinities or by a portion of them to the affinities of the carbon.[2] The main features of the doctrine of the "Linking of Atoms" (*Bindung der Atome*) were contained in those sentences of Kekulé's.

Almost at the same time Couper,[3] independently of Kekulé, arrived at similar views with respect to the mutual linking of several carbon atoms. Being definitely of opinion that Gerhardt's doctrine of types did not satisfy the claims required by a theory, he made the attempt to get at the constitution of chemical compounds by falling back upon the elementary atoms. He laid stress upon the point that, in addition to the affinity proper (*Wahlverwandtschaft*), the degree of that affinity (*Gradverwandtschaft*) of the small particles came into

[1] *Sich aneinander lagern.*

[2] Cf., for instance, *Ann. Chem.*, vol. cvi. p. 155.

[3] *Comptes Rendus*, vol. xlvi. p. 1157 ; *Ann. Chim. Phys.* (3) vol. liii. p. 469.

play in the formation of chemical compounds. For the atom of carbon the highest power of combination was expressible by the number 4. In general he adopted Frankland's doctrine of the varying saturation-capacities of the elements. Couper further laid great emphasis upon the capacity of the carbon atoms to unite with one another, and this in such a manner that a part of their own individual power of combination was thereby neutralised. This linking of the atoms, he illustrated by bars drawn between the chemical symbols of the combining particles; he thus laid the foundation of the so-called "structural formulæ."[1] The following examples will serve to illustrate this:—

CH_3 | $C\begin{matrix}H_2\\O-OH\end{matrix}$ Alcohol;	CH_3 | $C\begin{matrix}O_2\\O-OH\end{matrix}$ Acetic acid;	$C\begin{matrix}O-OH\\O_2\end{matrix}$ | $C\begin{matrix}O_2\\O-OH\end{matrix}$ Oxalic acid.

Both Kekulé and Couper expressed with absolute definiteness the axiom that the "atomicity of the elements" was to be made use of for arriving at the constitution of chemical compounds. The idea of the term "atomicity" had without any doubt been introduced by Frankland six years previous to this. The further development of the above axiom and its utilisation in the theory of the linking of atoms was carried out mainly by Kekulé, and in the succeeding years also by Butlerow and Erlenmeyer.

Before an absolutely certain knowledge of the atomicity or, better, the valency of the elements could be attained, perfect clearness has to be arrived at with respect to the magnitudes of the atomic weights; and, more particularly, the distinction between the atom and equivalent of polyvalent elements had to be clearly grasped. That was, however, by no means the case at this time. In writing the formulæ of chemical compounds, most chemists employed Gmelin's equivalents from force of habit; but, in making use of these,

[1] Wurtz manifestly forgot Couper's paper in the *Annales de Chimie et de Physique*, of which he (Wurtz) was one of the editors, for he took credit to himself as being the first to make use of these linking-bars (see his *Atomic Theory*, fourth English edition, p. 214, note).

the true chemical values of the atoms remained indistinct and only became apparent after the conversion of the equivalents into atomic weights. For instance, the functions of the simple atoms C and S were ascribed to the double equivalents C_2 and S_2 in the formulæ employed by Kolbe, while for hydrogen, chlorine, nitrogen and other elements, the equivalents were identical with the atomic weights.[1] And the disorder was increased by many chemists, Couper among the number, giving to carbon its correct atomic weight (12), while retaining the equivalent (8) for oxygen. It is true that Gerhardt had already attempted to bring order into the prevailing confusion, but his mode of procedure had not been logical enough.[2]

Thanks to the efforts of the Italian chemist Cannizzaro, a way was prepared in 1858 for the clearing up of this unsatisfactory state of matters, although those efforts received only tardy recognition. It was he who, by his criticism in a paper entitled *Suntò di un Corso de Filosofia Chimica* ("Outlines of a Course of Chemical Philosophy"),[3] threw light upon the methods employed for arriving at the relative atomic weights of the elements. He recognised, as especially reliable, the deduction of these values from the vapour densities of chemical compounds,—a method now in universal use. And he further showed to what extent the specific heats of the metals might be regarded as a trustworthy aid in the determination of their atomic weights,

[1] The meaning of this is at once apparent if we take Kolbe's old formula for acetic acid, $C_2H_3.\ C_2O_2.\ OHO$, and convert it into our present formula, $CH_3.\ CO.\ OH$, by changing the double atoms C_2 and O_2 into the single ones C and O.

[2] Cf. p. 295.

[3] *Nuovo Cimento*, vol. vii. p. 321. This paper was edited, with notes, by the late Lothar Meyer for Ostwald's *Classiker* (German by Miolati) in 1891. Stanislao Cannizzaro, born in 1826, first studied medicine, then chemistry under Piria, and subsequently filled in succession the chairs of chemistry in Genoa, Palermo, and (since 1871) in Rome. This last he still holds, while he is at the same time a Senator and a member of the High Court of Public Education (*Mitglied des obersten Rathes des öffentlichen Unterrichts*). His experimental researches, *e.g.* those on benzyl alcohol and on santonine and allied compounds, are of a very high order.

wrong values for many of these having come to be accepted as the result of Gerhardt's statements.

After the correct atomic weights of the elements had been established in this way, it became possible to build up the doctrine of the chemical values of the elements from a more general point of view than before. First it was applied to the compounds of carbon, whose constitution became the subject of the most ardent investigation. Kekulé in his text-book (begun to be published in 1859), and Butlerow and Erlenmeyer in various papers and subsequently in text-books also, endeavoured to explain the connection existing between the elementary atoms within the molecules, by setting out with the conception that a definite atomicity appertained to each element; carbon, hydrogen, oxygen and nitrogen came primarily into question here.

Butlerow was the first to express himself clearly upon the principle which underlay these efforts, and, with this, upon the nature of the Structure Theory (which received its name from him).[1] We must premise here that he took up his position on the valency theory founded by Frankland, according to which many of the elements possess a varying saturation-capacity. Butlerow defined the *structure* of a chemical compound as the "manner of the mutual linking of the atoms in a molecule;" he decisively rejected the idea that it afforded any information as to the position of the individual atoms in space. He advanced the opinion that the chemical character of a compound depended first upon the nature and quantity of its elementary constituents, and then upon its chemical structure. The latter had, to his mind, but one meaning; he could not agree with Gerhardt that several rational formulæ might be proposed for one and

[1] *Ztschr. Chem.* for 1861, p. 549 *et seq.*—Alexander Butlerow, who was born in 1828 and died a few years ago, became professor of chemistry in the University of Kasan in 1858, and in that of St. Petersburg in 1868. He contributed materially to the development of organic chemistry by many admirable experimental researches, and in a very special manner by his *Text-book of Organic Chemistry*; this latter, which appeared first in 1864 in Russian and in 1868 in German, has had a far-reaching influence.

the same chemical compound, one formula only appearing possible to him.

The more that the former adherents of the type theory came to feel the necessity for abandoning it, and, free from the yoke of this doctrine, of basing all considerations with respect to chemical constitution upon the "atomicity" of the elements, the more definitely ought the views upon the nature of this property of the elements to have shaped themselves.—The conclusion, deduced from numerous experiments, that the atoms of certain elements show a constant combining value and the atoms of others a varying one, came at that time into opposition with the opinion that this capacity of the elements was invariable.

Controversies respecting constant and varying Valency of the Elements.

Frankland, the originator of the doctrine of the saturation-capacity of elementary atoms, held aloof from the lively discussions to which it gave rise, more especially after the year 1870. This in all probability accounts for his service in developing such an important doctrine having been forgotten by many chemists, and precisely by those who have taken the most active share in the above discussions.[1] About the year 1860 Frankland's views regarding a saturation-capacity peculiar to the elements, which, under certain circumstances, might be a varying one, were accepted either tacitly or expressly by most chemists of standing. Even so early as 1856 Gerhardt had stated in his text-book that nitrogen was sometimes triatomic, sometimes pentatomic,—a view which coincided exactly with that of Frankland. Wurtz, Williamson and Couper also held this opinion, and not for nitrogen and its analogues alone, but also as being characteristic of many other elements; that Kolbe likewise agreed with Frankland on this point has been stated already. In the assumption that a constant valency was

[1] See Note 1, p. 310.

characteristic of a few elements and a varying one characteristic of many more; Kolbe merely saw another expression for the law of multiple proportion; this conception, as corresponding with facts, he considered necessary, because nothing was known of the real cause of valency.

This view, then, which had so many observations to support it, led to the conclusion that each element possessed a maximum saturation-capacity, but that lower stages of saturation might coexist along with this; Kolbe had expressed himself in this sense so far back as the year 1854.[1] Towards the beginning of the sixties, several chemists who took an active part in developing the structure theory gave utterance to the same opinion in a more definite manner. Erlenmeyer, in particular, maintained in various papers,[2] and afterwards in his *Lehrbuch der organischen Chemie*, that each element possesses a maximum valency, or that each is furnished with a definite number of *Affinivalenten* or affinity-points (*Affinitätspunkten*), only part of these, however, being in many cases combined with the affinity-points of other elements. In ammonia, for instance, only three of the five equivalents of the nitrogen atom come into play, while in chloride of ammonium all five are satisfied. Following this out, Erlenmeyer distinguished between saturated and unsaturated compounds. Strickly speaking, this is nothing else than Frankland's view.

At about the same time a lively discussion with respect to the atomicity of the elements went on between Wurtz and Naquet[3] on the one hand, and Kekulé[4] on the other. The two former declared for the assumption of a varying valency in the case of many of the elements, while Kekulé, on the other hand, expressed his opinion more definitely than before that the "atomicity of the elements is a fundamental property of the atoms, quite as unalterable as their atomic weights."

[1] Cf. *Lehrbuch der organischen Chemie*, vol. i. p. 22.

[2] *Ztschr. Chem.* for 1863, pp. 65, 97, and 609; for 1864, pp. 1, 72, and 628.

[3] *Ibid.*, p. 679.

[4] *Ibid.*, p. 689; *Comptes Rendus*, vol. lviii. p. 510.

In order to confirm this theorem of absolute or constant valency, and to reconcile it with conflicting facts, Kekulé was obliged to have recourse to hypotheses which laid themselves strongly open to criticism. A few examples may be given here to illustrate his view of the valency of each element being constant. According to him, nitrogen and its chemical analogues acted only as trivalent, sulphur, like oxygen, only as divalent, and chlorine, bromine and iodine as monovalent. In order, therefore, to explain the constitution of compounds, in which, upon the assumption of a varying valency, the elements just named had a higher saturation-value than he assigned to them, Kekulé had to presuppose a fundamental difference as existing between compounds of one and the same element. To his first hypothesis of absolutely constant valency he added the further one, that those compounds, in which the elements are present in their supposed normal values, are distinguished from the others by a more compact structure; the former he termed atomic, and the latter molecular compounds. The components of the latter, *e.g.* ammonia and hydrochloric acid in salmiac, phosphorus trichloride and chlorine in phosphorus pentachloride, were, according to his view, held together by forces of another kind to those which acted in the atomic compounds. In order to give expression to the looser connection between the molecules of these substances, he placed their components dualistically alongside of one another in writing the formulæ; thus he gave $PCl_3 . Cl_2$ as the formula of phosphoric chloride, and $H_3N . H_2S$ as that of ammonium hydrosulphide. He would not admit a variation in the saturation-values of nitrogen and phosphorus in compounds like those just named.

Other chemists were thus justified in asking what his grounds were for assuming such a distinction between the forces by which chemical constitution was conditioned; for, in both kinds of compounds the same atomic laws held good. Kekulé regarded the breaking up of compounds into their components at a somewhat high temperature as a criterion of their being *molecular compounds*, while *atomic compounds* were those which could be converted into the gaseous state without decomposition. But this distinction between the two categories could not be maintained in the face of known facts; it soon became evident that such an artificial partition only served to introduce confusion and bring about contradictions which were irreconcilable.

This theory of the constant valency of the elements could not therefore long withstand the critical examination to which it was subjected by Kolbe,[1] and more especially by Blomstrand,[2] not to mention others. The known facts could not by any possibility be brought into accordance with the assumption of saturation-capacity being invariable, and this helped more than anything else to cause the theory to be abandoned by its most zealous adherents. How, for instance, could the existence and behaviour of the organic ammonium bases, the sulphones and sulphoxides, perchloric and periodic acids, and many other compounds be explained by the aid of the above hypothesis? Other weighty arguments have recently been brought forward which must be regarded as incompatible with those urged shortly after the setting up of Kekulé's theory; to take compounds of one element only, we may refer here to the discovery of the isomeric triphenyl-phosphine oxides, in one of which the phosphorus must be pentavalent, and also to the proof given of phosphorus pentafluoride existing in the gaseous state. Such facts are not to be reconciled with the assumption of phosphorus being only trivalent.

[1] Cf. *Journ. pr. Chem.* (2), vol. iv. p. 241.

[2] In his work, *Die Chemie der Jetztzeit*, Blomstrand went carefully into the doctrine of the saturation-capacity of the elements, and by his comprehensive treatment of the question materially lightened the labours of other critics as to the share taken by different workers in its development.—C. Wilhelm Blomstrand, born in 1826, filled the chair of chemistry in the University of Lund in Sweden from 1854 until 1895; he died in 1897. His eminent researches in various branches of mineralogical and also of organic chemistry are distinguished by their thoroughness, and show the influence of Berzelius, whose doctrines Blomstrand endeavoured, in his book mentioned above, to reconcile and bring into close connection with the more recent views. From the electro-chemical basis, in especial, he was able to throw light upon the valency question, and to gain for it new points of view.

We may assert that in the course of the last thirty years the majority of chemists have adopted the opinion that the atoms of most of the elements possess a varying saturation-capacity, varying according to the conditions. The idea prescribed as essential at the time the theory of an unchanging valency was set up, viz. that this was a fundamental property of atoms, may be fully recognised without our being thereby forced to the conclusion that the valency of the elementary atoms must therefore be constant.

In connection with these weighty discussions upon the nature of valency, reference may be made here to a problem nearly related to it, which has given rise of recent years to frequent debate, and also to important experimental work, viz.—the question whether the individual affinity-units or valencies of one element are alike or different. If we only took into consideration some isolated facts, such as the dissimilar functions of the two atoms of oxygen or sulphur in carbonic acid and carbon disulphide respectively, we might be inclined to favour the assumption of a difference in two affinities of the carbon atom with respect to the other two. But the numerous investigations which have been made by Popoff, Schorlemmer, L. Henry, Röse and others, with the object of deciding this point so far as regards carbon, have led to the conclusion that its four affinities are alike.

The equality or inequality of the affinities of the sulphur and nitrogen atoms is still undecided, notwithstanding that many facts bearing on the point have been collected together. Among other researches we may mention the work of Krüger, which appeared to prove a difference in the valencies of sulphur; but, while his results have been corroborated on one side, they have been doubted on the other. The remarkable isomerism in the derivatives of hydroxylamine, which has been worked out by Lossen, seems quite compatible with the assumption of the affinities of nitrogen being different; more recent researches by Lossen, Beckmann, Behrend and Werner, however, point to another solution of the question on stereo-chemical lines.

The main directions which chemical investigation has taken, since these discussions with regard to valency came up, are characterised by the endeavour to gather from the chemical behaviour of compounds an insight into their constitution, by the aid of the assumption that the elements have a definite saturation-capacity; while at the same time efforts are being made to arrive at the mutual relations between the physical properties of compounds and their constitution as determined by chemical means. To this problem, which has only recently been assiduously attacked, although it has been projected for a long time, an analogous one has been added, viz. the elucidation of the connection which obviously exists between the relative atomic weights of the elements and their chemical and physical properties.

The further Development of the Structure Theory—The chief Directions taken by Organic Chemistry during the last thirty Years.

At a first glance it strikes one as strange that organic chemistry in particular should have been made the field for speculations as to the composition of chemical compounds, speculations which had the valency theory as their basis. The reason for this preference is undoubtedly to be sought for in the peculiarity of that element which is never wanting in the so-called organic compounds, carbon, even if we allow for the fact that it was from compounds of carbon—the organo-metallic ones—that the idea of the saturation-capacity of elements developed itself.

From the tendency of the atoms of carbon to unite with one another according to different degrees of affinity (*Gradverwandtschaft*), *i.e.* by the interchange of one, two, or three affinities, the production of the variously composed carbon compounds could be explained without difficulty. The addition of elements like hydrogen, oxygen, sulphur, nitrogen and chlorine to the complexes of carbon atoms was rendered intelligible in a similar manner by assuming that the individual affinities of the elements named were satisfied by a like number of affinities of carbon. The combination of the carbon atoms among themselves or with other elementary atoms, as illustrated in this way, was termed "linking" (*Verkettung*). From thenceforth the adherents of the structure theory came to grasp more clearly the problem of chemical investigation. They sought to combine the atoms of the various elements in question suitably with one another, according to their saturation-capacities, directing their efforts mainly to investigating the structure of the compounds of carbon, since inorganic substances, as being of much simpler composition, seemed to offer few or even no difficulties to the application of the above principle. The conceptions thus gained of the structure of organic substances were then tested with more or less minuteness by actual experiment, with the object of seeing whether the modes of formation and decomposition of the compounds in question, and their chemical behaviour generally, agreed with the theoretical hypotheses.

The readiness with which many chemists took to the construction of formulæ which were meant to express the mutual relations existing between the atoms of a compound, *i.e.* the structure of the latter, may in some cases have given rise to the belief that by the aid of such symbols an insight into the actual arrangement of the atoms in space might be obtained. Some investigators of eminence may have incited to such daring hopes and expectations by indistinct modes of expression and unhappily chosen comparisons and illustrations. In the minds of younger chemists, especially, it was easy for erroneous ideas regarding the problems of chemistry to effect a lodgment. We may recall here that Kekulé spoke of the carbon atoms as sliding over and adhering to one another,[1] and of the *other* side of a molecule, etc.; that in his text-book he brought forward graphic formulæ, in which the elementary atoms have different forms according to their saturation-capacities; and, further, that

[1] "*Von einem Zusammenschieben oder Aneinanderleimen der Kohlenstoff-atome.*"

the smallest particles of an element have been pictured by Naquet and Baeyer as furnished with small hooks, by which they catch hold of one another. Metaphors such as these tended, at any rate, to an over-estimation of the capabilities of the structure theory.

The more prudent advocates of the latter, with Butlerow at their head, dissented all along from the idea that such formulæ could furnish any picture of the arrangement of the atoms in space. On the other side Kolbe, in particular, protested with all his critical acumen against such exaggerations, as leading easily to error. He remained staunch to the point of view which he had laid down in 1854,[1] believing that no clear conception could ever be arrived at as to how the atoms of a compound were thus arranged.

Constitution of Organic Compounds according to the Structure Theory.

Although the structure theory was unable to realise the highly-pitched expectations which aimed at a knowledge of the spacial arrangement of the atoms, it possessed none the less great practical value. The development of organic chemistry since the middle of the sixties shows in fact that, through the aid of the structural hypothesis, the discovery of new modes of formation and decomposition of compounds, the recognition of the relations existing between various classes of bodies, and, especially, the interpretation of the constitution of numerous organic substances became possible. Kekulé's theory of the aromatic compounds (see below) forms the most striking proof of this.

The working out of the constitution of the so-called *saturated* compounds offered fewer difficulties than that of the compounds poorer in hydrogen,—the *unsaturated* ones. Kekulé was the first to express the definite opinion that in all fatty compounds the carbon atoms were united to one another by an affinity of each, a point which might

[1] *Lehrb. d. organ. Chemie*, vol. i. p. 13.

have been deduced from Couper's and also from Kolbe's rational formulæ, had the equivalents used by them been converted into the atomic symbols. The expositions given by Kekulé and also by Erlenmeyer, Butlerow, Claus and others in text-books of organic chemistry and occasional papers, with regard to the constitution of such compounds soon became the common property of nearly all chemists.

More difficult was the question—What was the function of the carbon atoms in organic compounds poorer in hydrogen? With respect to the constitution of these, Kolbe, Couper and Wurtz had already expressed the view that in them—*e.g.* ethylene, acrylic acid, acetylene, etc.—one or several atoms of carbon acted as divalent. Kekulé hesitated at first between two opinions. He was, on the one hand, inclined to assume a "more compact," *i.e.* a double or treble, linking of particular pairs of carbon atoms in the substances in question; while, on the other, his experimental researches upon unsaturated organic acids led him to prefer the idea that the affinities of certain of their carbon atoms were not completely saturated, and that these therefore show gaps (*Lücken*), by means of which the capability of further combination which such compounds possess can be explained. The latter of the two views coincided in the main with the one mentioned above, in which divalent carbon atoms were presupposed. Kekulé, it is true, never definitely admitted that he regarded the saturation-capacity of carbon as a varying quantity. Of recent years preference has been given to the conception of a double or treble linking of the carbon atoms, although the other view does not want for eminent adherents. Thus Fittig,[1] from his

[1] Rudolf Fittig, born 6th December, 1835, after working for several years on the teaching staff of the University of Göttingen, became Professor of Chemistry at Tübingen in 1869, and was called from thence to the University of Strasburg in 1876, where he still continues; the beautiful laboratory there was planned by him. His name will often be mentioned in the special history of organic chemistry, which he has greatly enriched by most admirable researches, more especially upon aromatic and unsaturated compounds. Wöhler's *Grundriss der organischen Chemie* ("Out-

work upon unsaturated acids, has expressed himself in favour of the assumption of carbon being divalent in some of these compounds.[1] But the question of the constitution of such compounds has not yet been conclusively answered; for numerous observations have been made which appear to show that the complete solution of this problem by the aid of structural-chemical hypotheses alone is impossible.

Theory of the Aromatic Compounds.

In Kekulé's hands the structure theory scored by far its greatest victory, in the deciphering of the constitution of the so-called aromatic compounds.[2] These were defined by him as derivatives of benzene; his first task therefore consisted in elucidating the structure of this long-known hydrocarbon, *i.e.* in explaining how the six carbon and the six hydrogen atoms were combined together. Here Kekulé took up again his previously expressed idea of a more compact linking of the carbon atoms, and discussed the possible cases of how the six in benzene could be connected together, setting out with the assumption that the carbon acted as tetravalent and the hydrogen as monovalent. While the compounds of the fatty series contained—in the language then and now current—an *open chain*, Kekulé assumed in benzene a *closed* one, and pictured each of the six carbon atoms present in the molecule as being united to two others. The structural formula which followed from this was the *hexagon*, since then so widely made use of, whose angles were formed of carbon atoms linked alternately to each other by one and two bonds, and also combined in every case with one atom of hydrogen, thus—

lines of Organic Chemistry"), entirely recast by him and published under the same title, has run through numerous editions; he supplemented it in 1872 by the companion volume, *Grundriss der anorganischen Chemie.*

[1] Cf. *Ann. Chem.*, vol. clxxxviii. p. 95.

[2] *Bull. Soc. Chim.* for 1865, p. 104; *Ann. Chem.*, vol. cxxxvii. p. 129.

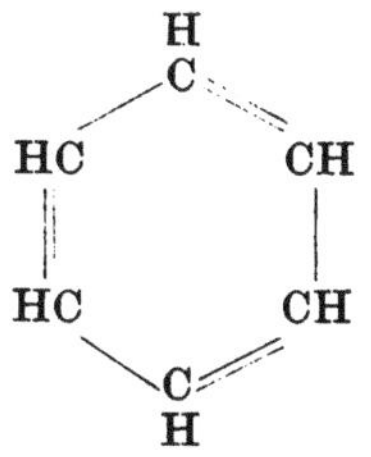

Kekulé and his pupils, together with many other chemists who had busied themselves with the derivatives of benzene after this view had been published, now directed their efforts to comparing all the known and rapidly increasing observations bearing upon this class of bodies with the deductions drawn from the above formula, and therewith to proving by actual experiment the admissibility of the assumptions on which the formula was based. An almost boundless number of facts were thus collected together, which, taken as a whole, were found to agree readily with Kekulé's hypothesis. The first inference to be drawn from it, viz. that the six hydrogen atoms which were distributed similarly among the six carbon ones were in every respect equal to one another, was confirmed by the observation, made over and over again, that only one and the same product resulted from the replacement of any one of the hydrogen atoms of benzene by a monovalent radical or element, and never a second isomeric compound. When two or three atoms of hydrogen became substituted, the case was otherwise. From his formula Kekulé deduced the number of isomers which were then to be expected; he stated his opinion that three isomeric compounds, and not more, would result in both cases through the replacement of two or three of the hydrogen atoms of benzene by the same substituent. If two dissimilar radicals took the place of two atoms of hydrogen, the number of possible isomers was not increased; these did augment, however, to a definite number when three hydrogen atoms were replaced by two or three different substituents. The truth of these and of other prognostications by Kekulé has since been verified in the most brilliant manner by a vast number of observations.

This happy interpretation of the constitution of benzene shed a great light over a hitherto neglected branch of the science. Not merely the immediate derivatives of benzene, but also substances much more distantly related to it, like naphthalene and anthracene, and more recently phenanthrene, fluorene and many other hydrocarbons, together with their numberless and often important derivatives, had their chemical constitution successfully investigated by the aid of Kekulé's hypothesis.

This hypothesis did not, however, completely satisfy a number of chemists, who considered modifications in it necessary. We need not enter here into the reasons which led to such modifications, but may just mention Ladenburg's [1] *prism formula* and Claus's [2] *diagonal* one (see appended figures), which were brought forward by those investigators as explaining more completely than Kekulé's hexagon formula the chemical behaviour of benzene.

[1] *Ber.*, vol. ii. p. 140; also his pamphlet, *Theorie der aromatischen Verbindungen.*—Albert Ladenburg, born at Mannheim on July 2nd, 1842, has been a notable contributor to organic chemistry by his excellent experimental work. His chief researches have been upon the organic compounds of silicon, the benzene derivatives, and more particularly the derivatives of pyridine and piperidine. His *Vorträge über die Entwickelungsgeschichte der Chemie in den letzten 100 Jahren* (1st edition, 1869, 2nd edition, 1887), is well known as a genuine historical work. He is editor of the chemical section of the *Encyklopädie der Naturwissenschaften* (published by Trewendt). Since 1890 Ladenburg has held the chair of chemistry at Breslau, having previously taught at Heidelberg and Kiel.

[2] *Theoretische Betrachtungen und deren Anwendung zur Systematik der organischen Chemie* (1867), ("Theoretical Considerations and their Application to the Systematising of Organic Chemistry"). Adolph Claus, born June 6th, 1840, studied under Kolbe and Wöhler, and now holds the professorship of chemistry at the University of Freiburg-im-Breisgau. His experimental work has been mainly in organic chemistry, of which he has systematically investigated various branches,—*e.g.* the derivatives of quinoline, the fatty-aromatic ketones, etc. He has further from time to time published papers giving his views on many important points of chemical theory (cf. the *Grundzüge der modernen Theorie in der organischen Chemie*, Freiburg, 1871; and also the *Journal für praktische Chemie* since 1888).

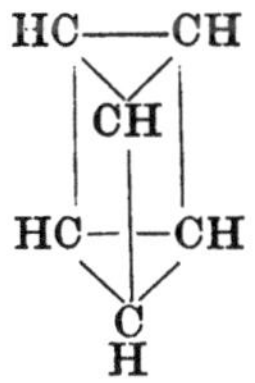

Ladenburg's formula.

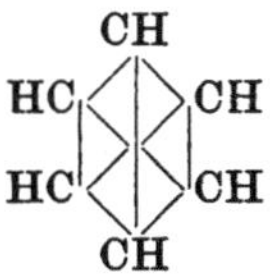

Claus's formula.

The discussions upon this point still continue; thus, the results of recent admirable investigations on the hydrophthalic acids, etc., by A. von Baeyer[1] had, he considered, given him grounds for disputing all the above hypotheses on the constitution of benzene, while Claus[2] maintained—and not without cause—that Baeyer's view was identical with his own. The latter has quite lately acknowledged[3] that Claus's formula agrees best with known facts, including those which cannot be made to harmonise with either Kekulé's or Ladenburg's hypothesis. The most recent discussions upon the constitution of benzene, naphthalene, quinoline, etc., can only be indicated here.[4]

But, notwithstanding all this, the fact must be fully recognised that Kekule's conception, even although it by no means affords a complete picture of the constitution of benzene, has borne many and rich fruits. Through the stimulus which was given by his theory of the aromatic compounds, the work of numberless chemists with this class of substances, work extending over a long period of time, received a particular stamp of its own; their chemical labours have been carried out entirely under the influence of the benzene theory.

The meaning of the term *Aromatic Compounds* has of

[1] *Ann. Chem.*, vol. ccxlv. p. 103; vol. ccli. p. 257; vol. cclviii. pp. 1 and 145.

[2] *Journ. pr. Chem.* (2), vol. xxxvii. p. 455.

[3] *Ann. Chem.*, vol. cclxix. p. 177.

[4] Cf. especially, in addition to the papers cited in note 3, p. 351, Ad. Claus, *Journ. pr. Chem.* (2), vol. xlviii. p. 576; vol. xlix. p. 505; W. Marckwald, *Ann. Chem.*, vol. cclxxiv. p. 331; Brühl, *Journ. pr. Chem.* (2), vol. xlix. p. 201; E. Bamberger, *Ann. Chem.*, vol. cclvii. p. 1; Collie, *Journ. Chem. Soc.*, vol. lxxii. p. 1013.

late years undergone a wide extension since the near relation of pyridine, quinoline and iso-quinoline and their derivatives to benzene has come to be recognised. The ardour shown in the investigation of these nitrogenous bodies, with their endless derivatives, has gone on increasing in proportion with the increasing surmise of a close connection existing between them and the vegetable alkaloids, and with the actual proof of this in some particular cases. Körner was the first to propound the important idea that pyridine may be regarded as benzene in which a methine (CH''') is replaced by the trivalent nitrogen atom.[1] The inferences drawn from this with respect to the derivatives of pyridine, like those deduced from the structure of benzene, have formed the subject of numberless experimental researches and theoretical discussions which are still proceeding. Reference will be made to some of the more important results of these investigations, and of others upon the nitrogen compounds termed polyazines, in the special history of organic chemistry.

The efforts to gain a clear conception—in the widest sense of the word—of the structure of benzene and its derivatives have also been of use in the case of other classes of compounds, especially for those analogous substances furfurane, thiophene and pyrrol, which are now universally regarded as being characterised by a closed five-membered ring containing four carbon atoms together with an atom of oxygen, an atom of sulphur, or the imido-group (NH) respectively. Victor Meyer's [2] splendid and thorough re-

[1] Dewar was the first to publish this view (*Journ. Chem. Soc.*, vol. xxiv. p. 145; or *Ztschr. Chem.* for 1871, p. 117), Körner having however already given utterance to it in his lectures.

[2] Viktor Meyer, born 8th September, 1848, after filling the post of professor of chemistry at Stuttgart and at Zürich, was called to the chief chemistry chair at Göttingen on Wöhler's death in 1885, and removed from there in 1889, to succeed Bunsen at Heidelberg; he died suddenly on August 8th, 1897. His comprehensive researches upon nitro-compounds of the fatty series, upon iso-nitroso compounds, and upon thiophene are among the very first of our time, and have contributed largely to increase our knowledge of organic chemistry. The method devised by him for vapour-density determinations has become a standard one, and has also been successfully applied to the solution of important theoretical questions (*e.g.* to that of

searches on thiophene and its derivatives [1] have before all others led conclusively to the recognition of the analogous composition of the above substances, and also to a more precise conception of the term *aromatic compounds.* According to Meyer,[2] it is the chemical behaviour of a substance with regard to nitric acid, sulphuric acid, bromine, and acid chlorides (in the presence of chloride of aluminium) which decides whether it has a claim to be ranked among those compounds. He lays here the greatest weight upon facts, whereas in previous determinations of the nature of this class of substances the existence of a closed ring of six carbon atoms was held to be a fundamental condition.

Those chemists [3] who have made a special study of the constitution of benzene, naphthalene, quinoline, etc., are at present inclined to think that the reciprocal linking of the carbon atoms may vary with the metamorphoses of the compounds in question, in such a way that the "central" bonds change into the so-called double bonds, and *vice versa*; an interchange of linkage is thus assumed. And although we have as yet no knowledge of the actual nature of these modes of linking, such speculations have a certain value, serving as they may do to a better understanding of many curious facts.

Application of Structural-chemical Conceptions to the Investigation of Isomerism.

Detailed reference has already been made to the significance which the investigation of the isomeric relations of organic compounds has for the question of their chemical

the valency of aluminium). Among his most recent researches were those on the iodo- and iodoso- compounds, and on the laws governing the esterification of aromatic acids. Lastly, Victor Meyer and Jacobsen's large *Lehrbuch der organischen Chemie* is a work of very great value. A short but appreciative memorial address on Victor Meyer by Liebermann is to be found in the *Berichte*, vol. xxx, p. 2157.

[1] Cf. his work, *Die Thiophengruppe* ("The Thiophene Group"), Braunschweig, 1888.

[2] *Ibid.*, p. 276.

[3] Cf. especially Ad. Claus, *Journ. pr. Chem.* (2), vol. xlii. pp. 24, 260, 458; vol. xliii. p. 321.

constitution.[1] Indeed, the efforts made during the last thirty years to prepare as large a number of isomers as possible, and to establish their *structure*, is a main feature of the mode in which organic chemistry has been and still is being studied. Before the derivatives of benzene had acquired that predominating interest for chemists which they afterwards came to do, the constitution of metameric substances was held to be sufficiently explained by a difference in the grouping of the atoms of the radicals. We have only to recall here the proof given of the rational composition of trimethylamine as opposed to that of the isomeric propylamine; the reason assigned for the metamerism of diethyl oxide and methyl-propyl oxide; and, lastly, to think of the secondary and tertiary alcohols or acids, whose constitution was predicted with perfect definiteness before they had been discovered (i.e. of the metamerism of dimethyl-carbinol with ethyl-carbinol, and that of trimethyl-carbinol with propyl-, iso-propyl-, or methyl-ethyl-carbinol),[2] &c.

To such satisfactorily explained cases of metamerism as these, the investigation of the aromatic compounds now added numerous others which, however, unlike the former, could not be referred back to a different grouping of the atoms in the radicals. Kekulé therefore sought to explain the similar composition of various benzene substitution products (*e.g.* of the three dibromo-benzenes, the three phenylene-

[1] Cf. p. 250 *et seq.*

[2] The rational formulæ will serve to illustrate the above cases of metamerism—

$$\underbrace{\left.\begin{matrix}C_3H_7\\H_2\end{matrix}\right\}N}_{\text{Propylamine}} \quad \underbrace{\left.\begin{matrix}CH_3\\CH_3\\CH_3\end{matrix}\right\}N}_{\text{Trimethylamine}} \quad \underbrace{\left.\begin{matrix}C_2H_5\\C_2H_5\end{matrix}\right\}O}_{\text{Diethyl oxide}} \quad \underbrace{\left.\begin{matrix}CH_3\\C_3H_7\end{matrix}\right\}O}_{\text{Methyl-propyl oxide}}$$

$$\underbrace{C^{(CH_3)_2}_{H}(OH)}_{\text{Dimethyl-carbinol}} \qquad \underbrace{CH_2(C_2H_5)\,(OH)}_{\text{Ethyl-carbinol}}$$

$$\underbrace{C(CH_3)OH}_{\text{Trimethyl-carbinol}} \qquad \underbrace{C^{C_3H_7}_{H_2}(OH)}_{\text{Propyl-carbinol}} \qquad \underbrace{C\begin{matrix}CH_3\\C_2H_5(OH)\\H\end{matrix}}_{\text{Methyl-ethyl-carbinol}}$$

dicarboxylic acids, &c.) from his conception of the structure of benzene, by assuming different *relative positions* of the substituents to one another. Such compounds were termed *position-isomers.* The question of the relative positions occupied by the entering substituents, or, as it was also called, the *determination of the chemical position* of the latter, was ardently studied from different sides, after the problem had been raised by Kekulé. Among the investigations which helped in a special degree towards the solution of this were those of Baeyer upon the constitution of mesitylene and its derivative isophthalic acid, those of Graebe upon naphthalene and phthalic acid, and that of Ladenburg on terephthalic acid. By the ingenious conclusions drawn from these and many other researches, the structure of the so-called Ortho-, Para-, and Meta-compounds was arrived at with considerable certainty. Some errors, however, did creep in here,—for instance, the wrong interpretation of the constitution of quinone from theoretical considerations, a point which gave rise to very great confusion before the mistake was finally put right. Körner's researches [1] have been of immense value for the determination of position; he introduced a new method here.

The investigation of these metameric relations among the derivatives of benzene materially lightened that of the still more complicated phenomena among the pyridine and quinoline bases which were referable to similar causes. The metamerism of the pyridine-carboxylic acids and other derivatives, which had been predicted on theoretical grounds from conceptions as to the structure of pyridine, was beautifully confirmed later on by the comprehensive researches of Weidel, Skraup and others; while considerations of the same kind have proved equally fruitful in the investigation of the derivatives of thiophene and pyrrol, and also of indole and other aromatic compounds, such as the poly-azines and poly-azoles.

But the certainty with which the constitution of metameric substances was supposed to have been established left

[1] *Gazz. Chim. Ital.*, vol. iv., p. 305.

much to be desired in many cases. The symbols employed to express the structure of such compounds were intended to have but one definite meaning; Gerhardt's view, that several formulæ might be used indifferently to picture the reactions of the bodies in question, was entirely abandoned. On the other hand, more organic compounds became known whose constitution could be illustrated equally well by two totally different formulæ, according to their chemical behaviour in different circumstances. Many of the reactions of aceto-acetic ether, for instance, cause us to give to it the constitution which is apparent in its name, but in others it behaves like the ether of an oxy-crotonic acid; indeed L. Claisen[1] has just proved that its sodium compound is derivable from the latter ether. Phloroglucin, which has been for long, and justly, looked upon as trioxy-benzene, may also be indicated, from some of its reactions, as a metameric tricarbonyl compound.[2]

The constitution of these, as well as of certain other compounds, *e.g.*, isatin, oxindole, carbostyril, cyanamide, etc., is therefore capable of two explanations. Opinions are still divided among chemists who have busied themselves with this question as to which of the two possible structural formulæ is the correct one for such compounds. Baeyer distinguishes between a stable (*stabile*) modification and an unstable (*labile*) one, the latter being termed the *pseudo-form*; for isatin, *e.g.*, the formula containing hydroxyl is the stable

[1] *Ann. Chem.*, vol. ccxcvii. p. 92.

[2] The *tautomerism* of the above compounds is seen from the following formulæ:—

$$\begin{array}{l} CH_2{\cdot}CO{\cdot}CH_3 \\ | \\ CO{\cdot}OC_2H_5 \end{array} \qquad \begin{array}{l} CH{=}C(OH)CH_3 \\ | \\ CO{\cdot}OC_2H_5 \end{array}$$

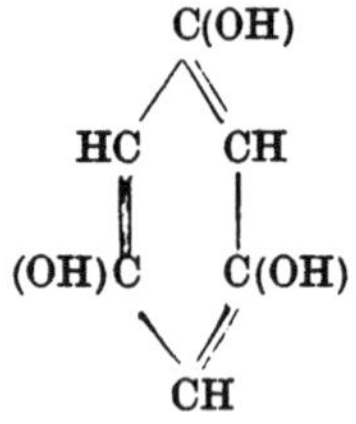

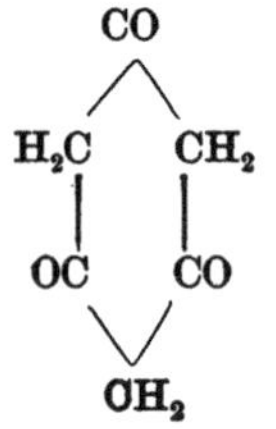

modification, while pseudo-isatin is unknown in the free (or unstable) state, only derivatives of it being capable of existence.

C. Laar,[1] who has discussed this question minutely, applies the name *tautomerism* to these phenomena. A "change in combination or position of hydrogen atoms"[2] is always involved here, as is readily seen in what is doubtless the simplest case of such a tautomerism—in hydrocyanic acid. The chemical behaviour of this acid leads on the one hand to the structural formula $H-C\equiv N$, and on the other to that of $C=N-H$ (in which the carbon is divalent); in the former case the hydrogen is linked with carbon, and in the latter with nitrogen. Laar imagines oscillatory conditions within the hydrocyanic acid molecule, which cause the hydrogen atom to take up the one and the other position alternately; he therefore presupposes the simultaneous existence of both modifications. Since all cases of tautomerism depend upon a change in the linking of the atoms of carbon, nitrogen and oxygen with respect to hydrogen, Victor Meyer and Jacobsen have proposed to replace the above indefinite term by the better one of *desmotropism*.

During the last few years experimental and speculative work have added largely to the number of known tautomeric compounds. In a lecture entitled "*Ueber Tautomerie*," delivered at Stuttgart last year (1897), W. Wislicenus gave an excellent *résumé* of the most important investigations in this field. Of special interest are those still rare cases in which the two tautomeric forms of a compound have actually been observed, by W. Wislicenus himself and by L. Claisen. Under these circumstances the former is justified in concluding that tautomeric phenomena are reversable intra-molecular changes which only lend themselves to observation in exceptional instances. According to F. Traube,[3] "tautomerism is a particular kind of isomerism in which we have to do with a state of equilibrium, excessively sensitive

[1] *Ber.*, vol. xviii. p. 648; vol. xix. p. 730.
[2] *Ein "Bindungs- oder Platzwechsel von Wasserstoffatomen."*
[3] *Ber.*, vol. xxix. p. 1723.

to outward conditions, of two isomers that change very readily the one into the other."

Here then we have an instance of the constitution of one and the same compound being expressible by two structural formulæ, either one of them apparently as correct as the other. In another group of metamers we find just the opposite conditions, *i.e.*, one and the same structural formula applying to two totally different chemical compounds of the same composition. J. Wislicenus[1] was the first to establish such an *identity in structure* (*Strukturidentät*) for two different substances—the fermentation- and para-lactic acids.[2] The structure theory is therefore insufficient to explain such cases of metamerism as this. Further instances of the same kind are found in crotonic and iso-crotonic, fumaric and maleïc, and mesaconic and citraconic acids. Wislicenus designated this species of metamerism *geometrical isomerism*, and Michael, who has likewise occupied himself for a long time with the study of this branch, *allo-isomerism*. These phenomena are now grouped under the term *Stereo-isomerism*, and the rapidly growing *Stereo-chemistry* now forms a distinct branch of the science.

J. Wislicenus[3] has attempted to explain phenomena of

[1] Johannes Wislicenus, born at Klein-Eichstedt, near Querfurt in Thüringen, on 24th June, 1835, became in 1885 professor of chemistry and head of the chief chemical laboratory in the University of Leipzig, after filling from 1872-85 the corresponding post at Würzburg, before which he taught at Zürich. After the death of Strecker, whom he succeeded at Würzburg, he re-edited the former's text-book of chemistry. His experimental researches, most of which have been published in the *Annalen der Chemie*, pertain almost exclusively to the domain of organic chemistry, in the special history of which we shall frequently have occasion to refer to them. The very important work which he did on the lactic acids impelled him even so early as 1873 to the conclusion that the cause of the difference between two of them must be sought for in the spacial relations of the atoms in the molecule. His quite recent speculations upon geometrical isomers are referred to above.

[2] *Ann. Chem.*, vol. clxvii. p. 343.

[3] Cf. *Die räumliche Anordnung der Atome in organischen Molekülen* (Leipzig, 1887), ("The Spacial Arrangement of the Atoms in Organic Molecules"); also the *Tageblatt der Naturforscherversammlung zu Wiesbaden*, 1887 ("Journal of the Assembly of Scientists at Wiesbaden, 1887").

this kind by the aid of an hypothesis propounded by van 't Hoff and Lebel.[1] According to this hypothesis, which was designed with the object of explaining the optical activity of isomeric compounds, the centre of gravity of an atom of carbon is supposed as in the middle of a tetrahedron, and its four affinities as at the four corners. When two atoms of carbon become linked together, with the subsequent neutralisation of one affinity of each, then van 't Hoff and, after him, Wislicenus assume that both are capable of rotating in opposite directions about a common axis; and the possibility of such rotation is supposed to cease with the double or triple linking of the carbon atoms. Wislicenus has made this hypothesis the basis of his discussions and his later experimental researches. An important aid to this conception is added in the supposition that, in the rotation of systems with carbon atoms linked together by one affinity of each, "specially directed forces, the affinity-energies," come into play, which regulate the spacial relations of the atoms to one another. Wislicenus believes that in these suppositions he possesses a means whereby "the establishing of the spacial arrangement of atoms in particular cases may be arrived at by experiment."

The theory which is based upon the presence of asymmetric carbon atoms in chemical compounds is in point of fact supported by many important observations. In the first place it is to be noted that all optically active organic compounds, whose constitution is established, contain one or more asymmetric carbon atoms. The observations which

[1] Cf. van 't Hoff's pamphlet, *Dix Années dans l'histoire d'une Théorie* (1887). Van 't Hoff first published his views on the subject in the small volume, *La Chimie dans l'Espace*, in 1875 (English edition by Marsh, under the title *Chemistry in Space*, 1891; and German, by Herrmann, 1877 and 1894). Le Bel also brought out the same hypothesis, independently of van 't Hoff, in the *Bull. Soc. Chim.* (2), vol. xxii. p. 337. Messrs. Longmans and Co. have just published (1898) a second revised and enlarged English edition of van 't Hoff's *The Arrangement of the Atoms in Space*, with a preface by Johannes Wislicenus, and an appendix entitled "Stereo-Chemistry among Inorganic Substances," by Alfred Werner. The book is translated and edited by Arnold Eiloart, who has made a special study of this branch.

have been made upon racemic, malic, mandelic and lactic acids, and upon a number of other substances, are in perfect accord with the above theory. The plan of breaking up inactive into active modifications, which was first followed by Pasteur[1] with such striking success, has since been applied in many other cases with equally good results. The theory has proved especially fruitful during the last decade, as applied by Emil Fischer[2] in his brilliant researches on the sugars.

Further, A. von Baeyer's important work upon the hydrophthalic acids,[3] whose isomerism is without doubt due to differences in the spacial arrangement of the atoms, constitutes a strong support for the theory of the asymmetric carbon atom in "ring-shaped" structures.

The investigation of the isomerism of certain compounds, in which the so-called "double linkage" of carbon is to be found, has proved exceptionally fruitful. The work undertaken by Johannes Wislicenus and his pupils[4] upon fumaric and maleic, crotonic and iso-crotonic, angelic and tiglic acids, and their halogen derivatives, with the object of getting at the root of these phenomena, has led to surprising results, which however do not harmonise with theory in many respects. In fact the investigations of A. Michael[5] and

[1] Louis Pasteur (born at Dôle on December 27th, 1822, died at Paris on September 28th, 1895) has proved a great pioneer in chemical as well as in the biological sciences. It was indeed his systematic work upon optically active compounds, especially the tartaric acids, which led him on to the treatment of biological questions,—to the isolation and artificial culture of pure ferments. His researches upon the alcoholic, lactic and acetic fermentations constituted him a chief founder of the new zymo-chemistry and bacteriology. The brewing industry is deeply indebted to him for the improvements which he brought about in it. Following on those researches we have his great work on inoculation against splenic fever, dysentery and hydrophobia. He belongs truly to the great benefactors of mankind.

[2] *Ber.*, vol. xxiii. p. 2114, vol. xxiv. p. 3997. See also the Special History of Organic Chemistry.

[3] *Ann. Chem.*, vol. ccxlv. p. 103; vol. ccli. p. 257; vol. cclvi. p. 1; vol. cclviii. pp. 1 and 145; vol. cclxvi. p. 169; vol. cclxix. p. 145.

[4] See the pamphlet already quoted; also *Ann. Chem.*, vol. ccxlvi. p. 53; vol. ccxlviii. pp. 1 and 281; vol. ccl. p. 224.

[5] Cf. especially *Journ. pr. Chem.* (2), vol. xlvi. p. 400, besides preceding numbers.

others [1] have shown that contradictions occur in them which throw doubt upon some of the theoretical hypotheses.

This idea of referring the cause of many cases of isomerism to the different geometrical arrangement of the atoms has had a most stimulating effect, and has led to the discovery of many hitherto overlooked relations existing between isomeric substances. The work done upon the dichlorides of tolane, the butylenes, the isomeric cinnamic acids, erucic and brassidic acids, and upon the alkyl-succinic acids deserve mention here.[2] Of recent years there have been numerous speculations advanced with the object of bringing conflicting phenomena into accord with theory, *e.g.* Victor Meyer and Riecke's [3] ideas upon the "constitution of the carbon atom," and Bischoff's[4] "dynamic hypothesis" of certain cases of isomerism.

All this work is due to the circumstance, of which there can no longer be any doubt, that *geometric-chemical* isomers do really exist. During the last few years there have been similar observations with regard to various nitrogen compounds, and efforts have been made to trace these cases of isomerism back to spacial relations,—to the configuration of the nitrogen atom. It is more especially in those compounds in which we have a double linkage between the carbon and nitrogen, *i.e.* $=C=N—$, that such isomers have been noticed. The theory of the stereo-isomerism of nitrogen compounds, —a theory due to a great extent to Werner and Hantzsch,[5] —is based upon the work of the late Victor Meyer and Auwers, of Beckmann, and particularly of A. Hantzsch himself on the oximes of aldehydes and ketones, together

[1] Skraup, Wiener *Monatshefte*, etc, vol. xii. p. 119; Anschütz, *Ann. Chem.*, vol. ccliv. p. 175.

[2] Cf. Special History of Organic Chemistry.

[3] *Ber.*, vol. xxi. p. 951.

[4] *Ber.*, vol. xxiii. p. 1467.

[5] *Ber.*, vol. xxiii. pp. 1 and 1243. For the literature on the subject, see Hantzsch's *Grundriss der Stereochemie* (Breslau, 1893). On p. 106 the vital part of this theory is expressed as follows:—"In the language of the valency theory, the geometrical isomerism of nitrogen compounds . . . . depends upon the three valencies of the nitrogen atom not being in the same plane in certain of these compounds."

with quite recent observations on the hydrazones and the carbo-di-imides by Overton and others. There is no question that a large number of important cases of isomerism have been in a way explained by the assumption of spacial differences in the relation of the nitrogen to the carbon atom. But whether stereo-isomerism is to be assumed in as many cases as Hantzsch is inclined to believe, is for the present doubtful (compare the diazo-compounds in the *Special History of Organic Chemistry*).

It is impossible to give a definite answer to the question whether the spacial arrangement of the atoms within a molecule actually corresponds with the configurations assumed by the above-named scientists, for no proof can be furnished of the correctness of these conceptions. The expectations raised by them—of obtaining a deeper insight into the mode in which the atoms are arranged in a compound—are possibly pitched too high. Criticism has indeed begun—as already indicated—to busy itself with the explanation of geometrical isomerism in particular cases,[1] but stereo-chemical theories[2] are not yet sufficiently advanced to give us a clear view of the whole subject. The time has not yet come for an objective historical account of stereo-chemistry, in which theory and fact shall have their true values assigned to them; the subject is but in its infancy.

[1] Ad. Claus has been especially active in disputing the correctness of the stereo-chemical view as applied to the isomeric oximes; cf. *Journ. pr. Chem.*, vol. xliv. p. 312; vol. xlv. pp. 1, 556; vol. xlvi. p. 544.

[2] Hantzsch's *Grundriss der Stereochemie* gives a good summary of the work done in this branch of chemistry up to the year 1893. Compare also Auwers' *Die Entwickelung der Stereochemie* (Heidelburg, 1890), and C. A. Bischoff and P. Walden's *Handbuch der Stereochemie* (vol. i, 1894), which goes minutely into the subject. In English there is Eiloart's book *A Guide to Stereo-Chemistry*.

The Development of Important Methods for investigating the Constitution of Organic Compounds.

The above-mentioned discussions upon isomers are sufficient to show us how materially these have aided the development of organic chemistry since the subject was zealously taken in hand. Hardly any other group of phenomena has furthered the solution of the question of chemical constitution in a more lasting manner, for the attempts to establish the constitution of isomeric bodies have coincided with those whose aim was to fathom the cause of isomerism. The methods followed during the last decade for investigating the rational composition of organic compounds have in great part developed themselves from others previously in use. The paths which have led towards the wished-for goals were smoothed by the indispensable preparatory labours of Liebig, Wöhler, Bunsen, Kolbe, Frankland, Dumas, Williamson, Gerhardt, Hofmann, Wurtz and others.

Synthetic Methods.

The mode of attaining to a knowledge of the constitution of organic compounds, which had been least worked out of any, was their artificial preparation from others of simpler composition. After Wöhler had published his memorable observation on the production of urea from its elements, and had therewith furnished a complete synthesis of it, years elapsed before any further successful work in this direction fell to be recorded. Referring the reader to the special history of organic chemistry, we need merely recall here the important discoveries during the fifties by Kolbe and Frankland,—the synthesis of acetic acid by the former, and the building up of hydrocarbons from substances of simpler composition by the latter.

The importance of synthetic research was from thenceforth recognised in an increasing degree;[1] indeed, it was from artificial modes of preparation that the constitution of many organic substances could first be deduced with certainty. Thus (to give only one or two instances) the rational composition of acetic acid was arrived at from its production from the methyl compounds—methyl cyanide and sodium-methyl. The constitution of hydrocarbons was inferred from their synthesis from halogen-alkyls with zinc or sodium, and that of the ketones through their formation from acid chlorides and zinc-alkyls. Light was thrown upon the true composition of the oxy-acids by their synthesis from aldehydes or ketones and hydrocyanic acid, and also from phenates and carbonic acid. And to what a wealth of synthetic reactions and discoveries of new compounds have not the sodium derivatives of certain acid ethers—*e.g.* aceto-acetic and malonic ethers—led![2]

In every section of the wide field of organic chemistry, great success has followed the application of synthetic methods; and the worth of these latter is not to be measured merely by the vast number of new compounds to which they have given rise, but by their own intrinsic value, which has shown itself in the knowledge thereby gained of the chemical constitution of organic compounds. The so-called *condensation* syntheses have proved themselves of especial value in this direction. This term "condensation" has, since Baeyer's explanation on the subject, been employed generally for those reactions in which several similar or dissimilar molecules coalesce together, with elimination of water, in such a manner that the carbon atoms become linked to one another. A classical instance of it (observed a long time ago by Kane, but first explained by Baeyer, as above) is given in the transformation into mesityl oxide—or into phorone—and then into mesitylene

[1] In 1889 chemical literature was enriched by an admirable systematic "Textbook of Syntheses" on a historical basis, in K. Elb's *Die Synthetischen Darstellungsmethoden der Kohlenstoffverbindungen.*

[2] With reference to these and other syntheses, cf. the special history of organic chemistry.

which acetone experiences under the influence of sulphuric acid. Similar reactions go on in the case of other ketones and of aldehydes—*e.g.* the condensation of acetic to crotonic aldehyde (Kekulé)—and that of a mixture of acetic and benzoic aldehydes to cinnamic aldehyde. Through these and other processes a bridge was thrown over the gap between the saturated and unsaturated compounds, while at the same time light was shed upon the constitution of the latter. The reaction discovered by and called after W. H. Perkin, sen., which depends on the condensation of aldehydes with fatty acids, formed the basis of some notable researches by Fittig, Claisen and others, while it likewise aided in clearing up the rational composition of unsaturated acids.

A. von Baeyer,[1] in conjunction with a large number of his pupils (E. and O. Fischer, v. Pechmann, Königs, Knorr, E. Bamberger, Paal, etc.), has minutely investigated this subject of condensation in the most admirable manner, as have also Kekulé, Fittig, Ladenburg, Victor Meyer, Hantzsch, Claisen, W. H. Perkin, Graebe, Liebermann, Collie, and, in fact, almost all chemists who have occupied themselves with organic chemistry of recent years; indeed, this study seemed for a time to be the chief feature of organic chemistry. The ardour for carrying out such syntheses increased more especially after it was seen that the chemical processes going on in plant organisms—*i.e.* the formation of compounds rich in carbon from carbonic acid,

[1] Adolf von Baeyer, born at Berlin on 30th November, 1835, became a pupil of Bunsen and of Kekulé, and applied himself under the stimulating influence of the latter to organic chemistry, which he has enriched by a wealth of admirable and important work. His untiring study of condensation reactions has led him to results of the highest value, which will frequently be referred to in the special history of organic chemistry. From his laboratory there has come forth much work of a fundamental nature; we need only recall here that of Graebe and Liebermann on alizarin, and that of E. and O. Fischer on rosaniline, etc. Since 1860, in which year Baeyer became assistant professor in Berlin, he has continued energetic as a teacher—first at the Berlin Technical College, then from 1872-75 in Strasburg, and lastly, from 1875, in Munich, where, as head of the University laboratory, which was built after his own plans, he has found a brilliant sphere of action.

water, and ammonia—were for the most part based upon condensation. The history of organic chemistry can tell of many results of efforts to imitate such natural processes, or at least to prepare products which occur in the vegetable kingdom (acids, colouring matters, alkaloids, carbohydrates, etc.) from substances of simpler composition. The most important of those vegetable acids which had long been known were prepared synthetically,—oxalic acid from carbonic, succinic acid from ethylene, malic and tartaric acids from succinic, and citric acid from acetone (which, like ethylene, could be built up from its elements); further, benzoic acid from benzene, cinnamic acid from benzaldehyde, and so on. By those observations, the list of which might be extended by numerous others on the artificial formation of acids occurring in the animal and vegetable kingdoms (*e.g.* the syntheses of chelidonic, vulpic, and uric acids), the chemical constitution of these substances was determined with greater precision than had hitherto been possible.

Similarly from the synthesis of vegetable colouring matters and other bodies—*e.g.* alizarin, purpurin, indigo blue, cumarin and vanillin,—trustworthy conclusions have been drawn with respect to their rational composition. The important problem of preparing the sugars and vegetable alkaloids artificially has been taken in hand with success,—witness the beautiful researches of Emil Fischer[1] upon carbohydrates, which have lately led to the artificial formation of grape sugar, and the ingenious synthesis of conine by Ladenburg.[2]

One may safely express the opinion that a clear idea of the chemical constitution of these and other difficultly accessible classes of compounds, whose proximate composition has as yet been but imperfectly worked out, will only be arrived at after they have been synthetised from simpler ones of known structure. The history of the synthesis of organic compounds has already proved the truth of this axiom in numerous instances.

[1] Cf. the special history of organic chemistry. [2] *Ibid.*

The chemical behaviour of organic compounds is in every case regarded as an aid of the first importance in working out their constitution, and has been valued accordingly ever since organic chemistry began to flourish. A short sketch only can be given here of a few of the more important methods which have been applied during these last decades, with the object of getting at the chemical constitution of organic compounds from their reactions, transformations and decompositions.

The general principle of such methods consists, in contradistinction to that of the synthetic, in investigating the products obtained by the chemical alteration of the compounds in question, and in deducing the constitution of the latter from these. In many cases of transformation the chemist keeps his attention fixed upon particular elements or atomic groups united to carbon, the carbon framework itself undergoing no change; in many others, on the contrary, carbon is separated as carbonic acid, carbonic oxide, or even in a more complex form. For those classes of substances which are among the best investigated, special reactions have been discovered which make it possible to decide whether a hitherto unknown compound belongs to this or that group. Of recent years great attention has been paid to the refinement of such reactions. To mention only one or two important steps in this direction:—Phosphorus pentachloride, acetic anhydride and hydriodic acid have been found of inestimable value for determining whether an organic compound contains hydroxyl, and, if so, what function that hydroxyl performs. Further, the transformation of nitro- into amido-compounds by reduction, and that of the latter into oxy-derivatives by oxidation, the conversion of cyanides into carboxylic acids, of hydrocarbons into acids, and of amido- into diazo-compounds, have all become typical reactions, which, when rightly interpreted, lead very quickly to the explanation of the constitution of such bodies. Lastly, we may recall here the beautiful method of V. Meyer and E. Fischer, by which the presence of the carbonyl group in a compound can be proved by means of hydroxylamine or phenyl-hydrazine. All the above and other similar reactions have for their aim the definite recognition of the *rôle* of elementary atoms or compound radicals in organic molecules, and with this the partial solution of the constitution of these latter; in numberless instances this aim has been accomplished.

The decompositions of organic substances into others poorer in carbon, which may be made use of for deciding the same point, are legion, and will just be touched upon here, in order to illustrate the principle of the method. This plan is the direct opposite of the synthetic; while by the latter the constitution of an organic compound is deduced from that of its components, the former leads to the same conclusion through a study of the resulting decomposition-products. To give only one or two examples:—Let us recall the important inferences drawn by v. Baeyer from the decomposition of derivatives of uric acid into simpler bodies; the constitution of those compounds thus deduced by him was subsequently confirmed by direct synthesis. The researches by Frankland, Geuther, Wislicenus and others on the modes of decomposition of aceto-acetic ether must also be mentioned, researches which, conjointly with other synthetic ones, cleared up the constitution of the latter. Further, carbonic acid, formic acid, etc., are very often eliminated from organic compounds, whose decomposition-products thus furnish a clue to their rational composition. The changes produced by oxidation in the case of numerous substances, such as the ketones, quinoline bases, naphthalene derivatives and unsaturated compounds, furnish excellent proof of the invaluable aid given by researches of this nature towards solving the question of chemical constitution. For further details on this point, the reader is referred to the special history of organic chemistry.

By this co-operation, by the use of the various methods which are now an integral part of organic chemistry, the problem of the rational composition of carbon compounds has been brought distinctly nearer to its solution.

The Main Currents in Inorganic and General Chemistry during the last Thirty-five Years.

The doctrine of the saturation-capacities of the elements, which has proved of such extraordinary importance for the development of organic chemistry, has not by any means found the same rapid and general application in inorganic. After Odling, so early as 1854, had applied Frankland's idea of valency to the oxides of a large number of the elements, remaining however at the same time enchained by the type theory (cf. p. 325), gradual attempts were made by a number of chemists, either in text-books or in their experimental researches, to engraft on inorganic compounds the ideas which had so quickly found acceptance with respect to the linking of carbon atoms among themselves or with other elements. The gain which arose from this was first apparent in the systematising of these compounds, which became classified into natural families according to the valencies ascribed to the individual elements. Similarity in saturation-capacity formed the common link which held the different members of such groups together. Thus Frankland had already recognised the analogy between nitrogen, phosphorus, arsenic and antimony, from the fact that they were all capable of acting either as tri- or as pentavalent. Alongside of carbon were ranged silicon, titanium and zirconium, as being in the main tetravalent elements, whereas boron, which had formerly been ranked along with carbon, was seen to be trivalent, and was relegated to another group. These and similar efforts to introduce clearness into the systematising of the elements, by classifying them according to their chemical values, soon led to the establishment of the important *Natural System of the Elements* (cf. p. 370).

The problem of interpreting the constitution of inorganic compounds similarly to that of organic, by getting at the relations which exist between their component elements, has not been treated with the same care as in the case of the latter. For substances of simple composition the difficulty of the point was usually under-estimated; this showed itself more particularly in the arbitrary attempts at explaining the constitution of inorganic compounds on the supposition that the valencies of the elements were invariable. Thus it was often overlooked that the chemical behaviour of a substance was not in accordance with the structural formula assigned to it. Sulphur chloride, for example, was given the formula,

$$\begin{array}{l} S - Cl \\ | \\ S - Cl \end{array},$$

without any heed being paid to the fact that one of its atoms of sulphur behaved quite differently from the other. And the constitution of phosphorus oxychloride could only be illustrated by the adherents of constant valency by the formula,

$$P \begin{array}{l} \diagup O - Cl \\ - Cl \\ \diagdown Cl. \end{array}$$

, a formula which indicated an (unproven) difference between one chlorine atom and the other two.

And how the ordinary rules were strained in order to indicate the composition of more complex compounds! According to Wurtz,[1] the constitution of bodies rich in oxygen could usually be explained by assuming the oxygen atoms to be linked to one another; take, for example, periodic anhydride, in which seven atoms of oxygen were linked together in a chain, with the two supposed monovalent iodine atoms at either end. This very one-sided assumption of a constant valency of the elements was however gradually superseded, a sounder view taking the place of such artificial explanations. But trustworthy methods of arriving at the constitution of complex compounds are scarcely yet developed in inorganic chemistry, although in organic much has already been done in this direction.

The researches of greatest value for inorganic chemistry which have been made during the last few decades are those upon particular elements, more especially upon such as had hitherto been imperfectly or even not at all investigated.

[1] *Leçons de Philosophie Chimique*, p. 157.

Thus the work of Roscoe[1] on vanadium, of Marignac[2] on niobium and tantalum, and of Zimmermann, Krüss, von der Pfordten, Moissan and others on uranium, gold, titanium, fluorine, etc., have enabled those elements to be put in their proper place among the others; this of course only became possible after their chemical character had been thoroughly examined. The same applies to the more recently discovered elements—thallium, indium, gallium, scandium, germanium, etc., which have likewise been investigated by their discoverers in a masterly manner.

All these researches, which will be referred to again in the special history of inorganic chemistry, have had the same ends in view, viz. the establishment of the chemical character, and, in particular, of the combining relations of the element in question, and the most careful possible determination of its relative atomic weight. In addition to all this, an increasing value has come to be laid upon the observation of its physical properties. Such investigations upon individual elements became more systematised after it was clearly seen that a close connection existed between their chemical and physical properties on the one hand and the magnitudes of their atomic weights on the other. Of course, when it came to a question of proving this intimate relation, the first thing was to determine the

[1] Sir Henry E. Roscoe, born in 1833, was a pupil of Bunsen's. For nearly thirty years he held the chair of chemistry at Owens College, Manchester, resigning in 1885. His work has been for the most part in inorganic and physical chemistry, the *Photochemical Researches* by Bunsen and Roscoe (London, 1858-1863) deserving mention here. He is also well known as the joint author of Roscoe aud Schlorlemmer's *Treatise on Chemistry*, as well as of other smaller text-books on the science. His *Lessons in Elementary Chemistry* has run through numerous editions, and has been translated into a great many different languages.

[2] J. C. Marignac, born at Geneva in 1817, retired several years ago from the professorial work to which he had devoted himself in his native city since 1842, and died there on April 15th, 1894. With the exception of some researches on the naphthalene derivatives, his most important work has been in the determination of the atomic weights of numerous elements, and in other subjects of inorganic chemistry. A detailed account of his life and his services to the science has been given by E. Ador in the *Archives des Sciences Physiques et Naturelles*, vol. xxxii, p. 5.

relative atomic weights as accurately as it was possible to do.

The efforts of many chemists had already for a long time been directed to improving as far as practicable the methods of determining atomic weights, before the importance of this question for the systematising of the elements had come to be recognised. The memorable labours of Berzelius were followed during the forties by those of Turner, Dumas, Marignac, Erdmann, Marchand and Pélouze, and were crowned by the classical researches of Stas[1] upon the atomic weights of oxygen, chlorine, bromine, iodine, nitrogen, sulphur, silver, etc. In Stas's case the extreme limit of accuracy was reached which was possible with the means at command. But this certainty with respect to the magnitudes of the relative atomic weights only extended to some of the elements, the values hitherto assigned to many (*e.g.* molybdenum, antimony, platinum, osmium, iridium, etc.) being exceedingly inaccurate. Much has, however, been accomplished in this direction of late years.[2]

The Periodic System of the Elements.

Prout's hypothesis, according to which the atomic weights of all the elements stand in a simple relation to that of hydrogen, acted for a long period like a ferment, in that it gave rise to continually renewed speculations upon the connection which existed between the elements and their atomic weights. The observed fact that chemically analogous ele-

[1] Jean Servais Stas, born at Löwen in Belgium in 1813, died in 1891 at Brussels, where he had occupied the chair of chemistry in the Military School for a number of decades. His unique services in the determination of the atomic weights of the elements are universally recognised. The various papers on this subject were published by him in a collected form in the well-known work, *Recherches sur les Rapports réciproques des Poids Atomiques*, and in the *Nouvelles Recherches sur les Lois des Proportions Chimiques*, etc. Organic chemistry and forensic analysis are also indebted to him for most important investigations (see Special History).

[2] Cf. the *Special History of Inorganic Chemistry*.

ments possessed either nearly equal atomic weights, or atomic weights separated from one another by definite numerical increments, afforded food for such theorising. For almost seventy years attention has frequently been drawn, with more or less emphasis and ability, to regularities of this kind; the discussions of the point by Döbereiner, L. Gmelin, Pettenkofer, Dumas, Kremers, Odling and others may be recalled here.[1] But it is only of comparatively recent years that a systematic classification has followed from those efforts to discover a connection between the atomic weights and the nature of the elements.

In the year 1864 Newlands[2] in England and Lothar Meyer[3] in Germany—independently of one another—arranged a number of the elements according to the magnitudes of their atomic weights,[4] and thereby observed that while, at a superficial glance, the elements following one another showed apparently no regularity in properties,

[1] Cf. L. Meyer's *Moderne Theorien* (fifth German edition), p. 133.

[2] *Chem. News*, vol. xxxii., pp. 21 and 192; also Newlands' pamphlet, *The Discovery of the Periodic Law* (London, 1884). Mendelejeff, in his *Grundlagen der Chemie*, p. 683, calls attention to the fact that, so early as 1862, some parts of the periodic law were enunciated by Chaucourtois.

[3] Lothar Meyer, born 19th August, 1830, filled from 1876 until his death on April 29th, 1895, the first chair of chemistry in the University of Tübingen, after having previously worked as an academic teacher in Breslau, Neustadt-Eberswalde and Karlsruhe. His first experimental researches dealt with questions of physiological chemistry; but he afterwards turned his attention more to theoretical and physico-chemical problems. The outcome of this was his valuable work, *Die Modernen Theorien der Chemie* (fifth edition, 1884), which has been translated into English by Professors Bedson and Carleton Williams under the title, *Modern Theories of Chemistry*; compare also his *Grundzüge der theoretischen Chemie* (1890). The efforts mentioned above, which he made with the object of firmly establishing the periodic system of the elements, led him on to a careful review of all that had been written on their atomic weights (cf. his and K. Seubert's meritorious work, *Die Atomgewichte der Elemente aus den Originalzahlen neu berechnet*, 1883) ("The Atomic Weights of the Elements newly Recalculated from the Original Numbers"). A detailed account of Lothar Meyer's life and work, from the pen of his pupil Seubert, is to be found in the *Berichte*, vol. xxviii, Ref. p. 1109; and another by Bedson in the *Journal of the Chemical Society* for 1896, p. 1403.

[4] Cf. *Moderne Theorien* (first German edition, 1864).

after the lapse of a certain *period* the chemical and physical behaviour of the elements now succeeding each other strongly recalled that of the previous group, in fact repeated it. The elements which resembled one another were therefore united into groups or *natural families*, and these in their turn were distinguished from the *periods*, which comprised the elements whose atomic weights lay between those of two successive members of a natural family. This attempt to classify the elements according to the magnitude of their atomic weights, and to deduce from this an important connection between the latter and the properties of the former, called forth at first more astonishment than recognition. Indeed, Newlands did not escape banter on the subject, being asked whether he would not try, with a similar result, to classify the elements according to the initial letters of their names.

After the year 1869 these very imperfect beginnings were soon greatly extended and improved by Mendelejeff[1] and Lothar Meyer,[2] quite independently of one another, the atomic weights of various elements having in the meantime been determined with greater accuracy than before. Mendelejeff made what was for that time the bold attempt to classify all the elements according to the magnitudes of their atomic weights, the correctness of some of which was extremely doubtful. He was thus able to show that the elements which belonged to a natural family, *i.e.* those which were chemically similar, followed one another in regular periods. In this way the elements were brought together into a *natural system*, as it was termed, in which

[1] *Ztschr. Chem.* for 1869, p. 405; and more fully, *Ann. Chem.*, Supplement, vol. viii., p. 133.—D. J. Mendelejeff, born at Tobolsk on February 7th, 1834, has devoted himself to researches on physical constants. *e.g.* specific volumes, expansion of gases, etc. He is best known by his famous treatise, *Die Periodische Gesetzmässigkeit der chemischen Elemente*, and also by his very original text-book, *Grundlagen der Chemie*. Since 1866 he has held the chair of chemistry in the University of St. Petersburg, having previously occupied that in the Technological Institute there.

[2] *Ann. Chem.*, Supplement, vol. vii., p. 354; and also in the recent editions of his *Moderne Theorien*.

however, there was much that was arbitrary because of the inaccuracy of many of the atomic weights. But the fundamental idea developed by the above investigators, viz. that the elements arrange themselves on the one hand into periods, and on the other into natural families, and that all their properties are periodic functions of their atomic weights, has been strengthened and verified in every direction by many subsequent investigations. The latter applies more especially to the chemical valency of the elements, the electro-chemical character, the atomic volume, the thermo-chemical behaviour and other physical properties, all of which stand in periodic dependence to the magnitude of the atomic weight.

These efforts, so invaluable for the systematising of the elements, have led to many important deductions. Thus, in virtue of the periodic system, definite values could be assigned to the hitherto uncertain atomic weights of various elements; for each element claims a place of its own in this system and an atomic weight corresponding with this place, the magnitude of the latter being calculable within certain limits. When, for example, only the equivalent of an element was known, the atomic weight could be deduced from its behaviour and from the position thus accruing to it in the natural system, as was actually done, *e.g.*, for beryllium and indium. Further, a choice could be made between different definite values for one and the same element, and the more suitable one taken, to be afterwards verified, of course, with the utmost care. In this way the periodic system has been applied in the happiest manner to correcting the atomic weights of molybdenum, antimony, cæsium, etc.

Other conclusions of a speculative nature have likewise been drawn with the best results from this classification of the elements into periods and natural families. The gaps shown by the system at the time it was brought forward, and in a lesser number to-day, were and are intended to be filled up by new and hitherto undiscovered elements.

Mendelejeff sought to predict from the positions of such blanks, not merely the existence of elements and their approximate atomic weights, but also their properties and chemical behaviour, together with that of some of the compounds which they would form. His prognostigations have been fulfilled in the most striking manner by the discovery of gallium, scandium and germanium, and by the verification of their behaviour as foreseen by him. On the other hand, the position in the periodic table of the elements argon and helium, recently discovered by Rayleigh and Ramsay, presents great difficulty, as these elements show no definite chemical reactions. (Cf. the *Special History of Inorganic Chemistry.*)

The perception of the fact that the physical and chemical properties of all the elements show a periodic dependence upon their atomic weights is therefore a result of this natural classification. But the discovery of the common cause which underlies these peculiar relations, and its formulation into a law, still remain tasks for the future. Some chemists have thought to lift this veil already by assuming that all the various elements, or at least those belonging to a natural family, may be referred back to still simpler ones. We perceive clearly here a reapproach to Prout's hypothesis, which threatened to exercise such an unfavourable influence on the rational development of the atomic doctrine, had not the ablest chemists of the time raised a protest against its admissibility. During the last few years Crookes has again brought up this ticklish question, whether the so-called elements are to be regarded as simple, and not rather as compound.[1] According to him, all the elements have resulted by gradual condensation from a primary material which he terms *protyle*, this view having been arrived at from his observations on the phosphorescence spectra of the

[1] Cf. *Chem. News*, vol. liv.; also Crookes' Presidential Addresses to the Chemical Society in 1888 and 1889, published in the Society's *Journal* for those years. Compare, too, W. Preyer's *Das genetische System der chemischen Elemente* (1893).

yttrium earths. The peculiar nature of these yttrium earths, which consist of a variety of substances, has conduced greatly to speculations of the kind.

But until the transformation of one element into another has been incontestably proved by experiment, chemists cannot give up the idea of indivisible elementary particles, *i.e.* the present atomic theory.

The General Significance of Physico-chemical Investigations.

The relations thus discovered between the atomic weights of the elements and their physical properties have materially contributed to enlarge our knowledge of the boundary-land between physics and chemistry. Many investigators had previous to this followed the example of H. Kopp [1] (who began his stimulating labours in the forties), in assiduously tracing out the connection existing between the chemical constitution of compounds and their physical behaviour.

[1] Hermann Kopp, born 30th October, 1817, at Hanau (at which place his father was an esteemed physician), after studying at Heidelberg, was drawn through Liebig's attraction to Giessen, where he became assistant professor of chemistry in 1841 and professor at a later date. At Giessen he remained until his removal to the University of Heidelberg in 1864, where he continued to work in full vigour until shortly before his death, which took place on February 20th, 1892. His services as a historian of chemistry have already been frequently referred to. All his historical works [*Geschichte der Chemie* ("History of Chemistry"), 4 vols. 1843-47; *Die Entwickelung der Chemie in der neueren Zeit*, 1873 ("The Development of Chemistry in Recent Times"); *Beiträge zur Geschichte der Chemie* ("Contributions to the History of Chemistry"); *Die Alchemie in älterer und neuerer Zeit.* ("Alchemy, Old and New")] are distinguished by their comprehensiveness and thoroughness. He possessed in a remarkable degree the gift of sympathetically tracing out the development of important ideas and hypotheses. The stimulus produced by his physico-chemical researches was a highly gratifying one (cf. the special history of physical chemistry). In addition to all this, he took a share in the editing of Liebig's *Jahresbericht* and of the *Annalen der Chemie und Pharmacie*, besides writing his *Lehrbuch der theoretischen Chemie* (1863) for the Graham-Otto series. Immediately after Kopp's death, A. W. von Hofmann delivered in his memory one of those biographical addresses, which none could do so well as he. (See *Ber.*, 1892, *Ref.* p. 505).

The advances made in this direction fall to be treated of in the special history of physical chemistry. Here it need merely be said that it has come to be more and more recognised, especially within the last thirty or forty years, that chemical investigation runs the danger of becoming one-sided without a free use of physical aids. Chemists have perceived the necessity for their science of physico-chemical methods.

Thus, what a wide application have not the latter found in the estimation of the molecular weights of elements and compounds! The determination of vapour density has proved its value for the attainment of this end in an infinite number of cases, and has been applied to the solution of the most important theoretical questions; thus, of recent years the saturation-capacities of numerous elements, *e.g.* tungsten, vanadium, beryllium, thorium, germanium, aluminium, etc., have been established by the aid of this process. The constant relations between the molecular weight of a substance on the one hand and its point of solidification and the vapour pressure of its solutions on the other, first definitely formulated by Raoult and de Coppet (independently), have rapidly become the basis of easily-carried-out methods for the determination of molecular weights. In fact, the investigations on the physical behaviour of solutions—*e.g.* their electric conductivity, osmotic pressure, vapour pressure, etc.—have during the last few years enlarged the boundaries of general chemistry to an extent that was undreamt of.

We need only refer here to the deduction of the atomic weights of elements from their specific heats, and of equivalents from the electrolysis of salts, in order to emphasise the significance of physical methods for establishing the most important of chemical values. Of the wealth of work which has been accomplished in the branches of spectrum analysis, thermo-chemistry, electro-chemistry, upon the doctrine of affinity (*Verwandtschaftslehre*), and in the investigation of the connection between optical properties and chemical constitution, an account will be given in the special section. The position of chemistry to physics will there come out more clearly than is possible at this point. Thanks to the labours of Ostwald,[1] van 't Hoff[2] and others, the physical conceptions of the transformation and conservation of energy have now come into general application in chemistry also, more especially in the explanation of affinity-phenomena.

Similarly, the relation of chemistry to other branches of science can only be properly represented by going into details. This will show itself in the case of mineralogy, which is

[1] Wilhelm Ostwald, born at Riga on September 2nd, 1853, taught first in the University of Dorpat, then at the Riga Polytechnic from 1880 to 1887, and has since the latter year held the chair of Physical Chemistry in the University of Leipzig. The amount of work which he has already done in this branch of the science has been very great, the subject having been immensely advanced both by his researches and his writings. Up to 1887 his papers were published in the *Journal für praktische Chemie*, but since that date in the *Zeitschrift für physikalische Chemie*, which he himself and van 't Hoff started, and continue to edit. Some papers have been brought out in the *Ber. der Konigl. Sächs Gesellschaft der Wissenschaften*. His large *Lehrbuch der allgemeinen Chemie* (2 vols.), of which two editions have so far appeared, and also his *Grundriss der allgemeinen Chemie*, have found universal acceptance. Part of the former has been translated into English by M. M. P. Muir, while the English edition of the latter is by James Walker. His *Hand- und Hülfsbuch zur Ausführung physiko-chemischer Messungen* (1893), also translated into English by Walker, is of great practical value. A smaller work, *Die wissenschaftlichen Grundlagen der Analytischen Chemie* (1894), is also highly original—in fact, the only book of its kind; it too has been translated (by G. McGowan). Lastly, his large work, *Elektrochemie, ihre Geschichte und Lehre* (1896), is thoroughly apposite at the present time, and deserves special mention. In addition to these, Ostwald's *Klassiker der exakten Wissenschaften*, being reprints of classical scientific papers, have made these accessible to any reader of German (the "Alembic" series is their analogue in this country).

[2] J. H. van 't Hoff was born at Rotterdam on the 8th of August, 1852. After studying at Delft, Leyden, Bonn, Paris and Utrecht, he became a lecturer in the Veterinary College of the last-named city in 1876, and Professor of Chemistry in the University of Amsterdam in 1878; in 1896 he was made an honorary professor of the University of Berlin. Van 't Hoff has shown himself to be a man of singularly original mind, and his work will be frequently referred to in the special sections of this book. In addition to his numerous papers contributed to scientific journals and his share in editing the *Zeitschrift für physikalische Chemie* along with Ostwald, he is the author of the following works :—*La Chimie dans l'Espace* (1875); *Dix Années dans l'Histoire d'une Theorie* (1887); *Ansichten über die organische Chemie* (1878-1881) ("Views upon Organic Chemistry"); *Études de Dynamique chimique* (1884); and *Lois de l'Équilibre chimique* (1885).

united to inorganic chemistry by a firm band. The connection with physiology is proved by the fact that organic chemistry is absolutely necessary to the latter. In fact, to whatever quarter we turn in the extensive range of the natural sciences, we find that chemistry is to most of them an indispensable aid, and to the remaining ones useful in a high degree. The history of the different branches of natural science shows in the most distinct manner this constantly recurring reciprocal action.

SPECIAL HISTORY OF THE VARIOUS BRANCHES OF CHEMISTRY FROM LAVOISIER TO THE PRESENT DAY

CHAPTER VI

SPECIAL HISTORY OF THE VARIOUS BRANCHES OF CHEMISTRY FROM LAVOISIER TO THE PRESENT DAY

Introduction.—In the general history of this period the attempt has been made to set forth the more important ideas and points of view which have led to the development of particular doctrines, and at the same time to give a description of the latter. In conjunction with these objective discussions, short sketches have been appended of the lives of those investigators who have exercised a permanent effect upon the development of chemistry, and more especially upon the systematising of it.

Up to the fourth or fifth decade of our century, the leading chemists were able to cover in their work a very large part of the ground which was either occupied by chemistry itself, or in which it was an indispensable aid; we have but to think of Berzelius and Liebig, and of their labours, which were at the same time both pioneering and fundamental, in analytical and pure chemistry, physiology and mineralogy. But during the later decades the tremendous growth of the science has necessitated a large subdivision of work, indeed an almost one-sided specialisation in research. This may even give rise to the apprehension that, with increasing specialisation, a danger is run of losing sight of general guiding principles. Organic chemistry may serve as an example of this subdivision of labour, particular branches of it having been opened up which in themselves alone are sufficient to absorb the full energies of large numbers of capable investigators; take for instance the chemistry of the aromatic compounds, and more especially that portion of it comprising the pyridine and quinoline bases and similar compounds richer in nitrogen. The new journals unmistakably reflect this subdivision of labour. While formerly *Poggendorff's Annalen*, *Liebig's Annalen*, the *Journal für praktische Chemie*, the *Journal of the Chemical Society*, &c., contained papers in every branch of the science, we now find journals exclusively devoted to inorganic, physical, physiological, agricultural, pharmaceutical and applied chemistry.

In the following special section of this book, which deals with the different branches of chemistry in succession, such facts and investigations are recorded as have contributed to the true advancement of the various parts of our science.

The history of analytical chemistry is placed first in order, since the latter is an indispensable aid to all chemical research, and therefore to all the other branches of chemistry, pure as well as applied. Following it comes the history of pure chemistry, which divides itself into inorganic and organic, although there is no natural partition between the two. Next to pure chemistry stands physical, with whose history that of the doctrine of affinity (*Verwandtschaftslehre*) is intimately bound up. It was the endeavour to discover relations between chemical and physical properties which led to the establishment of this important middle kingdom between chemistry and physics.

That chemistry is necessary for the healthy growth of other sciences is particularly shown in the history of mineralogical, physiological and pathological chemistry, which are also treated here according to their historical development. The opening up and cultivation of the fields of mineralogy, geology, and vegetable and animal physiology are indissolubly connected with the names of such distinguished chemists as Lavoisier, Vauquelin, Klaproth, Berzelius and Liebig.

Last in order comes the history of technical chemistry, which illustrates in the most brilliant manner the influence of chemical research upon the development of chemical industry. To give a historical account of the penetration of the scientific spirit and of chemical methods into this branch, a branch hitherto worked empirically, is a task which repays itself in a special degree.

As an appendix to the whole, an attempt has been made to picture within short space the growth which chemical instruction has undergone in the course of the present century.

HISTORY OF ANALYTICAL CHEMISTRY IN RECENT TIMES.

The main problem of chemistry, the investigation of the true composition of compounds, necessarily carries along with itself the constant endeavour to elaborate and perfect the means employed for arriving at this end. Thus, since the time of Lavoisier, analytical methods, which constitute the tools for the solution of this problem, have been and are being improved in a continuously increasing degree.

Qualitative Analysis of Inorganic Substances.

Even so early as during the phlogistic period, men like Boyle, Hoffmann, Marggraf, and especially Scheele and Bergman, had collected together a large number of valuable observations, by means of which it was possible to test with certainty for many inorganic compounds. In a knowledge of the various reagents which served for this end Bergman was the furthest advanced; he it was who first attempted to publish a system for the qualitative analysis of substances in the wet way (cf. p. 143). From the analytical course of procedure which he proposed, and which had for its aim the separation of different substances into particular groups by converting them into insoluble compounds, the methods in use at the present day have developed themselves. To the perfecting of this (previous to the time of Berzelius, who also worked with the greatest effect in this branch), Lampadius and Göttling materially contributed; the former published in 1801 his *Handbuch zur chemischen Analyse der Mineralien* ("Text-Book on the Chemical Analysis of Minerals"), and the latter his *Practische Anleitung zur prüfenden und zerlegenden Chemie* ("Practical Introduction to the Chemistry of Testing and Decomposing"),—works in which the best analytical methods of the time are given.

The many and varied observations collected by Klaproth, Vauquelin, Berzelius, Stromeyer and others in their analyses of minerals further helped to strengthen the qualitative method. The text-books of analytical chemistry by C. H. Pfaff and Heinrich Rose enable us to judge of the rate of its continuous development; alongside of the latter of those works, which became justly celebrated and ran through numerous editions, must be placed the well-known and highly prized *Anleitung zur qualitativen chemischen Analyse* ("Introduction to Qualitative Chemical Analysis") of R. Fresenius, which covers the whole ground on the subject, and is a marvel of thoroughness and accuracy. The procedure in qualitative analysis has undergone no material alterations since Fresenius first published his book, and is treated of in numerous works, most of which are intended to instruct the beginner in its principles.[1]

Qualitative analysis in the dry way has been perfected by the more general and improved use of the blowpipe, which Berzelius[2] and Hausmann were in a high degree instrumental in introducing into chemistry and mineralogy. This valuable little instrument has been employed with the greatest success, more especially for the detection of the constituents of minerals; Bunsen's important flame-reactions[3] have, however, enabled it to be dispensed with in a number of cases. Among the most noteworthy of dry reactions are the spectroscopic, which, thanks to their extraordinary delicacy and certainty, serve for the detection of the most minute quantities of many metals, and have rendered possible the discovery of a number of new elements. Spectrum analysis, through which we are able to deduce the nature of a glowing substance by examining the light which it emits,

[1] Out of the large number of such text-books, those of Beilstein, Birnbaum, Classen, Drechsel, Geuther, Medicus, Rammelsberg, Städeler-Kolbe, Will, Odling, Harcourt and Madan, Thorpe, Clowes, and Jones may be mentioned.

[2] His pamphlet, *Ueber die Anwendung des Löthrohrs* ("On the Application of the Blowpipe"), was first published in 1820; cf. also p. 144.

[3] *Ann. Chem.*, vol. cxxxviii. p. 257; also in a much extended form as a separate pamphlet.

was founded by the masterly researches of Bunsen and Kirchoff;[1] Talbot, Miller, Swan and others had before this investigated the spectra of coloured flames, without however applying their results with a definite aim to the analysis of substances. The first proposal to utilise the different flame colourations for distinguishing potash from soda salts was made long ago by Marggraf.[2]

Quantitative Analysis of Inorganic Substances.

The accurate investigation of the behaviour of bases, acids and salts towards different reagents, especially towards such as yield with them either sparingly soluble or insoluble precipitates, constituted the basis of the gravimetric estimation of individual substances. Before the time of Lavoisier few attempts had been made at quantitative analysis, but the path which it was bound to follow had been already clearly indicated by Bergman; for he was the first to enunciate the principle of converting the substance to be analysed into a convenient form of known composition, and then deducing from the weight of the compound thus precipitated or otherwise obtained that of the substance in question. At that date chemists either already knew or became acquainted with the precipitation of silver solutions by hydrochloric acid, of solutions of lime salts by oxalic or sulphuric acid, of lead salts by liver of sulphur or sulphuric acid, and many similar reactions. It was Klaproth who taught the ignition of precipitates before weighing them, in those cases where they did not suffer decomposition through this procedure, and he also co-operated largely with Vauquelin in developing the quantitative analysis of minerals. The observations of both of these chemists, especially of Klaproth (who directed his efforts to ascertaining correctly the composition of those compounds into which the constituents of the substances to be analysed were usually transformed),

[1] *Pogg. Ann.*, vol. cx. p. 161.

[2] Cf. p. 143. It has been already mentioned that Scheele made the same observation.

attained to a fairly high degree of accuracy; and this also applies to the analyses of salts carried out by Wenzel at an earlier date, although to these hardly any attention had been paid. Richter's endeavours to establish the quantitative composition of salts, and the success which followed them, have been sufficiently described in the general history of this period; in spite of the fact that his analyses were not particularly accurate, he understood how to draw important and correct deductions from them.

Lavoisier, who had from the outset of his scientific career clearly grasped the importance of proportions by weight, and with this of quantitative analysis, examined more particularly the composition of oxygen compounds, Thus he established with tolerable correctness (for example) the relation of carbon to oxygen in carbonic acid, but only approximated to that of hydrogen to oxygen in water, and was wide of the mark in the relation of phosphorus to oxygen in phosphoric acid. He also sought to apply the values which he had obtained for the composition of water and carbonic acid to establishing the composition of organic substances. Lavoisier, however, introduced no original methods for the quantitative analysis of inorganic bodies and their separation from one another.

Proust effected infinitely more in this branch, his analytical work leading, as has already been stated, to a clear grasp of the law of constant proportions, and of the alteration by definite increments in combining proportions. Quantitative analysis was also strengthened and extended by the establishment of stöchiometry (which found its perfect support in Dalton's atomic theory), since a check upon the results obtained was thereby rendered possible.

Endeavours were at that time mainly directed to the determination of the relative atomic or, to speak more correctly, combining weights. The splendid results obtained by Berzelius from his pioneering labours in this direction have already been detailed. He devised a large number of new gravimetric methods of estimation, and tested those already in use for the separation of substances, working out better modes for attaining to this end. His researches on the composition of chemical compounds embraced every element which was at all well known. Berzelius, far more than any other man, developed the principles by which atomic weights could be established; and the degree of accuracy at which he arrived in his analyses is seen from the tables of atomic weights published by him after the year 1818 (cf. pp. 218 and 224).

The great task of determining the atomic weights—the constants of the atomic theory—with the utmost possible accuracy, has led ever since the time of Berzelius to the development and improvement of gravimetric methods; for, what is required here is to establish by various procedures an unalterable value for each element, a value which shall form the basis for the composition of all the compounds of that element. The efforts and speculations to round off these numerical values in accordance with Prout's hypothesis were replaced by exact quantitative determinations. Among the latter the researches of Dumas, Erdmann and Marchand, Marignac, and Stas deserve special mention.[1]

The systematic development of quantitative analysis was thus mainly promoted by the investigation of mineral substances, since the chief requirement here was to find out modes for separating their constituents from one another. After the valuable preparatory labours of Bergman (with whom, for instance, the fusion of silicates with alkaline carbonates originated), and the researches of Klaproth, Vauquelin and Proust, it was Berzelius who worked out entirely new methods; we need only recall here his plan of decomposing silicates by hydrofluoric acid, and that of separating metals from one another by means of chlorine. He it was, too, who first employed far smaller quantities of substances than the large amounts recommended by Klaproth, who introduced the spirit-lamp which bears his name, thus facilitating the ignition of precipitates, and who taught how to incinerate the filter-paper and determine its

[1] Cf. Lothar Meyer and K. Seubert, *Die Atomgewichte der Elemente* (1883).

ash; in fact, to speak generally, he was the first to make use of a large number of practical contrivances and apparatus for the carrying out of analyses. His greater analytical researches, such as those upon platinum ores and on mineral waters, show Berzelius as a master in devising good methods of separation.

His pupils, more especially H. Rose[1] and Fr. Wöhler, worked up the valuable experiences of their teacher, extended them largely by wide-reaching observations of their own, and made analytical methods public property by their admirable books[2] on the analysis of minerals and chemical bodies generally. R. Fresenius,[3] until quite lately our chief exponent of analytical chemistry, likewise perfected and strengthened this branch of the science in all its various parts by collating and sifting the methods formerly in use,

[1] The brothers Heinrich and Gustav Rose belonged to a Berlin family which produced distinguished chemists for several generations. Their grandfather, Valentin Rose the elder, a pupil of Marggraf, and also their father, Valentin Rose the younger, were energetic pharmacists and chemists. Gustav Rose, who was born in 1798 and died in 1873 as Professor of Mineralogy at Berlin, was only connected with chemistry indirectly. But Heinrich Rose (born 1795, died 1864) was an ardent exponent of the science, and enriched it by most important work, especially in analytical and inorganic chemistry (see special history of these). He reciprocated fully and truly the affection of his master Berzelius, as is vividly shown in the beautiful memorial address which he gave of the latter (cf. p. 209). In his two-volume *Handbuch der analytischen Chemie*, H. Rose collected together in a masterly manner the best of the then known methods in qualitive and quantitative analysis.

[2] H. Rose, *Ausführliches Handbuch der analytischen Chemie* ("Detailed Textbook of Analytical Chemistry"); Fr. Wöhler, *Die Mineralanalyse in Beispielen* ("The Analysis of Minerals, illustrated by Examples").

[3] C. Remigius Fresenius, born at Frankfurt on the Maine in 1818, became assistant to Liebig in Giessen in 1841, and assistant professor there in 1843; in 1848 he opened his now universally known laboratory at Wiesbaden, which has undergone a continuous extension, and been frequented by students from all parts. His text-books of chemical analysis, of which the *Qualitative* appeared for the first time in 1841, and the *Quantitative* in 1846, have had an extraordinarily wide distribution, as their numerous editions in different languages prove. Every one who has used them systematically cannot fail to have been struck with their wonderful accuracy and at the same time great breadth. Fresenius died suddenly, while still in active work, on June 11th, 1897.

and, more especially, by working out many new ones. By founding in 1862 the *Zeitschrift für analytische Chemie*, Fresenius supplied a centre-point for the analytical branch of the science. It is impossible to enumerate here what other workers (among whom Liebig, Thomson, Stromeyer, Bunsen, Turner, Scheerer, Rammelsberg, Gibbs, Blomstrand, R. Schneider, Pélouze, and Winkler may be named) have done for the development of quantitative analysis.

We may, however, mention here that the galvanic current has of late years been called in to the service of analysis, the quantitative determination of many metals being rendered possible by its aid. After Gibbs (in 1865) had worked out the electrolytic determination of copper, and other chemists had subsequently busied themselves with similar investigations, Alexander Classen[1] rendered special service in the development of the method. This branch of chemical analysis is of the utmost use for metallurgy, in which even already it forms an important part of docimacy. The latter, originally confined to the determination of the noble metals in the dry way, has expanded into an important branch of analytical chemistry, particularly since C. Fr. Plattner's comprehensive researches and the publication of his classical book, *Die Probierkunst mit dem Löthrohr* Leipzig, 1835), ("Docimacy by Means of the Blowpipe").[2]

Volumetric Analysis.

Besides the analytical methods which have been touched upon above, volumetric ones have become developed within the last seventy years or so; these are of great use, particularly in manufacturing chemistry and pharmacy, and have therefore the widest application. Since in volumetric

[1] Cf. his work, *Handbuch der chemischen Analyse durch Electrolyse* ("Text-Book of Chemical Analysis by means of Electrolysis"). In the *Berichte*, vol. xxvii., p. 2060, there is a further paper by Classen, in which he gives very useful data regarding particular points in electrolytic determinations.

[2] Cf. Kerl's *Metallische Probierkunst* ("Metallic Docimacy"), (1886); Balling's *Probierkunde* ("Docimacy"), (1879), and his *Fortschritte im Probierwesen* ("Advances in Docimacy"), (1877).

methods no weighing is required after the standard solutions have once been made up, and the wished-for results are arrived at simply by reading off the amounts of the solutions used, much time is saved and at the same time sufficient accuracy attained, the requirements of technical analysis (more particularly) being thereby met.

Gay-Lussac must be regarded as the man who introduced volumetric methods into the science, and rendered them available for chemical industries; before him various investigators—of whom Descroizille and Vauquelin must be specially mentioned—had attempted to apply such methods empirically to comparative determinations of chemical products.

Gay-Lussac worked out with the greatest care his methods of *chlorimetry* (1824), of *alkalimetry* (1828), and of the determination of chlorine and silver (1832).[1] Notwithstanding the excellent results which those volumetric processes yielded, they received but slowly the recognition which was their due. The application of permanganate of potash to the estimation of iron by Margueritte in 1846, and, more particularly, Bunsen's process with equivalent solutions of iodine and sulphurous acid (by means of which a large number of different substances can be accurately estimated by one and the same reaction) are landmarks in the history of "titrimetry," which soon after this began to rank alongside of gravimetric analysis. One of the chief promoters of volumetric methods was Friedrich Mohr, who both improved old processes and introduced many new ones; he rendered great service by the publication of his *Lehrbuch der chemischen Titrirmethode* ("Text-book of Volumetric Analysis").[2] Among the many investigators who have enriched this branch of the science we may name J. Volhard,[3] who devised an exact method (the determination of silver by means of ammonium sulphocyanide) capable of numerous applications.

[1] Cf. his *Instruction sur l'Éssai des Matières par la Voie Humide* (1833).

[2] The latest edition of this is edited by A. Classen. Among other valuable books on volumetric analysis are those of Cl. Winkler, Medicus, and Fleischer in Germany, and of Sutton in England.

[3] Cf. *Ann. Chem.* vol. cxc. p. 1, *et seq.*

In organic chemistry volumetric analysis has not been able to take up anything like the same position that it has in inorganic, the methods as yet introduced being wanting in precision. Among the most noteworthy processes here are Fehling's for the determination of grape sugar, Liebig's for that of urea, the volumetric estimation of phenol by means of bromine,[1] etc.

Development of Methods of Gas Analysis.

The history of the volumetric analysis of liquids naturally leads us on to a description of the efforts to analyse gases qualitatively and quantitatively. It is worthy of note here that the systematic qualitative analysis of these was much later of being developed than their quantitative determination. The first attempts in this direction were made by Scheele, Priestley, Cavendish and Lavoisier, to be followed by those of Dalton, Gay-Lussac, Henry, de Saussure and others at the beginning of this century. But it has been through Bunsen's fundamental researches[2] that the quantitative analysis of gases has been brought to such perfection that those methods which depend upon the absorption or combustion of the gas under investigation are among the most exact of our science, having required but trifling modifications since he first published them.

In addition to Bunsen's methods, others have been worked out with a special view to technical gas analysis; although the same as the former in their main principle, these allow of determining the composition of the so-called industrial gases by the aid of simple apparatus within a short time, and with sufficient accuracy. Cl. Winkler and

[1] Cf. Degener, *Journ. pr. Chem.* (2), vol. xvii. p. 390; Koppeschaar, *Ztschr. Ann. Chem.* for 1876, p. 223.

[2] These researches of Bunsen's began about the year 1838, and were collected together under the title of *Gasometrische Methoden* (Brunswick, 1857; second edition, 1877); this most valuable work was translated into English by Roscoe. Kolbe, in the *Handwörterbuch* (under the article "Eudiometer"), had already brought the details of these methods before public notice so early as 1843.

W. Hempel have rendered great service here by materially simplifying the apparatus required and by generalising methods.[1] Among others who have done good work in gas analysis of recent years may be mentioned Frankland, Pettersson, Orsat, Coquillon and Bunte.

The qualitative analysis of gases has only quite recently been developed scientifically, and here, too, Winkler has laboured with success; by the systematised use of absorptives he has divided gases into different groups, thus proceeding in the same manner as is done in the analysis of substances in the wet way. The recent work by Ramsay and his collaborators in connection with the two new gases argon and helium must also be referred to here. The improvements in methods of gas analysis have drawn the attention of chemists to gases in an increasing degree, and have proved of the greatest benefit to theoretical as well as to practical chemistry.

The Analysis of Organic Substances.

The fact that animal and vegetable products, which came to be comprised under the term "organic," always contain carbon, usually hydrogen and oxygen, and frequently also nitrogen, was—as already stated—a long time of being recognised. Here again we have a brilliant proof of Lavoisier's far-seeing glance, and of his power of drawing general conclusions from detached observations. It had indeed struck previous experimenters, *e.g.* van Helmont and Boyle, that spirit of wine, wax, etc., form water when burned, while Priestley perceived that carbonic acid was produced at the same time; in fact Scheele stated in 1777 that both of these compounds were products of the combustion of oils. After it had become clear to Lavoisier that carbonic acid consisted of carbon and oxygen, and water of

[1] Cf. Clemens Winkler, *Anleitung zur chemischen Untersuchung der Industriegase*, Freiberg, 1876-77 ("Methods for the Chemical Examination of Industrial Gases"); the same author's *Lehrbuch der technischen Gasanalyse* second edition, 1892), ("Text-Book of Technical Gas Analysis"); and W. Hempel's *Neue Methode zur Analyse der Gase* (Brunswick, 1880), and *Gasanalytische Methoden* (1890).

hydrogen and oxygen, he went on to deduce the composition of organic substances. Thus, with the discovery of what were the most important elements of organic compounds, the first step in qualitative organic analysis was reached. The principle of arriving at the constituents of organic bodies by transforming them into compounds of known composition has ever since been retained. In the same way nitrogen, which Lavoisier himself recognised as being peculiar to many organic substances,[1] was detected by conversion either into ammonia (Berthollet) or sodium cyanide (Lassaigne), and phosphorus and sulphur by conversion into phosphoric and sulphuric acids respectively.

While the elementary constituents of organic compounds are thus easily arrived at, the detection of the compounds alongside of one another is a much harder task; only small beginnings have as yet been made at a systematic course of qualitative organic analysis, in the sense in which we apply the term to inorganic.[2] In many instances one has to depend upon isolated characteristic reactions of organic substances, *e.g.* in the investigation of colouring matters, alkaloids, proteïn substances, carbohydrates, etc.

The quantitative analysis of organic compounds has developed itself from the observation that carbonic acid and water are products of their combustion; the method, therefore, which served for the detection of the constituents carbon and hydrogen was applied in a perfected form to their exact determination. Lavoisier was again the first to point out the right path here; he attempted to burn the organic compound in question completely, and to estimate the resulting carbonic acid and water—the latter indirectly. In order to be able to deduce the amounts of carbon and hydrogen themselves, it was necessary to know the quantitative composition both of carbonic acid and of water; but, since the values obtained by him for these were not very

[1] How uncertain the tests for the elements of organic substances were at the beginning of this century is shown by the fact that Proust believed he had proved nitrogen to be an integral constituent of acetic acid.

[2] Cf. Barfoed's *Qualitative Analyse organischer Körper*; also Allen's *Commercial Organic Analysis*.

accurate,[1] it was impossible that the results of his analysis of an organic substance could turn out correct, and this all the more from the method of the combustion being such as to involve errors in itself.

Lavoisier's process for easily combustible substances was to burn a weighed quantity in a known volume of oxygen, contained in a receiver closed by mercury, and then to estimate the resulting carbonic acid together with the residual oxygen; from these data the amounts of carbon, hydrogen and oxygen were calculated. For difficultly combustible bodies, such as sugars and resins, Lavoisier (as we now learn from his recently published journals)[2] used, instead of the free gas, substances which yield up their oxygen upon being heated, *e.g.* red oxide of mercury and red lead; he thus adopted the plan which later on became the standard one, while at the same time he estimated the weight of the carbonic acid produced by this oxidation by means of a solution of caustic potash.

Had those researches become known at that time, organic analysis would doubtless have undergone a more rapid development than it actually did. The efforts of Dalton (1803), Saussure (about 1800–1803), and Thénard (1807) to arrive at the composition of organic compounds by exploding their vapours with oxygen and analysing the resulting products would never have been made. Gay-Lussac and Thénard[3] endeavoured to solve this problem in a more felicitous manner by the combustion of the organic substance with chlorate of potash; from the amounts of resulting carbonic acid and residual oxygen they calculated the percentage of carbon, hydrogen and oxygen in the substance under analysis, and attained in some instances at any rate to

[1] The following are Lavoisier's figures for the composition of carbonic acid and water (the correct values being given in brackets):—

Carbonic Acid . . .	Carbon	28	per cent.	(27·2)
	Oxygen	72	,,	(72·8)
Water . . .	Hydrogen	13·1	,,	(11·1)
	Oxygen	86·9	,,	(88·9)

[2] *Œuvres de Lavoisier*, vol. iii. p. 773.

[3] *Recherches Physico-chimiques*, vol. ii. p. 265.

serviceable results. Compared with this method, uncertain as it was on account of the violence of the combustion, the one followed by Berzelius showed a marked improvement;[1] for here the organic substance, mixed with chlorate of potash and sodium chloride, was gradually decomposed, and then not merely the resulting carbonic acid but also the water determined directly—the latter by means of chloride of calcium. A further advance was made by Gay-Lussac[2] in 1815, in the use of black oxide of copper as the oxidising agent. But the rounding off of the whole procedure by the introduction of a convenient bulb-shaped apparatus, and the consequent simplification of the manipulation required, is due to Liebig.[3] Since his time elementary organic analysis has not altered essentially, the modifications introduced having had reference to the combustion furnaces (now heated by gas instead of charcoal), and to the mode of carrying out the combustion; with respect to the latter, Koppfer's method[4] must be mentioned, a method by which the substance is burnt in a current of oxygen, with the aid of platinum black. Plans for the combustion of organic compounds in a stream of oxygen had before this been proposed by Hess, Erdmann and Marchand, Wöhler, and others.

Quite recently W. Hempel[5] has succeeded in carrying out the combustion of organic compounds in oxygen under pressure (*i.e.* in autoclaves), and has perfected the method so much that it is now possible to make accurate determinations, not merely of carbon and hydrogen, but also of nitrogen and sulphur. Messinger[6] has also been successful lately in estimating the carbon of organic compounds in the wet way, by oxidation with permanganate of potash.

The exact determination of nitrogen in organic compounds first became possible after Dumas[7] (in 1830) had

[1] *Annals of Philosophy*, vol. iv. pp. 330, 401.

[2] *Schweigger's Journ.*, vol. xvi. p. 16; vol. xviii. p. 369.

[3] *Pogg. Ann.*, vol. xxi. p. 1; also his pamphlet, *Anleitung zur Analyse organischer Korper* ("The Analysis of Organic Compounds").

[4] *Ber.*, vol. ix. p. 1377.

[5] *Ber.*, vol. xxx. p. 202.

[6] *Ber.*, vol. xxiii. p. 2756.

[7] *Ann. Chim. Phys.*, vol. xliv. pp. 133, 172; vol. xlvii. p. 196.

devised his admirable method. For many nitrogenous organic substances the process worked out by Will and Varrentrapp[1] at a later date, in which the nitrogen is estimated as ammonia, has proved itself thoroughly applicable. In addition to these, the recent method of Kjeldahl[2] must be mentioned, a method which is found to be of great use, especially in agricultural-chemical analyses (for the determination of proteïn). Since this method was devised, it has been materially improved.

Only a bare reference can be made here to the numerous methods for the determination of the halogens, sulphur, phosphorus and other elements which occur less often in organic substances.[3]

Analytical methods have found the most extended application in judicial cases, in questions of hygiene, and in all the branches of technical chemistry; a short historical account of them must therefore be given here. Forensic chemistry, whose task consists in the absolutely certain detection of poisons, could only reach its present stage of development after analytical methods in general had been placed upon a firm basis. Fresenius admirably depicted in 1844 the position and duties of a forensic chemist at that date.[4] The great progress which has since been made in the precision with which poisons can be detected is distinctly seen by an examination of the various works on legal-chemical analysis which have been published from time to time.[5] In addition to Fresenius—J. and R. Otto, Marsh, Graham, Stas, Mohr, Husemann, Dragendorff and others have rendered special service in working out good methods. The Stas-Otto process for the detection of individual alkaloids

[1] *Ann. Chem.*, vol. xxxix. p. 257.

[2] *Ztschr. anal. Chem.*, vol. xxii. p. 366; vol. xxiv. p. 199.

[3] Cf. Fresenius' *Quantitative Analysis.*

[4] *Ann. Chem.*, vol. xlix. p. 275.

[5] Reference may be made here to Otto's *Anleitung zur Ausmittelung der Gifte* ("Methods for the Detection of Poisons"), sixth edition, 1884; Christison's *Treatise on Poisons in relation to Medical Jurisprudence, Physiology, and the Practice of Physic*, which was first published in 1829 and which ran through numerous editions; and to Stevenson's new edition of *Taylor on Poisons.*

has proved of great importance for the development of this branch; since the discovery of the ptomaïnes,[1] it has had to undergo some modifications, as the resemblance between many of the reactions of these products and those of the vegetable alkaloids may easily give rise to most serious mistakes, and in fact has already done so.

A special branch of chemical analysis is represented by the methods of testing and investigating used in industrial chemistry. Since these have for their aim the attainment of a fair degree of accuracy within the shortest possible time, volumetric processes are the ones most frequently employed. The rapidity with which acids and alkalies, chlorine, many metals in their compounds, and other substances can be determined quantitatively by volumetric methods, has rendered it possible to exercise a continuous control over manufacturing processes,—with what benefit need not be said.

A glance into the most recent text-books of technico-chemical methods [2] is sufficient to convince us of the high degree of development to which these have been brought. A large number of processes have in the course of time been devised, more especially for the commercial analysis of organic products; we may recall here the estimation of sugar by polarisation, the rapid determination of the heating power of combustibles, the valuation of coal-tar dyes by test-colourations and by specific reactions, and the estimation of alcohol, fat, albumen and starch, not to speak of numerous other methods which have become standard ones in chemical technology.

For technical chemists, and in an equal degree for medical officers of health, the development of the analysis of articles of food and drink has been of the first importance; the pharmacist, too, frequently finds it needful to apply the methods which have approved themselves in such cases. By their aid the analyst is able to decide whether the products

[1] Cf. the special history of physiological chemistry.

[2] The works of Post (Brunswick, 1882), of Böckmann (Berlin, 1888), and of Sutton may be mentioned here.

are what they pretend to be, or, if they should be adulterated, the nature of such adulteration. The reader has but to recall to mind the quickly executed methods for analysing milk, butter, meal, feeding-stuffs, wine, beer, coffee, etc., in order to appreciate the true blessing of these applied analyses. The gradual but continuous work of numerous investigators has rendered possible the development, within a comparatively short period of time, of the processes which have become standard ones here. We cannot now refer in detail to the services rendered by single individuals in this branch. Full particulars are to be found in König's admirable work, *Die menschlichen Nahrungs- und Genussmittel* (Berlin, 1883), a book which furnishes a complete review of the subject, and at the same time indicates clearly the share which different chemists have taken in it. *The Bibliothek für Nahrungsmittel-chemiker*, edited by F. Ephraim, and published by Barth of Leipzig, forms an excellent summary of works of this class; and in C. Flügge's *Lehrbuch der hygienischen Untersuchungs-methoden* ("Text-Book of Methods of Hygienic Research") hygiene possesses a splendid guide for such investigations. Among English books on the subject, A. Wynter Blyth's *Foods, their Composition and Analysis* (4th Edition, 1896) must also be mentioned.

As the importance of the analysis of foods and drinks became by degrees better appreciated, the greater refinement of analytical methods increased the need for laboratories in which such investigations should be carried on continuously. The long-cherished wish of many that the State[1] should control these laboratories and their chemists has not yet been realised. But the importance which is now attached to this branch of analysis is shown in the increasing provision made by universities and technical colleges for instruction in it. Marked advances have also been made in this direction in Great Britain of late years, thanks to a considerable extent to the care and vigilance exercised by the Institute of Chemistry.

[1] The German Government is referred to here.

THE PROGRESS IN PURE CHEMISTRY FROM LAVOISIER TO THE PRESENT TIME

While only the main currents of chemistry have been depicted in the general history of this period, we have now in the following section to pick out, from the endless number of experimental researches made, those which have materially contributed to the extension of our chemical knowledge. This rich material is divided between the two great branches of inorganic and organic chemistry. If we glance back over the labours of the last fifty years, we recognise that organic chemistry has gone on preponderating more and more over inorganic; the former has outgrown the latter,—its elder sister. But inorganic remains nevertheless the basis upon which organic chemistry rests, although on the other hand we must not forget that important fundamental principles and doctrines (*e.g.* the doctrine of valency and the true conception of chemical constitution) were first fruitfully developed in the domain of organic chemistry.

SPECIAL HISTORY OF INORGANIC CHEMISTRY

The great revolution in ideas with regard to the constitution of many substances, which was brought about by Lavoisier's system, has been described in detail in the special part of this book. A large number of bodies, which had formerly been looked upon as compound, belonged from thenceforth to the elements; while many, which had been considered simple substances, were either proved to be compounds, or were to be regarded as such from their analogy to others. The clarifying process which Lavoisier had commenced went vigorously forward, thanks to the efforts of Klaproth, Vauquelin, Proust, Davy, Berzelius, Gay-Lussac and others. But we are still far from having attained to a clear and definite knowledge of the nature of all the elements and their compounds, new elements being from time to time added to the long series already known; and the relations of those to the others have to be established by an accurate study of their chemical behaviour. Emphasis has already been laid upon the great effect which the so-called periodic system has had on the classification of the elements.

Historical Notes on the Discovery of Elements—The Determination of their Atomic Weights.

The knowledge of the elements was increased to a very large extent soon after the death of Lavoisier (who had not himself discovered any), and this exactly in proportion as methods of chemical analysis became more perfect. While Lavoisier was able to bring forward twenty-six elements in his *Traité de Chimie*, the number of those whose existence has been definitely established has now extended to at least sixty-eight.

To the aid which was rendered by improved methods of analysis, other means specially effective for the discovery of new elements soon came to be added. Among these was the application of the galvanic current to the decomposition of chemical substances, the production of higher temperatures, and the breaking up of haloid compounds by means of the alkali metals; in spectrum analysis, lastly, chemistry now possesses an invaluable instrument, which has already led to the isolation of a number of most important elements.

After the establishment of the atomic theory, the original task of acquiring a qualitative knowledge of a new element and its compounds was supplemented by the further and higher one of determining its relative atomic weight, and explaining, on the basis of the atomic hypothesis, the constitution of the compounds which it forms with other elements.

For oxygen, which Lavoisier was the first to claim as a simple substance, the elementary nature was always afterwards maintained. Nitrogen, on the other hand, was temporarily regarded by Davy (1808) and by Berzelius[1] (1810) as a compound of an unknown element, *nitricum* with oxygen, because only in this way could they find an explanation of the basic properties of ammonia, in which they likewise assumed the presence of oxygen. Davy was the first of the two to give up this hypothesis in favour of the simpler one of nitrogen being an element, Berzelius only doing this in 1820. Pure nitrogen has just been subjected to a searching process of diffusion by Ramsay and Travers,[2] but the various fractions were found to have precisely the same density, *i.e.* the gas showed perfect homogeneity.

Hydrogen, too, was for a short time looked upon by Berzelius as being compound, *i.e.* as containing oxygen, and the same applied to sulphur and phosphorus, in which the presence of hydrogen and oxygen, besides that of other unknown elements, was conjectured. That many distinguished chemists were inclined to regard chlorine as the oxide of a hypothetical element has been already detailed, as has also the profound influence which this view exercised upon important sections of chemistry.[3] Before this idea had been abandoned by Berzelius, iodine—discovered by Courtois in 1811 in the ashes of marine plants—was shown to be an element analogous to chlorine, through the splendid researches of Davy, and still more those of Gay-Lussac.[4] Bromine, isolated by Balard[5] in 1826 from the mother liquor of sea-salt, and the investigation of which was materially promoted by Löwig's[6] labours in 1829, completed for a

[1] Cf. Kopp, *Gesch. der Chemie*, vol. iii. p. 218.

[2] *Proc. R. S.* for 1898.

[3] Cf. p. 239.

[4] *Ann. de Chimie*, vol. xci. p. 5 (1813).

[5] *Ann. Chim. Phys.* (2), vol. xxxii. p. 337.

[6] K. J. Löwig was born in 1803, and died at Breslau in 1890. After studying under L. Gmelin and Mitscherlich, he taught at Zürich from 1833 to 1853, in which year he became Professor of Chemistry at Breslau, continuing in that post until 1889. Among his publications we may name :—*Das Brom und seine chemischen Verhältnisse* (1829), and his *Lehrbuch der Chemie* (1832; second edition, 1849), which latter was long in use as a text-book. His most important experimental work is referred to in the special history of the subject, while a memorial notice by Landolt is to be found in the *Berichte*, vol. xxiii. Ref. p. 905.

long time the group of Berzelius' "halogen" elements. Fluorine, the acid constituent of hydrofluoric acid, has only quite recently been isolated for the first time by Moissan[1] (in spite of a great many previous attempts[2]), by the electrolysis of hydrofluoric acid under suitable conditions, and, as was to be expected, has been found to be a substance of the most violent chemical energy. Those researches of Moissan upon fluorine are among the most noteworthy in inorganic chemistry in modern times.

The atomic weights, those all-important constants, have been determined with great accuracy for the non-metallic elements already spoken of, and by various different methods in each case. For oxygen, nitrogen, chlorine, bromine and iodine, the classical researches of Marignac[3] and Stas[4] have yielded the most reliable values; for fluorine the determination by Christensen[5] may be regarded as the most exact. Of late years the ratio between the atomic weights of hydrogen and oxygen has been re-determined by a number of different methods, with the result that a value has been arrived at which is slightly different from that hitherto accepted,[6] viz., 1·0032 : 16. It is hardly necessary to say that the fixing of this constant deserves all the attention which has been paid to it; among the recent minute experimental researches on the subject, those of Cooke and Richards, Scott, Rayleigh, Morley, Keiser and Noyes may be mentioned.

Tellurium (chemically analogous to sulphur, which had been known for so long, but had first been characterised as an element by Lavoisier) was discovered by Müller von Reichenstein in 1782, and investigated by Klaproth[7] in

[1] *Ibid.* (6), vol. xii. p. 472 (1887); *Comptes Rendus*, vol. cix. p. 861; and *Ann. Chim. Phys.* (6), vol. xxiv. p. 224.

[2] Cf. Gore, *Phil. Trans.* for 1869, p. 173.

[3] Cf. *Ann. Chem.*, vol. xliv. p. 1; vol. lix. p. 284; vol. lx. p. 180.

[4] *Untersuchungen über die Gesetze der chemischen Proportionen* (Leipzig, 1887) ("Researches upon the Laws of Chemical Proportions").

[5] *Journ. pr. Chem.* (2), vol. xxxv. p. 541.

[6] Cf. Ostwald, *Lehrbuch der Allgemeinen Chemie*, vol. i. p. 43 *et seq.* (second edition). Ostwald holds that the atomic weights should be referred to oxygen, taken as 16, instead of to hydrogen taken as 1.

[7] *Crell's Ann.*, vol. i. p. 91.

1798; an intimate knowledge of it was however first arrived at through the investigations of Berzelius.[1] Selenium was discovered by Berzelius[2] in 1817, and, along with its more important compounds, examined by him in the most thorough manner. The atomic weights of the two last elements have only recently been settled, after great fluctuations, that of selenium[3] being now taken as 79·07 and that of tellurium[4] as 125, previous determinations having, for a long time caused the wrong value 127–128 to be ascribed to the latter. This higher value was, no doubt, due to the difficulty of freeing tellurium from other elements of higher atomic weight. Since the work of Stas on the subject, the number 31·98 for sulphur has been accepted as firmly established.

The discovery of the analogues of nitrogen,—phosphorus, arsenic and antimony, to which bismuth may be added, took place a long time ago; but it is only of late years that they, and more especially their compounds, have been accurately investigated.[5] For phosphorus, the correct atomic weight arrived at by Berzelius was confirmed by Dumas (who found the value 31·02); similarly his atomic weight for arsenic (75) was corroborated by Pélouze and Dumas. But on the other hand R. Schneider and Cooke have proved, by their researches, that the value assumed by Berzelius for antimony was much too high.

Boron was discovered simultaneously and independently by Gay-Lussac[6] and Davy,[7] both of whom isolated it from boracic acid, which already Lavoisier had regarded as the oxide of an unknown element. Guided by a similar view, Berzelius succeeded in 1810 in discovering the element

[1] *Pogg. Ann.*, vol. xxxii. p. 28.

[2] *Schweigger's Journ.*, vol. xxiii. pp. 309, 430.

[3] Eckmann u. Petterson, *Ber.*, vol. ix. p. 1210.

[4] Brauner, *Ber.*, vol. xvi. p. 3055; Brauner has more recently found a higher value than this, but he concludes from his experiments that, in those cases where the value obtained is greater than 125, this is due to the presence of some foreign substance (which has not yet been isolated) in the tellurium (cf. *Ztschr. Phys. Chem.*, vol. iv. p. 344).

[5] Cf. Thorpe and Tutton, *Journ. Chem. Soc.*, vol. lvii. p. 545.

[6] *Recherches Phys. Chim.*, vol. i. p. 276.

[7] *Phil. Trans.* for 1809, p. 75.

combined with oxygen in silica, although he was only able to prepare silicium pure for the first time in 1823 by the action of potassium on potassium silico-fluoride;[1] with this he devised an important method for the isolation of various elements.

The definite knowledge that diamond and graphite are modifications of the element carbon belongs to the beginning of the new period; in addition to the researches of Lavoisier in 1773 and those of Tennant in 1796, the proof furnished by Mackenzie—that equal parts by weight of graphite, charcoal and diamond yield equal amounts of carbonic acid on combustion—was of special importance for the recognition of the similar chemical nature of the three substances.

The phenomenon of allotropy, the term applied by Berzelius to the existence of one and the same substance in different modifications, has been observed with especial frequency among the non-metals. The oldest example of it was that offered by carbon, whose allotropic forms show the greatest conceivable differences among each other; experiments are still being made on these different modifications, more especially on the conversion of amorphous carbon into diamond,—[2] a feat which has been accomplished by Moissan. The most remarkable case of it, however, is afforded by the conversion of ordinary oxygen into the chemically active ozone, which was discovered by Schönbein,[3] although van Marum had a long time previously (in 1785) called attention to the peculiar change produced in oxygen by the electric spark. The beautiful investigations of Schönbein, Marignac and de la Rive established the substantial identity of ozone and oxygen, while those of Andrews[4] and, more especially, of Soret[5] proved that the

[1] *Pogg. Ann.*, vol. i. p. 165.
[2] Cf. Moissan, *Comptes Rendus*, vol. cxvi. p. 218.
[3] *Pogg. Ann.* vol. l. p. 616 (1840).
[4] *Phil. Trans.* for 1856, p. 1; or *Ann. Chem.*, vol. xcvii. p. 371. Andrews and Tait, *Phil. Trans.* for 1861, p. 113; or *Pogg. Ann.*, vol. cxii. p. 241.
[5] *Compt. Rend.*, vol. lxiv. p. 904; or *Ann. Chem.*, Suppl., vol. v. p. 148.

molecule of ozone was made up of three atoms of oxygen. Quite recently Shenstone has succeeded in obtaining a very much larger yield of ozone from oxygen than was formerly held to be possible. Reference must be made here, in passing, to the remarkable observations of van 't Hoff and Jorissen, Victor Meyer, Engler and others on the "rendering active" (*Activirung*) of oxygen by other substances, such as the phosphines, aldehydes, etc. It has been shown that the slowly oxydised substance renders exactly the same amount of oxygen active as it takes up itself.[1]

The allotropic modifications of sulphur were investigated by Mitscherlich, and those of selenium by Berzelius and, later, by Hittorff. The transformation of ordinary yellow phosphorus into red was also observed by Berzelius, but was first discovered with certainty by Schrötter[2] in 1845, and its conversion into the metallic modification by Hittorff. The discovery of "black" sulphur, and of two further modifications of that element—one of which is soluble in water—belong to the present time;[3] as does also that of a new form of phosphorus.[4] The proof that boron and silicon, already long known in the amorphous state, also exist in the crystalline form, is due to Wöhler. That allotropic forms of metals can also exist has been shown by the observations of Carey Lea, E. A. Schneider and others upon silver;[5] but there are still many points here which are obscure. Nevertheless, it remains an undoubted fact that a number of elements, especially among the non-metals, are capable of existing in more than one form. Lastly, reference may be made here to the discovery of allotropic modifications of chemical compounds, *e.g.* mercuric sulphide and iodide, arsenic trioxide, etc.

To the metals which were regarded as elements by La-

[1] Cf. *Ztschr. Phys. Chem.*, vol. xxii. p. 34; vol. xxiii. p. 668; *Ber.*, vol. xxx. p. 1669.

[2] *Pogg. Ann.*, vol. lxxxi. p. 276.

[3] Knapp, *Journ. pr. Chem.*, vol. xliii. p. 305; Engel, *Compt. Rend.*, vol. cxii. p. 866.

[4] Vernon, *Phil. Mag.*, vol. xxxii. p. 365.

[5] Cf. E. v. Meyer, *Journ. pr. Chem.* (2) vol. lvi, p. 241, for the literature on the subject.

voisier many new ones were subsequently added, so a short account of the isolation of these must be given here. The memorable discovery of potassium and sodium, together with the conjoined discussion upon the nature of chlorine, had such a deep influence on the development of important chemical doctrines, that it has already been referred to in detail in the general section of this book. The relative atomic weights of these two alkali metals were determined by Berzelius with fair accuracy, allowing for the fact that he assumed their values to be four times greater than those now assigned to them. Marignac, Dumas and Stas afterwards arrived at much the same figures in their investigations already referred to.

Lithium was discovered by Arfvedson,[1] a pupil of Berzelius, in 1817; he found it to be a constituent of various minerals, *e.g.* petalite, and recognised its analogy to potassium and sodium, but was unable to isolate the metal itself. The latter was first properly investigated in 1855 by Bunsen and Matthiessen,[2] who obtained it by electrolysis. The red colouration which its salts impart to the spirit-of-wine flame was noticed by C. G. Gmelin in 1818.

The discovery of rubidium and cæsium[3] in lepidolite and in the Dürkheimer mineral water by Bunsen and Kirchhoff, by the aid of spectrum analysis, was the first great gain which accrued to chemistry from this new method. Since the chemical reactions of the salts of these two alkali metals are very similar to those of the salts of potassium, their presence would undoubtedly have been overlooked but for the spectroscope. Indeed, several years before the discovery of cæsium, the careful analyst Plattner[4] had examined the mineral pollux, which is rich in that element, and had been unable to explain the deficiency in the results of his analyses, this being really due to his taking the cæsium sulphate present for a mixture of the sulphates of potassium and sodium. The atomic weights

[1] *Schweigger's Journ.*, vol. xxii. p. 93.
[2] *Ann. Chem.*, vol. xciv. p. 107.
[3] *Pogg. Ann.*, vol. cx. p. 167; vol. cxiii. p. 337; vol. cxviii. p. 94.
[4] *Ibid.*, vol. lxix. p. 443.

of cæsium and rubidium were correctly estimated by Bunsen, although too low a value was at first assigned to the former, in consequence of the supply of material at disposal being insufficient. The atomic weight of lithium was definitely determined by Stas as 7·01.

The metals barium, strontium, calcium and magnesium were isolated by Davy from their amalgams, which Seebeck had been the first to prepare; but for a long time previous to this baryta and lime had been regarded as the oxides of unknown metals. Strontia had been discovered by Klaproth and Hope, independently of one another, and had been characterised as being similar to lime. Berzelius, Marignac and Dumas determined the atomic weights of these four metals. Magnesium, which has of late years increased in importance for manufacturing purposes, has been found by Clemens Winkler [1] to be an excellent reducing agent for metallic oxides. His comprehensive researches were carried out with the object of learning how the various metallic oxides comported themselves to magnesium, and what capacity the reduced metal showed for combining with hydrogen. Space will not allow more than this brief reference to the valuable results from the above piece of work, which has at the same time added to our knowledge of many of the elements. Ramsay also found magnesium to be the best agent for taking up the nitrogen of the air in the preparation of argon.

Beryllium, whose oxide Vauquelin had discovered in 1798 in the mineral beryl, was first obtained by Wohler [2] in 1828, by acting upon its chloride with potassium. The atomic weight gave rise to important discussions, since it remained for a long time uncertain whether this amounted to twice or three times its equivalent number. The point was only decided by the recent researches of Nilson and Pettersson [3] on the subject, which proved that beryllium, as a diatomic element, possesses the atomic weight 9·1. Later work by Krüss and Moraht would make it appear, however, that this number is still a little too high.

[1] *Ber.*, vol. xxiv. pp. 873, 1969. [2] *Ibid.*, vol. xiii. p. 577.
[3] *Journ. pr. Chem.* (2) vol. xxxiii. p. 15.

Cadmium was first observed by Stromeyer in 1817, then subsequently rediscovered by others, and recognised as being similar to zinc in character; its atomic weight has lately been redetermined with great accuracy by Partridge. Thallium, isolated by Crookes [1] in 1861 from the selenious mud of the sulphuric acid manufacture, owes its discovery to the characteristic spectrum given by its salts. The chemical nature of this metal, which approximates on the one hand to lead and on the other to the metals of the alkalies, was mainly established by Lamy, while Crookes determined its atomic weight.

Aluminium was isolated for the first time by Wohler [2] in 1827, by the action of potassium upon its chloride, and thus the conjecture which had long been entertained, that alumina was the oxide of a metal, was confirmed. The production of the metal on a large scale from its abundantly occurring oxide, by means of the electric current, is a feat of modern manufacture (see Special History). The elements indium and gallium, which together with aluminium constitute a family, were only discovered comparatively recently, the first-named in 1863 by Reich and Richter,[3] as a constituent of the Freiberg zinc blende, and the second in 1875, also in zinc ores, by Lecoq de Boisbaudran.[4] Here again it was the characteristic spectra of the two metals which led to their discovery. Their atomic weights were determined by the discoverers, and that of indium with especial accuracy also by Cl. Winkler [5] and by Bunsen; [6] while the atomic weight of aluminium has been worked out with the utmost care by Mallet.[7]

The isolation of the metals which constitute the cerium and yttrium groups has presented unusual difficulties. Although the discovery of yttria—impure, it is true, from admixture with other earths—was accomplished by Gadolin nearly a hundred years ago, and investigators of the first rank have

[1] *Chem. News*, vol. iii. p. 193. [2] *Pogg. Ann.*, vol. xi. p. 146.

[3] *Journ. pr. Chem.*, vol. lxxxix. p. 444; vol. xc. p. 172; vol. xcii. p. 480.

[4] *Comptes Rendus*, vol. lxxxi. pp. 493, 1100.

[5] *Journ. pr. Chem.*, vol. cii. p. 282.

[6] *Pogg. Ann.*, vol. cxli. p. 28. [7] *Phil. Trans.* for 1880, p. 1003.

busied themselves with the question, the chemistry of the cerium metals is not even yet completely cleared up, and may possibly remain unsolved for a considerable time to come. After Klaproth and Berzelius had simultaneously prepared cerium sesquioxide from cerite, and the latter had recognised this as the oxide of a metal, Mosander discovered two new oxides in crude yttria, the metals of which—lanthanum and didymium—he isolated. A few years later (in 1843) he added to these two others, erbium and terbium, whose existence and nature is not yet however definitely settled, in spite of the admirable work which has been done on the subject. This has given us a better knowledge of yttrium, while yttria, which was formerly held to be a homogeneous substance, has proved itself a mixture of the oxides of various metals, of which, however, only one or two have as yet been isolated; witness the discovery of scandium by Nilson, and of ytterbium by Marignac. The most recent additions to our knowledge of the chemistry of this group of elements have been made by Krüss, Cl. Winkler, Bettendorf, Crookes, Brauner and others.

Cobalt and nickel, whose discovery belongs to a preceding era (cf. p. 149), have lately been the subject of important researches, more particularly because of the remarkable compounds which they are capable of forming (see p. 428). Winkler[1] has proved that Krüss' and Schmidt's view[2]—that another element, hitherto unknown, was present in nickel and cobalt prepared in the ordinary way—is erroneous; and he has also given us most careful re-determinations of their atomic weights.

The elements molybdenum, tungsten and uranium, which belong to the same group as chromium, were discovered like the latter itself in the first decades of the modern period; but their investigation is still being proceeded with, thanks to the extraordinary diversity of the compounds which they form with other elements (see below). Vauquelin discovered chromium in 1797 as a constituent of red lead spar, and he also contributed materially to a knowledge of its compounds; Klaproth pointed out independently at

[1] *Ber.*, vol. xxii. p. 890. [2] *Ber.*, vol. xxii. pp. 11 and 2026.

the same time that there was probably a new metal contained in that mineral. The presence of molybdenum and tungsten in their oxygen compounds was foreseen by Scheele and Bergman, the former being isolated in 1783 by Hjelm, and the latter by d'Elhujar. Uranium, lastly, or rather an oxide of it which was looked upon as the element, was detected by Klaproth in 1798 as a principal constituent of pitchblende; Péligot[1] was the first to correct this error by proving that the supposed element contained oxygen, and also by preparing metallic uranium itself. The atomic weights of chromium and uranium, as determined by Péligot, have been corroborated by the recent estimations of Cl. Zimmermann, Berlin, and especially Meineke. For molybdenum, a somewhat higher value than that obtained by Berzelius has been arrived at from the work of Dumas, Rammelsberg and others. The atomic weight found by Schneider, Marchand and others for tungsten—viz. 183·5—has maintained its ground.

The elements which resemble tin in character—viz. titanium, zirconium and thorium (to which germanium has within the last few years been added)—belong practically to the chemical history of this century; for, although the oxides of titanium and zirconium were discovered at the end of last century, the isolation of the elements themselves was first accomplished by Berzelius, by means of the method already mentioned—viz. the decomposition of the double fluorides with potassium. Berzelius[2] also discovered thoria (ThO_2) and thorium in 1828; the atomic weight of this element was, however, only definitely established at a much later period by Nilson,[3] the value then obtained by him being subsequently corroborated by the determination of the vapour density of thorium chloride.[4] Germanium, the youngest at present of the metallic elements, was discovered some years ago by Cl. Winkler,[5] and led in his hands to some admirable

[1] *Ann. Chim. Phys.* (3), vol. v. p. 5.
[2] *Pogg. Ann.*, vol. xvi. p. 385.
[3] *Ber.*, vol. xv. p. 2527.
[4] Nilson u Krüss, *Ber.*, vol. xx. p. 1671.
Clemens Winkler, born in 1838, has held the Chair of Chemistry at

experimental work,[1] which threw the clearest light upon its nature and that of its compounds. The impulse to look for a new element was given him by the analysis of a Freiberg silver ore, which invariably showed a deficit of about 7 per cent. This led to the surmise that some substance was present for which the analytical methods in use were inadequate, just as in the case of cæsium, already mentioned. The atomic weight of germanium, as determined by Winkler, agrees with the position which naturally falls to this element in the periodic system.

Vanadium, tantalum and niobium—elements nearly related to antimony and bismuth—have only become well known through comparatively recent researches. Vanadium, recognised as a constituent of certain lead ores by del Rio so early as 1801, but more definitely by Sefström in 1830, was isolated in the metallic state by Roscoe[2] in 1867, who proved that the substance hitherto taken for an element really contained oxygen and nitrogen. The chemical relations of this element and its compounds were admirably worked out by him, and the atomic weight determined with certainty.

The investigations of Hatchett, Ekeberg, Wollaston and Berzelius on the minerals columbite and tantalite, in the two first decades of our century, had already pointed to the presence of the elements which afterwards received the names of tantalum and niobium, although the elements themselves had not been obtained. Nor did the work of H. Rose[3] lead either to their isolation or to a correct knowledge of their compounds, for in this case, too, niobium dioxide (Nb_2O_2) was regarded as the element itself. It was the researches of Blomstrand[4]

the School of Mines in Freiberg, Saxony, since 1873, having previous to that been engaged in practical mining work for fourteen years. Inorganic and technical chemistry are indebted to him for some most admirable researches, which have frequently included new methods, valuable either in the laboratory or the manufactory. His most important papers are referred to in the special history of analytical and of inorganic chemistry.

[1] *Journ. pr. Chem.* (2), vol. xxxiv. p. 177; vol. xxxvi. p. 177.

[2] *Phil. Trans.* for 1868, p. 1; or *Ann. Chem.*, Suppl., vol. vi. p. 86.

[3] *Pogg. Ann.*, vol. xcix. p. 80; vol. civ. p. 432.

[4] *Journ. pr. Chem.*, vol. xcvii. p. 37.

and of Marignac[1] which first furnished definite standpoints for a review of the chemical behaviour of the two elements and their compounds, and for the fixing of their atomic weights.

The metals of the platinum group, with the exception of platinum itself,[2] have all been discovered during this century as constituents of platinum ores. Platinum was only obtained perfectly pure after suitable methods had been worked out for separating it from the accompanying metals. Its employment for making certain kinds of apparatus, so important for the development both of scientific and of technical chemistry, also belongs to the present period.

Palladium came in 1803 under its present name into commerce, as a new metal, without its discoverer being known; it was only at a later date that it was learnt to have been isolated by Wollaston from platinum ore.[3] The remarkable property which it possesses of combining with hydrogen was first observed by Graham.[4] The discovery of palladium led Wollaston[5] on to that of another of the platinum metals, rhodium, which he thus named because of the rose-red colour of its solutions. It was investigated carefully by Berzelius,[6] who made a minute study of the platinum metals generally, and by C. E. Claus;[7] the separation of rhodium from other metals is due primarily to Bunsen,[8] and to Deville and Debray. Tennant[9] was the first to direct the attention of chemists to iridium and osmium, as two new metals which were contained in the residues left from the solution of platinum ores; while to Deville and Debray[10] we mainly owe the method of preparing both elements (the heaviest substances as yet known) pure. Ruthenium, lastly,

[1] *Ann. Chim. Phys.* (4), vol. viii. p. 5. [2] Cf. p. 149.
[3] *Phil. Trans.* for 1804, p. 428. [4] *Phil. Mag.* (4), vol. xxxii. p. 516.
[5] *Phil. Trans.* for 1804, p. 419. [6] *Pogg. Ann.*, vol. xiii. p. 437.
[7] *Beiträge zur Chemie der Platinmetalle* (Dorpat, 1854), ("Contributions to the Chemistry of the Platinum Metals").
[8] *Ann. Chem.*, vol. cxlvi. p. 265.
[9] *Phil. Trans.* for 1804, p. 411.
[10] *Comptes Rendus*, vol. lxxxi. p. 839; vol. lxxxii. p. 1076.

was likewise discovered in platinum ores, as well as in osmiridium, by Claus,[1] who has further given us most of our knowledge of this element and its compounds, together with its atomic weight. Debray [2] also investigated ruthenium and its oxygen compounds comparatively recently.

Up to a short time ago the atomic weights of the platinum metals had only in part been determined with the requisite definiteness. For platinum itself the most reliable value was supposed to be that obtained by Berzelius, viz. 196·7, until Seubert [3] showed (in 1880) that this figure was too high by at least two units. The atomic weights of palladium, rhodium and osmium were also determined by Berzelius, but required further corroboration; this applied more especially to that of osmium. Of late years, however, K. Seubert [4] has re-determined the atomic weights of iridium, osmium and rhodium, Keyser [5] that of palladium, and von Joly [6] that of ruthenium, with the result that those important constants have been brought into order in the above two groups. At the same time the periodic system of the elements has celebrated a new triumph, for it is only now that the platinum and palladium metals accord with it,—*i.e.* they now occupy the positions which they ought to do according to theory.

Within the last four years the number of the elements has been increased by two of peculiar interest—viz. argon,[7] discovered in the air by Lord Rayleigh [8] and William

[1] *Ann. Chem.*, vol. lvi. p. 257; vol. lix. p. 284.

[2] *Comptes Rendus*, vol. cvi. pp. 100, 328.

[3] *Ann. Chem.*, vol. ccvii. p. 29; *Ber.*, vol. xxi. p. 2179; also Dittmar and Arthur, *ibid.*, vol. xxi. Ref. p. 412.

[4] *Ber.*, vol. xi. p. 1770; *Ann. Chem.*, vol. cclxi. p. 257; vol. cclx. p. 314.

[5] *Amer. Chem. Journ.*, vol. xi. p. 398.

[6] *Compt. Rend.*, vol. cviii. p. 946.

[7] *Chem. News*, vol. lxx. p. 87 (1894); *British Association Report* for 1894; *Phil. Trans.* for 1895, vol. A. part 2, p. 187.

[8] John William Strutt, third Lord Rayleigh, was born on November 12th, 1842. After a brilliant career at Cambridge, where he was senior wrangler of his year and Smith's prizeman, he succeeded the late Clerk Maxwell as professor of physics in the University of Cambridge (1879—1884). Since 1887 he has held the chair of Natural Philosophy in the Royal Institution.

Ramsay,[1] and helium discovered by Ramsay in the mineral clèveïte.[2] Helium had already been found to exist in the sun's chromosphere by Janssen[3] so long ago as 1868, *i.e.* he observed in the spectrum a new bright yellow line which he designated D_3. The spectrum was further studied that year by Frankland and Lockyer, who also gave to the new hypothetical element the name Helium.[4] The isolation of argon arose from and followed upon careful determinations of the atomic weight of nitrogen by Lord Rayleigh, who found that nitrogen obtained from air was always specifically heavier by a distinct amount (about ½ per cent.) than nitrogen prepared from chemical compounds. The original paper in the *Philosophical Transactions*, already quoted, will undoubtedly rank as a classic, the investigation having been a particularly brilliant one. It is interesting to recall that a hundred years ago, Cavendish, reasoning from the results of his own experiments, had with marvellous acuteness suggested the possibility of such a gas in the air, and had surmised its approximate amount. Thanks to its incapacity for combining with other elements, argon remains behind after all the other gases in the air have been got rid of.

Although a physicist in the first instance, some of his work has been of the highest importance for chemistry, notably that upon the density of nitrogen, which led him and Ramsay on to the discovery of argon; and, in a secondary degree, his determination of the ratio of the atomic weights of hydrogen and oxygen.

[1] William Ramsay, born at Glasgow on October 2nd, 1852, studied chemistry under the late Professor Anderson at Glasgow University, and then under Fittig at Tübingen. After acting for some years as chemical assistant at the Andersonian College and afterwards at the University of Glasgow, he was appointed in 1880 professor of chemistry in University College, Bristol, and also principal of that college in the following year. In 1887 he succeeded Williamson in the chair of chemistry at University College, London, which post he still holds. His earlier investigations were in organic chemistry, but for the last twenty years or so they have been mostly in physical and inorganic, both of which branches he has greatly advanced by a wealth of work. In addition to his numerous published papers, he is the author of the well-known and original text-book :— *A System of Inorganic Chemistry* (J. and A. Churchill, 1891), besides of smaller works, his last published volume being *The Gases of the Atmosphere* (Macmillan and Co., 1896).

[2] *Chem. News*, vol. lxxi. p. 151; *Journ. Chem. Soc.*, vol. lxvii. p. 1107.

[3] *Compt. Rend.* vol. lxvii. p. 838. [4] Lockyer, *Proc.*, *R.S.*, vol. xvii. p. 91.

Helium, first obtained by heating clèveïte with sulphuric acid, and since found in small quantity—often together with argon—in a good many other minerals, as well as in the gases from some mineral water springs, is equally indifferent. Up to now, in spite of persistent effort, no compound of either argon or helium has been prepared. And further, although most careful and laborious diffusion experiments with both gases have been carried out,[1] with the object of seeing whether they were really elementary, the densities of both have remained unaltered, *i.e.* it has been found impossible to subdivide them by diffusion into two or more components. From the ratio of the specific heats at constant volume and constant pressure, it follows that the molecule and atom are identical in both argon and helium, *i.e.* that the gases are monatomic.[2]

The above short survey of the discovery of elements during the present chemical period is sufficient to allow of our properly appreciating the extent of the achievements in this branch of the science. Since chemists have had before their eyes the task of assigning a definite place in the periodic system to each element, the discovery of a new one has possessed quite another charm, and also a far higher significance than was formerly the case. What is now aimed at is to determine the atomic weight of each with such accuracy, and to examine its chemical behaviour with such completeness, as to permit of its being classified in

[1] Ramsay and Collie, Proc. R.S., vol. lx. p. 206 (1896); Ramsay and Travers, *ibid.* vol. lxii. p. 316 (1897).

[2] Ramsay and Travers have just discovered another gas of characteristic spectrum, which is present in the air in very small quantity, and to which they have given the name krypton. It is monatomic and has a higher density than argon (the discoverers expect that its atomic weight will be found to be 81 or 82, in accordance with the periodic law). They found it in the gas from the last 10 c.c. left on evaporating about a litre of liquid air.—Further, by fractionating about 20 c.c. of crude liquid argon, they have succeeded in subdividing the latter into the following three elementary gases, each of which has a well-defined spectrum:—argon, metargon, and neon. The atomic weights of argon and metargon will probably be about 40, and that of neon about 20. Argon and metargon have been proved to be monatomic, but neon has not yet been tested for this; it will no doubt be found monatomic like the others. (*Proc. R.S.* for 1898.)

this system. In the case of none among the recently discovered elements have those efforts been followed with such signal success as in that of germanium. On the other hand, as argon and helium cannot be properly fitted into the periodic system, a suggestion has been made to set up a separate group of "inactive elements" (Lecoq de Boisbaudran; J. Thomsen.)[1]

We find in chemical literature many accounts of supposed new elements, which afterwards turned out either to have been prepared before, or to be mixtures of substances partly already known and partly unknown. A passing reference may be made here to the fantastic attempts of Winterl[2] at the end of last and beginning of this century, who imagined that he had decomposed several metals into different elements. But even investigators of eminence fell into errors which could only be explained by defects in the analytical methods of their day; thus Bergman (in 1781) looked upon iron phosphide, prepared from "cold-short" iron by means of hydrochloric acid, as a new metal, to which he gave the name of *siderum*, and Richter claimed impure nickel as an element, terming it *nickolanum*. Even Berzelius thought that he had discovered (in 1815) a hitherto unknown earth in some Swedish minerals, but he corrected the error himself by showing that the supposed new body was phosphate of yttria. The history of the cerium metals, to which yttrium belongs, and also of tantalum and niobium, shows more especially a great many such errors, while even at the present day a number of new elements are being brought forward whose homogeneous nature is in the highest degree doubtful, *e.g. decipium, mosandrium,* and *philippium.*[3] Similarly, so little is known yet of *masrium*,[4] discovered in 1892, and of the element still more recently obtained from bauxite, that nothing can be said about them; and the same remark applies to *lucium* and *russium* (Barrière; Chrustschoff).

[1] *Ber* vol. xxix. Ref. p. 830.
[2] Kopp, *Gesch. d. Chemie*, vol. ii. p. 282.
[3] *Comptes Rendus*, vol. lxxxvii. pp. 148, 559, 632, etc.
[4] *Mon. Scient.* for 1892. p. 514.

Extension of the Knowledge of Inorganic Compounds.

The general standpoints arrived at during the present chemical period for the comprehension of inorganic chemical compounds, more especially the opinions with regard to the constitution of acids, bases, and salts, have been entered into in detail in the first section of this book. It remains now to give an account of the development of special knowledge in this branch of the science. An exhaustive treatment of the subject is of course impossible here; only researches of particular moment, which have materially aided in extending the knowledge of chemistry, can be mentioned.

Hydrogen Compounds of the Halogens.

The remarkable behaviour of hydrogen with respect to chlorine,—the readiness with which those two gases combine, was first investigated by Davy and Gay-Lussac, and afterwards made the subject of important physico-chemical work by Roscoe and Bunsen.[1] The researches of Davy and Faraday [2] contributed greatly to a more intimate knowledge of hydrochloric acid, showing among other things how to condense the gas, while those of Roscoe and Dittmar [3] established the chemical relations existing between hydrochloric acid and water. Gay-Lussac and Balard investigated hydriodic and hydrobromic acids, while the fundamental researches of Gay-Lussac, Thénard and Berzelius contributed a knowledge of hydrofluoric acid in aqueous solution, and those of Gore [4] and Frémy [5] of this acid in the gaseous state, these latter thus establishing its composition. Nicklès fell a victim to the frightful action of anhydrous hydrofluoric acid in 1869. Ampère was the first to point out the analogy between fluorine and chlorine.

Oxygen Compounds of Hydrogen and of the Halogens.

The investigations which led to a knowledge of the composition of water have been already described; the first

[1] *Pogg. Ann.*, vol. c. p. 43; or *Phil. Trans.* for 1857, p. 355; *Ann. Chem.*, vol. xcvi. p. 357; cf. *History of Physical Chemistry*.

[2] *Phil. Trans.* for 1823, p. 164.

[3] *Ann. Chem.*, vol. cxii. p. 337.

[4] *Phil. Trans.* for 1869, p. 173.

[5] *Ann. Chim. Phys.* (3), vol. xlvii. p. 5.

quantitative determination of its constituents, to which but little exception could be taken, was made by Berzelius and Dulong.[1] The discovery of peroxide of hydrogen [2] by Thénard in 1818 showed that water was not the only oxide of that element, while the chemical behaviour of this peroxide, which was examined by Thénard, Schönbein, etc., and of recent years by Schöne [3] and Traube,[4] stamps it as one of the most remarkable of inorganic compounds. It also appears to play an important part in many of the processes of nature, and the interest in it is heightened still further by the value which it promises to have for technical chemistry. Wolffenstein has lately succeeded in preparing pure hydrogen peroxide without difficulty by distilling it *in vacuo*.[5]

The various stages of oxidation of chlorine, iodine and bromine have been the cause of much valuable work since the beginning of our century, *e.g.* that of Gay-Lussac on chloric acid, of Stadion on perchloric acid, of Davy and Stadion on chlorine peroxide, of Millon [6] on chlorous acid, and of Balard [7] on hypochlorous acid. The knowledge of some of these compounds was much enlarged by Pebal's latest researches,[8] which established the nature of the so-called euchlorine and of chlorine peroxide. The oxygen compounds of iodine became known through the investigations of Davy and Magnus; periodic acid (discovered by the latter) [9] and iodic acid led later on to a knowledge of several series of salts, from whose composition important conclusions as to the saturation-capacity of iodine, and therefore of the halogens generally, were drawn. Excepting argon and helium, fluorine is the only element which does not combine with oxygen.

[1] *Ann. Chim. Phys.*, vol. xv. p. 386; for later determinations of the ratio $H_2 : O$, see p. 403. [2] *Ann. Chim. Phys.*, vol. viii. p. 306 (1818).

[3] *Ann. Chem.*, vol. cxcii. p. 258 (Schöne gives here a review of the previous literature on the subject).

[4] Cf. *Ber.*, vol. xx. p. 3345; vol. xxii. p. 1496; vol. xxvi. p. 1471.

[5] *Ber.*, vol. xxvii. p. 3307. [6] *Ann. Chim. Phys.* (3), vol. vii. p. 298.

[7] *Ibid.*, vol. lvii. p. 225.

[8] *Ann. Chem.*, vol. clxxvii. p. 1; vol. ccxiii. p. 113.

[9] *Pogg. Ann.*, vol. xxviii. p. 514.

Sulphur, Selenium and Tellurium Compounds.

To the early known compounds of sulphur and oxygen, sulphurous and sulphuric acids (the anhydride of the latter having been discovered by Vogel and Döbereiner), others came to be added, viz. "hyposulphurous acid" by Gay-Lussac,[1] and dithionic acid by Welter and Gay-Lussac (in 1819). The constitution of the first of these, which is really thiosulphuric acid, was only made out at a much later date.[2] The thio-acids containing more sulphur, and nearly related to sulphuric acid, were discovered at the beginning of the forties by Langlois, Fordos and Gélis, and Wackenroder; the question as to whether the pentathionic acid of the latter really exists has recently been vigorously discussed.[3]

To the above sulphur acids there has of late years been added Schützenberger's hyposulphurous acid (H_2SO_2) the chemical behaviour of which is of great interest.[4] The two well-known oxides of sulphur also received an addition in R. Weber's sesquioxide, S_2O_3.[5] Lastly, mention may be made here of per-sulphuric acid, whose existence Berthelot showed to be probable, and for the anhydride of which he assumed the formula S_2O_7; recent researches by Elbs and others have solved its true composition, viz. HSO_4, which thus corresponds with that of permanganic acid. Neither of these compounds, however, (*i.e.* S_2O_7 and HSO_4) has yet been obtained pure. The enormous impetus given to chemical industries generally by the development of the sulphuric acid manufacture must also be referred to. It is only within the last few years that such simple derivatives of sulphuric acid as the amide and imide have become known; and

[1] *Ann. Chim.*, vol. lxxxv. p. 199; sodium hyposulphite (thiosulphate) was first prepared by Chaussier in 1799, and afterwards more carefully examined by Vauquelin.

[2] Cf. Schorlemmer, *Journ. Chem. Soc.* (2), vol. vii. p. 256.

[3] Cf. Curtius u. Henkel, *Journ. pr. Chem.* (2), vol. xxxvii. p. 37; Debus, *Journ. Chem. Soc.*, vol. liii. p. 278; or *Ann. Chem.*, vol. ccxliv. p. 76.

[4] *Comptes Rendus*, vol. lxix. p. 169.

[5] *Pogg. Ann.*, vol. clvi. p. 53.

the same thing applies to fluor-sulphuric acid and other compounds.[1]

The compounds of selenium with hydrogen and oxygen were investigated by Berzelius, and an account of them given in his memorable treatise. After him there came Mitscherlich, who discovered selenic acid, and therewith furnished a beautiful confirmation of the analogy between selenium and sulphur more especially from the isomorphism of the sulphates and selenates. This chemical similarity has not however been maintained in all respects, Michaelis [2] having recently shown that the salts of selenious acid probably possess a constitution different from that of sulphites.

The chlorine compounds of sulphur, selenium and tellurium, the study of which has helped to characterise these elements, have been examined at various times; by his investigation of tellurium tetrachloride Michaelis has lately furnished an excellent proof of tellurium being tetratomic.

Even if we desired to mention only the more important of the investigations which have aided in the discovery and elucidation of the hydrogen, oxygen and halogen compounds of nitrogen, phosphorus, arsenic and antimony, it would be necessary to record a long series. Among them were the researches of Davy, Berthollet and Henry, which made clear the composition of ammonia,—so long looked upon as containing oxygen. The discovery of phosphuretted hydrogen (PH_3) by Gengembre [3] in 1783, and the examination of it by Pelletier (who was the first to prepare it pure), only became fruitful after Davy's investigations; the last-named elucidated the composition of this gas, and pointed out its analogy to ammonia, this being emphasised still more sharply by H. Rose later on. Thénard [4] discovered liquid phosphuretted hydrogen, and recognised in it the cause of the spontaneous inflammability of the not completely pure gaseous compound.

[1] Cf. W. Traube, *Ber.*, vol. xxvi. p. 607; Thorpe and Kirwan, *Ztschr. anorgan. Chem.*, vol. iii. p. 63; or *Journ. Chem. Soc.* vol. lxi. p. 921.

[2] *Ann. Chem.*, vol. ccxli. p. 150.

[3] *Crell's Ann.*, vol. i. p. 450.

[4] *Ann. Chim. Phys.* (3), vol. xiv. p. 5. Compare also Gattermann and Haussknecht, *Ber.*, vol. xxiii. p. 1174.

Arseniuretted and antimoniuretted hydrogens, which are analogous to ammonia in composition, were first obtained in the pure state by Soubeiran [1] and Pfaff.[2] The former compound cost Gehlen his life in 1815, from his not suspecting its extreme poisonousness; and the same fate has recently befallen H. Schulze (of St. Jago). The great importance of the formation of arseniuretted hydrogen for the detection of minute quantities of arsenic in judicial-chemical analyses (Marsh's process) is well known.

The oxygen compounds of nitrogen played, as already described, an important part in the history of the atomic theory, even although the true composition of all these oxides was not at that time made out. The number of the oxides of nitrogen known in Dalton's time was supplemented by nitrogen peroxide, whose relation to the others was arrived at through the researches of Berzelius, Gay-Lussac and Dulong; and by nitric anhydride, discovered by St. Claire Deville. The various obscure points with respect to nitrous acid and nitrogen peroxide have been for the most part explained by the recent investigations of Hasenbach [3] Lunge,[4] Ramsay [5] and others. The discovery of hyponitrous acid,[6] the acid corresponding to nitrous oxide, enlarged still further the series of the oxy-acids of nitrogen. W. Wislicenus and Paal, independently of one another, succeeded in preparing hyponitrous acid by the interaction of hydroxylamine and nitrous acid,[7] and since then Hantzsch and Kaufmann have proved that its molecular weight corresponds to the formula $N_2O_2H_2$.[8]

Reference must also be made here to the important discovery of hydroxylamine,[9] which, from its value as a reagent,

[1] *Ann. Chim. Phys.* (2), vol. xxiii. p. 307.

[2] *Pogg. Ann.*, vol. xl. p. 135.

[3] *Journ. pr. Chem.* (2), vol. iv. p. 1.

[4] Cf. *Ber.*, vol. xviii. p. 1376; vol. xxi. p. 67.

[5] *Journ. Chem. Soc.*, vol. xlvii, pp. 187 and 672; vol. liii. p. 621; vol. lvii. p. 590; *Phil. Mag.*, vol. xxiii. p. 129; vol. xxiv. p. 196.

[6] Divers, *Proc. R. S.*, vol. xix. p. 425; Zorn, *Ber.*, vol. x. p. 1306.

[7] *Ber.*, vol. xxvi. pp. 771 and 1026. [8] *Ann. Chem.*, vol. ccxcii. p. 317.

[9] Lossen, *Ann. Chem.*, Suppl., vol. vi. p. 220.

gations and discussions.[1] Hypophosphoric acid,[2] $H_4P_2O_6$, has also lately been added to the above oxygen compounds.

The discovery of the halogen compounds of nitrogen and phosphorus has proved of particular interest, the latter being largely employed for the preparation of many other substances, because of the readiness with which they enter into reaction. Chloride of nitrogen was discovered by Dulong,[3] who suffered serious injury in consequence of some of it exploding unexpectedly; this dangerous substance, whose composition was hitherto uncertain, has of late been made the subject of important investigations by Gattermann,[4] who has succeeded in preparing the pure chloride, NCl_3. The analogously formed iodide of nitrogen was first prepared by Serullas,[5] while Bunsen, Stahlschmidt and, quite recently, Raschig [6] have contributed towards a knowledge of its composition. The chlorine compounds of phosphorus were prepared in the first decade of our century, the trichloride by Gay-Lussac and Thénard, and the pentachloride by Davy. The trifluoride of phosphorus has only recently been prepared by Moissan; the pentafluoride, isolated by Thorpe,[7] is of especial interest from its not decomposing even at high temperatures, unlike the other penta-haloid compounds of phosphorus. Wurtz discovered phosphorus oxychloride, which is of great value as a reagent in organic work, and H. Rose antimony pentachloride. The oxybromide of phosphorus has been known for some time, but it is only lately that Moissan has obtained the oxyfluoride.

The halogen compounds of boron and silicon were mainly investigated by Berzelius and, later, by Wöhler and Deville,[8] and they constituted the material from which those elements themselves and others of their compounds were prepared; the

[1] Cf. Wurtz, *Ann. Chem.*, vol. xliii. p. 318; vol. lxviii. p. 41.

[2] Salzer, *Ann. Chem.*, vol. clxxxvii. p. 322; vol. cxciv. p. 28; vol. ccxi. p. 1; vol. ccxxxii. p. 114; Sanger, *ibid.*, vol. ccxxxii. p. 1.

[3] *Schweigger's Journ.*, vol. viii. p. 302.

[4] *Ber.*, vol. xxi. p. 751.

[5] *Ann. Chim. Phys.*, vol. xlii. p. 200.

[6] *Ann. Chem.*, vol. ccxxx. p. 212.

[7] *Ann. Chem.*, vol. clxxxii. p. 201.

[8] *Ibid.*, vol. cv. p. 67 *et seq.*

above researches, in fact, greatly extended the knowledge of these substances generally. Among other points, the discovery of boron nitride and silicon hydride may be mentioned here.[1] To the careful investigation of volatile silicon compounds is due the definite establishment of the atomic weight of that element, and, with this, of the composition of silica, to which another formula than the present was previously given. In recent years there have been further important researches on the halogen compounds of boron and silicon[2] by Moissan, Besson, and Sabatier.

Of the simple compounds of carbon, which from long custom are assigned to inorganic chemistry, the greater number were discovered and examined at the beginning of this century. Details have already been given with respect to carbonic acid and carbonic oxide. The study of the phenomena of combustion, and particularly of the processes which go on in the flame of burning carbon compounds, in which the two gases just mentioned play a prominent part, was first taken up by Davy, who advanced the subject immensely by his beautiful researches. We must also refer here to the more recent investigations of Frankland, Blochmann, Heumann, Smithells, and Lewes, on the theory of luminous flames. The luminous acetylene flame has again within the last few years been the subject of much investigation and discussion among chemists, the object being to get at the cause of the luminosity. The results have proved that, in the main, Davy's old theory of luminous flames is correct.

Carbon oxychloride or phosgene, which has proved of exceptional value as a reagent in organic chemistry, was first prepared by Davy in 1811, but carbon oxysulphide only comparatively recently by von Than.[3] Carbon disulphide, on the other hand, was obtained by Lampadius so early as 1796,

[1] Wöhler u. Buff, *Ann. Chem.*, vol. cii. p. 120.

[2] *Compt. Rend.*, vols. cxii. and cxiii.

[3] *Ann. Chem.*, Suppl., vol. v. p. 236. The properties of the pure compound were first established by Klason (*Journ. pr. Chem.* (2), vol. xxxv. p. 64).

and more minutely investigated by Clément and Desormes in 1802 ; it is now an important product of chemical manufacture. Its composition was arrived at correctly by Vauquelin and Berzelius, after the most confused opinions had previously been expressed with regard to its containing hydrogen and nitrogen. The profound influence which the classical researches of Gay-Lussac on cyanogen and its compounds exercised upon the development of chemistry has already been referred to.

Extension of the Knowledge of Metallic Compounds.

From the endless number of investigations which have contributed towards a knowledge of the metallic compounds, and, with this, of the metals themselves, the most important must now be mentioned, if they have not already been touched upon in the general section of this book.

The discoverers of the alkali metals also aided largely in their investigation; thus to Davy is due our knowledge of the oxides of potassium and sodium, to Gay-Lussac that of the corresponding peroxides, and to Bunsen that of rubidium and cæsium compounds. Sodium peroxide is now manufactured in quantity. The enormous influence which these researches on the compounds of the alkalies exercised upon the development of chemical industries will be detailed under the history of technical chemistry.

The peroxides of barium and calcium were discovered by Gay-Lussac and Thénard. The knowledge of the nature of chloride of lime was advanced by the researches of Balard, who was the first to express the opinion—still held by many—that this substance was a double compound of calcium chloride and hypochlorite. Since that time numerous further experiments have led many to regard it as an oxychloride of calcium, which has given rise to a large amount of discussion.[1]

The investigations which led to a knowledge of the com-

[1] Cf. the work of Göpner, Wolters, Kraut, Lunge and others.

pounds of beryllium and thallium have been cited above.[1] New oxygen compounds of copper, in addition to the oxides already known, were obtained by Rose [2] and Thénard, while Wöhler discovered silver suboxide and peroxide; it must however be mentioned here that the existence of the former of these has been vigorously contested.[3] The application of silver salts for the fixation of light impressions (*i.e.* in photography), so pregnant in its results, will be treated under the history of physical chemistry. Those chemists who shared in the discovery and investigation of aluminium, indium and gallium, also contributed at the same time to a knowledge of their compounds. With respect to the compounds of alumina, pure chemistry has frequently been called upon to elucidate difficult points pertaining to the manufacture of ultramarine, porcelain, glass, &c.

The compounds of the metals which form the iron group have been the object of a very large number of investigations, among which we may mention those on the different stages of oxidation of manganese by Liebig and Wöhler,[4] Mitscherlich [5] and, recently, Franke.[6] The chlorine and fluorine compounds of manganese were studied by Christensen. To the two oxides of iron (FeO and Fe_2O_3), a knowledge of which we owe to Proust and whose composition was established by Berzelius, Frémy added ferric acid, which he also carefully investigated; the existence of this acid was surmised by Scheele. Light was thrown upon the nature of the cyanogen compounds of iron by the beautiful researches of Gay-Lussac, Berzelius, Gmelin (who discovered potassic ferricyanide), and Liebig, out of which the present views held with regard to these substances have developed themselves. The

[1] Cf. pp. 408 and 409.

[2] *Pogg. Ann.*, vol. cxx. p. 1.

[3] Wöhler, *Ann. Chem.*, vol. xxx. p. 1; Friedheim, *Ber.*, vol. xxi. p. 316. On the other hand, von der Pfordten, who at first believed that he had proved the existence of silver suboxide, has subsequently expressed himself in favour of a "hydrate of silver," the most probable formula being $Ag_4 \cdot H_2O$. (*Ber.*, vol. xxi. p. 2288.)

[4] *Pogg. Ann.*, vol. xxi. p. 584.

[5] *Ibid.*, vol. xxv. p. 287.

[6] *Journ. pr. Chem.* (2), vol. xxxvi. pp. 31, 166, 451.

nitroprussides, so nearly allied to the ferrocyanides, were first obtained by Playfair, but their constitution has yet to be satisfactorily cleared up.

There are few more remarkable metallic compounds than those recently discovered ones which carbon monoxide forms with iron and nickel, when the gas is allowed to pass over the hot and finely-divided metal.[1] Nickel tetra-carbonyl, $Ni(CO)_4$, is especially interesting both from a physical and chemical point of view, and it is to be hoped that by means of it the question of the true atomic weight of nickel will be ultimately solved. Whether, as Mond hopes, the metallurgical production of nickel by the aid of this carbonyl compound will be possible, remains to be proved.

The metallic carbides and several compounds of carbon with other non-metallic elements are likewise very striking substances. Although only discovered a short time ago, some of them have already acquired great technical importance, especially carbide of calcium, which is used at present for the production of acetylene gas, and which will assuredly play an even greater part in the future than it does now. Silicon carbide or "carborundum" far surpasses corundum as a polishing material for hard substances. Moissan's splendid work on the carbides—of which a few were known previously, but had only been superficially examined—has thrown much light on these compounds.[2]

The chemistry of the cobalt salts was enriched by the discovery of the remarkable and highly varied ammonio-cobaltic compounds, which, observed by Genth for the first time in 1851, were afterwards investigated by Fr. Rose, Gibbs, Frémy, and especially Jörgensen.[3] The last-named investigator has brought the extraordinarily difficult question of the chemical constitution of these bodies materially nearer to its solution, by systematically examining the ammonia

[1] Mond, Langer and Quincke, *Journ. Chem. Soc.*, vol. lvii. p. 749; *Ber.*, vol. xxiv. p. 2248; Berthelot, *Compt. Rend.*, vol. cxii. p. 1343.

[2] Cf. *Compt. Rend.*, vol. cxvii. p. 679; also vols. cxv. and cxvi.; but particularly his brochure, *Le Four Electrique* (Paris, 1897).

[3] Cf. *Journ. pr. Chem.* (2), vol. xxiii. p. 227; vol. xxxi. pp. 49, 262; vol. xxxix. p. 1; vol. xli. p. 429.

compounds of those other metals analogous to cobalt in this respect—chromium and rhodium.[1]

The various combining relations which the different members of the group of elements comprising molybdenum, tungsten and uranium show towards other elements, have only been fully understood of recent years. The admirable work of Berzelius on molybdenum compounds has been supplemented by that of Krüss[2] on the sulphides, and of Muthmann[3] on the oxides, as well as by the earlier investigations of Blomstrand, Debray, Liechti and Kempe on the halogen compounds of molybdenum. The chlorides of tungsten were examined in detail by Roscoe, who thereby advanced the knowledge of the saturation-capacity of this element. The complicated salts of tungstic acid were first studied by Margueritte, Scheibler, Marignac and v. Knorre, but their ultimate constitution, as well as that of the phospho-molybdic and phospho-tungstic acids, has still to be unravelled. Tungsto- and molybdo-vanadic acids belong to the "compound acids" which have recently been investigated by Friedheim. The chemical nature of uranium and its compounds has been worked out with most success by Cl. Zimmermann[4] whose able researches have largely supplemented the earlier ones of Péligot, Roscoe and others.

Of the compounds of tin and its chemical analogues, the isomorphous double fluorides[5] aroused especial interest from their proving the connection which exists between silicon, titanium, zirconium and germanium. The peculiar nature of titanium was elucidated in a striking manner by the discovery of its nitrogen compounds,[6] and more recently by the preparation of its various sulphides.[7]

To Roscoe's admirable work[8] is due most of our knowledge

[1] *Journ. pr. Chem.* (2), vol. xxv. pp. 83, 321; vol. xxx. p. 1; vol. xxxiv. p. 394.

[2] *Ann. Chem.*, vol. ccxxv. p. 1. [3] *Ibid.*, vol. ccxxxviii. p. 109.

[4] *Ann. Chem.*, vol. ccxiii. p. 285 (contains a historical review); vol. ccxxxii. p. 274; also Alibegoff, *ibid.*, vol. ccxxxiii. p. 117.

[5] Marignac, *Ann. des Mines* (5), vol. xv. p. 221.

[6] Wöhler, *Ann. Chem.*, vol. lxxiii. p. 43. [7] *Ann. Chem.*, vol. ccxxxiv. p. 257.

[8] *Phil. Trans.* for 1869, p. 679; or *Ann. Chem.*, Suppl., vol. vii. p. 70.

of vanadium, as he worked out correctly the different stages of combination of this element with oxygen, chlorine, &c., and set right the former erroneous assumptions with regard to the composition of these compounds. Gerland's investigations [1] on vanadyl salts and vanadic acids, and those of v. Hauer on the salts of the latter, have also been of assistance here.

Similarly niobium and tantalum, whose chemical nature had been completely misjudged, were given their proper position among the other elements by the investigations already cited, more particularly by the determination of the true composition of the chlorides of both and of niobium oxychloride,[2] and by the examination of niobium fluoride and hydride.[3]

Valuable work has also been done lately on the compounds of gold, by Krüss [4] more especially, which has materially amplified the earlier researches of Proust, Berzelius, Figuier, &c., and has served to establish the chemical character and the atomic weight of this element.

The literature on platinum and its compounds is very voluminous, and gives evidence of most excellent experimental work. Reference may be made here to the discovery of the peculiar reactions to which platinum can give rise in virtue of its condensation of oxygen (the absorption of oxygen by platinum and palladium has been proved to be true oxidation, for the same amount of heat is given out by this absorption as in the formation of the oxides PtO and PdO) [5]; and to the numerous investigations on the platinum-ammonium compounds, the first of which were prepared by Magnus, and whose peculiarities were studied by Gros, Reiset, Cléve, Thomsen and Blomstrand. The recently published work of Jörgensen: [6]

[1] *Ber.*, vol. ix. p. 874; vol. x. p. 2109; vol. xi. p. 98.

[2] Deville and Troost, *Comptes Rendus*, vol. lx. p. 1221.

[3] Krüss and Nilson, *Ber.*, vol. xx. p. 1676.

[4] Cf. Krüss, *Ann Chem.*, vol. ccxxxvii. p. 274 (contains a historical review); vol. ccxxxviii. pp. 30 and 241; *Ber.*, vol. xxi. p. 126; Thorpe and Laurie, *Journ. Chem. Soc.*, vol. li. pp. 565 and 866.

[5] Mond, Ramsay and Shields, *Phil. Trans.*, vol. clxxxvi. p. 657; vol. cxc. p. 129.

[6] *Journ. pr. Chem.* (2), vol. xxxiii. p. 489.

Zur Konstitution der Platinbasen ("On the Constitution of the Platinum Bases"), marks an important step in the recognition of the constitution of these bodies. The compounds which carbon monoxide forms with chloride of platinum, discovered by Schützenberger, have lately been carefully investigated by Mylius and Förster and by Pullinger, who have thereby contributed greatly to solving the problem of their constitution.[1]

The researches which have assisted materially towards a knowledge of the platinum metals have already been mentioned under the history of the individual elements.

If we throw a glance over the wide field of inorganic chemistry, with its seventy elements approximately and their endless compounds, we cannot fail to recognise the fact that the atomic theory has rendered the main service in their classification. The endeavour, too, to establish periodic relations between the properties of the elements and their atomic weights has introduced order among the motley array of the elements and their compounds. The question of the constitution of the latter allows in most cases of a simple and satisfactory answer; as soon, however, as the composition of inorganic compounds becomes complicated, the usual aids to the solution of such points no longer suffice. The consequence of this is that the rational composition of a large number of compounds, whose empirical composition has long been known, has not yet been cleared up; as examples of such we may refer to the metallo-ammonia compounds (*e.g.* those of cobalt and chromium), the poly-silicic acids, the tungstic acids, and the host of compound acids. Even the constitution of the carbonyl compounds of nickel, iron, &c. is still uncertain.

[1] *Ber.*, vol. xxiv. pp. 2291, 2434 and 3751.

SPECIAL HISTORY OF ORGANIC CHEMISTRY IN THE NINETEENTH CENTURY.

The development of organic chemistry during the first few years of this century has been already described under the general history of the period (cf. p. 246); there, also, much of the pioneering work accomplished in this branch of the subject has been discussed, in so far, that is, as it had a determining influence on the origin and growth of important theoretical investigations. In this section the attempt will be made to pick out from the superabundance of work done in organic chemistry that which has proved of greatest significance, and to arrange it according to its nature (not according to its sequence in point of time)—more especially such investigations as have contributed to solving the question of the chemical constitution of whole classes of bodies. The general points of view by which individual experimenters have been guided in those researches have already been examined at various times in the first section of this book.

Before organic chemistry could be in a position to develop itself independently, the following two conditions had to be fulfilled:—In the first place, the determination of the empirical composition of organic substances was necessary (how this question was solved is described under the history of analytical chemistry);[1] in the second, it had to be proved that organic compounds were subject to the same atomic laws as inorganic, and that they were not, as many formerly assumed, to be classed as totally distinct from the latter. To Berzelius, more than to any other man, is due the removal of this dividing barrier between the two.

The most important methods, which have ever since remained standard ones in organic chemistry, were created by the fundamental researches of Gay-Lussac on cyanogen and its compounds, of Liebig and Wöhler on benzoyl and

[1] Cf. p. 393.

uric acid, of Bunsen on the compounds of cacodyl, of Dumas and Péligot on wood-spirit, and by the investigations of Kolbe, Frankland, A. W. Hofmann, Williamson, Gerhardt, Wurtz, Kekulé and others during the fifties and sixties. Many of these researches have already been referred to in the general section, because of the influence which they exercised on the development of views regarding the chemical constitution of organic compounds; but it will not be altogether possible to avoid recurring to some of them in this portion of the book.

The recognition of the totally different behaviour of the so-called saturated, unsaturated, and aromatic substances was of the first importance for the systematising of organic compounds. A precise distinction between and definition of the above three classes, more especially of the two latter, has been gradually brought about in the course of the last few decades, as the knowledge of them has been extended. In the study of organic compounds, the investigation of physical properties has of late years acquired very great prominence; and this is easily intelligible when it is stated that such investigation has greatly advanced the solution of the question of chemical constitution.

Hydrocarbons and their Derivatives.

The hydrocarbons, from which as the simplest organic compounds all the others are derivable, have been, as befits their "typical" importance, the object of numberless investigations, which have led to the development of doctrines of the utmost weight. We have only to think of the determination of the composition of marsh gas and of ethylene, which led to the recognition of multiple proportions, and with this to the setting up of the atomic theory; of the importance of Faraday's researches on butylene for the evolution of what became known as polymerism; of the labours of Regnault and others on ethylene and its haloid compounds, which afforded such rich food for the theories of substitution; and lastly, of the work of Kekulé and his pupils on benzene and its derivatives—investigations on which most of the work in organic chemistry for the last thirty years has been based.

Mitscherlich's researches on benzene (which he then termed *Benzin*) sixty years ago, taught new methods of preparing hydrocarbons; the formation of this substance from benzoic acid, in consequence of the separation of carbon dioxide, became typical for a large number of similar reactions, *e.g.* the production of cumene from cumic acid, of methane from acetic acid, of chloroform from trichloracetic acid, &c. Of great theoretical importance, too, was Kolbe's mode of formation of hydrocarbons by the electrolysis of the alkaline salts of the fatty acids, and also that of Frankland by the action of zinc upon alkyl iodides; the latter investigations led to the discovery of the zinc alkyls, and opened up this especially fruitful field in the synthesis of organic compounds.[1] The researches of Wurtz,[2] which showed how the combination of different alkyl radicals from hydrocarbons might be effected by the action of sodium upon two alkyl iodides, bore much fruit subsequently among the aromatic compounds; for, with this reaction as a model, the homologues of benzene were prepared synthetically, while at the same time the simple mode of formation allowed of their chemical constitution being deduced.[3]

Another synthesis[4] of homologues of benzene, depending upon the peculiar interaction of aluminic chloride with mixtures of benzene and chlorine compounds (such as methyl chloride), has also proved itself of general application, as well as serviceable for the artificial production of other bodies, *e.g.* ketones, acids, &c. Notwithstanding the care with which these reactions have been studied, a conclusive explanation of the mode of action of the aluminium chloride has still to be given; this much, however, may be taken as proved—that their cause is to be sought for in the formation of

[1] Cf. p. 362. [2] *Ann. Chim. Phys.* (3), vol. xliv. p. 275.

[3] Cf. Fittig, *Ann. Chem.*, vol. cxxxi. p. 301.

[4] Friedel and Crafts, *Comptes Rendus*, vols. lxxxiv., lxxxv., &c.

peculiar intermediate compounds of the chloride with aromatic hydrocarbons (Gustavson).[1]

Berthelot's method[2] of forming hydrocarbons out of different organic compounds by the action of hydriodic acid upon them at rather high temperatures, must also be mentioned here, since it has led to important results in many cases. And reference must be made to the method, so frequently employed, of reducing oxygen compounds to hydrocarbons by heating them with zinc dust.[3] The work of Berthelot on acetylene, of Butlerow and others on the butylenes and amylenes, of Freund on trimethylene, of W. H. Perkin, jun., on the derivatives of tri- and tetra-methylene, of Liebermann on allylene, &c., has materially enlarged our knowledge of the unsaturated hydrocarbons. The remarkable processes of the *isomerisation* of such compounds have recently been cleared up by the valuable researches of Faworsky.[4]

Out of the extraordinarily large number of investigations on aromatic hydrocarbons, whose constitution has given rise to important discussions, there may be mentioned here (in addition to the above) those of Fittig[5] and Baeyer[6] on mesitylene, which was found to be "symmetrical" trimethylbenzene, and also those of Graebe[7] upon naphthalene, and of Graebe and Liebermann[8] upon anthracene. Important conclusions were drawn from the two last with respect to the chemical constitution of these already long-known hydrocarbons, which from thenceforth were regarded as standing in a simple relation to benzene.

Other coal-tar hydrocarbons of complex composition have likewise been satisfactorily investigated; thus phenanthrene, the isomer of anthracene, has been shown by Fittig and

[1] *Ber.*, vol. xi. p. 2751.
[2] *Ann. Chim. Phys.* (4), vol. xx. p. 392.
[3] Baeyer, *Ann. Chem.*, vol. cxl. p. 295.
[4] *Journ. pr. Chem.* (2), vol. xxxvii. pp. 382, 417, 532.
[5] *Ztschr. Chem.* for 1866, p. 518.
[6] *Ann. Chem.*, vol. cxl. p. 306.
[7] *Ann. Chem.*, vol. cxlix. p. 22.
[8] *Ibid.*, Suppl., vol. vii. p. 257.

Graebe [1] to be a diphenylene derivative of ethylene, fluorene by Fittig [2] to be diphenylene-methane, and chrysene by Graebe [3] to be phenylene-naphthalene-ethylene. To Bamberger [4] is due the elucidation of the chemical nature of retene and pyrene. Lastly, the important researches of Kraemer and Spilker [5] throw light on the question—how the individual compounds occurring in coal-tar may be formed during the distillation of the coal.

A wide field has been opened up within the last ten years by the discovery of the hydrides of aromatic hydrocarbons—compounds of remarkable character. It will be sufficient to refer here to the comprehensive work of Bamberger,[6] Baeyer [7] and Markownikoff [8] on the subject. The last-mentioned has proved that a long series of the constituents of petroleum,—the so-called *naphthenes*—belong to this class of hydro-compounds. Again, many of the latter show close relations to the terpenes, substances about which until recently very little was known, but which have been made more and more accessible by the admirable systematic work of Wallach.[9] By means of definite reactions it has been found possible to introduce order among the dire confusion of these "ethereal oils"; and some quite recent researches of A. von Baeyer [10] have thrown much light upon the constitution of the terpenes.

We must further refer to the admirable work of E. and O. Fischer, Zincke and others on the phenyl derivatives of methane, more especially triphenyl-methane; this last was

[1] *Ann. Chem.*, vol. clxvi. p. 361; vol. clxvii. p. 131.

[2] *Ibid.*, vol. cxciii. p. 134.

[3] *Ber.*, vol. xii. p. 1078.

[4] *Ann. Chem.*, vol. ccxxix. p. 102; *Ber.*, vol. xx. p. 365.

[5] *Ber.*, vol. xxiii, pp. 78 and 3266.

[6] Cf. especially *Ber.*, vol. xxii. p. 767; vol. xxiv. p. 2463.

[7] *Ber.*, vol. xxv. p. 2122; vol. xxvi. pp. 229, 820.

[8] *Journ. pr. Chem.* (2), vol. xlv. p. 561; vol. xlvi. p. 86 (this last gives the literature on the subject).

[9] *Ann. Chem.*, vol. ccxxv., ccxxvii., ccxxx., ccxxxviii., ccxxxix., ccxli., cclviii., cclxix., cclxxv. and cclxxvii.; also his lecture on the Terpenes, *Ber.*, vol. xxiv. p. 1525.

[1] *Ber.*, vol. xxvi. pp. 820, 2267, 2558 and 2861.

proved by E. and O. Fischer to be the mother-substance of exceptionally valuable aniline dyes, whose constitution was thus explained.

The continuous and increasing effort to express organic compounds as derivatives of the hydrocarbons is further shown by the nomenclature. With the object of systematizing this on a uniform plan, an International Commission[1] of chemists met a few years ago at Geneva. The system adopted by this Commission makes the hydrocarbons the basis of the nomenclature proposed. But whether all the branches of organic chemistry are at present sufficiently advanced to allow of a satisfactory solution of the question is highly problematical.

The Alcohols and Analogous Compounds.

The close connection existing between the alcohols and the hydrocarbons was clearly recognised when methyl alcohol (the first member of a long series of compounds of this nature) had been successfully prepared from methane, by converting the latter into methyl chloride, and then transforming this into the alcohol. Formerly regarded as the hydrated oxides of hypothetical radicals, the alcohols were after this characterised as hydroxyl derivatives of the hydrocarbons. What an influence Williamson's researches on the formation of ether and Kolbe's views on the constitution of the alcohols had upon the development of the opinions now held with regard to this point, has been already described.

Among the most important of the investigations which helped to establish our knowledge of the alcohols were those of Dumas and Péligot[2] on wood-spirit, whose analogy to

[1] Compare the *Rapport de la Sous-Commission nommée par le Congrès Chimique de* 1889, &c. (Paris, 1892); the report by Pictet in the *Archives des Sciences Physiques et Naturelles,* May 1892; Tiemann's report in the *Berichte,* vol. xxvi. p. 1595; and Armstrong's report in *Nature* for 1892, vol. xlvi. p. 56.

[2] *Ann. Chim. Phys.*, vol. lviii. p. 5; vol. lxi. p. 93.

ethyl alcohol they clearly recognised. The true composition of the latter was worked out by Saussure, who thus did away with the fundamentally erroneous ideas regarding it which had prevailed since the time of Lavoisier; the latter had indeed arrived at a correct knowledge of its constituents, but not of the proportions in which these were present. Equally important were the fact that æthal ($C_{16}H_{33}OH$), discovered by Chevreul, was characterised as an analogue of alcohol by Dumas and Péligot in spite of its unlikeness to the latter, and the corresponding proof by Cahours [1] for the amyl alcohol obtained from fusel oil, to which isobutyl alcohol [2] was afterwards added. The discovery of the secondary and tertiary alcohols, so memorable for the history of this class of compounds, was, as already stated, prognosticated by Kolbe. The series of the secondary carbinols was begun with isopropyl alcohol, isolated by Friedel, and that of the tertiary with Butlerow's trimethyl-carbinol. The modes of formation of these substances (that of isopropyl alcohol from acetone by the addition of hydrogen, and that of trimethyl-carbinol from acetyl chloride and zinc methyl) have since been extensively made use of for the preparation of analogous compounds.

Carbinols of other series were investigated by Cannizzaro, who discovered benzyl alcohol,[3] the simplest carbinol of the aromatic series, and by Cahours and Hofmann, who isolated allyl alcohol;[4] while an accurate acquaintance with various new primary carbinols of the fatty series was arrived at by the systematic researches of Lieben and Rossi.[5] The above-mentioned investigations were also of great importance for the development of the views upon chemical constitution, and more especially upon the isomerism of organic compounds.

The knowledge of the polyatomic alcohols had its beginning in the already-mentioned important researches of

[1] *Ann. Chim. Phys.*, vol. lxx. p. 81; vol. lxxv. p. 193.
[2] Wurtz, *Ann. Chem.*, vol. xciii. p. 107.
[3] *Ibid.*, vol. cxxiv. p. 324.
[4] *Ibid.*, vol. c. p. 356. [5] Cf. *Ibid.*, vol. clviii. p. 137.

Berthelot on glycerine, as representing the triatomic carbinols, and especially in those of Wurtz on the diatomic glycols. In connection with these we would call attention here to the notable discovery of the poly-ethylene alcohols, and of ethylene oxide (distinguished by the readiness with which it enters into reaction).

The discovery of the fact that certain sugars are polyatomic alcohols is of quite recent date; mannite, for instance, is a hexoxy-hexane and arabite, rhamnite and pentite are pentoxy-hexanes. The "carbohydrates," being aldehydes or ketones, are closely related to these.

The derivatives of the alcohols known as the simple ethers, with common ethyl ether at their head, have frequently been the subject of important investigations. The discussions upon the constitution of ether and its mode of formation—discussions which lasted for many years—were brought to an end by the work of Williamson and Chancel, which led to the discovery of mixed ethers.[2]

The knowledge of the compound ethers, frequently now also called *Esters*, has been greatly extended within the last sixty years. The recent observations of the late Victor Meyer and his pupils on the formation of esters of aromatic acids are of great interest here, the constitution of these acids determining the path which the synthesis follows.. To the neutral ethers of the acids, the number of which has gone on continuously increasing (but regarding which it is impossible to mention here even the more important researches), there have been added the so-called ether-acids, whose chemical nature has been cleared up by the work of Hennel, Serullas, Magnus and Regnault on ethyl-sulphuric and ethionic acids, of Pélouze on the ethyl-phosphoric acids, of Mitscherlich on ethyl-oxalic acid, and other more recent labours, *e.g.* that upon phenyl-ethyl-sulphuric acid by Baumann, and upon ethyl-oxalic acid by Anschütz.

Certain of the compounds prepared from ethyl alcohol

[1] *Comptes Rendus*, vol. xlviii. p. 101; vol. xlix. p. 813.

[2] Cf. p. 298.

[3] *Ber.*, vol. xxvii. pp. 1580, 3146; vol. xxviii. p. 2773.

and other carbinols have played an important part in the synthesis of organic substances, thanks to their capability of reaction; we have but to recall here the discovery of sodium ethylate by Liebig, that of chloro-carbonic ether by Dumas, and Debus' investigations of the products which result from the oxidation of ethyl alcohol by nitric acid.

The first step towards a knowledge of those compounds so nearly allied to the alcohols, which have received the generic name of *phenols*, was Laurent's investigation of carbolic acid and its derivatives.[1] Gerhardt was the first to point out the analogy between alcohol and phenol. Of great importance for the development of this class of compounds, and more especially for their technical production, was that mode of formation of phenol itself which was first observed by Kekulé[2] and Wurtz,[3] viz. by fusing benzene-sulphonic acid with potash. This reaction soon led to the discovery of a large number of mono- and poly-atomic *phenols*; the naphthols and other oxy-derivatives of naphthalene, the di- and trioxy-benzenes, etc. were isolated. The reactions of these compounds turned out to be of remarkable interest not merely from a technical, but also from a purely scientific point of view; one need but refer to the conversion of many phenols into quinones, and to the various transformations of these latter by chlorine and bromine. The comprehensive researches of Zincke[4] and his pupils on this subject are worthy of special mention here; the nature of the peculiar decomposition-products of the phenols allowed of conclusions being drawn as to the constitution of the original compounds.

[1] *Ann. Chim. Phys.* (3), vol. iii. p. 195. Runge was the discoverer of carbolic acid itself.

[2] *Lehrb. der. organ. Chemie*, vol. iii. p. 13.

[3] *Ann. Chem.*, vol. cxliv. p. 121.

[4] *Ber.*, vol. xxi. p. 3540; vol. xxii. pp. 1024, 1467; vol. xxiii. pp. 230, 1706, 2200, &c.; *Ann. Chem.*, vol. cclxi. p. 208.

Carboxylic Acids.

A field of immense extent and fertility became open to chemical research with the systematic investigation of the acids contained in animal and vegetable fats, as well as in other natural products. The important work on the fatty acids, suggested in the first instance by Liebig, and which was accomplished by his pupils Varrentrapp, Rochleder, Bromeis, Fehling, Redtenbacher and others, and that of Heintz[1] upon palmitic and stearic acids, not only materially supplemented the earlier investigations of Chevreul on the fats, but led to the discovery of new and wider domains. Important methods for the separation of the fatty acids resulted from these labours. The common link which unites the compounds of this class was only discovered when their chemical constitution came to be understood. The successful efforts of Kolbe, who was the first to recognise acetic as methylcarboxylic acid, and who established this view by direct experiment, have been already described in the general section. It has indeed been from acetic, as the most fully investigated of all the carboxylic acids, that our present ideas upon the constitution of the whole class of compounds have developed themselves. The recognition of the correct atomic composition of acetic acid by Berzelius in 1814, and of its relation to alcohol by Döbereiner in 1821, was of great importance for the solution of this problem.

After the constitution of the carboxylic acids had once been grasped, it became possible for Kolbe to predict the existence of other members of this class, as he had done in the case of the alcohols, and thus existing blanks could be filled up. Of special importance here was the discovery of isobutyric acid,[2] of the isomers of valeric acid—itself already long known, and of other acids richer in carbon, in the systematic investigation of which Lieben and

[1] *Ann. Chem.*, vol. lxxxiv. p. 297; vol. lxxxviii. p. 297; *Journ. pr. Chem.* vol. lxvi. p. 1.

[2] Erlenmeyer, *Ztschr. Chem.* for 1865, p. 651.

Rossi[1] and Krafft, among others, have rendered great service.

The knowledge of the polybasic saturated carboxylic acids, whose chemical constitution was likewise only thoroughly established by Kolbe's speculations, was greatly advanced by the work of Berzelius, Fehling and others on succinic acid (synthetised from ethylene cyanide by Maxwell Simpson[2]), by that of Arppe on adipic acid and homologous compounds,[3] and by the discovery and investigation of malonic acid,[4] &c. The ethers of this last acid have served for the synthesis of homologues of malonic and other polycarboxylic acids,[5] thanks to the facility with which they exchange hydrogen for sodium; while from aceto-acetic ether, which so closely resembles malonic, there have been prepared numerous compounds belonging to this class, to be afterwards systematically investigated. Drechsel's memorable synthesis of the simplest dibasic acid, oxalic, from carbon dioxide and sodium,[6] also deserves mention here. The synthesis of the mono- and polybasic acids has proved in most cases the best guide to their constitution.

The wide field of unsaturated carboxylic acids, some of which (*e.g.* acrylic, angelic, fumaric and maleic) were discovered at an early date, first became cultivated with success after a clear idea of the constitution of these compounds had been arrived at through Kekulé's admirable investigations[7] on the two last-named and on the pyrocitric acids, which explained the behaviour of these bodies to nascent hydrogen; and after Frankland and Duppa[8] had made their beautiful syntheses, which resulted in the

[1] Cf. *Ann. Chem.*, vol. clix. p. 75; vol. clxv. p. 116.

[2] *Proc. R. S.*, vol. x. p. 574; or *Ann. Chem.*, vol. cxviii. p. 373.

[3] *Ann. Chem.*, vol. cxv. p. 143; vol. cxx. p. 288.

[4] *Ibid.*, vol. cxxxi. p. 348.

[5] Cf. Conrad, Bischoff, and Guthzeit, *Ann. Chem.*, vol. cciv. p. 121; vol. ccix. p. 211; vol. ccxiv. p. 31.

[6] *Ztschr. Chem.* for 1868, p. 120.

[7] *Ann. Chem.*, vol. cxxx. p. 21; vol. cxxxi. p. 81; Suppl., vol. i. p. 129; vol. ii. p. 198.

[8] *Journ. Chem. Soc.*, vol. xviii. p. 133; or *Ann. Chem.*, vol. cxxvi. p. 1.

conversion of oxalic ether into unsaturated carboxylic acids. In fact this last investigation led Frankland to express the view that acrylic acid and its homologues were derivatives of acetic acid, and a simple explanation was given of their transformation into the latter (by means of potash). The more recent systematic researches of Fittig[1] and his pupils on the unsaturated carboxylic acids have contributed in great degree to round off and deepen our knowledge of this class of compounds. The latest observations on the molecular transformations of the so-called *α-β*-unsaturated acids[2] into the isomeric *β-γ*-acids, and *vice versa*, calls for particular mention. Remarkable results, too, have been obtained by A. Saytzeff and others on the oxidation of such acids by permanganate of potash. The discovery of tetrolic and propiolic acids[3] prepared the way for an acquaintance with the carboxylic acids derived from acetylene.

The discovery and careful investigation of peculiar isomers among the unsaturated acids has been carried out more particularly during the last ten years. The observations made on fumaric and maleic, crotonic and isocrotonic, angelic and tiglic acids led to the successful attempt—already spoken of on p. 356—to explain the constitution of these and similar isomers on *stereo-chemical* principles. Facts bearing on this subject are gradually accumulating, *e.g.* the discovery of the isomeric cinnamic acids by Liebermann,[4] the investigation of the relations existing between erucic and brassidic[5] acids by Holt,[6] Fileti and Saytzeff, &c; but we are not yet in possession of a theory which satisfactorily explains all the phenomena of this kind.

The class of the aromatic carboxylic acids, with benzoic

[1] *Ann. Chem.*, vol. clxxxviii. p. 87; vol. cxcv. p. 50; vol. cc. p. 21; vol. ccvi. p. 1; vol. ccviii. p. 37.

[2] Fittig, *Ann. Chem.*, vol. cclxxxiii. pp. 47 and 269.

[3] Geuther, *Journ. pr. Chem.* (2), vol. iii. p. 448; Bandrowski, *Ber.*, vol. xiii. p. 2340.

[4] *Ber.*, vol. xxiii. pp. 141, 512, 2510; vol. xxv. pp. 90, 950.

[5] $C_{22}H_{42}O_2$.

[6] *Ber.*, vol. xxiv. p. 4128; vol. xxv. p. 1961.

acid at their head, has been the subject of innumerable and fruitful researches. We have but to recall here the discovery of the peculiar mode of formation of these compounds from hydrocarbons by oxidation, as well as by the direct introduction of the elements of carbonic acid by means of aluminic chloride;[1] and the splendid investigations on the di-, tri-, and poly-carboxylic acids of benzene,[2] to the last class of which the already long-known mellitic acid was found to belong. The aromatic carboxylic acids of unsaturated character, like cinnamic acid, etc., proved particularly easy of examination after Perkin[3] had worked out the reaction now known by his name—a reaction which can be generally applied to their formation. Lastly, the isolation of phenylpropiolic acid[4] and its derivatives has led to results of importance.

The esters have in many cases proved serviceable for obtaining other important derivatives of the carboxylic acids; thus, by means of the reactions which have been investigated by L. Claisen and W. Wislicenus, ketones and ketonic acids, etc., have been prepared (see those compounds).

The discovery of the chlorides, anhydrides and amides of the carboxylic acids deserves particular mention here, since these classes of compounds fill an important place in the history of organic chemistry. The first organic acid chloride was benzoyl chloride, obtained by Liebig and Wöhler by the action of chlorine on oil of bitter almonds, in their classical research already so frequently referred to. The general method for the preparation of such compounds, *i.e.*, by acting upon organic acids with phosphorus pentachloride, is due to Cahours;[5] since then this reagent has been a

[1] Friedel and Crafts, *Comptes Rendus*, vol. lxxxvi. p. 1368.

[2] Baeyer, *Ann. Chem.*, Suppl., vol. vii. p. 1; vol. clxvi. p. 325; Fittig, *ibid.*, vol. cxlviii. p. 11; Graebe, *ibid.*, vol. cxlix. p. 18, etc.

[3] *Journ. Chem. Soc.*, vol. xxi. p. 53; or *Ann. Chem.*, vol. cxlvii. p. 229.

[4] Glaser, *Ann. Chem.*, vol. cliv. p. 140; Baeyer, *Ber.*, vol. xiii. p. 2258.

[5] *Ann. Chem.*, vol lx. p. 254. A. Cahours (1813–1891) filled the chairs of chemistry at the École Centrale and the École Polytechnique of Paris, and was at the same time Master of the Mint there. In addition to his

standard one in organic chemistry, and has proved its value in the most varied circumstances, but more especially for the replacement of oxygen or hydroxyl by chlorine. Phosphorus oxychloride was applied by Gerhardt,[1] and phosphorus trichloride by Béchamp[2] for the same purpose; these are however used but seldom in comparison with the pentachloride.

The great capability of reaction which the acid chlorides possess had already been shown by Liebig and Wöhler in the case of benzoyl chloride, from which they prepared the amide of benzoic acid with ammonia, the ether with alcohol, and the sulphide with sulphide of lead, thus introducing at the same time general modes of formation for these classes of compounds. The acid chlorides afterwards led Gerhardt[3] on to the important discovery of the acid anhydrides, which have likewise proved of great value for the synthesis of organic compounds; take, for instance, acetic anhydride, so often used for obtaining other acetyl compounds and condensation-products, and phthalic anhydride, an extremely reactive substance. Brodie[4] then prepared from some of those anhydrides the peroxides of the acid radicals, so remarkable in their behaviour, which have since been ranked alongside of peroxide of hydrogen. To the acid amides, a class which had been opened up by Dumas' discovery of oxamide, Gerhardt added the anilides, and thus gave the impulse to the sub-division of the former into primary, secondary and tertiary amides. The discovery of the aminic acids and the imides of polybasic acids must

Lécons de Chimie générale Élémentaire—a work greatly valued in France —he published numerous researches which helped materially to advance certain branches of organic chemistry; *e.g.*, papers upon amyl alcohol, cuminol, anisol, oil of winter green, the sulphines, arsines, stannines, and— conjointly with A. W. Hofmann—upon allyl alcohol. But the claim put forward by Étard in his obituary of Cahours (*Bull. Soc. Chim.* vol. vii. p. 1), that the latter was the discoverer of sulphines, is mistaken; the priority in this belongs to von Oefele.

[1] *Ann. Chim. Phys.* (3), vol. xxxvii. p. 285.

[2] *Comptes Rendus*, vol. xl. p. 944.

[3] *Ann. Chem.*, vol. lxxxii. p. 131; vol. lxxxvii. p. 151.

[4] *Proc. R. S.*, vol. xii. p. 655; or *Ann. Chem.*, vol. cxxix. p. 282.

also be mentioned here,—compounds which are closely related to the amides; oxamic acid was isolated by Balard, and succinimide by Fehling. And reference must be made, too, to the connection between the acid nitriles and the primary amides of the acids, the latter being converted into the former by the abstraction of the elements of water.

The investigation of certain derivatives of the carboxylic acids has led to results of very great moment, in that a thorough grasp has been gained of the relations existing between them and two other great classes of compounds—the oxy- and amido-acids. The distinct idea which is now associated with the terms "oxy-carboxylic acid" and "amido-carboxylic acid" has developed itself from lactic acid and alanin as oxy- and amido-propionic acids, and from those other compounds already known for such a long time before their constitution had been deciphered,—glycollic acid and glycocoll. The work of Wurtz,[1] and of R. Hofmann and Kekulé,[2] among others, upon those substances, and especially the decisive investigations of Kolbe, which furnished the key to a thorough explanation of the facts, laid the foundation of our present knowledge of these classes of compounds.[3]

Of great importance for the true recognition of the relations of the substances just named to one another, and to the carboxylic acids from which they are derived, was the transformation of the amido- into oxy-acids by means of nitrous acid (Piria, Strecker, etc.), and the conversion of the latter into the corresponding carboxylic acids by means of hydriodic acid. In this way the constitution of malic, tartaric, aspartic, lactic, and many other acids was definitely arrived at,[4] so that the method may be considered as a peculiarly valuable aid in elucidating the rational composition of many organic compounds. Wislicenus[5] has contributed in a very marked degree to a knowledge of the various

[1] *Ann. Chim. Phys.* (3), vol. lix. p. 171.

[2] *Ann. Chem.*, vol. cii. p. 11; vol. cv. p. 288. [3] Cf. p. 319.

[4] Cf. Schmitt, *Ann. Chem.*, vol. cxiv. p. 106; Kolbe, *ibid.*, vol. cxxi. p. 232; Lautemann, *ibid.*, vol. cix. p. 268.

[5] *Ann. Chem.*, vol. cxxviii. p. 11; vol. clxvi. p. 3; vol. clxvii. p. 302.

lactic acids, his work on the subject having helped greatly to extend the doctrine of isomerism. The idea of "physical isomerism," which originated in the different behaviour of substances of the same composition towards polarised light, has since developed itself more and more, Pasteur's memorable researches[1] on lævo- and dextro-tartaric acids, and on the inactive racemic acid produced by their combination, having previous to this thrown much light upon the subject. It has been already explained how the theory of the asymmetric carbon atom arose. The few isolated observations which led to its establishment have since been materially increased, and prediction has been verified by the discovery, after patient search, of two lactic, mandelic and malic acids, besides other compounds.

Once the constitution of many of the naturally occurring oxy- and amido-acids became known, the synthetic preparation of such compounds was merely a question of time; thus, lactic acid was prepared artificially from propionic acid as well as from aldehyde,[2] inactive tartaric acid from dibromo-succinic,[3] citric acid from acetone,[4] hippuric acid (first recognised as a definite compound by Liebig) from glycocoll,[5] and salicylic acid from phenol.

This last leads us to the aromatic oxy-acids, and to the important method of their formation from phenates and carbonic acid, discovered by Kolbe.[6] A complete explanation of this general reaction has only of late been given by Schmitt,[7] who has proved that the production of an isomer (sodium phenyl-carbonate, $C_6H_5O.O.CO_2Na$) precedes that of the sodium salicylate. The observation that the phenates behave very differently according to the nature of their

[1] *Ann. Chim. Phys.* (3), vol. xxiv. p. 442; vol. xxviii. p. 56; vol. xxxviii. p. 437.

[2] Wislicenus, *Ann. Chem.*, vol. cxxviii. p. 11.

[3] Kekulé, *ibid.*, vol. cxvii, p. 124.

[4] Grimaux and Adam, *Comptes Rendus*, vol. xc. p. 1252.

[5] Dessaigne, *Jahresber. d. Chem.* for 1857, p. 367.

[6] Cf. *Ann. Chem.*, vol. cxiii. p. 125; vol. cxv. p. 201; *Journ. pr. Chem.* 2), vol, x. p. 95.

[7] *Journ. pr. Chem.* (2), vol. xxxi. p. 397.

alkali,—that, for instance, phenol-potassium and carbonic acid yield the isomeric para-oxy-benzoic acid instead of salicylic,—deserves to be noted here as especially important. Ost's discovery of the phenol-di- and tri-carboxylic acids, which result from the same reaction at a higher temperature, must also be recalled.

Of late years a special group has been formed of a peculiar class of oxy-acids which readily change into the so-called *lactones* or intra-molecular anhydrides, with separation of water. Fittig,[2] in conjunction with his pupils, has investigated this remarkable class of compounds systematically, and has largely contributed towards a knowledge of the relations between the lactones and the corresponding acids, and also of their constitution, which formerly received a different interpretation; thus, the simplest member of the series, butyro-lactone, was previously held to be the aldehyde of succinic acid. The relation of many lactones to unsaturated acids is particularly interesting. Numerous lactonic acids have also been examined, and found to be carboxylic derivatives of the lactones.

Aldehydes.

The knowledge of the aldehydes, so important from many different points of view, has gone on steadily increasing ever since bitter almond oil or benzoic aldehyde was first investigated by Liebig and Wöhler, and ordinary aldehyde also by the former; the latter compound, first obtained by Fourcroy and Döbereiner, was carefully examined by Liebig. The chemical constitution of the aldehydes and of the nearly allied ketones was first definitely grasped and given expression to by Kolbe. Both classes of compounds acquired special importance after their capacity for combining with other organic bodies became known; they were thenceforward largely utilised for the synthesis of compounds richer in carbon.

[1] *Ibid.*, (2), vol. xiv. p. 95.

[2] Cf. *Ann. Chem.*, vol. ccviii. p. 111. ; vol. ccxvi. p. 27 ; vol. ccxxvi. p. 322 ; vol. ccxxvii. p. 1 ; vol. cclv. pp. 1, 257 ; vol. cclvi. p. 50 ; vol. cclxviii. p. 1.

Liebig[1] was the first to explain the relation of the aldehyde of acetic acid to alcohol on the one hand, and to acetic acid on the other, whereupon Berzelius pointed out clearly the analogy existing between aldehyde and acetic acid and bitter almond oil and benzoic acid respectively. The mode of formation of the aldehydes, by oxidation of the alcohols, has since then remained the general one. It was only discovered at a much later date that members of this class of compounds could be prepared from the salts of the acids, by heating these with sodium formate.[2] Still more recent is the discovery of the method of preparing aromatic aldehydes from phenols, chloroform and alkali (*i.e.* nascent formic acid), a reaction which has led to the isolation of some curious compounds.[3] And quite lately Gattermann[4] has made the remarkable observation that aldehydes are formed by the union of a hydrocarbon with carbon monoxide in presence of a mixture of hydrochloric acid, cuprous chloride and chloride of aluminium. The aldehyde of formic acid, the first member of its series, was prepared by A. W. Hofmann,[5] the simplest representative of the di-aldehydes, glyoxal, having already been obtained long before by Debus (1856) as one of the products of the oxidation of alcohol. With regard to aldehydes of complex composition, many of these were long ago isolated from various ethereal oils, *e.g.* oil of cinnamon, oil of cumin, etc., and recognised as analogues of ordinary aldehyde. The agreeable odour which many aldehydes possess rendered their artificial production desirable, and so in this way vanillin, heliotropin and others were synthetised, and their constitution established.

Ordinary aldehyde has been ever and anew the subject of important investigations, more especially since Liebig and Fehling observed its tendency to polymerise (into para- and meta-aldehydes).[6] Liebig's observation that benzoic aldehyde

[1] *Ann. Chem.*, vol. xiv. p. 133; vol. xxii. p. 273.

[2] Piria, *Ann. Chem.*, vol. c. p. 114; Limpricht, *ibid.*, vol. ci. p. 291.

[3] Reimer, *Ber.*, vol. ix. p. 423; Tiemann, *ibid.*, vol. ix. p. 824; vol. x. p. 63.

[4] *Ber.*, vol, xxx. p. 1620.

[5] *Proc. R. S.*, vol. xvi. p. 156.

[6] *Ann. Chem.*, vol. xxv. p. 17; vol. xxvii. p. 319.

changed into the polymeric benzoïn in presence of cyanide of potassium must also be mentioned here; it was the origin of further work which led to the discovery of such interesting compounds as benzile, benzilic acid, etc. And those researches gained an increased interest through the discovery of aldol [1] (a condensation product of aldehyde, of the same percentage composition with it), and of its nearly allied compound, crotonic aldehyde; [2] the perception of the constitution of the last-named substance was of importance in that it led to an explanation of this "condensation," and therefore also of other similar processes.

It was thus from aldehydes that a knowledge was gained of the peculiar chemical reactions now known generally under the above name of *condensations*. The aldehydes possess in a superlative degree the capacity for combining with other compounds of similar or dissimilar nature—*e.g.* acids, ketones, amines, etc.—water being eliminated (cf. p. 363). They are thus of exceptional value for the synthesis of organic compounds. A. v. Baeyer was the first to point out that formic aldehyde—the simplest member of the series—played a prominent part in the building up of carbohydrates, acids, and other compounds in plants.

The numerous investigations which have been made with the object of explaining such reactions of aldehydes with other compounds, under elimination of water, cannot be given in detail here. Reference can only be made to those of W. H. Perkin, sen., who showed how the condensation of aromatic aldehydes with fatty acids might be effected,—a reaction which, developed as it has been, still continues to yield rich fruit; [3] and to the researches of L. Claisen, who has systematically examined the manifold condensation processes of which the aldehydes and ketones are capable.[4]

While an extraordinary number of new and important

[1] Wurtz, *Comptes Rendus*, vol. lxxiv. p. 1361.

[2] Kekulé, *Ann. Chem.*, vol. clxii. pp. 92, 309.

[3] Cf. *Ann. Chem.*, vol. ccxvi. p. 115; vol. ccxxvii. p. 48, etc.

[4] Cf. *Ibid.*, vol. clxxx. p. 1; vol. ccviii. p. 121; vol. ccxxiii. p. 137; vol. ccxxxvii. p. 261; *Ber.*, vol. xxi. p. 1135.

compounds has been obtained in this way, the energies of many workers have also been devoted for a long time to the preparation of others resulting from the action of ammonia upon the aldehydes (especially benzoic aldehyde), and, more recently, to the compounds similarly obtained with hydroxylamine and phenyl-hydrazine, *i.e.* the aldoximes and hydrazones.

The thio-aldehydes were first observed a long time ago, but have only been investigated minutely of late years, more especially by Baumann; remarkable cases of isomerism have been discovered among them, to explain which recourse is being had to stereo-chemistry.—Mention must also be made of the discovery and gradual examination of the aldehyde-alcohols, aldehyde-acids, oxy- and amido-aldehydes, and the acetals,—these last being closely related to the aldehydes. Like aldehyde itself, these various substances have proved of much service for the synthesis of many important compounds.

Ketones and Ketonic Acids.

The work done upon the ketones, compounds so closely allied to the aldehydes, has also been most fruitful. The simplest member of this class of bodies, acetone, had already been known for a long time and had been the subject of frequent investigation when Liebig[1] definitely established its composition. Important points in the further history of the ketones were (1) the discovery of their mode of formation from acid chlorides and zinc alkyls,[2] and (2) the preparation of mixed ketones by distilling the lime salts of the two carboxylic acids together.[3] The formation of those peculiar compounds, mesityl oxide, phorone and mesitylene, from acetone was observed a long time ago, but it was only completely explained after similar processes depending upon the condensation of aldehyde had been correctly interpreted.

[1] *Ann. Chem.*, vol. i. p. 223.

[2] Freund, *Ann. Chem.*, vol. cxviii. p. 1.

[3] Williamson, *Journ. Chem. Soc.*, vol. iv. p. 238 ; or *Ann. Chem.*, vol. lxxxi. p. 86.

The remarkable method, discovered by Friedel and Crafts, of synthetising ketones from aromatic hydrocarbons and acid chlorides in presence of chloride of aluminium,[1] threw open the wide field of fatty-aromatic ketones. The behaviour of these last towards oxidising agents, especially permanganate of potash, has been largely investigated and has led to very curious results.[2]

The transformation of ketones into secondary carbinols by the addition of hydrogen has been already spoken of.[3] Equally worthy of notice was the conversion of acetone into pinacone,[4] a diatomic alcohol, and that of the latter into pinacoline; those reactions, extended to other—especially to aromatic—ketones, have led to important results.[5]

The analogy of the ketones to the aldehydes is very clearly shown by the fact that the former also react with hydroxylamine and phenyl-hydrazine to produce oximes and hydrazones, the investigation of which has likewise proved of great value. (See below.)

Entirely new fields have been opened up by the investigation of the di-ketones, to which acetyl- and benzoyl-acetones, acetonyl-acetone, naphthoquinone, anthraquinone, and, as recent researches have shown, benzoquinone and similar compounds belong,—substances whose nature has been elucidated by the labours of Graebe, Liebermann, Fittig, Zincke, Claisen, Paal, Combes and others.

The beautiful condensation of esters with ketones, discovered by Claisen,[6] has made known to us the so-called β-diketones, these being totally distinct from the corresponding α- and γ-compounds. The quinones are now regarded with special interest, on account of a quinonic constitution being assumed for many dyes. Indeed Armstrong, Nietzki

[1] *Ann. Chim. Phys.* (6), vol. i. p. 449; or *Ber.*, vol. xvii., Ref. p. 376.

[2] Cf. Popoff, *Ann. Chem.*, vol. clxi. p. 289; Claus, *Journ. pr. Chem.* (2) vol. xli. p. 396; and especially Wagner, *ibid.*, vol. xliv. p. 257 (this last gives the literature on the subject).

[3] Cf. p. 438.

[4] Fittig, *Ann. Chem.* vol. cx. p. 25; vol. cxiv. p. 54.

[5] Cf. Zincke, *Ber.*, vols. x. and xi.

[6] *Ber.*, vol. xxii. pp. 1009, 3273, etc.

and others suppose that this actually determines the dye-character of the compounds in question.

The acids known as croconic acid, carboxylic acid ($C_{10}H_4O_{10}$), etc., prepared from potassium carboxide, were obtained a long time ago by Will and Lerch; the beautiful researches of Nietzki[1] have shown that some of them are related to benzoquinone, while others are derived from a compound (not yet isolated) containing a ring-shaped molecule of five carbon atoms. The obscurity hitherto surrounding the constitution of these remarkable bodies has thus been dispersed. They are now known as poly-quinones.

The so-called ketonic acids, certain of which (*e.g.* pyroracemic) have been known for a long time, have of late years awakened the interest of a large number of investigators, and rightly so; we have but to think of the splendid results, more especially from the synthetic point of view, which have been achieved with aceto-acetic ether,[2] levulinic acid,[3] acetone-dicarboxylic acid,[4] benzoyl-carboxylic acid[5] (which has become of importance through its relation to isatin), and other similar compounds. These ketonic acids acquire a still greater theoretical interest from the circumstance that they show a double chemical behaviour, their constitution, as judged from certain reactions, being that of hydroxyl compounds, and as judged from certain others, that of carbonyl ones.[6] Thanks to the reaction discovered by Claisen and W. Wislicenus, of which mention has already been frequently made, the synthesis of the ketonic acids has been carried out most thoroughly. These compounds have proved of the greatest interest in many respects; to mention only a few,—take the production of oxalo-acetic and formyl-acetic esters[7]

[1] *Ber.*, vol. xviii. pp. 499 and 1833; vol. xix. pp. 293 and 772.

[2] Cf. Wislicenus, *Ann. Chem.*, vol. clxxxvi. p. 161 (contains a historical review).

[3] Conrad's investigations showed this to be β-aceto-propionic acid (*Ann. Chem.*, vol. clxxxviii. p. 223).

[4] v. Pechmann, *Ber.*, vol. xvii. p. 2542; *Ann. Chem.*, vol. cclxi. p. 151.

[5] Claisen, *Ber.*, vol. x. p, 430.

[6] See General Section, p. 354.

[7] *Ber.*, vol. xx. pp. 2931, 3392. Claisen and v. Pechmann have lately

and the remarkable transformations which these are capable of undergoing, the synthesis of chelidonic acid[1] from oxalic ether and acetone, and that of hydro-chelidonic acid and others. And if, in addition to these points, we recollect that a large number of interesting compounds like camphor, menthone, dehydracetic acid, pyrone derivatives, etc., belong to the family of ketones, we can form some idea of the extent of the field, and of the variety of results to be obtained from it. Camphor, especially, with its almost endless allied compounds and derivatives, has for some years past been investigated minutely and also with good results, but the views of different experimenters as to its constitution are still far from concordant. Among those who have done most here, A. v. Baeyer Beckmann, Bredt, Bourcault, Friedel, Tiemann and Wallach must be named.

Carbohydrates and Glucosides.

The sugar varieties, which are so widely distributed in nature, and many of which have been known from an early age, belong partly to the alcohols and partly to the aldehydes and ketones. Just as the practical importance of many of these bodies has increased in an extraordinary degree, so has also their purely scientific interest augmented with an advancing knowledge of the close relations which exist between the sugar varieties and compounds whose constitution has been already worked out. Thus, many of the hexoses have been transformed into mannite, which is now known to be primary hexyl alcohol containing six hydroxyl groups in place of five hydrogen atoms; the rational composition of saccharic, mucic and levulinic acids, which are more or less intimately related to the sugars, has been arrived at; and the acid ethers of the latter have been obtained, &c. Such observations as these give support to the assumption that those carbohydrates which are comprised under the term

proved that this so-called formyl-acetic ester is really oxy-acrylic (*Ber.*, vol. xxv. p. 1040).

[1] *Ber.*, vol. xxiv. p. 111.

glucoses, or—better—hexoses, are to be regarded as derived from hexatomic alcohols, from which two atoms of hydrogen have been withdrawn in such a manner that they contain the formyl of the aldehydes or the carbonyl of the ketones (Baeyer, Fittig, V. Meyer).

The investigation of the individual sugars—of their chemical behaviour and the products of their decomposition—has been participated in by a great number of chemists; among those who have actively busied themselves with the subject we may mention Brouchardat, Brown and Heron, Kiliani, v. Lippmann, O'Sullivan, Salomon, Scheibler, Soxhlet, Tollens [1] and, especially, Emil Fischer.[2] Fischer's beautiful investigations, published in the *Berichte* [3] during the last ten years or so, have given us a deep insight into the constitution of the sugars. They have not only corroborated the assumption that the latter are partly aldehyde-alcohols (*aldoses*), and partly ketone-alcohols (*ketoses*), but have also paved the way for the stereo-chemical elucidation of the numberless isomers which exist among them.

Phenyl-hydrazine (p. 365) has proved itself of the greatest value for characterising individual sugars; and, by means of the *osazones* produced by this interaction, the conversion of one carbohydrate into another can be effected. The aldehydic or ketonic nature of these compounds was established by this reaction, by the formation of addition-compounds with hydrocyanic acid,[4] and by other means. To crown all, various

[1] Cf. Tollens' *Handbuch der Kohlenhydrate* ("Text-Book of the Carbohydrates," second edition).

[2] Emil Fischer, born on the 9th of October, 1852, at Enskirchen in Rhenish Prussia, was a pupil of A. von Baeyer. He has done an immense amount of brilliant work in organic chemistry, much of which will be referred to in the special section. After filling successively the chairs of Chemistry at Erlangen and (after 1885) at Wurzburg, he was called in 1892 to Berlin as successor to the late A. W. von Hofmann. His *Anleitung zur Darstellung organischer Präparate* has established itself as a laboratory manual,

[3] Victor Meyer and Jacobsen's *Lehrbuch der organischen Chemie*, p. 876 *et seq.*, contains a very clear account of the chemistry of the sugars, besides giving the literature on the subject. E. Fischer's lecture on the Sugar Group (*Ber.*, vol. xxiii. p. 2114) and his *résumé* of sugar syntheses (*Ber.*, vol. xxvii. p. 3189) should also be read.

[4] Kiliani; E. Fischer.

sugars (partly new ones, partly sugars occurring in nature) have been built up artificially from such simple compounds as formic and glyceric aldehydes; in this way E. Fischer has succeeded in synthetising fruit and grape sugars.

That the systematic arrangement of the carbohydrates has become infinitely clearer from these researches requires no demonstration. The mono-saccharides are now distinguished from the poly-saccharides (cane sugar, starch, cellulose, etc.), the former including not only the (6-carbon) glucoses or hexoses, but also compounds of similar chemical character containing no more than 3, 4 and 5 atoms of carbon in the molecule.

A great deal of work has also been done upon starch, dextrine, etc., among others by Brown and Heron,[1] Brown and Morris,[2] and O'Sullivan. But, notwithstanding this, our knowledge of the poly-saccharides, which are regarded as ethereal anhydrides of the glucoses, is very imperfect indeed in comparison with that of the mono-saccharides.

The glucosides,[3] which stand in the most intimate relation to the glucoses, and whose occurrence in the vegetable and animal kingdoms awakened the interest of chemists of the highest eminence at a very early date, have been the subjects of important work ever since the memorable investigation of Liebig and Wöhler on amygdalin, and that of Piria on salicin. Among other researches we would refer here to those of Will[4] on myronic acid, of Tiemann and Haarmann on coniferin, of Will on æsculin, and lastly of Tiemann and de

[1] *Journ. Chem. Soc.*, vol. xxxv. p. 596; or *Ann. Chem.*, vol. cxcix.

[2] *Journ. Chem. Soc.* vol. lv. p. 473.

[3] Cf. the article *Glycoside*, by O. Jacobsen, in Ladenburg's *Handwörterbuch der Chemie.*

[4] Heinrich Will (1812—1890), after working for some time with a pharmacist, studied chemistry under L. Gmelin. Coming subsequently into contact with Liebig, he became *Docent* at Giessen, succeeding to Liebig's chair there when the latter was called to Munich, and soon making his mark as a teacher. Besides producing a great quantity of admirable experimental work, mostly in organic chemistry, but partly in analytical, the results of which were published in the *Annalen der Chemie*, his literary labours were of very high value, notably his collaboration in editing Liebig's *Jahresbericht* and the *Annalen.*

Laire on iridin, the glucoside of the Florentine iris root—researches which resulted in the elucidation of the decomposition-products of the glucosides named, and which laid the foundation for a knowledge of the constitution of these and other compounds of the same class, so widely distributed in nature. The expectation that those natural products will ultimately be obtained artificially has been brought within measurable distance by E. Fischer's recent discovery of a simple method for preparing the glucosides of the alcohols.[1]

Haloid Derivatives of the Hydrocarbons and other Compounds.

As an appendix to the results of the investigations referred to above, investigations which have largely increased our knowledge of the hydrocarbons, alcohols, carboxylic acids, aldehydes and ketones, some others must be mentioned here which bear upon the haloid and other similar derivatives of those compounds.

Hand in hand with the examination of the hydrocarbons went that of their haloid- and nitro-derivatives, for in some cases these were easily obtained from the hydrocarbons, while in others they often served for the preparation of the latter. The formation of chlorine and bromine compounds from hydrocarbons was the subject of highly important discussions, arising from the experiments upon substitution-reactions made and suggested by Dumas and Laurent, and for the explanation of which special theories were advanced; take, for example, the first investigations made in this direction—those upon the action of chlorine on naphthalene, ethylene and ethylene chloride.

Other views began to prevail when, with the setting up of a new theory of the aromatic compounds, the difference between the hydrogen atoms of the benzene molecule and those belonging to the substituting radicals which had entered it came to be recognised. This difference was markedly ap-

[1] *Ber.*, vol. xxvi. p. 2400; vol. xxvii. p. 1145.

parent in the case of the halogens, and was clearly demonstrated by the work of Kekulé, Fittig, Beilstein and others.[1] Further, the study of the remarkable isomeric relations, predicted on theoretical grounds by Kekulé for the derivatives of benzene, led to the thorough examination of the haloid substitution-products of the aromatic hydrocarbons.

After substitution by chlorine has been more or less investigated, attention was directed to the action of bromine and iodine upon organic compounds. And here it was soon recognised that the presence of certain reagents such as phosphorus, iodic acid and mercuric oxide had a wonderful effect in facilitating the replacement of hydrogen by these elements.

Closely connected with this were the researches on the so-called "halogen carriers," which include a large number of the elements—those, namely, whose compounds with the halogens are capable of partially yielding up the latter again; this explains their action as halogen conveyers. The above action has been examined more especially in the case of the aromatic hydrocarbons; without entering into details, we would refer here to the investigations[2] on the subject carried out at L. Meyer's suggestion by Aronheim, Page, Scheufelen, Schwalb and others, and to those of Willgerodt.[3] The earliest observations on this point were made by H. Müller in 1862, when he noticed how chlorine was conveyed by iodine in the action of the former upon benzene and its homologues.

Two classes of peculiar iodine-oxygen compounds have lately been added to the aromatic group by Willgerodt[4] and Victor Meyer[5] respectively. Corresponding in composition with the nitroso- and nitro-compounds, they have been

[1] Cf. *Ann. Chem.*, vol. cxxxvi. p. 301; vol. cxxxvii. p. 192; vol. cxxxix. p. 331.

[2] Cf. *Ann. Chem.*, vol. ccxxxi. p. 152 (contains a historical review).

[3] *Journ. pr. Chem.* (2), vol. xxxiv. p. 264; cf. also Neumann, *Ann. Chem.*, vol. ccxli. p. 33 ("Sulphuric Acid as a Carrier of Iodine").

[4] *Ber.*, vol. xxv. p. 3494; vol. xxvi. pp. 357, 1307, 1532.

[5] *Ber.*, vol. xxv. p. 2632; vol. xxvi. p. 1354; vol. xxvii. pp. 1592, 2326; vol. xxviii. p. 83.

named accordingly (*e.g.* iodoso-benzene, C_6H_5IO, and iodo-benzene, $C_6H_5IO_2$). The interesting *iodonium* bases must also be included here.

Attempts, which have been to some extent followed with success, have also been made to determine the laws governing the substitution of definite hydrogen atoms by halogens; in connection with this the recent systematic experiments of Victor Meyer and his pupils deserve mention.[1] An infinity of work has been done in this direction with aromatic compounds, the object being to determine the order in which the hydrogen atoms of benzene and its homologues and of their derivatives are thus replaced.

The numerous researches on the combination of halogens with unsaturated hydrocarbons were of very great moment, the first example of such an addition being afforded by ethylene. It would be out of place here even to mention only the more important investigations bearing upon reactions of this nature; but it may be stated generally that our present views with respect to the constitution of unsaturated compounds have resulted in great degree from the behaviour of such hydrocarbons to the halogens and halogen hydrides. These addition-reactions have, besides, proved of unexpected value in the explanation of cases of stereo-isomerism (cf. p. 358).

The modes of formation of haloid derivatives of the hydrocarbons are typical, *i.e.* are also applicable to other classes of compounds, *e.g.* acids, ketones, etc. And the same holds good for the chemical behaviour of such compounds, this having been in most cases first established for the haloid derivatives. To mention only one or two of the researches which have advanced our knowledge of the subject—take the discovery and investigation of trichloracetic acid by Dumas,[2] that of chloral by Liebig and Dumas,[3] and that of monochlor-acetic and monochloro-propionic acids, from whose chemical behaviour the constitution of the corresponding oxy-

[1] Cf. V. Meyer and Fr. Müller, *Journ. pr. Chem.* (2), vol. xlvi. p. 161.
[2] *Ann. Chem.*, vol. xxxii. p. 101.
[3] *Ibid.*, vol. i. p. 189; *Ann. Chim. Phys.*, vol. lvi. p. 123.

and amido-acids was established by Kolbe. It is impossible to record here even the most important work of recent years in this direction, but a passing reference must be made to the production in theoretical quantity of aromatic halogen derivatives from the corresponding diazo- or amido-compounds.[1]

Mention must lastly be made of the important part which the halogen compounds have played in organic syntheses; take, for example, their interactions with sodio-aceto-acetic ether, sodio-malonic ether, and the zinc alkyls, besides many other synthetic reactions.

Organic compounds of fluorine have repeatedly been the object of research, notwithstanding which our knowledge of them is still limited. Although six decades have passed since methyl fluoride was described by Dumas and Péligot, it is only within the last few years that the systematic study of these fluorine compounds has been taken in hand by Moissan, Meslans and others.

Nitro- and Nitroso-compounds.

Mitscherlich's discovery and investigation of nitro-benzene[2] paved the way for a knowledge of the nitro-compounds; the formation of this substance from benzene and its relation to the latter were, however, only clearly understood after the adoption of Dumas and Gerhardt's view that nitro-benzene was a substitution-product of benzene. Since then the group nitroxyl (NO_2) has been ranked as a substituent alongside of the halogens. There is scarcely any reaction which has been more frequently applied among the aromatic compounds than the action of nitric acid upon them; take, for instance, the discovery of nitro-naphthalene, of di- and tri-nitrobenzenes, and of the nitro-derivatives of benzoic acid, benzoic aldehyde, phenol, etc. Picric acid, which was so much earlier known than nitro-benzene, was first characterised as trinitro-phenol by Gerhardt. It may be

[1] P. Griess and also O. Sandmeyer, *Ber.*, vol. xvii. pp. 1633, 2651; vol. xxiii. p. 1880.

[2] *Ann. Chem.*, vol. xii. p. 305.

taken for granted that nitro-derivatives of every aromatic compound are known, or at any rate can be prepared. Attention will be called later on to the history of some of the classes of compounds proceeding from these nitro-derivatives, *e.g.* the amines and azo-compounds, which have been destined to play such a prominent part in industrial chemistry.

The first nitro-derivatives of saturated compounds date from the year 1872, when Kolbe discovered nitro-methane[1] and Victor Meyer nitro-ethane.[2] The modes of formation of these substances were particularly calculated to arouse the reflection of chemists, since it was to have been expected here that compounds of quite other constitution—ethers of nitrous acid—would have been obtained instead. The thorough investigation and explanation of the chemical nature of nitro-ethane is due to V. Meyer. Those splendid researches[3] of his resulted further in the discovery (by himself) of other remarkable compounds, which include the nitrolic acids and nitrols. It must however be mentioned that the constitution of the nitro-paraffins hitherto assumed, viz. $R(NO_2)$, has of late been called in question more than once, on the ground of the chemical behaviour of these compounds.[4]

The nitrolic acids and nitrols have been proved to be representatives of the two classes of isonitroso- and nitroso-compounds, which have repeatedly, and more especially of late years, awakened the interest of chemists. It was those investigations of Victor Meyer and his pupils which established the constitution of the isonitroso-compounds, and showed how they were formed by the action of hydroxylamine upon substances containing the radical carbonyl. Thanks to this perfect reaction, so universally applicable, many substances which were formerly numbered among the nitroso-compounds have since been recognised as really belonging to the class of their isomers. On the other hand,

[1] *Journ. pr. Chem.* (2), vol. v. p. 427.
[2] *Ber.*, vol. v. pp. 399, 514.
[3] *Ann. Chem.*, vol. clxxi. p. 1; vol. clxxv. p. 88; vol. clxxx. p. 111.
[4] Nef, *Ann. Chem.*, vol. cclxxx. p. 263.

the above reaction has proved itself a convenient means of testing whether or not compounds contain the radical carbonyl.[1] From those simple researches there have thus been drawn valuable conclusions with respect to the constitution of whole classes of compounds, *e.g.* of the quinones.[2]

The compounds obtained by the action of hydroxylamine on the aldehydes and ketones—the *aldoximes* and *ketoximes* —have for a number of years back been much studied by the late Victor Meyer, Beckmann, Behrend, Hantzsch, Auwers and others, on account of the remarkable cases of isomerism that they show. In fact, the investigation of isomeric oximes and the peculiar chemical behaviour of these substances, form the basis of the stereo-chemistry of nitrogen (cf. p. 359). Only a few chemists (Claus, Minunni and Nef) have brought forward arguments against this view, seeking to explain these isomers on structural grounds.

Organic compounds containing the group phosphyl (PO_2) were also prepared a few years ago;[3] the constitution of these is analogous to that of the nitro- and of the iodo-compounds (p. 459—460).

Development of the Knowledge of Sulphur Compounds.

The examination of organic sulphur compounds has proved of great value for the development of our views upon the constitution of organic compounds generally, and more especially upon the saturation-capacity of the sulphur group of elements. Their investigation has led to the abandonment of the one-sided opinion that sulphur, selenium and tellurium can only act as divalent elements, by furnishing proofs that they may also be tetra- or hexa-valent.

The earliest known of those compounds, which contain sulphur combined in the same manner as the alcohols, carb-

[1] In phenyl-hydrazine E. Fischer discovered an analogous and equally serviceable reagent for carbonyl compounds, which has proved of the utmost value in establishing the constitution of a very large number of substances —the sugars, for instance (cf. p. 455). [2] Cf. p. 452.

[3] Michaelis and Rothe, *Ber.*, vol. xxv. p. 1747.

oxylic acids, ethers, etc., contain oxygen, was mercaptan, discovered by Zeise; its true constitution, as a hydrosulphide corresponding to alcohol, was recognised by Liebig.[1] To this there were soon added ethyl sulphide and its polysulphides, whose analogy to the sulphides of the metals was obvious. The similarly constituted selenium and tellurium compounds were to a great extent worked out by Löwig[2] and Wöhler.[3]

Of organic acids which contain sulphur in place of oxygen, thiacetic acid,[4] discovered by Kekulé, was the first known, although benzoyl sulphide had previous to this been regarded as the "thio-anhydride" of such an acid. Since then the number of these acids and their corresponding aldehydes has been greatly extended (cf. p. 450). Thioglycollic acid (analogous to glycollic) and its analogues have been investigated mainly by Klason.[5]

By the action of powerful reagents on many of the compounds containing divalent sulphur, which have just been spoken of, it has been found possible to prepare others in which the sulphur present possesses a higher valency—compounds which are comparable with sulphurous and sulphuric acids, and which can be derived and in part prepared from the latter. The earliest known of these were the sulphonic acids and sulphones, whose first representatives—phenyl-sulphonic acid and diphenyl-sulphone (*Sulphobenzid*)—were obtained by Mitscherlich,[6] by acting upon benzene with sulphuric acid. These compounds, however, only came to be fully understood after Kolbe had shown them to be derivatives of sulphuric acid and its anhydride. Previous to this (in 1844) he had enlarged the then existing knowledge of the sulphonic acids by his work upon methyl-sulphonic acid and its chlorine derivatives. The important discovery[7] of the transformation of hydrosul-

[1] *Ann. Chem.*, vol. xi. pp. 2, 11. [2] *Pogg. Ann.* vol. xxxvii. p. 552.

[3] *Ann. Chem.*, vol. xxxv. p. 111; vol. lxxxiv. p. 69.

[4] *Ibid.*, vol. xc. p. 311.

[5] Cf. *Ibid.*, vol. clxxxvii. p. 113.

[6] *Pogg. Ann.*, vol. xxix. p. 231; vol. xxxi. p. 628.

[7] Löwig, *Pogg. Ann.*, vol. xlvii. p. 153; Muspratt, *Ann. Chem.*, vol. lxv. p. 251.

phides, disulphides and sulphocyanides into sulphonic acids furnished a general method for the preparation of the latter.

In a similar manner the conversion of the alkyl sulphides into sulphones, which contain two atoms of oxygen more in the molecule, was effected.[1] Kolbe was again the first here to point out definitely the analogy between sulphones and ketones, and sulphonic and carboxylic acids. There have been added lately to the di-ketones the di-sulphones and the sulphone-ketones (products intermediate between the two), in whose investigation R. Otto[2] has done more than any one else. The di- and tri-sulphonic acids, which correspond to the poly-carboxylic, have been known for a long time, Hofmann and Buckton[3] having been the first to investigate them.

The mercaptals, sulphur analogues of the acetals, were prepared by Baumann[4] by the action of aldehydes upon mercaptans; and the mercaptols were got in the same way, ketones being substituted for aldehydes. Among the di-sulphones produced by the oxidation of the mercaptols is the well-known soporific sulphonal.

Von Oefele's discovery of the sulphines[5] was particularly pregnant in its results, because the existence of these compounds stood in contradiction to the assumption that the sulphur atom was invariably divalent. And the same applies to the investigation of the sulph-oxides by Saytzeff,[6] and to that of the sulphinic acids, whose formation and chemical behaviour was cleared up by the work of Kalle, Otto, Klason and others. Mention must also be made here of the remarkable conversion of sulphinates into sulphones,[7] and of sulphites into sulphonic acids[8] by means of alkyl iodides, those reactions having led to conclusions respecting

[1] Von Oefele, *Ann. Chem.*, vol. cxxxii. p. 80.
[2] *Journ. pr. Chem.* (2), vol. xxx. pp. 171, 321; vol. xxxvi. p. 401.
[3] *Ann. Chem.*, vol. c. p. 133.
[4] *Ann. Chem.*, vol. cclxxiv. p. 173; *Ber.*, vol. xxvi. p. 2155.
[5] *Ann. Chem.*, vol. cxxvii. p. 370; vol. cxxxii. p. 82.
[6] *Ibid.*, vol. cxliv. p. 148.
[7] Otto, *Ber.*, vol. xiii. p. 1274.
[8] Strecker, *Ann. Chem.*, vol. cxlviii. p. 90.

the constitution both of the sulphinic acids and the sulphites. By the discovery and careful investigation of the thionylamines, Michaelis has added another class to the list of sulphur compounds.—Organic compounds of selenium and tellurium corresponding to the above-mentioned sulphur ones are as yet but sparingly known.

Organic Nitrogen Compounds.

An exceptionally wide field in organic chemistry was opened up by the discovery of the nitrogenous bases corresponding to ammonia. When their connection with the latter was found out, the question of their chemical constitution in general was solved. A. W. Hofmann's classical researches[1] on the substituted ammonias and ammonium bases, whose salts result from the action of alkyl iodides upon ammonia, deserve the first mention here, since they led to the true perception of the constitution of these bodies, and established a basis upon which they might be systematised. His splendid work upon aniline and its numerous substitution-derivatives (*e.g.* cyan-aniline), begun in 1843,[2] and on the addition products of this base, immensely enriched organic chemistry. These investigations resulted in the discovery of a wealth of new and striking facts, *e.g.* the observation of the influence exerted by halogens entering the aniline molecule upon the chemical character of the resulting compounds.[3] Upon the basis of those labours, which prepared the way for a knowledge of the aromatic bases, the aniline colour industry has since developed itself in the most brilliant manner. From a theoretical point of view, also, these researches on the di- and tri-amines and on the corresponding ammonium bases (obtained from ethylene bromide and ammonia) were of special importance; Hofmann, in fact, worked out and explained organic nitrogen compounds generally as no other

[1] *Ann. Chem.*, vol. lxxiv. p. 117; vol. lxxv. p. 356; cf. also p. 296 of this book.

[2] *Ibid.*, vol. xlvii. p. 37, and numerous later papers.

[3] *Ibid.*, vol. liii. p. 1; cf. also p. 284 of this book.

man has done. His investigations on the formation of substitution-products of ammonia contributed more than anything else to the establishment of the "typical" theory towards the end of the forties (*cf.* p. 295 *et seq.*).

The observation that the organic ammonias result from the nitro-compounds by reduction[1] was a point of special significance in their history, this step having been first effected by Zinin[2] in the conversion of nitro- into amido-benzene. The above reaction has proved itself of the greatest use as a general method, has served for the preparation of di- and tri-amines, and has since been applied with success in innumerable instances, besides having been extended to the later discovered nitro-compounds of the fatty series. The mode of formation of the primary amines from the cyanic ethers, discovered by Wurtz,[3] must also be referred to here as of historical importance, since the simplest organic ammonia, methylamine, was first prepared in this way.

From the vast number of observations on the chemical behaviour of the classes of compounds in question, we can but pick out a very few, such, namely, as have led to the elucidation of their constitution and to the discovery of new and important groups. To what an unlooked-for significance the action of nitrous acid upon amines and similar bodies (a reaction which had already been studied by Hofmann and others) attained in the hands of P. Griess, who demonstrated the conditions under which diazo-compounds were formed, and examined these with the utmost success! To the latter there were afterwards added the azo-compounds and hydrazines, classes which are of such importance as also to merit a detailed description (see below). The transformation of aromatic amines into valuable dyes by oxidation, observed by W. H. Perkin, sen., A. W. Hofmann, and others, marked the commencement of a new era in chemical industry.

[1] Bamberger and Wohl have found that the first product of a moderated reduction (in neutral solution with zinc dust) is phenyl-hydroxylamine—a compound of great reactive power (*Ber.*, vol. xxvii. pp. 1348, 1432).

[2] *Journ. pr. Chem.*, vol. xxvii. p. 149.

[3] *Ann. Chim. Phys.* (3), vol. xxx. p. 443.

Only a passing reference need be made here to the conversion of the organic ammonias into quinoline, acridine, quinoxaline and other basic substances by similar processes of condensation, since these reactions will be considered further on, especially in their connection with the pyridine and quinoline bases, and the relations of the latter to the alkaloids.

Great advances have been made in the artificial production of naturally-occurring nitrogenous substances, by suitable transformations of ammonia or amines. The important work effected by Hofmann on the mustard oils brought out clearly the relation existing between this class of compounds and the amines, and furnished a firm basis for arriving at their constitution.

Oil of mustard itself (allyl iso-thiocyanate), which is obtained from the seeds of the black mustard, was prepared from allylamine, and also by converting allyl iodide into the thiocyanate, which changes on heating into the isomeric iso-compound. Hofmann's investigation [1] of the chemical behaviour of the mustard oils and their isomers the thiocyanates left no doubt as to the constitution of these two classes.

After the base which was isolated from herring brine had been recognised as identical with the artificially prepared trimethylamine, further researches led to the synthesis of the physiologically important compounds choline and neurine from trimethylamine and ethylene-chlorhydrin,[2] and also to that of betaïne, a substance found in the juice of beet. And just as trimethylamine served for the formation of the latter, so from methylamine and monochloro-acetic acid sarcosine (found naturally in the juice of flesh) was obtained; further, by assimilating the elements of cyanamide, sarcosine was converted into creatine. These reactions left the constitution of the compounds perfectly plain.[3] Reference must also be made here to the synthesis of many com-

[1] *Ber.*, vol. i. p. 176.

[2] Wurtz, *Ann. Chem.*, Suppl., vol. vi. pp. 116 and 197.

[3] Volhard, *Ann. Chem.*, vol. cxxiii. p. 261; *Jahresber. d. Chemie* for 1868, p. 685.

pounds nearly related to urea, *e.g.* guanidine,[1] and parabanic, oxaluric and barbituric acids,[2] which were known as derivatives of uric acid long before they were prepared of set purpose from urea. Uric acid itself was synthetised a few years ago,[3] after many unsuccessful attempts. Indeed urea and guanidine—compounds of such great physiological importance—have proved themselves singularly suited for building up complex "condensed" compounds, for instance, with the ketonic esters and di-ketones. The constitution of the *guanamines*, substances obtained by the action of organic acids upon guanidine, has lately been worked out by Bamberger.[4] The remarkable and highly nitrogenous compounds prepared from amido-guanidine[5] must also be mentioned.

The study of the amides, which include urea and several others of the compounds just mentioned, has gone on simultaneously with that of the amines. Here we can only refer to the important conversion of these substances into cyanides (by means of phosphorus pentoxide), and their re-formation from the latter; and to the interesting behaviour of the substituted amides with phosphorus pentachloride, a reaction which has been studied more especially by Wallach,[6] and which has led to a knowledge of certain peculiar bases, the oxalines. Hofmann[7] worked out the curious transformation of amides into amines containing an atom of carbon less in the molecule by subjecting them to the action of bromine in alkaline solution. The corresponding thiamides, investigated by Cahours, Hofmann, and many others, have on their part been converted into other nitrogenous compounds, *e.g.*, the amidines,[8] the study of which has likewise yielded many

[1] *Ann. Chem.*, vol. cxlvi. p. 259.

[2] Ponomareff, *Bull. Soc. Chim.*, vol. xviii. p. 97; Grimaux, *ibid.*, vol. xxxi. p. 146.

[3] Behrend and Roosen, *Ann. Chem.*, vol. ccli. p. 235.

[4] Nencki, *Ber.*, vol. vii. p. 776 and 1584; Bamberger, *Ber.*, vol. xxv. p. 534.

[5] Thiele, *Ann. Chem.*, vol. cclxxiii. p. 133.

[6] *Ann. Chem.*, vol. clxxxiv. p. 1; vol. ccxiv. p. 193.

[7] *Ber.*, vol. xv. p. 765.

[8] Wallach, *Ann. Chem.*, vol. clxxxiv. pp. 5 and 91; Bernthsen, *ibid.*, vol. clxxxiv. p. 321; vol. cxcii. p. 1.

useful results. The exhaustive researches by Pinner[1] are worthy of special notice here; he has prepared the amidines from the highly reactive imido-ethers, and has thoroughly studied their chemical behaviour.

Through the discovery and investigation of the organic compounds of phosphorus, antimony and arsenic, the connection existing between those three elements themselves and also their relation to nitrogen were proved in the clearest manner, so that here, as well as in other cases, the study of organic compounds has thrown a brilliant light upon particular branches of inorganic chemistry. The phosphines and phosphonium bases first became known through the classical and comprehensive researches of A. W. Hofmann,[2] and the corresponding compounds of the aromatic series through those of Michaelis.[3] The organic compounds of phosphorus were thenceforth recognised as derivatives of the well-known inorganic ones,—phosphuretted hydrogen (PH_3) and phosphonium iodide, and phosphorus tri- and penta-chlorides. The study of the organic compounds of arsenic and antimony, the former of which were admirably investigated by Bunsen, and at a later date by Cahours, Baeyer and Michaelis,[4] and the latter by Löwig, Landolt, Michaelis[5] and others, likewise led to the conclusion that those substances were derivable from the inorganic compounds of the elements. The influence exercised by some of these researches upon the development of the doctrine of valency has been already sufficiently referred to in the general section.

The field comprising the organic compounds of nitrogen is by no means exhausted with the description of the classes which have been shortly alluded to above. A number of

[1] Compare his monograph, *Die Imidoäther und ihre Derivate*; or, failing that, *Ber.*, vol. xvi. p. 1654; vol. xvii. p. 2520; vol. xviii. p. 759.

[2] *Ber.*, vol. iv. p. 605; vol. v. p. 104; vol. vi. p. 306.

[3] Cf. *Ann. Chem.*, vol. clxxxviii. p. 275.

[4] For the literature on the subject, cf. *Ann. Chem.*, vol. cci. p. 184.

[5] Cf. *ibid.*, vol. ccxxxiii. p. 39; *Ber.*, vol. xxvii. p. 244.

others must be referred to here, with regard to the chemical constitution of which much has also been accomplished; many of these are now of great technical importance.

Of the azo-compounds, azo-benzene was the first to be discovered (by Mitscherlich),[1] while much later there came azoxy-benzene by Zinin[2] and hydrazo-benzene by A. W. Hofmann.[3] The now universally accepted views held with regard to these three kinds of azo-compounds are due to Erlenmeyer,[4] and still more to Kekulé,[5] who assumed in azo-benzene two doubly-linked nitrogen atoms, and in oxyazo- and hydrazo-benzene two singly-linked ones. The ready production of these and similar substances from diazo-compounds has greatly tended to advance our knowledge of them. The investigations of Griess, Kekulé, Victor Meyer, H. Caro, Witt and others, which showed how diazo- could be converted into azo-compounds, have led to the establishment of a flourishing industry—the manufacture of azo-dyes. The doctrine of isomerism has also been enriched by a wealth of observations arising out of these labours. The remarkable molecular transformations of hydrazo-compounds into the isomeric diamido-derivatives of diphenyl and its homologues, and of diazo-amido- into amido-azo-compounds, also fall to be mentioned here. In this class it was the amido- and oxy-derivatives of azo-benzene and its homologues which first found employment as dyes. The view held by many,—that the above substances are to be looked upon as derivatives of quinone or quinone-imide respectively,[6] is of importance for understanding the connection between chemical constitution and dye-character.

The diazo-compounds, so remarkable for their reaction-capacity, were discovered by Griess[7] and investigated by him

[1] *Pogg. Ann.*, vol. xxxii. p. 324. [2] *Ann. Chem.*, vol. lxxxv. p. 328.

[3] *Jahresber. d. Chemie* for 1863, p. 424.

[4] *Ztschr. Chem.* for 1863, p. 678.

[5] *Lehrb. d. Chem.*, vol. ii. p. 703.

[6] Cf. Goldschmidt, *Ber.*, vol. xxv. p. 1324.

[7] Peter Griess (1829—1888), a pupil of Kolbe's, became assistant to A. W. Hofmann in London, but relinquished that post after a short time, on receiving in 1862 the appointment of chemist to Messrs. Allsopp and Sons,

in a long series of admirable researches, which disclosed their most important characteristics. Griess showed how they were formed by the action of nitrous acid on aromatic amido-compounds,—a reaction which had previously been studied under other conditions, and had not led then to the discovery of those bodies. In a number of papers[1] dating from the year 1859, which followed one another with great rapidity, the above-named investigator made the chemical world acquainted with the diazo-derivatives of phenol, aniline and benzoic acid, and with their remarkable properties. The view accepted by most chemists with respect to the constitution of these bodies, according to which two atoms of nitrogen are linked together as in the azo-compounds, originated with Kekulé.[2] Another view, in which one of the nitrogen atoms is assumed to be pentavalent and the other trivalent, was expressed by Blomstrand,[3] who brought forward arguments in its favour.

The existence of diazo-compounds in the fatty series has only been proved comparatively recently by the exhaustive researches of Curtius[4] on diazo-acetic and diazo-succinic ethers. The first of these, obtained by the action of nitrous acid on amido-acetic ether, shows certain points of resemblance to the aromatic diazo-compounds, but also many differences; its power of combining with other substances, nitrogen being eliminated, is more strongly marked than in its aromatic

at their well-known brewery at Burton-on-Trent. Although continuing there engaged in this branch of technical chemistry until his death, he at the same time carried out a number of most valuable scientific researches. His brilliant discovery and investigation of the diazo-compounds led him on to the azo-dyes; he was thus the father of this now enormous industry. Griess's work generally was marked by great refinement of execution, as well as great power of observation. A. W. von Hofmann has left us a full and sympathetic account of his life, while E. Fischer and H. Caro have told of his services to science (*Ber.*, vol. xxiv. *Ref.* pp. 1007, 1058).

[1] *Ann. Chem.*, vol. cxiii. p. 201; vol. cxvii. p. 1; vol. cxxi. p. 257; vol. cxxxvii. p. 39.

[2] *Ztschr. Chem.* for 1866, p. 689.

[3] In his *Chemie der Jetztzeit*, p. 272; cf. also *Ber.*, vol. viii. p. 51; and Strecker, *ibid.*, vol. v. p. 786.

[4] *Journ. pr. Chem.* (2), vol. xxxviii. p. 401.

congeners. Diazo-acetic ether is therefore of very great value for the synthesis of other compounds. Diazo-methane (the simplest of the diazo-compounds), discovered by v. Pechmann,[1] is also of the greatest interest.

Another class of bodies, the hydrazines, which stand in a near relation to the diazo-compounds, was discovered in 1875 by E. Fischer[2] and carefully investigated by him.[3] Phenyl-hydrazine—the first of the series to be discovered—has proved of the greatest value both as a specific reagent and as an aid in the synthesis of complex compounds. Its relation to diazo-compounds was definitely proved by Fischer, through its formation from diazo-amido-benzene or diazo-benzene chloride and its conversion into diazo-benzene imide. The importance of phenyl-hydrazine and similar bases for the preparation of hydrazones and osazones has already been referred to; they have also been of material aid to the theory of stereo-isomerism.

The production of derivatives of pyrazolone and indole (besides other condensed compounds) by the aid of phenyl-hydrazine must be mentioned here. The latter substance is also now used in large quantity for the manufacture of the well-known febrifuge antipyrine. The simplest member of this series—hydrazine itself—whose discovery is noticed in the history of inorganic chemistry, likewise reacts with the greatest readiness with aldehydes, ketones, and similar substances, and hence has also proved of signal service in extending the domain of nitrogen compounds (cf. the papers by Curtius and his pupils on hydrazides and azides of organic acids, *Journ. pr. Chem.* (2) vol. l. and succeeding volumes).

Within the last few years some remarkable reactions have been carried out with diazo-compounds, which have led either to hydrazones or to so-called formazyl derivatives; the reader is referred to the latest papers on the subject, which

[1] *Ber.*, vol. xxviii. pp. 855 and 1624; cf. also Bamberger, *ibid.* p. 1682.

[2] *Ber.*, vol. viii. p. 589.

[3] *Ann. Chem.*, vol. cxc. p. 67; vol. cxcix. p. 281; vol. ccxii. p. 316.

explain these reactions.[1] Von Pechmann's researches on the oxidation of diazo-compounds and on their constitution are also worthy of note.

After the discovery of the iso-diazo-compounds by Schraube and Bamberger, the constitution of the diazo-compounds became a burning question. The addition of new facts has greatly enlarged the chemistry of the subject. Blomstrand's view (p. 471) has again come into favour; Hantzsch, on the other hand, adheres strongly to the opinion that certain series of isomeric diazo-compounds are stereo-isomeric; while Bamberger can find in the experimental results no proof for Hantzsch's idea, but contends that a structural isomerism is probable. This controversy has now gone on for some years without any definite result being arrived at, although Blomstrand, not long before his death, expressed himself in favour of Bamberger's theory.[2]

Since Scheele's discovery of hydrocyanic acid, the cyanogen compounds have been the subject of frequent investigation by the most able chemists, so that the knowledge of them has been immensely increased. The development of this branch of organic chemistry is in a great degree due to the marked property possessed by most of these compounds of changing into isomers, and also of combining with other substances to yield new compounds.

The composition of prussic acid and of many of the cyanides was worked out by Berthollet and Ittner, and especially by Gay-Lussac in his classical researches, in which he discovered cyanogen and recognised its analogy to the halogens. He it was, too, who assumed in yellow prussiate of potash (a substance already known for a long time) the presence of the radical *ferrocyanogen,* while Berzelius, adhering strictly to the dualistic theory, explained it as being a double salt of iron protocyanide and cyanide of

[1] V. Pechmann, *Ber.*, vol. xxv. p. 3175; vol. xxvii. p. 219. Bamberger, *Ber.*, vol. xxv. pp. 3201, 3539; vol. xxvi. p. 2978. W. Wislicenus, *Ber.*, vol. xxv. p. 3459.

[2] Cf. *Ber.*, for the years 1894—1897; also Blomstrand, *Journ. pr. Chem.* 2), vol. liii. p. 169; vol. liv. p. 305; vol. lv. p. 481.

potash. The discovery of potassium ferricyanide by L. Gmelin in 1822, and that of the so-called nitro-prussides by Playfair [1] extended the knowledge of cyanogen compounds of complex composition, in which, at Graham's suggestion, the radical tri-cyanogen was assumed.

Sulphocyanic acid, together with its salts, was discovered by Porret, and subsequently investigated by Berzelius, who established its composition; Liebig succeeded in isolating cyanogen sulphide in 1829, and he also showed what remarkable products were obtained from the decomposition of ammonium sulphocyanide, viz. mellone, melame, melamine, etc.[2] Of recent years Reynolds, Volhard, Delitzsch and more especially Klason,[3] among others, have advanced our knowledge of this class of compounds.

Cyanic acid, whose chemical behaviour and relation to its own isomers gave rise to important discussions respecting the constitution of all of them, was first isolated by Wöhler,[4] who was led during the investigation of its salts to his memorable discovery of the artificial formation of urea.[5] Cyanuric acid, obtained by Serullas from the solid cyanogen chloride which he discovered, was recognised by Liebig and Wöhler as being of the same percentage composition as cyanic acid. The influence which this observation, taken in conjunction with that of the isomerism of both of these compounds with fulminic acid, had upon the doctrine of isomeric substances, has already been discussed in the general section of this book. The haloid compounds of cyanogen have been known for a long time, cyanogen chloride having been obtained by Berthollet, and the iodide by Davy; but cyanamide, which was destined to become of so much importance for the synthesis of organic compounds,[6] was first prepared in 1851 by Cloëz and Cannizzaro.[7]

[1] *Phil. Trans.* for 1849, vol. ii. p. 477. [2] *Ann. Chem.*, vol. x. p. 11.

[3] Cf. more particularly *Journ. pr. Chem.* (2), vol. xxxvi. p. 57; vol. xxviii. p. 366. [4] *Pogg. Ann.*, vol. xv. p. 619; vol. xx. p. 369.

[5] Cf. p. 252.

[6] Cf. Volhard's, Strecker's, and Drechsel's researches, more especially *Journ. pr. Chem.* (2), vol. xi. p. 284.

[7] *Comptes Rendus*, vol. xxxi. p. 62.

Owing to the readiness with which they unite with other substances, the cyanogen compounds as a whole have been of great service for opening up new branches of the science, and for advancing our knowledge of these; take, for example, the formation of guanidine and its derivatives from cyanamide or cyanogen chloride and ammonia, and also the formation of derivatives of the last-named compound.[1] The tendency shown by hydrocyanic acid to combine with aldehydes and ketones has already been mentioned; this property has rendered it possible to synthetise a large number of oxy-carboxylic acids.

The compounds of cyanogen as well as of thiocyanogen with organic radicals have, thanks to their diversity and capacity for transformation, yielded an almost inexhaustible material for investigation. The alkyl cyanides or nitriles, with methyl cyanide at their head, were first prepared by Dumas[2] from the ammonium salts of the fatty acids, by acting upon these with phosphoric anhydride; the amides afterwards replaced the ammonium salts of the acids for this purpose. The exceptionally important connection which exists between the nitriles and the fatty acids was demonstrated by Frankland and Kolbe[3] when they converted the former into the latter by treatment with caustic potash. Another passing reference may be made here to the generalisation of this reaction, and the consequent production of an immense number of carboxylic acids and their derivatives from simpler compounds, even although it was spoken of when those compounds themselves were being described. The investigation of mandelic acid,[4] resulting from oil of bitter almonds and hydrocyanic acid in presence of hydrochloric, gave the first impetus to the study of the compounds obtained under similar conditions from other aldehydes and ketones. The simplest nitrile of the aromatic series, phenyl

[1] Cf. Erlenmeyer, *Ann. Chem.*, vol. cxlvi. p. 253; A. W. Hofmann, *ibid.*, vol. cxxxix. p. 111; *Ber.*, vol. i. p. 145, etc.

[2] *Comptes Rendus*, vol. xxv. pp. 383 and 442.

[3] *Ann. Chem.*, vol. lxv. p. 269.

[4] Liebig, *Ann. Chem.*, vol. xviii. p. 319.

cyanide or benzo-nitrile, was first observed by Fehling.[1] Since then the number of these nitriles has been enormously extended, all those which correspond to the important carboxylic acids being known. The amidoximes—derivatives of the nitriles—which were discovered by Tiemann,[2] are of particular interest. The imido-ethers, which result from the nitriles by the addition of one molecule of an alcohol, are also worthy of note, because of the ease with which they yield the amidines,[3] compounds of great reactive power. The cyanogen compounds corresponding to the halogen fatty acids are also nearly related to the nitriles; the simplest members of this series, viz., cyano-carbonic and cyan-acetic acids, have led on to important transformation-products, thanks to the ease with which they enter into reaction.

The isocyanides, isonitriles, or carbamines, which are isomeric with the nitriles, were discovered simultaneously by A. W. Hofmann[4] and Gautier,[5] by different procedures, their existence having previously been foreseen by Kolbe. The perception of the cause of the isomerism existing between these two classes of compounds marked an important advance in theoretical chemistry. The conclusive explanation of the similar isomerism between the alkyl thiocyanates and the mustard oils, of which mustard oil proper (allyl iso-thiocyanate) was the earliest known, is due to Hofmann; the latter succeeded both in preparing the iso-thiocyanates artificially, and in proving at the same time their chemical constitution from their various decompositions.[6] The difference in constitution between the thiocyanates and the mustard oils was especially seen in their transformations. Hand in hand with the acquirement of the above knowledge went the gradual establishment of the views upon the analogously constituted cyanic and isocyanic ethers; and

[1] *Ann. Chem.*, vol. xlix. p. 91.

[2] *Ber.*, vol. xvii. pp. 126, 1685, etc.

[3] Cf. Pinner, *Ber.*, vols. xvi. and xvii., and especially his monograph:—*Die Imidoäther und ihre Derivate* (Berlin, 1892).

[4] *Ann. Chem.*, vol. cxliv. p. 144; vol. cxlvi. p. 107.

[5] *Comptes Rendus*, vol. lxv. pp. 468 and 862.

[6] *Ber.*, vol. i. pp. 26 and 169; vol. ii. pp. 116 and 452.

here again Hofmann acted as the pioneer with his researches, after the simplest compounds of this nature had been obtained by Wurtz and Cloëz. The ease with which the isocyanic ethers and the corresponding mustard oils assimilate the elements of ammonia and the amines led to the discovery of the extensive class of the substituted ureas;[1] the simplicity of the reaction, upon which the formation of these substances was based, allowed of the explanation of the numerous cases of isomerism which occur here.

The question of the chemical constitution of the polymeric cyanogen compounds presented far greater difficulties, the number of these having increased to an extraordinary extent after it was proved that cyanuric, fulminic and cyanic acids had all the same percentage composition. It is only comparatively recently (*i.e.*, since 1884) that a certain degree of clearness has been arrived at with regard to the constitution of the cyanuric and isocyanuric compounds, and this has been due more particularly to the admirable investigations of A. W. Hofmann and of Klason, and also to those of Rathke, Weddige, Bamberger and others. These researches have proved that isocyanuric acid and isomelamine are not in themselves capable of existence, although derivatives of both are. The doctrine of stable and unstable modifications, already referred to,[2] was developed and strengthened mainly from observations made upon these polymeric compounds. The obscurity surrounding the compounds of this nature, as well as those decomposition-products of ammonium sulphocyanide known under the names of mellone, melame and meleme, and the bases resulting from the nitriles by polymerisation (cyan-ethine, etc.), is now beginning to vanish, and a knowledge of their constitution is being gradually acquired. The recent work of Otto and Voigt, Weddige and Krafft has introduced us to the true alkyl cyanurates, the isomeric cyan-alkines (which are obtained directly from the nitriles by the action of sodium or sodium ethylate) possessing a totally different constitution. E. v. Meyer's

[1] Cf. Wurtz, *Ann. Chem.*, vol. lxxx. p. 346; A. W. Hofmann, *ibid.*, vol. xxxiii. p. 57.

[2] Cf. p. 354 *et seq.*

investigations[1] on this subject have proved that the cyanalkines are to be regarded as amido-miazines or amido-pyrimidines; the mode in which they are formed is an instructive case of polymerisation, this being brought about by the migration of hydrogen atoms. The formation of the di-molecular nitriles,[2] which from their behaviour are to be classified as imido-nitriles, depends upon a similar reaction, but one which does not go so far.

The rational composition of fulminic acid and allied compounds, *e.g.*, fulminuric acid and other isomers, is now becoming much better understood, thanks to the pioneering researches of Liebig,[3] and the investigations of Kekulé,[4] Schischkoff,[5] and, more recently, of Steiner, Carstanjen, Ehrenberg,[6] and especially Nef,[7] the last-named chemist assuming a divalent carbon atom; the point, however, has not yet been made absolutely clear.

Historical Notes on Pyridine and Quinoline.[8]

An extensive group of nitrogen compounds—the pyridine and quinoline bases—has only been worked at with success of quite recent years, although these substances were in part discovered during the earliest decades of the century; their investigation has been carried on with the utmost zeal ever since it came to be recognised that the vegetable alkaloids were among their derivatives. The researches of Anderson [9] on the volatile bases of bone oil, those of Williams [10] on the

1 *Journ. pr. Chem.* (2), vol. xxxix. p. 262, besides preceding numbers.

2 E. v. Meyer, *ibid.*, vol. xxxviii. p. 336; vol. xxxix. p. 188.

3 *Ann. Chem.*, vol. xxvi. p. 146.

4 *Ibid.*, vol. cv. p. 279.

5 *Ibid.*, vol. ci. p. 213.

6 *Journ. pr. Chem.* (2), vol. xxv. p. 232; vol. xxx. p. 38.

7 *Ann. Chem.*, vol. cclxxxvii. p. 265.

8 With regard to the sources of the following notes, cf. the pamphlets of Metzger, Hesekiel and A. Pictet on these bases, and Calm-Buchka's work, *Die Chemie des Pyridins und seiner Derivate.*

9 *Phil. Trans. E.*, vol. xvi. p. 4, and vol. xx. (2), p. 247; *Phil. Mag.* (4), vol. ii. p. 257; *Ann. Chem.*, vols. lx., lxx., lxxv., lxxx. and lxxxiv.

10 *Phil. Mag.* (4), vol. viii. p. 24; *Phil. Trans. E.*, vol. xxi. (2), p. 315, etc.

similar bodies contained in coal tar, and Gerhardt's observation on the production of quinoline from quinine [1] were the first beginnings in the cultivation of this field, which has since been worked with such wonderful success. The investigation of these substances received a special impetus from the recognition of the similarity between the pyridine bases and quinoline, and of the distinct analogy between these substances and the aromatic compounds. The earliest attempt to explain the constitution of pyridine and quinoline was due to Körner,[2] and it bore the richest fruit; he assumed these bodies to be benzene and naphthalene respectively, in which a methine group $(CH)'''$ was replaced by the trivalent nitrogen atom. This hypothesis was applied to the facts already known, to which a large number of new ones were being continually added, with the result that they were without difficulty made to accord with it. The theory of the aromatic compounds, which had by this time become strongly developed, gave those endeavours a more or less secure basis to go upon, especially when it came to criticising and sifting the rapidly augmenting number of isomers among the pyridine and quinoline derivatives.

The connection of pyridine and quinoline with benzene and naphthalene, assumed in the above hypothesis, was clearly proved by a succession of beautiful researches. We may refer here to the analogous behaviour with regard to oxidising agents shown by the alkylated pyridines and the alkyl derivatives of benzene. The investigation of these relations, more especially those of the isomeric methyl- and ethyl-pyridines and the pyridine mono-carboxylic acids, we owe to the admirable work of Weidel, Skraup, Ladenburg and Wischnegradsky. Just as the admissibility of the hypothesis respecting the constitution of benzene was arrived at from the number of its substitution-products which could actually be prepared, so in like manner a similar deduction was drawn for pyridine, viz. that only the theoretically possible methyl-pyridines and pyridine-carboxylic acids were capable of preparation, and no more.

[1] *Ann. Chem.*, vol. xlii. p. 310.

[2] Cf. p. 350.

Among the experimental researches which have furnished further support for the above view must be mentioned those of Königs, Ladenburg and A. W. Hofmann, which distinctly proved the connection between pyridine and piperidine (the latter containing six atoms of hydrogen more in the molecule than the former). The analogy between this compound and pyridine on the one hand, and hexahydro-benzene and benzene on the other, thus became at once apparent. Ladenburg found that sodium, acting on an alcoholic solution of the particular substance in question, was a most excellent reducing agent for pyridine bases, and since then it has been used with good effect in numberless instances. We have but to think of the conversion of trimethylene cyanide into piperidine and pentamethylene-diamine; this last compound, produced from the above-mentioned cyanide by the addition of eight atoms of hydrogen in the molecule, was proved to be identical with the ptomaïne, cadaverine.

The different modes of formation of pyridine bases from substances of simpler composition likewise assisted towards a knowledge of their constitution. We may refer here to the synthesis of one of the collidines from aldehyde-ammonia, as well as from ethylidene chloride and ammonia; to that of a chloro-pyridine from pyrrol-potassium and chloroform; to the researches of Hantzsch, which resulted in the artificial production of lutidine; and to the production of β-methyl-pyridine from glycerine (Stoehr).

The synthetic investigations of quinoline and its derivatives have proved themselves extraordinarily fruitful; they have served more particularly to confirm the constitution ascribed to those compounds, this being also deducible from the products of decomposition of the latter. Out of the great amount of work done in this branch, only one or two researches can be mentioned here, viz. those of Skraup, who (doubtless stimulated by the previous investigations of Königs and Graebe) discovered the general method of preparing quinoline and its derivatives, by the action of glycerine on the aromatic amines; Baeyer's beautiful investigations on the formation of quinoline, oxy-quinoline, etc., by the condensation of *o*-amido-phenyl compounds; the synthesis of quinoline and its homologues from a mixture of *o*-amido-benzaldehyde and other aldehydes by Friedländer; and that from aniline and aldehyde by v. Miller and Döbner. The syntheses of homologues of quinoline and of quinoline-carboxylic acids effected by C. Beyer and W. Pfitzinger are also closely connected with the above modes of formation.

While these syntheses have made clear the constitution of quinoline, other investigations have established its connection with pyridine; thus it was seen that the quinolinic acid obtained by oxidising quinoline was a pyridine-dicarboxylic acid, the formation of which was in every respect analogous to that of benzene-dicarboxylic acid from naphthalene.

The minute study of the derivatives of quinoline has led to a systematic investigation of the whole field, the researches of Ad. Claus[1] and his pupils on the halogen derivatives and sulphonic acids of quinoline deserving special mention. In this way other compounds of analogous constitution have been isolated, *e.g.*, the naphtho-quinolines and anthra-quinoline. The discovery of iso-quinoline and its preparation from derivatives of naphthalene (Gabriel, Bamberger, and Zincke) also calls for notice.

The bases known as the di- and tri-azines, which have been investigated with much care during the last few years, stand in the closest relation to pyridine and quinoline, just as these do to benzene and naphthalene. In this connection the work of Stoehr and of L. Wolff on pyrazine and piperazine derivatives, and that of Pinner on pyrimidine must be mentioned. The latest researches on cyanuric compounds have shown these to be derivatives of triazine. Among the highly nitrogenous compounds which proceed from quinoline, the quinoxalines (Huisberg and others), which are analogous to the pyrazines, and the quinazolines (Weddige, Paal, Widmann and others), analogous to the pyrimidines, must be named.—Specialization in organic chemistry has of late years increased to such an extent that we have now detailed works

[1] *Journ. pr. Chem.* from 1888 onwards.

dealing with branches of it that were either unknown or disregarded only a short time ago.[1]

A still greater interest than that aroused by the discovery of the compounds just named was awakened by the proof (gradually arrived at from a long series of admirable researches) of the intimate connection existing between pyridine, quinoline and iso-quinoline and various vegetable alkaloids, whose constitution was thereby explained. Wischnegradsky and then Königs were the first to express the opinion that the alkaloids were derivatives of pyridine or quinoline. They grounded this view upon the conversion of pyridine into piperidine, which is a decomposition-product of the alkaloid piperine contained in pepper, and on the retransformation of piperidine into pyridine; to this was added later on the precisely analogous conversion of conine into conyrine, a propyl-pyridine.[2] Quickly following the recognition of this last important fact came the further one [3] that this alkaloid of hemlock was the dextro-rotatory modification of *a*-propyl-piperidine.

Ladenburg's ingenious synthesis of conine [4] consisted in the preparation of *a*-allyl-pyridine, the conversion of this (by means of sodium) into *a*-propyl-piperidine, and the subdivision of the latter optically inactive substance into its active components.

The complete synthesis of other vegetable alkaloids is without doubt merely a question of time; some of them have already been partially built up from their decomposition-products, *e.g.*, atropine from tropine and tropic acid (Ladenburg), [5] and cocaïne from ecgonine, benzoic acid and methyl iodide (Merck).[6] In the case of most of the alkaloids,—nicotine, piperine, the alkaloids of opium, hydrastine, quinine, strychnine, etc., the nature of their products of decomposition

[1] Cf. O. Kühling's admirable *Handbuch der stickstoffhaltigen Orthokondensationsprodukte* ("Text-book of the Nitrogenous Ortho-condensation Products." Berlin, 1893).

[2] A. W. Hofmann, *Ber.*, vol. xvii. p. 825.

[3] Cf. Ladenburg, *Ann. Chem.*, vol. ccxlvii. p. 80 (1888).

[4] *Ber.*, vol. xxii. p. 1403.

[5] *Ann. Chem.*, vol. ccxvii. p. 74.

[6] *Ber.*, vol. xviii. p. 2952.

affords a basis for conclusions with respect to their constitution. The subject is too wide to be entered upon in detail here. But it may just be stated that in most cases the degradation-products show that a close connection exists between the alkaloids and pyridine, quinoline, or iso-quinoline as their nitrogenous nucleus.[1] In addition to conine, the constitution of the following important members of this class has now been fairly well established:—nicotine (Pinner, Blau); pitocarpine (Hardy and Calmels); cocaïne (Einhorn, Merling); and papaverine (Goldschmidt).

The above very short summary of but a few of the many investigations which have been carried out in this branch is of itself sufficient to show how necessary is a knowledge of the chemical nature and constitution of the pyridine and quinoline bases for the proper understanding of the alkaloids, and what a rich harvest may still be expected here.

Certain non-nitrogenous compounds also, which are naturally related to the alkaloids, viz. meconic, comenic, pyromeconic and chelidonic acids, whose constitution remained quite obscure although the substances themselves had long been known, have been shown, more particularly by the recent researches[2] of Ost and of Lieben and Haitinger, to be naturally connected with pyridine. Light was thrown upon their constitution, as also upon that of the similarly constituted compounds obtained from citric and malic acids,[3] by the important observation that they are converted by ammonia into oxypyridine-carboxylic acid. And Lieben and Claisen's successful synthesis of chelidonic acid[4] has finally solved the problem.

[1] The literature on this branch of the science is already voluminous; the reader is specially referred to Pictet's admirable monograph:—*Die Pflanzenalkaloïde*, etc. (Berlin, 1891).

[2] *Journ. pr. Chem.* (2), vol. xxvii. p. 257; vol. xxix. p. 57; *Ber.*, vol. xvi. p. 1259.

[3] A. W. Hofmann, *Ber.*, vol. xvii. p. 2687; v. Pechmann, *ibid.*, vol. xvii. p. 936; vol. xix. p. 2694.

[4] *Wiener Monatshefte*, vols. iv., v., and vi.; *Ber.*, vol. xxiv. p. 111.

Pyrrol and Analogous Compounds.

Another group of compounds, of which pyrrol, furfurane and thiophene are the representatives, has been the subject of the most ardent investigation during recent years, with the result that the constitution of these substances and also that of many of their derivatives has been cleared up. The analogy existing between those compounds gradually came to be recognised; they all contain the same nucleus, consisting of four atoms of carbon and four of hydrogen, this being combined in pyrrol with the imido-group (NH), in furfurane with one atom of oxygen, and in thiophene with one atom of sulphur. Their similarity to benzene became more apparent the better they came to be known, and was shown in a particularly striking manner in the investigation of thiophene (discovered by Victor Meyer) and its derivatives. The work which has been done upon this class of bodies is amongst the most brilliant of our time.[1]

The artificial formation of thiophene from succinic acid and phosphorus trisulphide,[2] that of pyrrol from succinimide by means of zinc dust, and the conversion of pyrrol into the compounds richer in hydrogen—pyrroline and pyrrolidine (Ciamician)—are reactions of special importance, which helped greatly to elucidate the constitution of these bodies. Pyrrol, which was observed by Runge in coal tar and named by him, and first isolated by Anderson, has with its rapidly-augmenting host of derivatives been closely and comprehensively examined by Ciamician, Dennstedt, Paal and others of late years, Schwanert[3] a long time ago having made the fundamental observation that pyrrol could be produced from ammonium mucate.

The work done upon furfurane (which was discovered by Limpricht[4]) is to be taken in conjunction with that upon pyromucic acid (first observed by Scheele, and recognised as a distinct compound by Labillardière) and its aldehyde furfurol (discovered by Döbereiner and examined by Stenhouse,

[1] Cf. pp. 350—351.
[2] *Ber.*, vol. xviii. p. 454.
[3] *Ann. Chem.*, vol. cxvi. p. 278.
[4] *Ibid.*, vol. clxv. p. 281.

Fownes and others). The analogy in behaviour of the latter to benzoic aldehyde was proved more especially by Baeyer and E. Fischer,[1] and the close connection between pyromucic and maleic acids by Hill.[2] Paal's beautiful investigations have shown that derivatives of furfurane, thiophene and pyrrol are produced from γ-diketones and γ-diketonic acids,[3] and have thus contributed in a marked degree to solve the constitution of these compounds (*i.e.*, of pyrrol, etc.).

Among the aromatic compounds proper, to which the substances just named show a great similarity in chemical behaviour, indole (discovered by Baeyer) was recognised by him as being an analogue of pyrrol, and was made the basis of important researches which resulted in showing its relation to the compounds of the indigo group, more particularly to isatin, oxindole and dioxindole. Various derivatives of indole have lately been prepared by a method discovered by E. Fischer,—*i.e*, from the condensation of phenyl-hydrazine with aldehydes and ketones.[4] Cumarone, obtained by Fittig and Ebert from cumarine, has been designated by Hantzsch[5] the "furfurane of the naphthalene series," and he has confirmed this view by some ingenious syntheses of its derivatives. The analogy existing between the three compounds furfurane, thiophene and pyrrol, and diphenylene oxide, sulphide and imide (carbazole) respectively, was perceived about the year 1885.

For some years past the attention of a large number of investigators has been given to the study of compounds which are related to pyrrol and its analogues as pyrazine and pyrimidine are to pyridine, or quinazoline to quinoline (*cf.* p. 481). Those remarkable compounds the azoles (pyrazole, glyoxaline, triazole, etc.) are pyrrol derivatives of this kind, which have been made known to us by the researches of Marckwald, v. Pechmann, Bladin and others.

[1] *Ber.*, vol. x. p. 13.
[2] *Ibid.*, vol. xiii. p. 734; *Journ. Chem. Soc.*, vol. xl. p. 36.
[3] Cf. Paal's monograph on the subject (Würzburg, 1890).
[4] *Ann. Chem.*, vol. ccxxxvi. p. 116.
[5] *Ber.*, vol. xix. p. 1290; also vol. xx.

Pyrazolone, iso-pyrazolone and their derivatives have proved of special interest in the hands of Knorr and his pupils, Curtius and von Rothenburg, etc. The thiazoles and oxazoles, derived from thiophene and furfurane respectively, have been studied by Hantzsch, Claisen and others.

Organic-metallic Compounds.

After it had come to be seen that not only hydrogen, oxygen, nitrogen, sulphur and the halogens could combine directly with carbon, but also arsenic as well—a point which Kolbe was the first to indicate in his interpretation of cacodyl,[1]—new fields in organic chemistry became opened up in rapid succession. Frankland's discovery[2] of the action of zinc on methyl and ethyl iodides, in which the metal breaks up the iodide in order to combine with the alkyl radical, led to a knowledge of the organo-metallic compounds. Thanks to the readiness with which these enter into reaction, they have been destined to aid in the development of organic chemistry to an unlooked-for extent, more especially as regards synthetic methods. With the aid of the zinc-alkyls many other organo-metallic compounds were prepared and minutely investigated in due course, *e.g.* the ethyl compounds of tin, mercury, lead, sodium, aluminium and other elements.[3] Among the last were those non-metals of which organic compounds had not previously been known; boric methide and other similar substances were prepared by Frankland,[4] and the important alkyl compounds of silicon by Friedel and Krafts, the composition of these latter proving the complete analogy between that element and carbon. To the

[1] Cf. p. 324.

[2] *Journ. Chem. Soc.*, vol. ii. p. 263; or *Ann. Chem.*, vol. lxxi. p. 171 (1849).

[3] Cf. the papers of Buckton, Odling, Frankland, Cahours, Ladenburg, etc., in the *Philosophical Transactions*, *Journal of the Chemical Society*, and *Annalen der Chemie*.

[4] *Proc. R. S.*, vol. xii. p. 123; or *Ann. Chem.*, vol. cxxiv. p. 129. For aromatic compounds of Boron, see Michaëlis and others, *Ber.*, vol. xxvii. p. 244.

organo-metallic compounds of the fatty series, various others belonging to the aromatic have since been added, the first of these having been mercury di-phenyl.[1] Magnesium, bismuth and thallium alkyls have also been prepared within the last few years. The peculiar compounds of nickel, iron and platinum with carbonic oxide, which find a place alongside of the organo-metals, have already been spoken of under the metals themselves.

The short description which has just been given of the development of organic chemistry is sufficient, notwithstanding its incompleteness, to allow of our recognising the main currents which have prevailed, and which still do so, in this branch of the science. The review of the numberless organic substances, which have been investigated during the last fifty or sixty years, is materially facilitated by the general points of view which have become gradually established from the classification of those compounds and from the deduction of their chemical constitution. A prominent place in this respect is to be given to the gradually growing perception that organic compounds might be looked upon as derivatives of inorganic, and to the increasing certainty with which their constitution could be defined on the basis of the saturation-capacities peculiar to the atoms of the various elements.

[1] R. Otto, *Ann. Chem.*, vol. cliv. p. 93. Cf. more especially Michaëlis' work on the phosphenyl compounds, etc.

HISTORY OF PHYSICAL CHEMISTRY IN RECENT TIMES[1]

The influence which certain branches of physics have exercised on the development of chemical doctrines cannot be estimated too highly. It was through the introduction of physical methods, more particularly through the application of weighing, measuring and calculating to chemical problems, that chemistry first became an exact science. The importance of those methods, in so far as they have had a determining influence on the chemical tendency of the present period, has already been entered into in the general section of this book. From the time of Lavoisier onwards, it came to be more and more clearly seen that an intimate connection existed between the chemical and physical properties of substances. Definite relations were found to hold good both between the proportions by weight of substances which enter into chemical combination and between the volumes of combining gases (Avogadro, Gay-Lussac). Investigators sought to determine the more important physical constants of compounds in their various states of aggregation, *e.g.* the specific gravity, specific heat, etc., as well as the changes in physical properties which were brought about by chemical reactions, and thus to arrive at general relations from which the chemical constitution and physical behaviour of different substances could be elucidated. To the efforts at solving such problems as these, physical chemistry owes its origin and gradual development.

Although Lavoisier, in conjunction with certain eminent physicists (Laplace, in particular), took up some of the above

[1] With regard to the sources of information on which this and the following sections are based, the reader is referred to W. Ostwald's admirable *Lehrbuch der allgemeinen Chemie*, 1st. edition in two volumes, 1885—7; 2nd. completely revised edition, of which two volumes have so far been published, 1890—1897 (cf. note 1, p. 377). W. Nernst's *Theoretische Chemie* (1893), which has been translated into English by C. S. Palmer of Colorado, is also a book of originality; and this remark likewise applies to the earlier work with the same title by Horstmann.

problems, and Gay-Lussac at a later period established the relations which exist between the volumes of different gases and their chemical composition, while Dulong and Petit pointed out the connection between the specific heat and atomic weight of the elements, the boundary land between physics and chemistry was first systematically explored by Hermann Kopp; with the investigations of the last-named chemist on the relations between atomic weight and specific gravity, on the laws which regulate the boiling temperatures of liquids, and so on, the history of physical chemistry is intimately bound up. The attention paid to physico-chemical questions has gone on steadily increasing during the last three or four decades, and this applies in a special degree to such as bear upon the relations between the thermo-chemical, optical, and electro-chemical behaviour of substances and their chemical constitution. All this work in physical chemistry has found a rallying point in the *Zeitschrift für physikalische Chemie*, which was projected by Ostwald in 1887, and which has throughout been edited by himself and van 't Hoff.

But there is another allied branch also, viz. that of chemical affinity (*Verwandtschaft*), which has been greatly benefited by the investigations just referred to. With the aid of physico-chemical methods, and the calculations requisite for these, a beginning is being made towards the solution of the old problem respecting the cause and nature of chemical affinity. It will therefore be appropriate to speak of the history of the doctrine of affinity while describing the development of physico-chemical researches. Through both of these branches there runs the continuous endeavour to make chemical reactions capable of mathematical treatment.

The behaviour of gases and vapours has been, almost more than anything else, the subject of fruitful physico-chemical investigations, doubtless because the physical properties of a substance in the gaseous state are observable with fewer complications than in any other, and hence definite relations between these properties and the chemical constitution of the compound are more readily apparent.

Determination of Vapour Density.

The laws of Boyle and Mariotte and of Gay-Lussac, which expressed the connection between the volume of a gas and its temperature and pressure, prepared the ground for a knowledge of other relations. Gay-Lussac's law of volumes, which has already been treated of,[1] was the first result in this branch which benefited chemistry in an exceptional degree. The recognition of the intimate connection between the specific gravity of a gas and its molecular weight we owe to Avogadro,[2] although it was a long time of taking root in the science; this "law of Avogadro," which expresses the above relation, still governs chemical research, and is an indispensable aid in the determination of the molecular weights of many chemical compounds.

The due appreciation of its value has led to continuous endeavours towards simplifying and refining the methods for determining the specific gravity of gases and vapours. Dumas, as already mentioned, was the first to devise a generally applicable method for vapour density determinations,[3] and by this he achieved great results. Another plan, according to which the volume of vapour produced from a given weight of substance is accurately estimated, was worked out by Gay-Lussac and afterwards modified by Hofmann.[4] And to the above methods there was added in 1878 that of Victor Meyer,[5] which depends upon the measurement of the air (or any other indifferent gas) which is expelled from the apparatus by the vapour resulting from a given weight of the substance in question. The improvements which those methods have undergone since their introduction cannot be entered into here, but emphasis must be laid upon the point that through their means the all-important knowledge of the relative weights of the atoms and molecules of elements and compounds has been immensely advanced.

The determination of the specific gravity of vapours has proved in certain cases the most reliable means of decid-

[1] Cf. p. 214. [2] Cf. pp. 215 and 294.

[3] *Ann. Chim. Phys.*, vol. xxxiii. p. 341.

[4] *Ber.*, vol. i. p. 198. [5] *Ibid.*, vol. xi. pp. 1867 and 2253.

ing between the values arrived at by different methods, stöchiometric or otherwise, and so getting at the correct atomic weights of the elements. To give only some more or less recent instances of this, we would refer to the deduction of the atomic weights of silicon, beryllium, thorium and germanium from the vapour densities of their chlorides. Starting with Avogadro's hypothesis—that the vapour density is proportional to the molecular weight—chemists have been able to deduce from the specific gravities of gasified elements most striking conclusions with respect to the number of atoms in their molecules at different temperatures. One has but to think of the results of Dumas' and Mitscherlich's investigations [1] on the vapour densities of sulphur, arsenic, phosphorus and mercury, the molecules of which contain different numbers of atoms, as was deduced at a later date from the specific gravity of their vapours after the revivification of Avogadro's law. The reader is further referred to the important work of V. Meyer and of Nilson and Pettersson on the vapour densities of compounds, more especially of such as show a varying composition with changing temperature. Aluminium chloride, for instance, has the simplest molecular weight which is possible (that expressed by the formula $AlCl_3$) at a temperature sufficiently high, but one double as great (Al_2Cl_6) at lower temperatures; and the same applies to stannous chloride ($SnCl_2$ or Sn_2Cl_4), etc. The latest efforts of workers in this field are being directed to the decomposition of molecules into their elementary atoms, by making use of exceedingly high temperatures (V. Meyer).

These few examples are sufficient to illustrate what has just been said above. The significance which is attached to the results of vapour density determinations is most strikingly shown in the fact that such estimations are held to be the most reliable means of getting at the valency of an element, The amount of care, however, which is requisite here, is proved by the different results obtained by different experimenters, and is particularly apparent in the behaviour of aluminic chloride, from whose vapour density the

[1] Cf. p. 225.

conclusion was drawn (and held to until quite recently) that aluminium was tetravalent, although the whole behaviour of the element pointed to its tri-valency; this has now been confirmed by the determination of the normal density of vapour of the chloride.

Dissociation.

From the observations made upon what are known as anomalous vapour densities, the cause of which has been recognised in a gradually increasing decomposition of the compound with rise of temperature, the doctrine of dissociation—so important for physical chemistry—has developed itself; the name "dissociation" was first made use of by H. de St. Claire Deville to express decompositions of this nature. He was the earliest (from the year 1857) [1] to work systematically at this branch of the science, which has also been made the subject of important investigations by others since, *e.g.* Debray, Cahours, Wurtz, Horstmann, Isambert and A. Naumann. Most of these experimenters did not confine themselves to cases of so-called abnormal vapour density alone, but studied generally the gradual increase in decomposition of chemical compounds under an increasing temperature. Of late years the assumption that every electrolyte is dissociated in solution has come prominently forward (see below).

The Liquefaction of Gases.

The investigation of the transition of gases and vapours into the liquid and solid states has given rise to work of exceeding importance. We have but to recall the comprehensive researches of Faraday [2] on the liquefaction of gases which were at that date held to be uncondensable, and especially the remarkable investigations of R. Pictet,[3] Cailletet,[4] Wroblevsky and Olzevsky,[5] which proved that

[1] Cf. *Comptes Rendus*, vol. xlv. p. 857.

[2] *Phil. Trans.* for 1823, p. 160; and for 1845, p. 1.

[3] *Comptes Rendus*, vol. lxxxv. p. 1214; also in subsequent volumes of the *Archives des Sciences Naturelles*.

[4] *Comptes Rendus*, vol. lxxxv. p. 1213 (1877).

[5] *Ann. Phys.*, N. F., vol. xx. p. 243, etc.

there was no known gas that could withstand the combined effect of sufficiently high pressure and low temperature. Nitrogen, oxygen, hydrogen and argon were thus all reduced by these experimenters to the liquid, and nitrogen and argon to the solid states, and their boiling temperatures determined,—observations of very great moment. The recent work of Linde and Hampson has greatly simplified the process for the liquefaction of air, which will probably in the future prove of importance from a technical point of view, *e.g.* may cheapen the production of oxygen. Dewar has just succeeded in obtaining a measurable amount of liquid hydrogen (about 50 c.c. at one time). Liquid hydrogen is clear and colourless, it shows no absorption spectrum, and the meniscus is as well defined as in the case of liquid air. The boiling point, determined by Dewar with a platinum resistance thermometer, is $-238°C.$, and the density of liquid hydrogen at its boiling temperature is 0·07 approximately; it is thus by far the lightest liquid known. Hydrogen does not possess in the liquid state the characters of a metal.[1]

Thirty years ago Andrews[2] had made a thorough study of the conditions under which a gas can be liquefied, and had established the important conceptions of "critical temperature" and "critical pressure," Mendelejeff[3] having some time before this made certain fundamental observations on the subject.

Light was thrown upon the behaviour of gases to liquids in the first decade of the century by the investigations of Henry and Dalton, which established the fact that the amount of absorption of a gas or of a mixture of gases by a liquid is dependent upon the pressure, and this law was afterwards confirmed by Bunsen's classical researches.[4]

The Kinetic Theory of Gases.

The thorough investigation of gases, of their physical behaviour in particular, led to the setting up of a theory by

[1] *Journ. Chem. Soc.* for 1898, p. 528.

[2] *Phil. Trans.* for 1869, p. 575; or *Pogg. Ann.*, Suppl., vol. v. p. 64 (1871).

[3] *Ann. Chem.*, vol. cxix. p. 11.

[4] *Ibid.*, vol. xciii. p. 1 (1855).

means of which the various phenomena exhibited by them—specific heat, diffusion and friction—have been brought together under one common standpoint and explained in a satisfactory manner. The fundamental idea that a gas was an assemblage of moving particles had previously been put forward by D. Bernoulli in 1738 and by Herapath, and Joule had in 1851 made a great step in advance by calculating the mean translational velocity of these particles. This idea, in the hands of Krönig and more especially Clausius (in 1857), gave birth to the modern kinetic theory of gases, which has been so splendidly worked out by Clausius and Maxwell, and since then perfected in detail by Boltzmann, O. E. Meyer, van der Waals and many others. It may be regarded as springing from the mechanical theory of heat.[1]

Spectrum Analysis.

The examination of the optical behaviour of glowing gases and vapours has exercised a most profound influence upon physical chemistry. Spectrum analysis has grown out of some apparently insignificant and disconnected observations made by Marggraf, Scheele, Herschel and others upon the light emitted by flames coloured by certain salts. The spectra of such flames were investigated by various physicists, among whom Talbot, Miller and Swan deserve first mention; but it was only after Kirchhoff[2] (in 1860) had made and proved the definite statement—that every glowing vapour emits rays of the same degree of refrangibility that it absorbs,—that spectrum analysis became developed by Bunsen and himself into one of the great branches of our science. Its importance for analytical chemistry has already been touched upon.

The application of the spectroscope to the determination of the composition of the heavenly bodies, and with this the firm establishment of stellar-physics, must be mentioned

[1] For an account of the development of the above theory, see O. E. Meyer's work, *Die Kinetische Theorie der Gase* (Breslau, 1877); also Watson's *Kinetic Theory of Gases*.

[2] *Pogg. Ann.*, vol. cix. p. 275.

here. With respect to general chemistry, the efforts to arrive at harmonic relations between the lines of the spectrum themselves, and at a connection between those lines and the atomic weights of the elements which give rise to them, appear to be well founded, as is seen from the work of Maxwell, Balmer, Stoney, Soret and Lecoq de Boisbaudran.[1] A complete theory of the spectral phenomena peculiar to gases remains still a problem for the future, although much admirable preparatory work has been done on the subject.

Atomic Volumes of Solids and Liquids.

The endeavour to establish relations between the physical properties of solid and liquid bodies and their chemical composition has given rise to a large amount of investigation, of which the most important must be mentioned here. H. Kopp was the first to work out in a through manner the connection between the specific gravity of elements and compounds and their atomic composition, Dumas, Herapath, Karsten, Boullay and Ammermüller having previously given some attention to the subject. After establishing the atomic or specific volumes of these latter, Kopp succeeded in discovering a number of relations, and, more particularly, in working out the specific volumes of the elementary atoms in compounds; it thus became possible to calculate the atomic volumes of complex compounds.[2]

The work done of recent years in this branch, among which that of Thorpe, Lossen, Staedel and R. Schiff may be mentioned, has for the most part been carried out upon the principles laid down by Kopp; it has resulted in bringing out many new points of view, and has led to a number of modifications in the values arrived at by him. The formerly accepted opinion—that the atomic volumes of the elements in their compounds are mostly invariable—has been greatly shaken by this later work. Among the numerous researches

[1] Cf. Ostwald's *Lehrbuch*, 2nd. edition, vol. i. p. 260 *et seq.*

[2] Cf. Kopp's pioneering researches, *Ann. Chem.*, vol. xli. p. 76; vol. xcvi. pp. 153 and 303. The last piece of work which he carried out dealt with the *Molecular Volumes of Liquids* (*Ann. Chem.*, vol. ccl. p. 1).

(in addition to H. Kopp's) which have been made with the object of discovering a connection between the volumes of solid compounds and their atomic composition, those of Schroeder are especially worthy of note. He assumes volume units [1] of chemically analogous elements, and believes that he has in this found the key to the solution of the above problem (the doctrine of "Parallelosterism"). But here again we are still far from a knowledge of any law governing the atomic volumes of solid or liquid compounds, whereas, in the case of gases, the simple relations existing between specific gravity and composition were worked out a long time ago.[2]

Laws regulating the Boiling Temperature.[3]

Kopp was likewise the first, in his classical researches,[4] to point out a connection between the boiling temperature and the composition of compounds (more especially of organic ones), in so far that he drew from his results the deduction that approximately equal differences in boiling-point correspond to equal differences in the composition of organic substances. And even although this supposed regularity turned out to be only applicable to certain compounds, and could not be relied upon for other series, still Kopp's work gave a powerful impetus to the search after actual relations —expressible by figures—between boiling-point and chemical composition.

The question arose,—in what manner does the different chemical constitution of isomeric and chemically analogous compounds exercise an influence on their boiling temperatures—to be subjected to examination by Kopp.[5] Other more recent and more extended investigations, *e.g.* those [6] of

[1] These units he terms *Steren.*

[2] Cf. pp. 214 and 490.

[3] For the literature on this subject, cf. the article *Siedepunkt* in Fehling's *Handwörterbuch* by Nernst and Hesse.

[4] *Ann. Chem.*, vol. xli. pp. 86 and 169; vol. lv. p. 166, etc.

[5] *Ann. Chem.*, vol. l. p. 142; vol. xcvi. p. 1.

[6] Cf. A. Naumann, *Allgemeine und Physikalische Chemie* ("General and Physical Chemistry"), (1877), p. 553 *et seq;* Ostwald, *Lehrbuch der allgemeinen Chemie*, 2nd. edition, vol. i., p. 330 *et seq.*

Linnemann, Schorlemmer, Zincke, Naumann and others, have resulted in showing that there are a number of definite relations here also, without, however, having rendered it possible to formulate a precise law setting forth the dependence of boiling-point upon chemical constitution; but it has been clearly established that there is a distinct connection between them. It is possible that a closer knowledge of the intimate relation sought for may be arrived at rather from the occasionally observed anomalies (*e.g.* the lowering of boiling temperature with increasing molecular weight, as in the case of the glycols and certain chlorine compounds, etc.) than from regularities. The efforts to establish definite formulæ for the relationship of the vapour pressure of liquids to the temperature have been followed with great success; they have resulted in the laws worked out by Dühring, Winkelmann, and Ramsay and Young.

There have not been wanting zealous endeavours also to discover regular relations between the temperatures at which solid substances become liquid and their composition, but no definite results have been arrived at in this way. Of more importance, however, have been the researches made with the object of determining melting-point and heat of solidification, *e.g.* those of Pettersson and Nilson, and those on the influence of pressure upon melting-point (James Thomson, Bunsen). Krafft and Weilandt[1] have recently made some very remarkable observations on the great reduction of the boiling-point in a very high vacuum.

Specific Heat of Solid Bodies.

The work which has been done upon the specific heat of elements and compounds is among the most important in the whole field of physical chemistry, the dependence of this property on the atomic composition having been definitely established. We would recall here the Dulong-Petit law of the approximate equality in the specific heats of solid

[1] *Ber.*, vol. xxix. p. 1316.

elements, the significance of which for the development of the atomic theory has already been detailed in the general section;[1] the extension of this law by Neumann; and its enlargement by Regnault's classical researches, as well as by those of H. Kopp, Weber and others, which proved that the specific heat varies with the temperature at which it is determined. And even if the confidence felt in the applicability of the Dulong-Petit law was shaken by the marked deviation from it shown by certain elements, still its usefulness in a very large number of cases and the great value of its principle remained; as Berzelius had predicted, it formed "the foundation of one of the most beautiful pages in chemical theory." The investigation of the specific heat of liquids has not led to conclusions of such a general nature as have resulted in the case of solids.

Optical Behaviour of Solids and Liquids.

A long series of excellent experimental researches has been induced by the endeavour to discover definite relations between the optical behaviour of solid and liquid substances and their chemical composition. The earlier labours of Becquerel, Cahours and Deville, and the later ones of Gladstone and Dale, Landolt, Brühl, Kanonikoff and others have led to conclusions of importance respecting the connection between the constitution of a substance and its power of refracting a ray of light.[2]

The working out of the refraction-equivalents pertaining to the individual elementary atoms within their compounds has led to the discovery of stöchiometric regularities with respect to refraction. Of special interest is the proof that the varying function or mode of combination of the elements, carbon in particular, has a determining influence on the molecular refraction. If the latter is accurately known,

[1] Cf. p. 220.

[2] For the literature on this subject, cf. Landolt and Börnstein's *Physikalische-Chemische Tabellen*, p. 220; and Ostwald's *Lehrbuch*, 2nd. edition, vol. i. p. 415 *et seq.*

then conclusions may be drawn from the refractive power as to the constitution. Deductions of this kind have been applied more especially to solving the question of the constitution of benzene. The great class of keto-compounds has also been investigated optically, with the view of making certain of the constitution.[1]

Only a passing reference can be made here to the importance to crystallography of the observed relations between light refraction and crystalline form, and to the pioneering work of Brewster and Fresnel on the subject.

Another optical property of many substances, more especially organic, has greatly excited the interest of chemists in quite recent years, viz. circular polarisation, which it has been attempted, and with success, to connect closely with the chemical constitution of the compounds in question. After the first memorable investigations of Arago, Biot and Seebeck had been made, the observation—that certain substances, whether in the solid or liquid state, are capable of turning the plane of polarisation of light—was held to be of importance for physics alone. It has only been since Pasteur's beautiful researches[2] on the optically active tartaric acids, and the inactive racemic acid produced by their combination, that relations between optical activity and crystalline form have been discovered, and deductions drawn from these as to chemical constitution.

The desire to gain light upon this point produced in 1874 a theory, which was given out at the same time and independently by Lebel[3] and van 't Hoff,[4] and which is based upon the hypothesis that the cause of this optical activity is to be sought for in the presence of one or more asymmetric carbon atoms, *i.e.* a carbon atom which is linked to four other different atoms or radicals. Should this

[1] Cf. particularly the latest investigations by Brühl, *Journ. pr. Chem.* (2), vols. xlix and l; and the most recent volumes of the *Berichte*, especially vol. xxix. p. 2902.

[2] *Comptes Rendus*, vol. xxiii. p. 535 (1848); vol. xxix. p. 297; vol. xxxi. p. 480.

[3] *Bull. Soc. Chim.* (2), vol. xxii. p. 337.

[4] *Ibid.* (2), vol. xxiii. p. 295.

assumption become fully demonstrated (and it has this in its favour,—that an asymmetric carbon atom has been found in every optically active substance whose constitution has been determined with the necessary accuracy), then it may with confidence be stated that there is an intimate connection between this physical property and chemical constitution.

We may again refer shortly here to van 't Hoff's spacial conception of the distribution of the four valencies of the carbon atom (represented as in the middle of a tetrahedron, with its four affinities at the four corners), and to the extension of this hypothesis by Wislicenus, who has explained by its means the constitution and formation of geometrical isomers, *e.g.*, fumaric and maleic acids, and the crotonic acids, with their derivatives,[1] and also the chemical behaviour of these compounds. Such speculations have very quickly proved themselves fruitful, in that they have led to the perception of relations which had been hitherto overlooked.

In addition to what has just been said with regard to circular polarisation, mention must be made here of the work done upon the rotation of the plane of polarisation by a magnet, since stöchiometric regularities, *i.e.*, relations between magnetic polarisation and chemical constitution, have been brought to light in this case also by the careful investigations of W. H. Perkin sen.[2]

Diffusion, etc.

The properties of liquids which are comprised under the designation "capillarity," together with the friction [3] and diffusion of liquids and of solutions of solids in liquids, have given rise to numerous and valuable researches. Ramsay and Shields [4] have proved that by the measurement

[1] Cf. p. 356.

[2] *Journ. pr. Chem.* (2), vol. xxxi. p. 481; or *Journ. Chem. Soc.* vol. xlv., p. 421; also *Journ. pr. Chem.* vol. xxxii. p. 523.

[3] Internal friction is treated historically in Ostwald's *Lehrbuch*, 2nd. edition., vol. i. p. 550, where one also finds an elegant method for its quantitative determination.

[4] *Journ. Chem. Soc.* for 1894, p. 1089.

of the molecular surface energy of liquids, the latter can be divided into two classes, viz., those of which the molecules are as simple in the liquid as in the gaseous state (and this applies to the generality of liquids), and those—such as water and the alcohols—in which the molecules form complices. This complexity of liquid molecules has been confirmed by Guye.[1]

Graham's memorable researches[2] gave a powerful impulse to the investigation of fluid friction and diffusion; here, too, relations have been found between these phenomena and chemical composition. Mention must be made, in conjunction with this, of his division of substances into crystalloids and colloids, according to their behaviour on diffusion. The reader is also referred to the work upon osmose (so nearly connected with diffusion, and of such great importance for physiology) by Jolly, C. Ludwig, Pfeffer and Brücke. Pfeffer's observations on osmotic pressure have proved of the first importance for the dissociation theory of solution (see below).

Theory of Solution ; Electrolytic Dissociation.[3]

For about a dozen years past a number of eminent investigators, who have devoted themselves to physical chemistry, have been occupied with the question of solution; among those who have done most to extend this subject, van 't Hoff, Arrhenius, Ostwald, Fr. Kohlrausch and Planck must be named. The fundamental idea underlying this work was that substances in highly dilute solution are in a state which is comparable with that of gases. While this idea was not a new one, van 't Hoff was the first to make the following definite statement, and to bring forward strong arguments in its support, viz. that the osmotic pressure of a

[1] *Ann. Chim.* vol. xxxi. (6), p. 206.

[2] *Phil. Trans.* for 1850, 1851, and 1861; or *Ann. Chem.*, vols. lxxvii., lxxx., and cxxiii.

[3] For the historical development of these speculations, see Ostwald's *Lehrbuch*; Nernst's *Theoretische Chemie*; van 't Hoff, *Ber.*, vol. xxvii. p. 6; Horstmann, *Naturwissenchaftliche Rundschau* for 1892, p. 465.

solution (*e.g.* a solution of sugar in water) is equal to the pressure which the same quantity of dissolved substance would exert if it were in the state of gas and filled the space at present occupied by the solution.

Similar relations to those observed in the determination of osmotic pressure had been found by various experimenters (Blagden, Rüdorff, de Coppet, and Raoult) when they established the facts that the freezing point of a solution is dependent on the concentration and the nature of the dissolved substance, and that the lowering of the vapour presure of a solution or the raising of its boiling point also depends on the amount of substance dissolved.

Raoult was the first to point out the great significance of these laws—laws which are theoretically deducible from van 't Hoff's axiom—for the determination of the molecular weight of a dissolved compound.[1] From those laws, therefore, the deduction was immediately drawn that equi-molecular solutions (*i.e.* solutions which contain, in equal volumes of the solvent, quantities of different substances proportional to their molecular weights) show the same osmotic pressure, freezing point, vapour pressure and boiling point. And, thanks to the facility with which freezing and boiling temperatures can be determined, methods were quickly devised by which the molecular weights of substances in solution could in this way be arrived at. E. Beckmann has rendered signal service in the practical elaboration and the scientific testing of such methods, while Raoult, Auwers, Eykman and others have striven to make this procedure for molecular weight determination applicable as far as possible to every case.

Of quite exceptional importance were the deductions drawn when the osmotic pressure, freezing point and boiling point of solutions of salts, acids and bases were viewed in the light of the above-mentioned theory. The marked deviations which were observed in this case found a simple explanation in the assumption that these compounds underwent *dissociation* when the solution became very dilute,—an

[1] *Ann. Chim. Phys.* (6), vol. ii. p. 92.

assumption which is in accordance with the numerous observations on the electric conductivity of such solutions. Arrhenius was the first to attempt an explanation of this behaviour of the dissolved electrolyte by assuming an *electrolytic dissociation*, according to which every electrolyte in aqueous solution undergoes dissociation into its ions, the degree of dissociation depending on the dilution and on the nature of the electrolyte itself. Although this hypothesis has met with great opposition in many quarters, and although it may seem at first sight to be far-fetched, there is no denying its exceeding usefulness for the explanation of numberless chemical processes; it has proved itself of special importance for electro-chemistry, analytical chemistry, and the doctrine of affinity.[1]

Electrolysis of liquid or of dissolved Substances.

The importance of the first work which was done upon this subject for the development of the electro-chemical theory has already been shortly touched upon in the general section.[2] The connection, so early assumed between electricity and chemical action, received the most brilliant confirmation from Faraday's electrolytic law, according to which equal amounts of electricity, when passed through different electrolytes, set free equivalent quantities of analogous substances at the two poles.[3] This law was vigorously contested by Berzelius, because it appeared to him to imply that all the components of the substances decomposed by the current were held together in these by equal affinities. Later experimental researches have corroborated the validity of this law in its full extent, and permit of our hoping for a definite solution of the important problem of chemical equivalents, and, with this, of the true saturation-

[1] Cf. Küsser's very able paper in the *Zeitschrift für Elektrochemie*, vol. iv. p. 105, entitled *Ueber Ionenreaktionen und ihre Bedeutung für die Elektrochemie*; Nernst's *Die Elektrolytische Zersetzung wässriger Lösungen* (*Ber.*, vol. xxx. p. 1547); Dampier Whetham's "Solution and Electrolysis" (1895); and Ostwald's recent work "The Scientific Foundations of Analytical Chemistry."

[2] Cf. p. 229 *et seq.*

[3] Cf. p. 228.

capacities of the elements; the reader is here reminded of Renault's investigations[1] on the various "electrolytic equivalents" of one and the same element, according to the nature of the compounds in which it is contained.

These and other observations, together with the above-mentioned conceptions regarding the nature of solution, have helped to make clearer the process of electrolysis itself, in so far that they have shown the intimate mutual relations existing between chemical and electrical energy. In the light of this Faraday's law appears as the expression of the fact that equal quantities of electricity require equivalent amounts of *ions* in their passage through different electrolytes. Electric conductivity and its relations both to physical properties and chemical composition have frequently been made the subject of investigation, among others by G. Wiedemann, Lenz, Long, and W. Ostwald. The recent work of the last-named chemist, and of Walden and others of Ostwald's pupils, more especially, has proved that a close connection exists between the conductivity of acids and their affinity for bases. Electro-chemistry has been advanced in an extraordinary degree by the attempts to solve the problem of electrolytic dissociation; and there can be no question that both theoretical chemistry and electrical manufactures have been and will continue to be greatly benefited by such researches. Organic chemistry, too, has of recent years come into much closer contact with electro-chemistry, as may be seen in the important experimental work of Gattermann, Elbs and others, and in the comprehensive reports of Lob, Neumann and Elbs.[2]

It has been attempted, too, to connect magnetism with chemical properties. The researches of Plücker, and especially those of G. Weidemann,[3] have, in fact, resulted in showing that there are certain definite relations between the intensity of the magnetism of compounds and their chemical nature.

[1] *Ann. Chim. Phys.* (4), vol. xi. p. 137.
[2] Cf. *Ztschr. für Elektrochemie*, vol. iv. p. 81.
[3] *Pogg. Ann.*, vol. cxxvii. p. 1; vol. cxxxv. p. 177.

Isomorphism, etc.

The investigation of the connection between the forms of solid bodies and their composition has been of great importance for the development of chemical doctrines. The growth of crystallography benefited mineralogy in the first instance, but it also led to the discovery of isomorphism, which—as already stated in the general section[1]—exercised great influence upon the atomic theory. The services rendered here by E. Mitscherlich, to whom even his friend G. Rose owed much, may again be recalled at this point. Mitscherlich did away with the erroneous conceptions which ascribed the crystalline form of a substance to the presence of minute quantities of other bodies, and proved irrefutably the connection existing between crystalline form and chemical composition. The deduction drawn both by himself and by Berzelius, viz. that in true cases of isomorphism of several substances, the chemical constitution of all became known as soon as that of any one of them was made out, because similarity of crystalline form is "a mechanical consequence of similarity in atomic constitution,"—this deduction was soon overthrown by observations of a contrary nature. It was found that dissimilarly constituted substances might be isomorphous, and analogously constituted ones heteromorphous; Mitscherlich himself added to his brilliant discovery of isomorphism that of *dimorphism* and *polymorphism*, while Scherer pointed out cases of the so-called *polymeric* isomorphism, which proved that elementary atoms might be replaced by atomic groups without change of crystalline form.

These and other similar observations have resulted in the view that isomorphism is only to be applied with great caution as a means for determining chemical constitution, otherwise false conclusions are unavoidable. A passing reference may be made here to the later researches of H. Kopp upon the relations between isomorphism and atomic volume, and to those of Schrauf, Pasteur and others upon the

[1] Cf. p. 221.

phenomena of *isogonism*. The problem—what changes of crystalline form are produced through the substitution of particular atoms by other atoms or radicals—has been systematically attacked by P. Groth[1] in the case of certain groups of organic compounds; the phenomenon of the partial alteration of crystalline form, in consequence of such substitution, he terms *morphotropism*. But much study is still required for the investigation of this newly opened out branch of the science. Among recent researches in this field, those of Retgers, published in the *Zeitschrift für physikalische Chemie* must be mentioned; apart from his most admirable work, he has critically examined that of other investigators like Dufet, Bodländer and Wyrouboff.

The so-called *allotropism* of elements and compounds is probably closely connected with polymorphism, *i.e.* with the fact that the same chemical substance can exist in different forms. A most important distinction between the two kinds of phenomena consists, however, in this,—that we have in the former case chemical as well as physical differences. Reference has been already made, under the history of the elements, to the discovery of certain of the more striking "allotropic modifications" of these.[2] But it may be mentioned at this point that material progress has recently been attained in this branch through the investigation of the physical constants of such allotropic bodies, *e.g.* their specific heat, heat of combustion, atomic volume, etc.[3]

Speaking generally, chemists lean to the idea that the same cause underlies both allotropism and polymerism, and that therefore the former is to be explained by assuming that different numbers of atoms (of one and the same element) are grouped together into dissimilar molecules; as has been stated already, the molecular weights of oxygen and ozone have been established, and thus the difference between them explained.

[1] *Pogg. Ann.*, vol. cxli. p. 31.

[2] Cf. pp. 405–406.

[3] Cf. the work of Hittorff, Lemoine, and others.

Thermo-Chemistry.

It is now a long time since the first attempts were made to determine the amounts of heat liberated during and in consequence of chemical reactions, with the object of thereby arriving at a measure of the affinities active in those processes. But the efforts of Laplace and Lavoisier, Davy, Rumford and others in this direction remained incomplete, their methods for the estimation of heat quantities being too inexact.

Thermo-chemistry only became firmly established with the exact measurement of the thermal changes accompanying chemical reactions. Of the earlier investigations, those of Favre and Silbermann on heat of combustion deserve special mention, because the calorimeter was materially improved by these chemists. Emphasis must also be laid here upon the almost forgotten labours of G. H. Hess,[1] who deduced from his own observations the all-important principle of the *Constanz der Wärmesummen* (*i.e.* that the heat evolved in the formation of a given compound is always the same), and thus taught in 1840 the application of the first law of the mechanical equivalent of heat to chemical reactions, before the law itself had been brought forward.

From this principle Hess[2] established the point that the amount of heat evolved in any chemical reaction was always the same, whether the reaction was consummated at once or by degrees in separate instalments. This law, taken in con-

[1] To Ostwald belongs the merit of having referred with emphasis, in his *Lehrbuch der allgemeinen Chemie*, to the services of the St. Petersburg chemist, Hess, as the founder of thermo-chemistry. On p. 12 of vol. ii. in the 1st edition of his book (among other passages) Ostwald expresses himself as follows: "In his fate we find a repetition of that which befel Richter, the importance of whose work for stöchiometry was for so long overlooked. Hess himself (*Journ. pr. Chem.*, vol. xxiv. p. 420) assigned to the latter his proper position by correcting the mistake of confounding Richter with Wenzel, which was due to Berzelius. It is now again needful that the same loving service should be rendered to him, who on his own part did justice to an investigator wrongly criticised and too little esteemed in his own day."

[2] *Pogg. Ann.*, vol. l. p. 385 (1840).

junction with the principle at which Lavoisier and Laplace had arrived fifty years before—viz. that the decomposition of a compound into its constituents requires exactly the same amount of heat as is evolved during its formatiom from the latter—constitutes the basis of thermo-chemistry.

Since the conception of heat as energy of motion found perfect expression in the mechanical theory, and especially since the development of the term *energy*, the above principles appear as self-evident deductions from that theory. The earliest application of the mechanical theory of heat to thermo-chemical processes was made by Julius Thomsen,[1] who has devoted himself to investigating thermo-chemically the more important chemical reactions, *e.g.*, the formation of salts, oxidation and reduction, and the combustion of organic compounds. This branch of the science has been enriched by him in an extraordinary degree, both by the working out of new methods and by the systematic investigation of numerous chemical processes. In addition to Thomsen, Berthelot[2] and especially (since 1879) F. Stohmann[3] have contributed

[1] Julius Thomsen, born at Copenhagen in 1826, where he continues to work as Professor at the University, has since 1852 applied himself with the utmost ardour to building up and developing thermo-chemistry. The large number of scattered papers, which contain the records of his comprehensive researches, were some years ago collected together and published by him in four volumes under the title *Thermochemische Untersuchungen* ("Thermochemical Researches").

[2] M. P. E. Berthelot, born in Paris in 1827, became professor in the Collège de France there, and recently held for short periods the posts of Minister of Education and Foreign Minister; he first made himself known by the beautiful researches, already spoken of, entitled, *Sur les Combinaisons de la Glycérine avec les Acides.* He soon directed his attention to the synthesis of organic compounds, which at that time had been but little studied, and in his comprehensive work, *Chimie Organique fondée sur la Synthèse* (1860), gave a detailed account of the observations and discussions in this branch of the science. Later on he turned with all his energy to the experimental solution of thermo-chemical problems, which he collected together in the two-volume book, *Mécanique Chimique fondée sur la Thermochimie* (1879); while last year (1897) he published a large work in two volumes entitled *Thermochimie.* To him we also owe a number of valuable historical works, dealing more particularly with the development of alchemy and with the oldest chemical writings of the Middle Age (Cf. p. 23, Note 2).

[3] Cf. his papers, published in the *Journ. pr. Chemie* since 1879.

in conjunction with their pupils a large number of important observations in thermo-chemistry, and have materially assisted in the refinement of calorimetric methods.

The efforts of these investigators were mainly directed to the discovery of relations between the thermo-chemical values (which, calculated upon the molecular weights of the reacting substances, were termed *molecular heats*) and the chemical constitution of compounds. The heats of combustion, in particular, furnished much food for speculations of this nature. But although regularities of various kinds became apparent, *e.g.* with respect to the heats of combustion and heats of formation in homologous and other series, very great caution requires to be exercised in forming deductions as to constitution from calorific values; this has lately been clearly shown by Brühl,[1] in a critique upon such attempts. A salutary limit has thus been placed upon the too great extension and over-valuation of the conclusions drawn from thermo-chemical work, a temperate criticism (on the part of Lothar Meyer and others) having previously done away with the erroneous view that an absolute measure of affinity was furnished by the heat evolved or absorbed in the formation or decomposition of chemical compounds. In spite, however, of this failure, thermo-chemical investigations will certainly prove to be indispensable for the perfected doctrine of affinity of the future. It should be added that the results of Stohmann's researches, *e.g.* those on the heat of formation of the various hydrides of benzene, promise to throw light upon the constitution of these compounds.

Photo-chemistry.

This short account of the growth of physical chemistry would be incomplete if nothing were said respecting the

Friedrich Stohmann, born in 1832, is Professor of Agricultural Chemistry in the University of Leipzig, having previously filled a similar post at Halle, while before that he was Director of the Agricultural Experimental Station at Brunswick. He is well known by his numerous and fundamental works, *e.g. Handbuch der Zuckerfabrikation*, *Handbuch der Stärkefabrikation*, etc.; and by his editorship of the *Encyklopedisches Handbuch der technischen Chemie*.

[1] *Journ. pr. Chemie* (2), vol. xxxv. pp. 181, 209.

chemical action of light. The latter, a particular form of radiant energy, gives rise to various chemical reactions, of which the great process of assimilation in plants was the earliest to attract the attention of chemists. The detailed treatment of this process, first observed towards the end of the last century, belongs to the recently developed science of vegetable physiology.

The earliest superficial observations on the action of light upon compounds of silver were made by Schultze so long ago as at the beginning of the eighteenth century; indeed, Boyle had noticed the blackening of chloride of silver, but had ascribed it to the influence of the air. The fundamental experiment which called photo-chemistry into life was made by Scheele, who thus proved himself a pioneer in this as in other branches of the science; he studied the action of the solar spectrum upon paper covered with silver chloride, and established the point that the effect begins first and is strongest in the violet portion. We must recall here too the experiments of Ritter, who observed the action of the ultra-violet rays; and, especially, the epoch-making discoveries of Daguerre and Talbot, who succeeded, after many attempts, in permanently fixing light-pictures.[1] This gave birth about 1839 to the art of photography, so enormously developed of recent years.

[1] The following notes may be added here upon the history of photography (cf. Schiendl's *Geschichte der Photographie*, published by Hartleben, Vienna): Nièpce had associated himself with Daguerre in his work, but did not live to see the perfecting of the Daguerreotype process. Talbot replaced Daguerre's silver plates by paper rendered sensitive to light. Among the further advances made in photography may be mentioned the production of negatives upon glass and the application of substances for attaching the chloride of silver, *e.g.* albumen and collodion (Nièpce de St. Victor—the nephew of the Nièpce mentioned above—and Legray, 1847); the multiplication of photographic pictures through pressure by means of the so-called photo-lithography, heliography, and the phototype method, which in time became superseded by the splendid autotype process (Meisenbach) and the heliotype one (Obernetter); and, lastly, the preparation of plates particularly sensitive to light (bromo-gelatine, etc.), or, to speak generally, the introduction of the so-called dry-plate process. Much interest has been aroused within the last few years by the discovery of colour-photography by Lippmann and others, but the subject is as yet in its infancy.

The foundation of comparative photo-chemistry, which is termed actinometry, was laid by the memorable researches of Bunsen and Roscoe,[1] Draper[2] having previously made important experiments in a similar direction. These investigators, along with others, *e.g.* B. H. W. Vogel, made clear the laws to which the actinic rays are subject. Especially remarkable were the results of the observations on the absorption of chemically active rays, and upon photochemical induction, a term employed by Bunsen and Roscoe to designate the process by means of which the substance sensitive to light was brought into such a condition that it underwent decomposition proportional in amount to the intensity of the light. In addition to the above, mention must be made here of the remarkable researches of Tyndall upon vapours and gases sensitive to light, in whose decomposition the action of the light is shown; thus he proved that the vapour of amyl nitrite (to give an instance) was decomposed by the actinic rays.

The phenomena, whose investigation has just been discussed, come properly speaking under the doctrine of affinity, whose task it is to show that chemical reactions, *i.e.* the formation and decomposition of chemical compounds, are the results of definite measurable forces. True, this important branch of the science is still far from attaining to such a goal; but the development of the doctrine of affinity,[3] a short sketch of which now falls to be given, shows that much zealous work is being done with the view of solving the difficult problems involved here.

[1] *Phil. Trans.* for 1857, p. 355, and for 1863, p. 139; or *Pogg. Ann.*, vol. c. p. 43 (1857); vol. cxvii. p. 531.

[2] *Phil. Mag.* for 1843.

[3] Compare the admirably clear *résumé* given by Ostwald in his *Lehrbuch der allgemeinen Chemie*, 2nd edition, vol. ii. p. 1.

Development of the Doctrine of Affinity since the Time of Bergman.

In a previous section of this book an account has been given of the earlier efforts to arrive at a knowledge of the phenomena of affinity. Through most of the speculations upon this question, ever since the time of Boyle, there runs the assumption that the so-called force of chemical affinity is in the main identical with that of gravity; only in that the former is exerted within very small distances, whereby the form of the material particles has to be taken into account, are differences between the two forces apparent. The attempts to estimate the affinity of substances for one another remained at that time (*i.e.* previous to Berthollet) very imperfect, because it was sought to determine qualitatively the relative intensities of the affinities under arbitrary conditions, without taking physical considerations into account. This period, from about the time of Geoffroy (1718) to that of Berthollet (1800), is characterised by the bringing out of "Tables of Affinity" (*Verwandtschaftstafeln*).[1]

Bergman's doctrine of chemical affinity and his determinations of the latter belong in part to this evolutionary stage, although he paid more attention to the influence of temperature upon the phenomena investigated by him than his predecessors had done. The reaction proper against the merely empirical conception of these latter is, however, to be found in Berthollet, whose *Essai de Statique Chimique* was a protest against the neglect of physical conditions during chemical processes.

Bergman's Doctrine of Affinity.[2]

Although the work of this investigator belongs to the phlogistic period, his doctrine of affinity can only be conveniently treated of here, in order that it may be compared or rather contrasted with that of Berthollet.

[1] Cf. p. 138.

[2] Cf. Bergman's *Opuscula phys. et chem.*, vol. iii. p. 291 (1783).

Bergman's conception of the phenomena of affinity, or perhaps it would be more correct to say his method of designating these phenomena, came into such general adoption that it is to be found even now, at least portions of it are, in many text-books.

The chief law of his doctrine states that the value of the affinity between two substances which act chemically upon one another is constant under similar conditions, and therefore that it is independent of the masses of those substances. Bergman assumed the universal force of gravity as the cause of affinity, this being, however, greatly modified by the form and position of the small particles of the reacting bodies. Partly from his own speculations with regard to affinity, and partly from the incorrectly determined composition of neutral salts, he drew erroneous conclusions with respect to the magnitudes of the affinities of bases to acids, and *vice versa*; he thus set up the tenet that an acid has the strongest affinity for that base of which it saturates the largest quantity, in order to form a neutral salt. Berthollet, as will presently be shown, deduced precisely the opposite from his own assumption,—that mass-action comes into play in chemical processes. It is noteworthy that Bergman recognised the impossibility of carrying out absolute affinity-determinations, and that he devoted his entire energies to making relative ones (by decomposing one compound by another), and then collating these in "affinity tables."

Berthollet's Doctrine of Affinity.

Against Bergman's ideas, and especially against the assumption that affinity is independent of the masses of the interacting substances, Berthollet raised a lively opposition. Setting out, like Bergman, with the hypothesis that affinity is identical with gravity, he went on to emphasise the undeniable conclusion that the forces of chemical affinity, like those of general attraction, must be proportional to the masses of the acting substances. The further deductions from this principle he worked out with masterly clearness in his *Essai de Statique Chimique.*

These views of Berthollet did not at the time receive the recognition which they merited, mainly, no doubt, because their author came into collision with the established facts of chemistry by carrying his deductions too far. His fundamental law of the dependence of chemical action upon the masses of the substances concerned in it led him to regard the "chemical effect" of any body as the product of its affinity and mass. From this he drew the further conclusion that the formation and composition of a chemical compound depended substantially upon the masses of the acting constituents which went to produce it. According to this view, any two substances must combine with one another in constantly varying proportions; with this deduction, however, Berthollet found himself in a serious dilemma.

But, if he went too far here, he so immensely advanced the doctrine of affinity and followed up its true aims by a more discreet application of his fundamental principle, that the errors into which he fell may well be forgotten. He pointed out with perfect clearness that it was impossible to determine the absolute values of chemical affinities, seeing that these were necessarily dependent to a great extent upon the physical properties of the substances which were formed or decomposed by the chemical reactions in question. According to him, such determining (and opposite) properties were *cohesion, i.e.* the mutual attraction of the small particles of any substance for one another, and *elasticity, i.e.* the tendency of those particles to occupy the greatest possible space. He saw in the greater or lesser insolubility of substances a measure of cohesion, and in their volatility a measure of elasticity, and by means of such conceptions conclusively explained chemical changes in which the separation of a precipitate or the escape of a gas or vapour had a determining influence on the course of the reaction. In fact, he stated distinctly that a complete rearrangement

(*Umsetzung*) of substances can only take place if cohesion or elasticity comes into play, and never by the mere action of affinity alone. He thus brought forward entirely new points of view, which have borne much rich fruit.

The Supplanting of Berthollet's Opinions by other Doctrines.

The first good which resulted from Berthollet's conception consisted in the recognition of the uselessness of tables of affinity, in so far as these were supposed to give the relative affinities of different substances. The important fundamental idea of his doctrine of affinity, viz. that the chemical action of a body is proportional to its mass, and is therefore to be expressed by the product of this into the affinity (*i.e.* by a factor still to be determined), led Berthollet to conclusions which were directly opposed to many known facts, and to numerous other data worked out at that time by Proust. The controversy between these two men, which turned upon the question whether chemical compounds are built up of elements in proportions which only alter in amount by certain definite increments, or in proportions which continually vary, has already been discussed in the general section (cf. p. 185 *et seq.*).

In bringing forward his theory Berthollet either neglected to pay sufficient heed to the stöchiometric relations known at that time, or else his knowledge of these was incomplete. It is precisely to the circumstance that he carried his theory of mass-action too far, and made it the starting-point for the most far-reaching deductions, that we have to ascribe the miscredit into which his principles—notwithstanding their clearness—fell, in fact they were held to be totally erroneous. It was thus that Bergman's doctrine, although based upon wrong assumptions and therefore leading its author to false conclusions, kept for so long a time the upper hand, and this all the more readily since it could be better made to accord with the atomic theory. The revival of Berthollet's principles was reserved for quite recent times, after various isolated experimental researches had furnished proof of their admissibility.

After Berthollet's temporary overthrow, the rapidly developing atomic theory formed the main subject of interest for chemists; and hand in hand with its development went that of the electro-chemical doctrines, whose object it was to show that the closest connection existed between electricity and the force termed affinity.

The doctrine of affinity now sought to perfect itself through the development of electro-chemistry; Berzelius' theory caused Berthollet's to be neglected. The successful work which has since been accomplished, with the object of getting at the actual relation between electrolysis and affinity, enables us to perceive now that in those efforts the investigators of that time were carried too far.

These endeavours could only result in showing the qualitative differences in the affinities of different substances; in fact, the electro-chemical theories reached their culminating point in the proof of an analogy between the electrical and chemical properties of substances. Faraday's electrolytic law, which threw light upon the quantitative side of electrolytic processes, did not give any information as to the relative magnitudes of the affinities of the substances in question.

The fortunes of the most important of the electro-chemical theories, that of Berzelius, have already been described. Blomstrand's ingenious attempt [1] to bring it back to life again has indeed shown how valuable it is for the explanation both of chemical processes and of the constitution of compounds; but it was unable at that time to aid materially in penetrating the obscure domain of the phenomena of affinity.

New prospects were opened out for the doctrine of chemical affinity by the thorough investigation of thermo-chemical processes, whose importance for physical chemistry has already been referred to. But in this case also, as

[1] Cf. his work, *Die Chemie der Jetztzeit* (1869).

in the application of electro-chemical conceptions to the problems of affinity, the worth of thermo-chemical determinations very soon became greatly over-estimated. Thus, even Julius Thomsen, who was for a long period the most eminent worker in this field, regarded the heat evolved or absorbed in chemical reactions (more especially in the formation and decomposition of compounds) as an absolute measure of the affinity; in his view the work of affinity was transformed into measurable heat.

But although the inadequacy of thermo-chemistry for the solution of the problems of affinity has now been made manifest, its present and future significance must not be depreciated. On the contrary, by the careful application of thermo-dynamic principles to the interpretation of chemical processes, great benefits have already accrued to the doctrine of affinity.

The Revival of Berthollet's Doctrines.

The most powerful impulse to a further healthy development was given to the doctrine of affinity by the revivification of Berthollet's theory. This was accomplished in its fullest extent by the publication in 1867 of the work of two Scandinavian investigators, Guldburg and Waage.

Several years previous to this H. Rose had proved with absolute clearness the mass-action of water in many reactions, *e.g.* in the decomposition of alkaline sulphides and of potassium bisulphate, and in the formation of basic salts. The attention of such distinguished workers as Rose, Malaguti, Gladstone and others had further been directed to the study of the mutual decomposition of two salts, whether those were soluble or one of them was insoluble. In fact, attempts were made to work out in various ways the relative affinities of particular substances, and thus to solve a problem which Berthollet had sketched out theoretically.

The ideas of the latter received, lastly, valuable experimental confirmation from the extremely important researches of Berthelot and Péan de St. Gilles[1] on the

[1] *Ann. Chim. Phys.* (3), vols. lxv. lxvi. and lxviii.

formation of compound ethers and ether-acids from an alcohol and an acid. In subsequent theoretical discussions, these and the more recent valuable experiments of Menschkutkin[1] (which furnished information with regard to the chemical equilibrium existing between different substances and to the time-rate of reaction) were applied with success to proving and confirming the correctness of Berthollet's axioms.

The observations on chemical equilibrium in *reciprocal* processes especially contributed to the general adoption of those doctrines of Berthollet; it was thought then (and still is) that the values thus obtained offered the surest data for arriving at the relative affinities of substances taking part in a reaction. With regard to the ideas held respecting such states of equilibrium, the opinion prevailed for a long time that a statical equilibrium must be assumed. A reversal of this was prepared for by the view originated and propounded by Williamson[2] in 1850, which was also worked out independently by Clausius several years afterwards, viz. that the atoms of substances are in a state of continual motion, not merely during chemical reactions but also when the substances are apparently at rest. A dynamical equilibrium thus took the place of a statical, *i.e.* an equilibrium of the opposing reactions. Pfaundler has of late ingeniously applied such speculations to the explanation of the phenomena of dissociation and of reciprocal reactions generally.

But although Williamson emphasised the point that his speculations were in accord with Berthollet's principles, a sufficiently secure and broad basis was still wanting, upon which they could at that time be further developed. Such a foundation for the building up of the doctrine of affinity was furnished by the above-mentioned work of Guldberg and Waage,[3] who took Berthollet's axioms as their immediate

[1] Cf. *Ann. Chem.*, vol. cxcv.; *Journ. pr. Chemie* (2), vols. xxv. xxvi. and xxix.

[2] In a paper read before the British Association at Edinburgh; *Ann. Chem.*, vol. lxxvii. p. 37.

[3] *Études sur les Affinités Chimiques* (1867); this was published in German in the *Journ. pr. Chem.* (2), vol. xix. p. 69.

starting-point, reanimated these anew, and proved their agreement with facts.

Like Berthollet, the investigators just named stated the chemical action of a substance as being proportional to its *active amount*,[1] the latter being given by the quantity contained in unit of space. The intensity of the interaction of two substances is expressed, according to them, by the product of the active amounts; but a coefficient[2] still remains to be determined which shall express the dependence of the reaction upon the nature of the substances taking part in it, the temperature, and other factors. By the aid of such hypotheses the relations existing between the amounts of the reacting substances and their actions[3] can be deduced mathematically. Important conclusions have also been drawn from them with respect to time-rate of reaction and chemical equilibrium, and these have been found to agree sufficiently well with the results of actual experiment.

The latest Development of the Doctrine of Affinity.

Guldberg and Waage's theory, based as it was upon Berthollet's principles, has had an extraordinarily stimulating effect. It has led in particular to the successful determination of the specific affinity-coefficients of different substances, especially of bases and acids; and these experimentally-determined constants have been made use of to test the correctness of the theory itself. Among the work done with this aim in view, that of Ostwald[4] deserves special mention; he has determined by different methods, volumetric and optical, the manner in which a base is distributed among different acids present in excess, and has deduced from this the specific affinity-coefficients of the latter. Julius

[1] ". . . . *seiner wirksamen Menge proportional.*"

[2] Such affinity-coefficients have hitherto only been determined in particular cases, and then only approximately.

[3] *Wirkungen.*

[4] Published in the *Journ. pr. Chem.* since the year 1877.

Thomsen[1] had previously attempted to solve the same problem by thermo-chemical methods.

Ostwald[2] further sought, some years ago, to deduce the affinity-coefficients of acids from reactions which go on with a measurable velocity under the influence of those acids, *e.g.* the decomposition of acetamide and of methyl acetate, and the inversion of cane sugar; in this case too, the results obtained have shown a sufficiently near agreement with calculation. The reader is referred, lastly, to the remarkable relations—already spoken of —which have been discovered by Arrhenius, and also by Ostwald, between the affinity-coefficients and the capacity for (chemical) reaction of acids and bases on the one hand, and their electrical conductivity in dilute solution on the other. Ostwald's researches[3] have thrown a surprisingly new light upon the chemical relations —especially upon the constitution—of the compounds investigated, showing as they do that the affinity-coefficients of substances alter definitely according to the constitution of the latter. At the same time it has turned out that the position or function of the atoms has a determining influence upon these coefficients, this important fact being most apparent in the case of isomeric compounds, *e.g.* the oxy-benzoic and chloro-propionic acids, etc.[4] The general conclusion to be drawn from these and other allied researches is that the specific chemical actions of acids depend upon their hydrogen ions and those of bases upon their hydroxyl ions.

The limits, within which this short account of the development of the doctrine of affinity is necessarily confined, would be widely overstepped were the results of other investigations—even taking only those of importance—to be described. Merely a passing reference can be made to the work of Wilhelmy, which has led to a better knowledge

[1] *Pogg. Ann.*, vol. cxxxviii. p. 575.

[2] Cf. *Journ. pr. Chem.* for 1884 and 1885.

[3] Cf. *Journ. pr. Chem.* (2), vol. xxxii. p. 300; and especially *Ztschr. phys. Chem.*, vol. iii. pp. 170, 241, and 369.

[4] Cf. also Raoult's work bearing on affinity-coefficients, as developed by Planck and others.

regarding time-rate of reaction, and to that of Menschutkin, van 't Hoff, Horstmann and others, which has resulted in making clear the conditions of chemical equilibrium in various reactions.

The hypothesis that the small particles of substances are in continual motion, not merely during chemical reactions, but also when the whole system is in a state of equilibrium, is now held to be indispensable for the new doctrine of affinity. The clear comprehension of the various kinds of energy, and especially of the relation of chemical energy to the other forms, such as electric and thermic energies, etc., has resulted in greatly developing this branch of the science.[1] The chief aim of the doctrine of affinity is to convert chemistry into a branch of applied mechanics—an aim which Berthollet and Laplace, notwithstanding the imperfection of the appliances at command in their day, had the prescience to designate as the highest possible.

[1] Compare the able treatment of energetics in Ostwald's *Lehrbuch der allgemeinen Chemie*, 2nd edition, vol. ii.

A SKETCH OF THE HISTORY OF MINERALOGICAL CHEMISTRY DURING THE LAST HUNDRED YEARS.[1]

Mineralogy only attained to the rank of a science after it had recognised the fact that chemistry was indispensable to it for ascertaining the composition of minerals. It is true that even in this century Mohs,[2] who did so much for minerals physics, almost denied that the chemical characters of minerals had any signification; but the system which he set up was only temporarily adopted by a few scientists. The benefits which accrued to mineralogy from the application of chemical aids were so obvious that the latter could never again be dispensed with. Mineralogy has been brought to its present high position by the joint assiduous work of mineralogists and chemists together. The beautiful aim —of making clear the connection which exists between the physical and chemical properties of individual minerals— has firmly retained its place for the mineral chemist ever since the labours of Berzelius, Mitscherlich, G. and H. Rose, and others were consummated.

The first modest attempts to gain a knowledge of the chemical composition of minerals were made in the seventeenth and first half of the eighteenth centuries, but these did not extend beyond mere superficial observations of a few qualitative reactions. In the second half of last century, however, there was much important preparatory work done,

[1] Cf. Kopp, *Geschichte der Chemie*, vol. ii. p. 84 *et seq.*; v. Kobell, *Geschichte der Mineralogie* (1650-1860), more especially p. 303 *et seq.*

[2] Mohs set up the axiom that a mineralogist had merely to consider the natural-history properties of minerals, *i.e.* crystalline form, specific gravity, hardness, and so on. If their chemical behaviour is taken into account, then, he expressly states, mineralogy oversteps its legitimate bounds and entangles itself in difficulties. This renunciation of the most important aid to mineralogical research is certainly characteristic. Berzelius was fully justified in comparing such a mineralogist to a man who objects to use a light in the dark, on the ground that he would thereby see more than he actually requires to do.

which helped materially to found the science of mineralogy. Mineral chemistry had its distinguished exponents in Bergman, and, a little later, in Klaproth and Vauquelin, whose services in devising methods for the analysis of inorganic substances have already been referred to.[1] The chemical investigation of minerals was carried on at that time, upon the principles which they laid down, by numerous other workers, among whom we may name Lampadius, Bucholz, Wiegleb, Westrumb, Valentin Rose the younger, Kirwan, Gadolin and Ekeberg.

The extraordinary benefit which accrued to mineralogy from the introduction of the blowpipe by Cronstedt, and its subsequent use by Gahn, Bergman, Rinman, and particularly Berzelius, may again be emphasised at this point.[2]

Even before the gradual development of a mineral chemistry, and also simultaneously with it, Romé de l'Isle, Werner, Haüy and Bergman had recognised crystallography as being essential to the study of mineralogy, and had applied themselves to it. Haüy, in particular, achieved wonderful results in this branch; he referred back the various crystalline forms to a few primary ones, and took account of chemical as well as of physical properties in classifying minerals. That he carried his deductions too far here is seen from his well-known axiom that difference in crystalline form signifies also difference in chemical composition.

The endeavours made to classify minerals during that period are for the most part characterised by the desire to recognise their chemical as well as physical properties. If this had only a subordinate signification in Cronstedt's, Haüy's and especially Werner's systems, it was on the other hand put prominently forward by Bergman[3] as an essential aid to the classification of minerals, so far as this was possible with the then existing chemical knowledge. But few of the mineralogists of that day, however, subscribed to Bergman's principles, most of them giving in their

[1] Cf. p. 384 *et seq.* [2] Cf. p. 385.

[3] In his *Sciagraphia Regni Mineralis*, etc. (1782).

adhesion to Werner's system, in which only a very modest place was assigned to mineral chemistry.

A new life began for mineralogical chemistry when Berzelius turned himself to its study. Basing his arguments upon his own comprehensive labours, which had for their aim the exact determination of the composition of minerals and artificial inorganic compounds, he was enabled to show that the doctrine of chemical proportions (and therefore the atomic theory) was applicable in its fullest extent to minerals also.[1] He was the first to characterise these latter as being in every respect "chemical compounds." At the same time this gave him occasion to classify them similarly to substances prepared artificially, and thus arose his Chemical System,[2] in which he gave definite expression to the view that mineralogy should only form a part of, or an appendage to, chemistry. The order of the minerals in his system was determined by the position of their electro-positive constituents in the so-called "tension series." Ten years later [3] Berzelius altered his principle of classification, in so far that he came to look upon the electro-negative constituents as primarily determining this, and he arranged the minerals accordingly. For his two main classes he took non-oxidised and oxidised substances, and between these he divided minerals with a marvellous perspicacity. All previous attempts at classifying minerals according to chemical principles were thrown into oblivion by Berzelius' system.

The development of this system, whose main features were subsequently reproduced in later classifications, was influenced in the highest degree by an observation made by N. Fuchs, viz. that certain substances can replace each other in minerals, and still more by the extension of this doctrine through Mitscherlich's discovery of isomorphism.[4] The results of the analyses of minerals hitherto obtained were henceforth regarded from entirely new points of view and were in many cases simplified to an unexpected extent. A

[1] Cf. p. 205. [2] *Schweigger's Journ.*, vols. xi. and xii. (1814).
[3] *Leonhard's Zeitschrift für Mineralogie*, vol. i. [4] Cf. p. 221.

high, perhaps too high, significance was now attributed to crystalline form in its connection with chemical composition. This over-estimate quickly became manifest after Mitscherlich discovered the first cases of dimorphism,—to be extended later on to tri- and polymorphism. Haüy's principle—that a difference in crystalline form also means a difference in chemical composition—was thereby overthrown; and, in spite of the opposition of this distinguished investigator, the doctrine of isomorphism took its place triumphantly in mineralogy.

The various mineralogical systems which were brought forward after that of Berzelius, *i.e.* after the year 1824, are almost all characterised by the endeavour to classify minerals according to their chemical composition, a greater or lesser signification being at the same time attached to their physical properties. In addition to G. Rose's classification of mineral bodies, which rested upon a purely chemical basis, the mixed systems of Beudant, C. F. Naumann, and Hausmann may be named here as having become best known.

The nomenclature of minerals has by no means kept equal pace with their strictly scientific investigation. The empirical principle still prevails here, this being apparent from the way in which minerals are named after their discoverers, or after localities in which they are found, or according to their physical properties, etc., instead of the name expressing or at least indicating their chemical composition.

Mineralogy owes its present flourishing condition to the immense development of mineral chemistry. Berzelius and his pupils, among whom Chr. Gmelin, E. Mitscherlich, Wöhler, H. and G. Rose, Svanberg and Mosander may be mentioned, were the first to really open up the ground which Bergman, Klaproth, Vauquelin and others had prepared. It is impossible to give a detailed account here of the wealth of new methods which have been devised for the analysis of minerals, and for the separation of their individual constituents. The almost inexhaustible field of minerals has ever since then been investigated chemically by numberless workers. To the problem which naturally comes first, viz. the establishment of their empirical composition, the further and higher one was added of getting at their chemical constitution. The silicates in particular, on account of their extraordinary variety, have given rise to continually renewed investigations.[1]

The limits of this short account of the development of mineralogical chemistry do not permit of citing even a few examples of the services rendered to this branch of the science by such men as Stromeyer, Th. Scheerer, Rammelsberg,[2] Bunsen and others. Among other chemists who have done good work for mineralogical chemistry the following may be named:—v. Bonsdorff, O. L. Erdmann, Marignac, Th. Thomson, Blomstrand, Deville, v. Hauer, Hermann, Th. Richter, Sandberger, Smith and Brush, Streng, Cl. Winkler, P. Jannasch, Th. Petersen; to these many more names might be added.

[1] Efforts have not been wanting to apply specially to minerals the more recent chemical views which have been arrived at with respect to the constitution of organic compounds. Wurtz was the first to do this, by comparing the poly-ethylene alcohols (discovered by himself) with the poly-silicic acids. That such attempts to explain the structure of the most complex silicates have often overshot the mark, and have therefore remained unfruitful, is due to the circumstance that the methods employed for gaining an insight into the constitution of organic compounds cannot as a rule be applied to inorganic.

[2] C. Rammelsberg, born in Berlin in 1813, worked from the year 1840 partly at the Technical College (*Gewerbeakademie*) and partly at the University there, and was from 1874 until a few years ago head of the second chemical laboratory of the latter; he is since dead. His researches, which greatly enriched inorganic and especially mineralogical chemistry, appeared for the most part in *Poggendorff's Annalen.* He rendered very great service by the publication of his *Handbuch der Mineralchemie* (second edition, 1875), and of his *Krystallographisch-physikalische Chemie* (1881-82).

The Artificial Production of Minerals[1]*—Beginnings of Geological Chemistry.*

To the older analytical method, which was the one naturally first followed in the investigation of minerals, the synthetic method has in recent times been added, with the result that mineralogical chemistry has been enriched by an extraordinary number of new facts and has led to the development of geological chemistry. The endeavour to imitate and to explain the natural production of minerals, by preparing them artificially under various conditions, has been the cause of many memorable researches, of which a short account must be given here.

After Berzelius had defined minerals as chemical compounds whose composition was dependent upon the same laws as that of compounds artificially produced, the problem at once arose of preparing mineral substances from their components. But several decades passed by, during which mineral chemistry was developed by improved analytical methods, before the synthesis of minerals was definitely taken in hand with this conscious aim in view. Only isolated observations on the artificial formation of such substances, *e.g.* of calc-spar and arragonite by G. Rose, and some experiments made by Gay-Lussac, Berthier and Mitscherlich, fall to be recorded during the first half of this century;[2] the brilliant development of this branch of mineralogical or geological chemistry only began in 1851 with the memorable labours of Ebelmen, Durocher, Daubrée and Sénarmont. These investigators elaborated a series of methods which led to the production of minerals under conditions similar in part to those found in nature. It was justifiable to draw careful deductions with respect to naturally occurring pro-

[1] Cf. *Die Künstlich dargestellten Mineralien*, etc. ("Artificially-prepared Minerals"), by C. W. C. Fuchs (Haarlem, 1872); and the *Synthèse des Minéraux et des Roches*, by Fouqué and Michel Lévy (Paris, 1882).

[2] The earliest observation of this nature was doubtless that made by James Hall upon the transformation of chalk into marble in 1801.

cesses from these methods of formation; at any rate, hypotheses which were brought forward to explain the formation of minerals and rocks could be put to the test in this way. Geology thus gained a firmer foothold, and found in chemistry an indispensable helpmeet.[1]

Reference may be made here to Bunsen's beautiful investigations[2] upon the geological conditions of Iceland, and especially upon the geysers, and to those on the formation of volcanic rocks, all of which were productive of new views; and also to the labours of G. Bischof,[3] who was indefatigable in advancing chemical geology.

Among the distinguished array of investigators who made further advances in this direction, and, in particular, who discovered new modes of formation of minerals, H. St. Claire Deville and Troost, Becquerel, Debray, Hautefeuille, Wöhler, Rammelsberg, R. Schneider, and especially Fouqué and Michel Lévy, stand out pre-eminent. Of recent years Friedel, Sarasin and Moissan have carried out important syntheses of minerals.

The chief founders of the synthetic method in mineralogical-geological investigations have been Frenchmen, and so reference is with perfect justice made to a *French school* in this branch, the five gentlemen last named being its principal exponents of recent years.[4]

The modes of formation of minerals observed by them

[1] Sénarmont expressed himself in the following significant words with regard to the necessity of chemistry for geology: "*C'est à la chimie minéralogique, que la géologie doit l'utile contrôle expérimental de ces conceptions rationelles. Les minéraux cristallisés ont, en effet, une origine toute chimique, et c'est l'expérience chimique qui doit servir d'appui à la géologie, si elle veut faire un pas de plus dans l'étude des roches, qui en sont composées.*"

[2] *Ann. Chem.*, vol. lxii. p. 1; vol. lxv. p. 70.

[3] Cf. his *Lehrbuch der chemischen Geologie.*

[4] Fouqué and Michel Lévy consider that the cause of this pre-eminence in the above field is to be found in the "nature of the French national character." The argument with which they support this assumption (see p. 5 of the work, *Synthèse des Minéraux*, etc.) is so characteristic, that it may find a place here: "*Notre génie national répugne à l'idée d'accumuler un trop grand nombre de faits scientifiques, sans les coordonner, et si cette tendance nous entraine quelquefois à des hypothèses hasardées, elle a, d'autre part, le mérite, de nous induire aux expériences synthétiques.*"

vary greatly, the processes being partly wet and partly fusion ones. To mention only one or two of the more important, take the production of many natural minerals by the slow mutual decomposition of two salts in solution, *e.g.* the formation of quartz and calc-spar from gypsum and silicate of potash in presence of carbonic acid; the deposition of artificial minerals from solution (formation of gypsum); the production of calc-spar or arragonite according to the conditions prevailing; the decomposition of various substances by water under increased pressure (formation of quartz, wollastonite, apophyllite, etc.); and, lastly, the production of numerous minerals by processes requiring fusion and a white heat—processes similar to those which go on in volcanoes (formation of tridymite, olivine, and other silicates).

The synthesis of the numerous sulphides of copper, iron, zinc and cadmium, partly in the dry and partly in the wet way, also deserves mention, and this applies too to the artificial production of gems, *e.g.* of the ruby by Frémy and of the diamond, the latter having been obtained by Moissan in minute crystals, by suitably cooling a carboniferous iron from an excessively high temperature.

Since nature but seldom allows her workshops to be spied into, the numerous experiments on the production of minerals, made in imitation of natural processes, and which have been carried to a successful issue, possess the highest significance for the explanation of those processes. The repeated proofs that one and the same mineral can be artificially prepared in the most diverse ways, by wet as well as by fusion methods, has rendered the former one-sided conception of geological processes (*i.e.* the view that rock-masses have been produced *either* in the wet way *or* by igneous action) almost impossible now. The synthesis of minerals has riveted still more firmly than before the already long-established link between mineralogy and chemistry.

DEVELOPMENT OF AGRICULTURAL AND OF PHYSIOLOGICAL CHEMISTRY

The history of these branches of chemistry is primarily associated with the work done by Liebig, of which a short description has already been given in the General Section. It is true that this gifted investigator had many predecessors, who found out various isolated chemical facts of great importance for vegetable and animal physiology; but it was he who first with far-seeing glance collected such facts together under general points of view, and conjoined them with still more important observations of his own. The ideas of a Palissy upon the necessity of mineral substances for plant life;[1] the investigations which towards the end of the seventeenth century led Malpighi and Mariotte to definite conclusions with respect to the nutrition of plants through their leaves and roots; the bold and comprehensive speculations of Lavoisier[2] regarding metabolism in plants and animals,—his conviction that the life processes are made up of a series of chemical reactions; lastly, the work of Fourcroy, Vauquelin, Proust, Berzelius and Chevreul upon products of the animal body; —all these, together with other labours, served to prepare the ground upon which Liebig afterwards raised the edifice of chemistry in its relation to agriculture, physiology, and pathology.

Those branches of chemistry are most closely interlaced with organic, for one of their main problems consists in isolating compounds of an organic nature and establishing the composition of these. To this is added the further task of elucidating the *rôle* which such substances fill in the organism. Vegetable and animal physiology are especially indebted to chemistry in questions of nutrition.

[1] Cf. p. 90.

[2] These are set forth in a paper written in 1792, but only published in 1860 (in vol. iv. of the *Œuvres de Lavoisier*).

Agricultural Chemistry and Vegetable Physiology.[1]

The work done in physiological chemistry towards the end of last century and the beginning of this by Priestley, Ingen-Houss, Senebier and Th. de Saussure had led to many important results with respect to the nutriment of plants. One might now suppose that, from the analysis of the ashes of plants, a distinct connection between the plants and the soil would have been apparent. The decomposition of carbonic acid by the leaves, which was observed by those workers, ought, one might further suppose, to have pointed to carbonic acid as the main source of the organic matter of plants. In like manner the early made observation that salts of ammonia were highly conducive to the growth of vegetables,[2] might have found an explanation in the recognition of ammonia as the source of their nitrogenous constituents.

These deductions, however, which now appear to us self-evident, were not drawn, and it was sought to credit humus as being the universal nutrient of plants, without paying any heed to those older fundamental observations which have just been mentioned. The processes of nutrition of plants were thus entirely misunderstood, for, according to this doctrine, they fed like animals upon organic matter.

This assumption, which dominated agricultural chemistry for many decades, found its chief advocates in Germany and France in Albrecht Thaer [3] and Mathieu de Dombasle respectively. In their opinion inorganic salts, the importance

[1] For the literature consulted on this subject (in addition to the books and papers cited below), see the *Geschichte der Botanik*, by J. Sachs; *Lehrbuch der Pflanzenphysiologie*, by Pfeffer; *Lehrbuch der Agrikulturchemie*, by W. Knop; *Chimie et Physiologie appliquées à l'Agriculture*, etc., by L. Grandeau; *Neues Handwörterbuch der Chemie*, vol. ii. pp. 119 and 1012; and Ville's *Artificial Manures. . . . . .* etc. (English Edition by Crookes). See also Storer's *Agriculture in some of its relations to Chemistry* (2 vols.).

[2] Nicolas Leblanc pointed out the importance of salts of ammonia in this respect so long ago as at the end of last century.

[3] Cf. his work, *Grundsätze der rationellen Landwirthschaft* ("Principles of Rational Husbandry"). Even Saussure, the originator of the doctrine of plant nutrition, fell into the humus theory error.

of which could not be absolutely denied, acted merely as stimulants, and not as if they were essential to the growth of the plant.[1] Indeed, Thaer held that the formation (*i.e.* creation) of earths in plants through their vital forces was possible. In this assumption he followed the opinion of Schrader, who so early as the year 1800 imagined that he had proved by actual experiments the generation of the ash-constituents of plants by the vital forces.[2]

Liebig put an abrupt end to this period of unscientific attempts at explaining the process of plant nutrition by his critical demolition of the humus doctrine. Taking his stand upon a large number of investigations carried through by himself and his pupils, in conjunction with earlier work done by others, he brought out in 1840 his book, *Die Chemie in ihrer Anwendung auf Agrikultur und Physiologie*[3] ("Chemistry in its Application to Agriculture and Physiology") ; in this he did battle with the arbitrary axioms of the humus theory, and completely undermined the foundations of the latter, hitherto looked upon as secure. The following sentences by Liebig constitute the quintessence of his doctrine; they already contain the complete programme of the agricultural chemistry which has been created since that time. "The nutritive materials of all green plants are inorganic substances."... " Plants live upon carbonic acid, ammonia (nitric acid), water, phosphoric acid, sulphuric acid, silicic acid, lime, magnesia, potash and iron; many of them also require common salt." . . . " Dung, the excrementa of the lower animals and of man, does not act upon plant life through (the direct assimilation of) its organic elements, but indirectly through the products of its decomposition- and putrefaction-processes, *i.e.* by the transformation of its carbon into carbonic acid, and of its nitrogen into ammonia or nitric acid. Organic manure, which

[1] Several writers have ascribed to Sprengel, who achieved so much for botany, the merit of having proved the indispensability of the ash-constituents for plants, but this is incorrect.

[2] This erroneous view was first combated upon good grounds by Saussure, and then by Davy.

[3] The incitement to this work came from the British Association for the Advancement of Science.

consists of portions or *débris* of plants and animals, may be replaced by the inorganic compounds into which it breaks up in the ground."[1] From these axioms Liebig drew the all-important conclusion that the soil must be replenished with whatever constituents have been withdrawn from it by the culture of plants, if its exhaustion is to be provided against.

In the further development of this pregnant doctrine, whose victory over the old system was soon complete, distinguished pupils of Liebig took part as well as himself. Indeed nearly every agricultural chemist since that time has come either directly or indirectly from Liebig's school. Boussingault[2] strove independently after similar goals, and the services which he rendered in carring out researches on the nutrition of plants by new methods must be emphasised here. The now world-famous field experiments of Lawes and Gilbert at Woburn in Bedfordshire, begun more than half a century ago, and which are being continued with unabated vigour, will always hold a distinguished place in the history of agricultural chemistry. And the service which the late Georges Ville rendered to this branch of the science by his work in France should also be borne in mind.

Definite researches were first made in order to explain the chemical conditions existing in the soil, from which plants are supplied with their purely mineral constituents. These included the investigation of the processes involved in the weathering of rocks, through which soil is produced. Liebig, Boussingault, Déherain, Dietrich and others showed by their investigations what were the parts played by the

[1] Liebig himself carried out practical experiments in manuring, and succeeded in changing a sandy piece of ground in the neighbourhood of Giessen into a productive garden by the aid of mineral manures alone.

[2] J. B. Boussingault, who was born in 1802 and who died in 1886, first became known through his adventurous journeys in South America, where he turned his catholic knowledge to brilliant account. After returning to France he devoted himself more and more to agricultural-chemical questions, which he treated partly in experimental researches, and partly in his detailed works, *Économie Rurale ; Agronomie ; Chimie Agricole et Physiologie* (1864).

active agents here,—water, carbonic acid and oxygen; they also came to the conclusion that free nitrogen as such was not directly assimilated by plants, but this view has been overthrown by the work of Ville, Hellriegel and others (Cf. below). It is only after rocks have been "weathered" that the inorganic substances necessary for the nutrition of plants are brought into such a condition that they can be assimilated by these. The valuable experimental work done by E. Wolff, Henneberg, W. Knop, F. Stohmann, Zöller, Lehmann and Nobbe, among others, upon the composition of different soils must be mentioned here, and also the closely allied experiments by them on the nutrition of plants in sterile soils and in solutions of salts,—*dry culture* and *water culture.* These methods have served to solve the most important questions regarding plant nutrition.

These researches all went to prove that the same substances as are found in the ashes of plants are the true nutrients of the latter, and are absolutely indispensable to them. But they did more than this, in showing the significance—indeed, the determining influence—as regards nutrition, not merely of the nature of the nutritive materials contained in the soil, but also of the form in which these are present, and of their action upon the other constituents.

The earliest series of experiments on the absorption by different soils of the mineral constituents which serve as food for plants was due to Liebig, while similar work by Henneberg and Stohmann, Peters, Knop, Zöller, etc., must also be recorded; these observations were likewise of great importance for the explanation of the action of manures.

A few words must be added here about nitrification in soils and the assimilation of free nitrogen by plants,—the most important discoveries in agricultural chemistry of recent years. So long ago as 1849 the late Georges Ville, then director of the Agricultural Experiment Station at Vincennes, proved by actual experiment that certain plants could and did assimilate free atmospheric nitrogen; but at the time his conclusions were strongly disputed, being directly opposed to those of Boussingault and Liebig, and also to subsequent investigations by Lawes, Gilbert and Pugh in 1857. An important experiment bearing on the point and extending over many years was begun in 1855 by Herr Schultz of Lupitz in Altmark, Germany. He grew lupines on very poor soil with the addition of non-nitrogeneous manures only, and found that notwithstanding this, the soil became richer in nitrogen year by year. The next step towards the solution of the question was the discovery in 1877 of the now well-known process of *nitrification* in soils by MM. Schloesing and Müntz, this nitrification being the work of definite microbes, some of which have been isolated by Winogradsky, Warington, and P. Frankland; while the recent work of Hellriegel and Wilfarth (in 1888), Frank, Schloesing, Berthelot and others has proved that the direct assimilation of atmospheric nitrogen by leguminous plants is brought about by the agency of certain micro-organisms originally present in the soil, which enter the root at a very early period of the plant's growth. At the place where the micro-organism enters, a disturbance is set up and a nodule or tubercle formed, in which the micro-organism multiplies rapidly. These nodules are highly nitrogenous substances, and through their agency the plant is somehow enabled to assimilate the free nitrogen of the air and to convert it into albuminous compounds; but how this is actually brought about has still to be explained. Cultures of these specific bacteria are now prepared on a manufacturing scale, under the name of *nitragins*, for application to soils naturally deficient in them; but whether they will actually be of value on ordinary arable land remains to be proved.[1]

Notwithstanding, however, that an immense number of new facts have been brought to light through these and other labours, the fundamental principles of Liebig's doctrine have undergone no alteration since he first gave them to the world in his pioneering work of 1840. He clearly recognised in all its broad features how plants draw their nutriment from the constituents of the air and the soil.

[1] Cf. A. P. Aitken, *Transactions of the Highland and Agricultural Society for* 1898, p. 299.

Upon this he based his doctrines of rational husbandry, which have already borne the richest fruit, and in the elaboration of which scientific and practical men are still engaged.

Development of Phyto-Chemistry.

After the importance of various inorganic substances for the life of plants had come to be recognised, the pressing question arose for physiologico-chemical investigation—How and in what phases is the formation of organic substances from carbonic acid, ammonia and water consummated? The problem to be solved here consists in isolating the chemical compounds present in the various organs of plants, and in establishing their physiologico-chemical relations to one another,—a magnificent task, and one which has already occupied many able investigators.

The conversion of carbonic acid into organic compounds under the influence of water and light, the process of the assimilation of carbon, which was already correctly apprehended in its main outlines by Saussure,[1] has naturally formed the subject of numerous investigations. Thus recent researches by Lommel, Pfeffer, N. J. C. Müller, Engelmann, and others have elucidated the nature of the light rays which are active here. Much valuable work too has been done upon chlorophyll, although the opinions of men like Sachs, Pringsheim, etc., differ as to the part which

[1] Cf. his *Recherches Chimiques sur la Végétation* (1804). Previous to this Ingen-Houss had observed the assimilation of carbonic acid and water by the leaves of plants, but, being enchained by the phlogistic theory, had not perceived that the oxygen thereby liberated came from this carbonic acid. The above relation was first made clear by Senebier, and became a certainty after Saussure's masterly researches, through which the balance between the substances absorbed and eliminated was approximately ascertained. Ingen-Houss, too, and Saussure still more definitely, recognised that the converse of this assimilation process (*i.e.* a breathing in of oxygen and giving out of carbonic acid) goes on in various parts of plants. Saussure and, after him, Dutrochet and others further observed the evolution of heat which accompanies respiration in plants, and thus established a noteworthy analogy between the processes in the vegetable and animal organisms.

this substance plays in the assimilation of carbon. Speculation has still, however, pretty free play in the answering of the questions—What is the organic compound which is in the first instance produced from the carbonic acid, and what are the intermediate products in the formation of starch, cellulose, albumen, etc.?

A. v. Baeyer's view—that formic acid is produced in plants by the reduction of carbon dioxide, and is then converted into carbohydrates by numerous condensations,—has been corroborated to some extent by the laboratory experiments of O. Loew, Bockorny, E. Fischer and others. This assumption is at any rate the simplest that could be brought forward to explain the nutrient action of carbonic acid.

The multifarious substances produced by plants have been the objects of ardent investigation, more especially since the stimulus which was given to the subject by Liebig's work; the chemistry of plant life has been developed alongside of that of animal life, particularly since the close of the forties. Reference must be made here, in passing, to Rochleder's researches in this field (so important from the chemical point of view), upon caffeine, various glucosides, tannic acids,[1] and other vegetable products. The attention of phyto-chemists has been directed in a special degree to the nitrogenous compounds which are formed in plants, *i.e.* to the albumens in the first instance, and then to the compounds produced by the breaking up of these. After Mulder had pointed out the similarity of the former to animal albumen, they were investigated by Liebig and his pupils, and they have formed the subject of excellent work by Ritthausen during recent years. The hope that conclusions might be arrived at with regard to the constitution of the albumens from the nature of their decomposition-products, more especially from the amido-acids like leucine, asparagine, glutamic acid, etc., has not indeed been realised; but, from the point of view of vegetable physiology, the researches on the nitrogenous compounds which are formed during the

[1] Kraus's monograph :—*Grundlinien zu einer Physiologie des Gerbestoffes* (1889) shows the importance of the tannic acids in vegetable physiology.

germination of seeds and other processes have furnished much valuable preparatory work for the future development of that branch of the science.[1]

There are, besides, many other vegetable products containing nitrogen which have occupied the attention of chemists as well as of physiologists, *e.g.* various glucosides such as myronic acid and amygdalin, and, in particular, the great class of the alkaloids,—compounds whose importance for chemistry has already been discussed.

The carbohydrates in their signification for the life of plants have likewise been much investigated, with regard both to the conversion of some of them into others by chemical means, and to their physiological modes of formation; but here again the necessary link is often wanting between particular products. The reader is referred to the pioneering investigations of Brücke, Nägeli, Sachs and others upon starch and the substances formed before it, *e.g.* dextrose, and upon the connection which exists between the formation of starch and the activity of chlorophyll; to the excellent work of Cross and Bevan and others on cellulose; to the numberless researches on the sugar varieties, especially dextrose and cane sugar, the occurrence of the latter in beetroot and its technical production from the same having created a chemistry of its own; and to the laborious work which has been and still is being done with the object of elucidating the chemical nature of the glucosides and their peculiar behaviour to ferments. The most important of the investigations upon vegetable fats, ethereal oils, and various other (vegetable) compounds belong in the main to organic chemistry proper, and have been referred to under the history of this.

[1] Cf. the investigations of E. Schulze and others.

Development of Zoo-Chemistry.[1]

The physiological chemistry of the animal body, zoo-chemistry, has made extraordinary progress since the early investigations of Fourcroy and Vauquelin, Chevreul, Berzelius and others were made. From the examination of the chemical constituents of animal organs, secretions, etc., an advance was made to the infinitely more difficult problem—Under what conditions are those substances formed in the organism, and what are their relations to one another? From the chemical investigations which arose from this, animal physiology was first constituted into the science as we now know it. And this applies in a special degree to the important question of nutrition, and, speaking generally, to the modern views of the metabolic processes of the animal body. Chemical investigation has thus been the means of dispelling the obscurity in which so many erroneous views grew and flourished.

Since the publication of the above-mentioned researches, the most distinguished physiologists and chemists have co-operated in the development of zoo-chemistry, in so far as this has aimed at a knowledge of the substances of which the animal body is composed. From the large number of excellent investigations of this kind, only one or two can be touched upon here. Reference must first be made to the work of v. Bibra, Mulder, Frémy and Heintz upon the constituents of bones, through which the true composition of these was established. Schmiedeberg's investigations, made in 1891, have been the first to throw light upon the nature of the substances present in bone cartilage.

The question as to the nature of the albumens has given rise to many important researches, especially since Mulder first proved the presence of compounds of this kind in plants, and Liebig and his pupils strove to arrive at their composition;

[1] The numerous sources of physiologico-chemical investigations are to be found in Hoppe-Seyler's *Lehrbuch der physiologischen Chemie.* Cf. also Bunge's *Lehrbuch der physiologischen und pathologischen Chemie.* Only in a few instances have direct references been given here.

but they have not as yet led to a knowledge of the true constitution of these bodies. Among those who have worked at this subject may be mentioned A. Schmidt, Graham, Brücke, Johnson, Hoppe-Seyler, Kühne, Hammarsten, Lehmann, Schützenberger, Nencki, Drechsel and Harnack. To the physiologist the question of the behaviour of albumen in the animal body (in particular, the changes which it undergoes during digestion, etc.) is of more importance than its rational composition. Some investigations will be referred to later on, in which an answer to such physiological questions is attempted.

The most important of the researches which led gradually but ultimately to a true explanation of the composition of fats have already been spoken of.[1] The part played by fats in metabolism has only been satisfactorily worked out of recent years, and the same remark applies to the carbohydrates.[2] The pathological occurrence of those substances has also given much occupation to chemists, who, by furnishing definite tests for sugar, albumen, etc., have in many cases lightened, and even rendered possible, the diagnosis of a disease by the physician.

As in all the other branches of chemistry, so too in physiological and pathological, have special methods of a zoo-chemical analysis gradually developed themselves and become indispensable.

The investigations that have been made with the object of elucidating the chemical processes which go on in the animal organism, and with this the processes which condition or accompany life, are almost innumerable. Our present knowledge of the various animal fluids which take part in such processes has only been attained by the most arduous labours. To mention but one or two of these, reference may be made in the first instance to the more important of the researches on the secretions which promote digestion. The classical investigations of C. Ludwig, Brücke and Cl.

[1] Cf. p. 441.

[2] With regard to the chemical importance of the carbohydrates, see p. 454 *et seq.*

Bernard proved that the secretions from the glands were to be looked upon as resulting from essentially chemical processes. The importance of the saliva for digestion was also shown by its chemical investigation; Leuchs, in 1831, discovered the ferment *ptyalin* which saliva contains, and which has the power of transforming starch into sugar, and the chemistry of the saliva has since been materially advanced by the later work of O. Nasse, C. Ludwig, Brücke, Herter and others.

Many scientists of repute have occupied themselves with the investigation of the gastric juice; thus the work of C. Schmidt, Bidder, Beaumont, Frerichs, Lehmann, v. Wittich and others has resulted in establishing the composition of this secretion, and also the peculiar nature of *pepsin*, the ferment which it contains. The excessively important part played by the latter in the digestion of the albumens, which are thereby converted into *peptones*, has been mainly arrived at through the labours of Lehmann, Hofmeister, Henninger, and more recently Neumeister, Kühne and Chittenden.

Our knowledge of the pancreatic fluid and of its powerful influence on the digestive process, which is due to the presence in it of particular ferments, we owe to W. Kühne, Hüfner and others.

The chemistry of the bile, lastly, which originated with Strecker's memorable work[1] on the bile-acids and their decomposition-products, has been subsequently extended by Städeler, Frerichs, Gorup-Besanez, Maly, etc.

The present knowledge of the chemical composition of the blood and of its various constituents (so difficult to separate from one another), together with the chemical behaviour of these, is the outcome of an infinite number of laborious investigations; and it is still very far from being complete. Reference must be made here to the pioneering work of Al. Schmidt upon the causes of the coagulation of blood; to that of C. Schmidt, Hoppe-Seyler, Hüfner, Preyer and others on hæmoglobin and oxy-hæmoglobin, and the behaviour of these to gases, and also to the successful application of the

[1] *Ann. Chem.*, vols. lxi. lxv. lxvii. and lxx.

spectroscope here; further, to the memorable researches which finally established the composition of the blood-gases and especially the difference existing between arterial and venous blood in this respect. The services rendered by C. Ludwig deserve to be particularly emphasised, the investigations which he carried out along with his pupils from the year 1858 far surpassing the earlier ones of Magnus and of L. Meyer in accuracy.

The numerous researches, by means of which the quantitative relations between the air inhaled and exhaled by animals were exactly determined, have been of the utmost value for a knowledge of the metabolic processes of the animal body. We have only to recall here the experiments carried out on a large scale by Pettenkofer and by Regnault and Reiset since the year 1862, and the important observations by C. Ludwig, and by Pettenkofer and Voit, on the effect of muscular exertion upon the consumption of oxygen and the production of carbonic acid.

The exceedingly numerous researches on the substances which occur in blood serum, on the inorganic constituents of blood, and on the pathological changes which the latter undergoes, cannot be entered upon here.

Milk has been the subject of frequent investigation ever since Chevreul, Lerch, Heintz and others established its principal constituents. Much attention has been paid in more recent work to the process of coagulation, to the changes which milk undergoes in the organism, to the nature of the albuminous compounds which it contains, and so on; witness the important researches on the subject by Soxhlet, Hammarsten, Hoppe-Seyler, and especially Lehmann.

Much excellent chemical and physiological work has been done upon urine—the secretion of the kidneys. Take, for instance, the observations on the artificial production of urea, of such moment from a chemical point of view, and those upon uric acid and its manifold transformation-products, the synthesis of which has already been achieved.[1] Then there are, too, the important physiological and pathological investi-

[1] Cf. *The History of Organic Chemistry*, pp. 467–468.

gations by Liebig, Voit, Bischoff, Fick and Wislicenus on the separation of urea in its bearing upon metabolism; the researches on the formation of hippuric acid, by Wöhler, Liebig, Dessaignes and Meissner; on that of the phenol-sulphuric acids by Baumann; on the formation of sugar, albumen, glycuronic acid, cynurenic acid (an oxyquinoline-carboxylic acid) and indole; and on the separation of all of those substances just named in the urine.

The explanation of the manner of origin of these and other substances, which are partly found under normal conditions and partly under pathological, has long been recognised as constituting an important problem of physiological chemistry. From the results of a large number of observations, a systematic method of analysing urine has gradually been developed,[1] and this daily stands the practising physician in good stead; for, from the occurrence or accumulation of certain substances in the urine, the latter can recognise particular diseases with greater precision than by any other sign.

The work which has been done upon the chemical composition of flesh,[2] a subject to which peculiar difficulties are attached, can only be briefly referred to. Liebig's classical researches on "the constituents of the fluids of flesh," [3] and the nearly allied ones of his pupils Schlossberger, Scherer, Strecker and Städeler, prepared the way for later and even more ambitious labours; we would refer here to the observations of Helmholtz, Ranke, Brücke and others on the effect of muscular action upon the chemical processes which go on in muscle-substance,—observations to which the first incitement may have been given by Liebig's ingenious and far-reaching speculations. The important part which glycogen plays in these, as well as in other processes (*e.g.* the processes of the liver), was arrived at through the admirable work of Brücke, Cl. Bernard, Külz, v. Mering, Voit, etc.

[1] Compare Neubauer and Vogel's book:—*Anleitung zur Analyse des Harns.*

[2] Cf. (*e.g.*) Falk's book, *Das Fleisch* (1880).

[3] *Ann. Chem.*, vol. lxii. p. 257.

From the rich material of facts relating to the chemical composition and physiological importance of particular parts of the animal organism, which have thus been accumulated, the views regarding the metabolic processes of the animal body have been developed, and indeed completed, in certain of their details. The establishing of the laws which govern the nutrition of animals was long ago felt to be of the first importance. And here again Liebig gave the powerful impulse to the first, even if incomplete, solution of this question from the chemical standpoint.

The service which he rendered with regard to the development of the doctrine of metabolism appears especially great when one recalls to mind how erroneous were the opinions of physiologists respecting the chemical processes going on in the animal body, before he set forth his views on nutrition and other physiological processes in his standard work, *Die Thierchemie oder die Organische Chemie in ihrer Anwendung auf Physiologie und Pathologie* (1842), (" Animal Chemistry, or Organic Chemistry in its Application to Physiology and Pathology"). The most eminent physiologists of that time, Tiedemann, Burdach and others, were by no means fully convinced of the necessity of chemistry for their science; to explain the processes in the organism they had recourse to " vital forces," many of them indeed flatly refusing the aid of chemistry. It was left to Liebig to form a truer estimate of the problems of physiology and of the means to be used in solving these; the opinion which he expressed—that it must adopt the methods of physics and chemistry—coming as this did with the full weight of his authority, was quickly taken to heart. And what a change came over physiology in consequence!

The powerful influence exercised by Liebig on the development of the doctrine of metabolism has already been frequently referred to. But a short *résumé* may be given here of the main conclusions of his comprehensive work and ingenious speculations. He endeavoured to establish the various importance of different nutritives for the animal body, in so far that he defined the albumenoids as *plastic* compounds which served mainly for building up the tissues and as the source of muscular power, and the fats and carbohydrates as *respiratory* compounds, which went for the most part to produce the animal heat. It was he in fact who first drew sharp distinctions between nutritive substances among themselves, and between these and other substances which, while not directly nutrient, bring about metabolic changes in the organism.[1] And he also successfully determined the relative values of the former by direct experiment.

The potent effect of Liebig's ideas respecting nutrition and metabolism showed itself during the succeeding years in the splendid work which was done by Bidder and Schmidt, Bischoff, Voit, Pettenkofer, Frerichs and others, as the result of his stimulus. By the aid of improved methods and, especially, by the use of larger respiration apparatus, Liebig's views were subjected to a sharper scrutiny, and thus underwent many corrections, more particularly with respect to the *rôle* of albumen and to the formation of fat. But in all essential points he was right. To the elucidation of the functions and actions of particular nutritives in the animal body, the classical researches[2] of Voit and Pettenkofer, together with those of their pupils (among whom were Ranke, Forster, Rubner, Falck, Fr. Hofmann and Renk) upon nutrition, and therefore upon metabolism, have contributed in an especial degree. An important deduction drawn from these researches, viz. that fat is produced from albuminous matter, has lately been disputed by Pflüger[3] as having no sufficient basis. This eminent physiologist is further of opinion that it is not the carbohydrates and fats but the albumens which are the sources of muscular power.

The aims of the above branch of physiological chemistry are so intimately connected with those of hygiene that the two overlap at this point. Hygiene may indeed be looked upon as a branch of chemistry, having found in the latter science the most powerful of all aids to her development.

[1] *Genussmittel.*

[2] Most of these were published in the *Zeitschrift für Biologie.*

[3] *Pflüger's Archiv für Physiologie*, etc., vol. xli. p. 229.

Reference has already been made in the history of analytical chemistry[1] to the continuous improvement in the methods of analysis of foods and drinks, a point of such immense importance to the community in general.

Fermentation; Putrefaction.[2]

The various processes by which ferments are set in action, and by which their action is conditioned, have now attained to such a supreme importance for hygiene and for physiology as a whole, that a few words must be said here with regard to the development of our knowledge of the processes of fermentation and putrefaction during recent years.

It is a long time since the vinous fermentation first attracted the attention of chemists, but Lavoisier was the earliest to recognize that the two main products resulting from it—alcohol and carbonic acid—came from the sugar present; at the same time he attempted to work out the quantitative relations between the latter and the two former compounds. As to the reason for the breaking up of sugar in the presence of yeast, no views were expressed at that time which were at all tenable. Before it was known that yeast consisted of living cells, Liebig's mechanical-chemical theory of fermentation[3] gained many adherents. This theory, which was propounded in the year 1839, attempted to explain alcoholic fermentation and other similar processes from one common point of view. Liebig here regarded ferments in general as easily decomposable bodies, from which the stimulus to the decomposition of fermentable substances proceeded. This view recalls that which Stahl and Willis had brought forward long before, for they also assumed a transference of the motion of fermenting

[1] Cf. p. 398.

[2] For the literature consulted here, see the articles "*Fermente*" and "*Gärung*" in the *Handwörterbuch der Chemie;* A. Mayer, *Lehrbuch der Gährungschemie;* and Schützenberger, *Gährungserscheinungen.*

[3] Cf. *Ann. Chem.*, vol. xxx. pp. 250 and 363.

particles to a large number of others. Some investigators had contented themselves with attributing to yeast a "catalytic" action, but this simply meant the employment of a word to cover their ignorance of the subject.

In 1836, *i.e.* shortly before Liebig had brought out his theory, Cagniard de Latour, Schwann, and Kützing made simultaneously and independently of one another the important discovery that yeast consists of low organisms which are self-propagating. The subsequent comprehensive researches of Pasteur[1] entirely confirmed the correctness of these observations. From all this the vitalistic theory of fermentation followed as a necessary consequence, although its recognition was retarded by the force of Liebig's great authority; according to this theory the decomposition of the sugar is dependent upon the vitality and consequent activity of the yeast fungus.

Other processes of fermentation were now investigated from the standpoint thus obtained, with the result that low organisms were found to be the cause of the action in their case also. We would refer here to the splendid researches of Pasteur upon the acetic and lactic fermentations, of equal importance physiologically and chemically; to the discovery of the particular fission fungi which gave rise to various fermentations; and to the work of Rees, de Bary, Brefeld, A. Mayer, Fitz and others, the object of which was to elucidate the conditions of the life and especially of the nutrition of organised ferments (more particularly yeast and its connection with fermentation), and also the products of these latter.[2] E. Chr. Hansen's wide-reaching investigations in this branch have been of the utmost value, more especially to the technical side of the brewing industry.[3]

[1] Cf. his large works, *Études sur la Bière,—sur le Vin,—sur le Vinaigre.*

[2] C. Schmidt found succinic acid, and Pasteur glycerine, among the products of the vinous fermentation. It is only of comparatively recent years that sufficient attention has been paid to the various alcohols contained in fusel oil, which are now recognised as products of secondary fermentations.

[3] Hansen, *Untersuchungen aus der Praxis der Gährungsindustrie* (Munich, 1890).

Much vigorous discussion has ultimately led to agreement upon the most important of the disputed points, with regard to which the views of different workers were formerly far apart. A twofold growth of the yeast cells is now established, viz. (1) a growth in presence of oxygen, which is not followed by fermentation, and (2) one in absence of oxygen, through which fermentation is produced.[1]

The difference between organised and unorganised ferments, the latter of which are termed *enzymes*, came to be clearly recognised, this being mainly due to Pasteur's work. The extraordinarily important functions of these unorganised ferments in the animal and vegetable organisms has led physiologists and chemists of the highest eminence to devote their close attention to the subject, but as yet no satisfactory theory of the action of such ferments has been brought forward; in conjunction with this, reference must be made here to the work of Nasse, Hüfner, Traube, Hoppe-Seyler, Nencki, Al. Schmidt and Wurtz. Quite recently Büchner[2] has isolated an enzyme which is capable of inducing the alcoholic fermentation in the absence of yeast cells.

The phenomena of putrefaction, which were placed by Liebig in the same category with the processes of fermentation (both being brought about, in his view, by similar mechanical-chemical causes), acquired a heightened physiological interest after it was perceived that they were connected with the presence of certain peculiar organisms. Here again the researches of Pasteur and also of Nencki, Hoppe-Seyler, etc., stand out pre-eminent. The chemical examination of the products of putrefaction has led to

[1] Liebig maintained an antagonistic attitude to the vitalistic theory of fermentation; he did not indeed contest the organised nature of yeast, but would not acknowledge that the latter itself gave rise to fermentation through its life processes. Instead of this he assumed in yeast the presence of an albuminous ferment, which, on the death of the former, he imagined to bring about the decomposition of the sugar into alcohol and carbonic acid. Nägeli's attempt to explain the phenomena of fermentation may be looked upon as an effort to reconcile the vitalistic and mechanical theories (cf. his *Theorie der Gährung*, 1879).

[2] *Ber.*, vol. xxxi. p. 568.

remarkable results, which have also a high importance for the chemist. Most interest has been centred in the nitrogenous compounds which originate from the decomposition of animal albuminous substances by putrefaction; thus we would recall here the discovery of various amido-acids, of indole and its homologues, and, particularly, of the so-called *ptomaïnes*.[1] The formation of these powerful poisons, which have also been called *corpse alkaloids*, because of their likeness to the alkaloids from plants, is of the first importance to the forensic chemist,[2] seeing that cases have occurred in which the ptomaïnes have been confounded with the true alkaloids, on account of similarity in reaction. The Italian toxicologist Selmi was the first to clearly recognise the important *rôle*, from a forsenic point of view, of these putrefaction bases, and he it was who gave them the generic name by which they are now known,—the ptomaïnes.

In addition to Selmi—Otto, Husemann, Dragendorff, Kobert, Brieger and others have rendered good service in extending our knowledge of these substances. Brieger, in especial, and also Nencki, Étard, Gautier, Guareschi and Mosso have succeeded in characterising certain ptomaïnes chemically. The constitution of some of them has been recently established, witness the beautiful syntheses of cadaverine[3] and of putrescine,[4] which have been respectively shown to be penta- and tetra-methylene diamines.

The Relation of Chemistry to Pathology and Therapeutics.

The phenomena of putrefaction possess the highest interest for pathologists, because such processes lie at the root of many diseases. An increasing knowledge of the causes of these proceeses has thus resulted in the establish-

[1] For a historical notice of these peculiar compounds, cf. Beckurts' *Ausmittelung giftiger alkaloïde* ("Detection of Poisonous Alkaloids"), (*Archiv Pharm.* for 1886, p. 1041); also Armstrong, *Journ. Chem. Ind.*, vol. vi. p. 482. [2] Cf. pp. 397–398. [3] Cf. p. 480.

[4] *Ber.*, vol. xx. p. 2216; vol. xxi. p. 2938.

ment of a close connection between chemistry and pathology, the former having now become indispensable to the latter. And this necessity for chemistry has shown itself not merely in the investigation of the products of putrefaction; through its means the more delicate tests for the recognition and distinction of disease-producing bacteria have been elaborated, and it has thus been instrumental in helping to found the new science of *bacteriology*. This subject cannot, however, be entered into here.

Above all, it has been reserved for chemistry to direct the attention of physicians to remedies for counteracting the pathological processes induced by micro-organisms. Only a passing reference can be made here to the wonderful results which have been achieved in medicine and surgery, and also on the large scale in the preservation of food and drink, by the use of antiseptics. One is probably not wrong in assuming that the old practices of smoking flesh and of dipping wood into tar drew attention to the carbolic acid which the latter contains, and the antiseptic action of which has now found such world-wide application in Lister's method of treating wounds. The discovery of the anti-fermentation and anti-putrefaction powers of salicylic acid by Kolbe originated in the idea that this compound tended to break up into carbolic and carbonic acids in its passage through the organism. The last decade has introduced us to a large number of new antiseptics, which are now used more or less in medical and hygienic practice; these are mostly substances which stand in a near chemical relation to phenol, *e.g.*, the homologous cresols and thymol, the sulphonic and carloxylic acids of these, the iodo-derivatives of phenol- and oxy-quinoline-sulphonic acids, etc. The assumption made by various investigators—that antiferments and antiseptics act by precipitating or chemically altering the readily decomposable albuminous substances—explains the *rôle* of these in a sufficiently satisfactory manner; for, when those bodies are got rid of, the ferments are deprived of their necessary nutriment.

The nearly allied question of the great benefit which

chemistry has conferred upon medicine [1] by enlarging its stock of remedies can only be touched upon very briefly, as any detailed treatment of the subject here would overstep the limits of this work. With the history of medicine in the earlier ages the conditions were quite otherwise; for, in the iatro-chemical as well as in the phlogistic periods the latter was in the main conjoined with the history of chemistry, whereas now chemical investigation pursues totally distinct aims.

To mention only one or two of the specially important services which chemistry has rendered to medical science, take the introduction of narcotics and anæsthetics—chloroform, ether, nitrous oxide, chloral, bromide of potassium, sulphonal, etc. A number of other chemical compounds have been proposed as anæsthetics during the last few years, but none of them have entered into serious competition with chloroform, ether and nitrous oxide. And the same remark applies to the substances newly recommended as soporifics, *e.g.* urethane, para-aldehyde, aceto-phenone, etc; compared with sulphonal and chloral, these have but little importance.

Reference must also be made to the success with which naturally occurring sedatives and febrifuges have been replaced by others artificially prepared, *e.g.* quinine by antipyretic remedies like salicylic acid, acetanilide, antipyrine, phenacetine, etc. It has already been shown [2] how, with the acquisition of the knowledge that the alkaloids are derivatives of pyridine or quinoline, a firmer foothold was gained for the artificial formation of these natural products—an object which has been striven after for so long.

[1] H. Thoms' work :—*Die Arzneimittel der organischen Chemie* gives an excellent summary of the rapidly extending list of artificially prepared medicines; compare also Beckurts' reports on Pharmaceutical Chemistry in the *Jahrbuch der Chemie*, vols. i—v.

[2] Cf. pp. 482–483.

The Relation of Chemistry to Pharmacy.

With the rapid enlargement of the medical treasury, the problems which confront the pharmacist have likewise grown in a very high degree. If the latter is to do justice to the demands which are made upon him, he must be equipped with a catholic and thorough knowledge of chemistry. The development of pharmaceutical chemistry in recent years is for the most part concurrent with that of particular branches of the pure and applied science. The discoveries of inorganic and organic compounds which have proved of importance for pharmacy have likewise been of great value for chemistry itself.[1]

In the domain of analytical chemistry we see the assiduous and scientifically educated pharmacist striving after similar aims with the chemist. The former ought to have a thorough knowledge and be master of the approved analytical methods which are required for the testing and examining of officinal drugs as well as of food and drink, and should also be prepared for legal cases where chemistry comes into play.[2]

Pharmaceutical chemistry is in fact connected in the most intimate manner with pure chemistry, for both have the same foundations. If we would convince ourselves of this, we have but to look through the numerous recent text-books of the former branch, to perceive that in contents and arrangement they are much the same as those of the pure science. So long ago as 1844 H. Kopp[3] expressed himself pertinently on the subject as follows: "Since the end of last century pharmaceutical chemistry has deviated more and more from the direction which it still followed during the earlier decades of the latter, when it merely borrowed from the investigations of scientific chemistry those results which had a bearing upon the preparation of medicines. It became more and more nearly allied to purely scientific chemistry; pharmaceutical

[1] Cf. *The History of Pure Chemistry*, p. 400 *et seq.*

[2] Cf. pp. 397–399.

[3] *Geschichte der Chemie*, vol. ii. p. 119.

text-books, which formerly were mere collections of empirical recipes, came to have a genuine scientific character, while the journals originally brought out for pharmacy became important miscellanies for pure chemistry."

At the close of last century and beginning of this one the relation of chemistry to pharmacy was, however, different from what it is now. Then the latter was an *Alma Mater* for the former, whereas now these positions are exactly reversed; pharmacy enjoys to-day the fruits of a highly developed chemistry. In earlier times the study of pharmacy was in truth the only road to that of pure chemistry, and this is why the most eminent chemists from the end of last century until well on in this one came from the pharmaceutical school. We have but to recall here the names of Scheele, Rouelle, Klaproth, Vauquelin, Liebig, H. Rose and many others.

The pharmaceutical institutes which began to spring into life at the close of last century were of great value for the education of chemists who wished at the same time to become pharmacists, for in these any young man who was anxious to learn received a course of systematic instruction. The Trommsdorff Institute in Erfurt, founded in 1795, deserves special mention in this connection. And good text-books of pharmacy were not wanting then either, *e.g.* Hagen's *Apothekerkunst* (" The Art of Pharmacy," 1778), Göttling's *Handbuch der Pharmazie* (1800), Hermbstädt's, Trommsdorff's, Westrumb's, and Buchholz's text-books, etc. The Pharmaceutical Society of London dates from 1841.

A historical account of how pharmacy proper has developed along with chemistry during the present century is unnecessary here, for the reasons already given.

HISTORY OF TECHNICAL CHEMISTRY DURING THE LAST HUNDRED YEARS.[1]

The immense development of large chemical industries and, in fact, of all the branches of chemical technology during the present century is the natural consequence of the great advances in chemical knowledge, and the rational application of these to technical processes. The light of scientific research has thus been shed upon the latter, and new branches of industry have been grounded upon exact investigations. The history of technical chemistry offers a continuous series of examples of this beneficial action of theory upon practice. On the other hand, numerous questions have arisen in the course of technical working which have given rise to investigations of the highest value for pure chemistry.

The great advances which have been made in chemical technology only became possible with the development of analytical chemistry, which allowed of a clear insight into the composition of the original, intermediate, and final products of technical processes. Since the beginning of this century methods of research have gradually become more perfect, methods which more and more meet the requirements of the technical chemist, and which have constituted and still constitute the most important aids to the development of chemical industry. Many of these methods have already been referred to in the history of analytical chemistry,[2] but the reader may also be reminded at this point of their use with respect to the wants of everyday life. The testing and examination of articles of food and drink are now carried on

[1] For the literature on the subject, see *Wagner's Jahresberichte* and his *Lehrbuch der Technologie* ("Annual Reports" and "Text-Book of Technology"); A. W. Hofmann's *Bericht über die Entwickelung der Chemischen Industrie* ("Report on the Development of Chemical Industries," etc., 1875-77); Karmarsch, *Geschichte der Technologie*, etc. ("History of Technology," etc.); and the text-books referred to in the succeeding pages.

[2] Cf. pp. 390, 392 and 398.

in a very large number of laboratories, the methods employed here having been elaborated from purely chemical investigations. This applies in a special degree to the analysis of water, which is of such enormous importance alike from a hygienic and an industrial point of view. We have only to think how necessary it is to establish the chemical composition of a water before employing it for any manufacture; and the various processes of purification, too, to which it has to be subjected, before it can be used for many purposes, are based upon rational chemical researches and observations. Another benefit which water analysis has conferred upon the community at large consists in its having rendered possible the artificial production of mineral waters, and thus called a flourishing industry into life; the great services rendered in respect to this by F. A. Struve (1820) deserve to be recalled here.

In the following pages mention will be chiefly made of such work as has either led to the introduction of important novelties into chemical technology or to the opening up of new branches of the latter.

It is hardly possible to estimate the benefit to the national well-being which has accrued, more especially in Germany, England, France, Switzerland and Belgium, from the growth of chemical industries. Take, for example, the coal-tar colour manufacture in Germany, which has arisen upon foundations of purely scientific work, and the alkali and sulphuric acid manufactures in Great Britain. The former illustrates in the most perfect manner the principle of the refinement of matter, a troublesome and almost worthless waste product—tar—being now worked up by chemical processes into a vast number of valuable substances. And the same applies in greater or less degree to the chief chemical industries of all the countries mentioned above; in every case men are striving to bring individual chemical processes to the highest state of perfection by utilising all the waste products. The soda industry of to-day offers a specially good instance of this, for in it we find competing processes successfully carried on, simply because they have called to their aid every means of rational chemical investigation. There is indeed hardly any branch of chemical manufacture of which the same may not more or less be said.

Reference may also be made here to the development of technical instruction,[1] which has of course contributed immensely to the advancement of chemical industries. Technical schools and colleges belong for the most part to the present century. The earliest of those on the continent of Europe were the *École Polytechnique* of France, founded in 1794, and called at first the *École Centrale des Travaux Publics*, the Vienna Polytechnic Institute (1815), and the Berlin Technical College (1821). The chemical laboratories of the above and other similar institutions in Dresden, Darmstadt, Hanover, Stuttgart, Munich, Zurich, etc. have continued to increase in importance as aids to the furtherance of chemical manufactures. As every one knows, Great Britain is by no means so well equipped with technical schools and colleges as many of its neighbours on the Continent, but public opinion is now becoming awakened on the subject, and the want is being gradually supplied.

The literature on technical chemistry has sprung from insignificant beginnings. Hermbstädt's works on *Dyeing, Bleaching, Distilling*, etc., which were published in and after the year 1820, deserve mention on account of their value at that time. During the last fifty years immense strides have been made in this respect, as is witnessed (*e.g.*) by the excellent encyclopedias of Prechtl and Karmarsch, Muspratt-Stohmann-Kerl, Bolley, Ure, Watts and Thorpe, and also by the text-books upon chemical technology, among others those of Dumas, Payen, Knapp, Wagner and Ost, in which the results of theory and practice are given together. In addition to these, the weekly and monthly journals, among which Dingler's *Polytechnisches Journal*, Wagner's *Jahresberichte* (now edited by F. Fischer) and the *Journal of the Society of Chemical Industry* may be named, supply us with information upon the results of current chemico-technical investigation.

[1] Cf. the excellent historical, critical and statistical work of Egon Zöller :—*Die Universitäten und technischen Hochschulen* (Berlin, 1891).

By such means the closest connection between chemical industry and the pure science is permanently maintained.

The Progress of Metallurgy.[1]

Although the production of iron and steel, as carried on in the phlogistic period, gave rise to chemical work through which the mutual relations of cast-iron, wrought-iron and steel were in some measure explained, there still remained a variety of problems in connection with these to be solved at a later date. The improvement of analytical methods rendered it possible to detect and estimate the various impurities in iron,—silicon, phosphorus, sulphur, arsenic, etc.,—and at the same time to recognise their influence in modifying the properties of the metal. The blast furnace process was explained by the excellent investigations of Gruner, Tunner, L. Rinman, and others, the analyses of the furnace gases by Bunsen[2] and Playfair[3] aiding in a special degree towards the elucidation of the reactions which go on in it. The determination of the composition of pig-iron—the proof that a chemical compound of iron and carbon exists—was also conducive to the establishment of a theory of the blast furnace process. The Bessemer process for the production of steel (1856) was the result of the clear perception of the connection existing between iron and steel, while the chemical investigation of the products which are formed during its various stages greatly assisted its development.

The Thomas-Gilchrist process for dephosphorising iron, introduced about the year 1878, has been a wonderful success. Light was shed upon the theory of it by various analytical researches, *e.g.* those of Finkener;[4] while, on the other hand scientific experiments by A. Frank, P. Wagner, and others have led to the utilisation of the phosphoric acid which

[1] Compare the works on metallurgy by B. Kerl, Stölzel, Balling and others.

[2] Cf. *Pogg. Ann.*, vol. xlvi. p. 193.

[3] *Brit. Assoc. Reports* for 1845, etc.

[4] Cf. *Wagner's Jahresber.* for 1883, p. 136.

accumulates in the slag produced in the process—the *Thomas* slag,—so that this latter has now become an artificial manure of the first importance, being sold in a fine state of division under the name of "basic slag." The ingenious application of the spectroscope to the examination of the Bessemer flame, whereby the end point of the reaction can be clearly distinguished,[1] and the introduction of the Martin process must also be referred to.

As another example of the utilisation of by-products, we may take the successful working up into iron of iron pyrites from which all the sulphur possible has been driven off.[2] The desire to waste no material of any value is also shown in the process of manufacturing copper from pyrites whose sulphur has been already utilised,—a process elaborated from chemical researches.

The metallurgy of nickel has developed rapidly since German silver began to be prepared upon a rational system, and especially since its employment as an ingredient of coins; the German nickel coinage dates from 1873. Nickel has, however, been long known to the Chinese, and used by them for making a variety of articles. An alloy of nickel and iron is now employed for armour-plating ships of war. A passing reference may also be made to the remarkable attempts to separate nickel from its ores in the form of the volatile compound with carbon monoxide,[3] and to regenerate the monoxide from this.

Numerous improvements have been made in respect to the production of silver, among others the Augustin and Ziervogel extraction processes, and the Pattison and Parkes processes for the desilverisation of lead; while the metallurgy of gold has also been facilitated by the introduction of good methods for separating the latter from other metals, *e.g.* by that of d'Arcet (1802), that of Plattner, and especially the now well-known cyanide process. The most important additions to the technology of platinum were made by Deville and

[1] Roscoe, *Chem. News* for 1871.

[2] Gossage, *Chem. Centr.* for 1860, p. 783.

[3] Mond, *Mon. Scient.* for 1892, p. 785; or *Nature* of July 7th, 1892.

Debray after the year 1852, in the fusion of large quantities of the metal and the introduction of methods which gave a larger yield.

The galvano-plastic process, *i.e.* the precipitation upon one metal of a thin layer of another one by means of electricity, has proved itself of great importance. The original observation in this direction was made by de la Rive in 1836, and this was followed by the publication in 1839 by Jacobi, and a little later by Spencer, of the process from which the more perfect electro-metallurgy of to-day has developed itself. The share taken by the late Werner Siemens in this development should not be forgotten.

Among the metals which have been isolated during the present century, aluminium was first made available for technical purposes by the assiduous and successful labours of H. St. Claire Deville,[1] while the Stassfurt mineral carnallite has proved itself a convenient source from which to prepare magnesium. The methods by which those metals are actually produced have grown out of the work of their discoverers.[2]

The application of electricity[3] for the extraction of metals from their compounds, *i.e. Electro-metallurgy*, has made very great progress during recent years, *e.g.* for the production of copper, zinc, gold, and especially aluminium. Sodium, which was before this used in such large quantity for the manufacture both of aluminium and magnesium, is now consequently of much less technical importance; but as sodium peroxide is coming into vogue for bleaching wool, silk and feathers, there is probably a fresh field of usefulness in store for this metal. The production of carbide of calcium, already referred to, must also be mentioned in connection with electro-metallurgical processes.

[1] *Comptes Rendus*, vols. xxxviii. xxxix. and xl.

[2] Cf. *The History of Pure Chemistry*.

[3] Compare E. Gerland's report in the *Chemiker Zeitung* for 1893, No. 30; Cl. Winkler, *ibid.* 1892, No. 22; and especially Borcher's *Elektrometallurgie*, 1891. See also a short paper by Thos. Ewan on *The Industrial Applications of Electro-Chemistry* (*Nature* for June 2nd, 1898).

Numerous improvements have also been made in the course of the century in the manufacture of alloys of every kind. Thus, from zinc and copper there have been prepared malleable brass, similor, etc., and from aluminium and copper, aluminium bronze, besides a great many alloys and amalgams of tin, including type metal; this last used to be made from antimony and lead only, but to these tin is now added.

This century has also witnessed the production of all sorts of metallic compounds, among which mineral pigments take a prominent place. The most important improvement in the manufacture of white lead was due to Thênard (1801), Scheele having before this made some fundamental observations on the subject (p. 147). Zinc white, which was made on an experimental scale by Courtois so long ago as at the end of last century, was first brought into general repute by Leclaire in 1840, after which it came to be produced on the large scale. The introduction of chrome colours, especially of chrome green and chrome red, both of which are so highly valued for enamelling, belongs to the present century. Schweinfurt green, a double compound of cupric arsenite and acetate, was discovered by Sattler in 1814; it was greatly in vogue for a long time, but is now superseded by other colours on account of its poisonous nature. The extended application of many metallic salts, formerly prepared in small quantities only, to new purposes (*e.g.* of nitrate of silver in photography, and of the yellow and red prussiates of potash in dyeing) has led to the rise of entirely new branches of manufacture. There are now but few salts of any of the more plentifully occurring metals which have not some use on the large scale; for instance, stannous and stannic chlorides and various salts of aluminium, iron and manganese in dyeing, and compounds of mercury, bismuth, antimony, zinc, etc., chiefly in pharmacy.

Development of the Great Chemical Industries.

The great chemical industries are a product of our own time, their growth having gone hand in hand with the growth of pure chemistry. The manufactures of sulphuric acid and soda, which may be looked upon as the basis of all the others, and which are naturally followed by those of hydrochloric acid, bleaching powder, chlorate of potash and other salts of potassium, nitric acid, etc., only attained to their full vigour after the various processes involved had been explained by chemical investigation, and after the most favourable conditions for those processes had been worked out. The introduction of easy methods of analysis into technical industries has also been of the utmost service to them.

Important practical improvements were made in the manufacture of sulphuric acid[1] so early as the beginning of the present century, *e.g.* the amount of steam required was regulated, and the process was made continuous (the latter by Holker). The first attempt to explain this remarkable chemical process of the formation of sulphuric acid from sulphurous acid, air, water and nitrous gas was made by Clément and Desormes,[2] who recognised the important part played by the nitric oxide. Later researches by Péligot, and more especially by Cl. Winkler,[3] R. Weber,[4] Lunge, Schertel and others, have served to elucidate the reactions which go on between the above-mentioned substances, and have therefore been of the utmost value in respect to the manufacture of the acid; they have led, for example, to an exact knowledge of disturbing conditions, which can therefore now be provided against. To Reich is due the merit of having brought the technical process under due control, by his analysis of the chamber gases; and, ever since Cl. Winkler called technical gas analysis into life, this

[1] Cf. Lunge's *Manufacture of Sulphuric Acid and Alkali.*
[2] *Ann. de Chimie*, vol. lix. p. 329.
[3] Cf. *Hofmann's Bericht*, etc., vol. i. p. 382.
[4] *Journ. pr. Chem.*, vol. lxxxv. p. 423; *Pogg. Ann.*, vol. cxxvii. p. 543.

has been a regular part of the operation. How essential for the manufacture the observations on the chemical behaviour of nitrous acid to sulphurous and sulphuric have been, is sufficiently evidenced by the introduction of the Gay-Lussac and Glover towers to which they gave rise, and which have made the process into one complete whole.

But if scientific chemistry has thus proved itself so necessary for technical, the latter has likewise done much to advance the former; for many important discoveries, *e.g.* those of selenium and thallium, have been rendered possible by its aid, and researches of high value, such as those of Lunge upon the various stages of the oxidation of nitrogen, have arisen from technical questions.

The preparation of sulphuric anhydride from sulphur dioxide and oxygen, which was formerly merely a lecture-room experiment, has been converted into a technical process through the admirable researches of Cl. Winkler,[1] and thus an important reagent has been made available for many branches of chemical industry. Sulphurous acid, whose sole technical application (practically speaking) for a long time was in the manufacture of sulphuric acid, is now condensed on the large scale and used for the bleaching of wool and silk, and as a refrigerant, and it has also recently found an extensive employment in the production of the so-called sulphite-cellulose and in the precipitation of lime from sugar juice. The utilisation of sulphurous acid for these purposes is all the more striking when we remember that in the roasting of sulphides it used often to be allowed to escape into the air, to the great detriment both of human beings and of vegetation.

The Soda Industry.—The transformation of common salt, which occurs so abundantly in nature, forms the foundation of this immense industry, whose history commences with the beginning of the present chemical period. Nicolas Leblanc[2] was the first to succeed in converting salt into soda,

[1] *Wagner's Jahresber.* for 1879 and 1884.

[2] This remarkable man, who was born at Issoudun (Indre) in 1742 (and not, as usually stated, in 1753), derived no pecuniary benefit from his

with sodic sulphate as an intermediate product, Malherbe and De la Metherie having some time previously attempted to utilise the latter substance in the same way, but without material success. It was in 1791 that Leblanc commenced the actual manufacture of soda, but political conditions and other circumstances hindered its growth for a long time, the chief difficulty being the high duty on salt. In the year 1823 Muspratt began the erection of his alkali works at Liverpool; his name deserves a foremost place in connection with the development of the soda industry. The advantages which have accrued to the manufacture of soda from chemical investigation are incalculable, but space will not allow of entering minutely into them here. The simple analytical methods which supplied the necessary information as to the composition of the raw, intermediate, and final products were and are still of the first importance for the regulation of the technical process. The formation of soda from the sulphate, by fusing the latter with coal and limestone, was ultimately so far explained by exact chemical experiments[1] (after various unsuccessful speculations on the subject by Dumas and others), as to allow of a tenable theory of this fusion process being brought forward.

Scientific researches have also given rise to numerous important improvements in the soda manufacture, *e.g.* to the beautiful process of Hargreaves and Robinson (by which sulphate of soda is prepared directly without the previous production of sulphuric acid), to the introduction of revolving soda furnaces, and to many processes for utilising and rendering harmless the unpleasant alkali waste. With respect to the last, we would refer here to the work of Guckelberger, Mond, and Schaffner and Helbig, who succeeded in making various laboratory reactions practicable on the large scale. But the greatest advance of all in this direction is

great labours. He died in the utmost poverty in 1806, his death being due to despair. A monument has recently been erected at his birthplace to his memory.

[1] Cf. Dubrunfaut in *Wagner's Jahresber.* for 1864, p. 177; Scheurer-Kestner, *ibid.* 1864, p. 173; and, especially, Kolb, *ibid.* 1866, p. 136.

the recent and exceedingly simple process of Chance,[1] by which nearly all the sulphur in alkali waste can be recovered at a very cheap rate; the result of this has been to enable the Leblanc process to compete on more equal terms with the younger ammonia-soda and electrolytic processes (see below).

Purely chemical observations have also led to what was, until quite recently, unquestionably the most important of all the innovations in the soda industry, viz. the conversion of common salt into carbonate of soda, without the intermediate formation of sulphate at all, by the ammonia-soda process.[2] Although the reaction upon which this method is based is extremely simple, it took a very long time before the most favourable conditions for it were established, and before it was made into a practical success; but this was ultimately achieved by E. Solvay. The manufacture of "ammonia soda" and of artificial manures has grown so enormously of late years that the demand for salts of ammonia has increased proportionately; but this requirement has in its turn been met by the introduction of improved apparatus for the working up of gas liquor, and by the attempts to extract the nitrogen of fuel in the form of ammonia,[3] at the same time that the heat from the fuel is itself being utilised. Here again the mutual influence of one branch of manufacture upon another is apparent, and also the benefits accruing to these from scientific investigations.

The production of "ammonia soda" has now attained to such a height that the manufacture of "Leblanc soda" has been greatly prejudiced. For many years back chemists have been striving to solve the problem—how to obtain hydrochloric acid or chlorine from the waste products of the ammonia soda process; should this be ultimately accomplished on the practical scale, then it is hardly conceivable that the Leblanc process can continue to exist. The numerous patents referring to the processes carried out on the large scale by

[1] *Journ. Chem. Ind.*, vol. vii. p. 162.

[2] For the history of this, cf. *Hofmann's Bericht*, vol. i. p. 445.

[3] Cf. Mond, *Chemiker Zeitung* for 1889, Nos. 81 and 82: or *Journ. Chem. Ind.*, vol. viii. p. 505.

Weldon and Pechiney, Solvay and others show that no efforts are being spared to overcome this difficulty.

Chemical labours have exercised a less profound influence upon the manufacture of hydrochloric acid, which is necessarily produced in such quantity in the Leblanc process, although laboratory researches have led to important improvements with regard to its condensation by water, and to its purification from admixed substances. It may be mentioned here, as a curious point in chemical history, that this acid, which is at present so cheap and which has at times been almost worthless, was in Glauber's time the most costly of the mineral acids.

The manufacture of chloride of lime, which uses up large quantities of hydrochloric acid, has also derived great benefit from chemical research, in fact it may be said to have arisen from the latter. Berthollet's experiments upon the bleaching action of chlorine and the chlorides (*i.e.* hypochlorites) of the alkalies led to the manufacture of the bleach liquor known under the name of *Eau de Javelle.* Chloride of lime was first produced by Messrs. Tennant and Co. in Glasgow in the year 1779. Weldon's beautiful process[1] for the recovery of the manganese dioxide, required in the preparation of chlorine, from the otherwise worthless chlorine waste—a process which has been in practical working since 1867—grew out of exact laboratory experiments; at the same time its development gave rise to a rich harvest of scientific results. Deacon's method of producing chlorine[2] directly from hydrochloric acid likewise originated in apparently trivial observations; a strictly scientific explanation of the action of the copper salt on the mixture of hydrochloric acid and air in this process has, however, still to be given.

The recent rapid development of electro-chemistry is strikingly shown in the electrolytic production of chlorine, caustic alkali, hypochlorites and chlorates from the chlorides of the alkalies. This method has already entered into serious

[1] *Chem. News* for September, 1870.
[2] *Journ. Chem. Soc.* for 1872, p. 725.

competition with the older processes for obtaining these substances.[1]

Bleaching powder itself has been the subject of numberless investigations, made with the object of arriving at its constitution. It may, in fact, be said that there is no other substance of equally simple composition regarding the nature of which so much doubt still prevails, notwithstanding all the efforts which have been made to clear this up.[2]

The two other halogens, bromine and iodine, also became in due course important from a technical point of view, although their much lesser abundance in nature, and consequent less extended practical application, cause them to be produced in small quantities as compared with chlorine. The manufacture of these is based upon the original work of Gay-Lussac and Balard. Laboratory experiments have also led to the production of iodine from mother liquors which were formerly looked upon as valueless, *e.g.* those from Chili saltpetre and from phosphorite after its treatment with acid. To A. Frank[3] is due the merit of having made bromine available for technical purposes, by preparing it from the Stassfurt waste salts. Large quantities of both of these halogens, especially bromine, (in combination with silver) are now employed in photography.

Nitric acid also plays an important part in chemical industries, especially since the development of the manufacture of explosives on a large scale. Potassium nitrate, which has been known and valued for so long, is still an indispensable ingredient of black gunpowder. Since the introduction of the nitrate of soda from the Chili deposits, nitric acid has been prepared from it (instead of from the more expensive nitrate of potash) by the old process of distillation with sulphuric acid, the latest step in advance here being the distillation of the nitric acid in a vacuum (Valentiner). At the same time nitrate of soda is now largely converted into the

[1] Cf. Oettel's *Entwickelung der elektrochemischen Industrie* (Stuttgart, 1896).

[2] Cf. *The History of Inorganic Chemistry*, p. 426.

[3] *Hofmann's Bericht*, etc., vol. i. p. 127.

potash salt by double decomposition with chloride of potassium. This process, so simple from a chemical point of view, could however only be carried out on an extensive scale after the rich deposits of potash salts at Stassfurt had been discovered; and it required careful chemical investigation to make those salts available,[1] for their composition had to be worked out, and proper methods for separating them from one another had to be devised. The large quantities of potassium chloride which occur in the Stassfurt mines have led in certain instances to the carrying out of the Leblanc process with it instead of with common salt, and to the consequent production of carbonate of potash, or *mineral potash*, as it was called (H. Grüneberg, 1861). The extensive use of the Stassfurt potash (and other) salts in the manufacture of artificial manures may also be referred to here; an immense new industry has thus been developed concurrently with the increased production of nitrate of soda, superphosphate of lime and salts of ammonia.

A reference to the history of gunpowder, and of explosives generally,[2] must not be omitted here, and this all the more because the discovery and use of the latter are connected in the most intimate manner with the development of the chemistry of the time. It is known that the Chinese and Saracens made use long ago of mixtures similar to gunpowder for fireworks, while in Europe it has been employed for the propulsion of projectiles since the beginning of the fourteenth century. But five hundred years passed before the chemical reactions, which go on during the combustion of powder, were in some degree understood. That its effect was due to the production of gas was stated by van Helmont; but it was only through the exact experiments of Bunsen and Schischkoff[3] upon the composition of powder gases and residues that the foundation was laid for a theory of its combustion, this being

[1] Cf. A. Frank, *Hofmann's Bericht*, etc., vol. i. p. 351; also Pfeiffer's *Kaliindustrie* ("The Potash Industry," 1887).

[2] Cf. the lecture given by Lepsius before the *Gesellschaft Deutscher Naturforscher* at Halle in 1891, entitled, *Das alte und das neue Pulver*, p. 17.

[3] *Pogg. Ann.*, vol. cii. p. 53.

further developed by the later work of Linck, Karolyi, Abel and Noble, Debus and others.

The explosives (with the exception of gunpowder), whose preparation now forms such a great industry, have all been made available for practical use by chemical investigations. The epoch-making discovery of gun-cotton by Schönbein, Böttger, and J. Otto (independently) in 1846 must be recalled here; its chemical nature and reaction upon ignition were cleared up by the laborious work of Lenk, Karolyi, Heeren, Abel and others. Nitro-glycerine had been known as a chemical preparation, discovered by Sobrero, for fifteen years before it began to find extended application in 1862, as the result of Nobel's researches. The careful investigations of Abel, E. Kopp and Champion upon its modes of formation and chemical behaviour immensely facilitated both its own manufacture and that of its various preparations,—dynamite, etc. Since 1888 an important forward step has been made here, in that nitro-glycerine and gun-cotton—up to then only applicable as explosives—were brought by the process of "gelatinising" into a condition in which they might be used with safety in guns. The "smokeless powder," which is now so much employed, but which varies widely in composition from the various methods used in its preparation, is also to be placed in the same category as the explosives just mentioned, since it contains nitro-cellulose. Systematic chemical investigation has now rendered it possible to prepare this powder with a definite ballistic value. Reference must also be made again at this point to the famous researches of Liebig and other chemists upon the fulminates, which rendered the manufacture of fulminate of mercury and its use in the preparation of fuses possible.

The whole match industry likewise owes its enormous development to the increased knowledge of chemical preparations and processes. What a contrast there is between the "chemical tinder" of 1807—*i.e.* matches containing a mixture of chlorate of potash and sulphur, which were ignited by dipping them into sulphuric acid—and our present friction matches! Those prepared with phosphorus were introduced in 1833 by Irinÿi of Pesth, Romer of Vienna and Moldenhauer of Darmstadt; they have since then undergone many improvements, the most important of these being subsequent to the discovery of amorphous (non-poisonous) phosphorus, which has been used since the year 1848, although for a long time only in small quantity, either in the match itself or in the material of the surface upon which the match is rubbed. Phosphorus, which last century was still a chemical curiosity, has been manufactured on the large scale for about fifty years. Scheele's process for its preparation was improved upon by Nicolas so far back as 1778, and has been materially modified in recent years, *e.g.* by Fleck.

Hand in hand with the development of the soda industry went the expansion of other branches of chemical manufacture, prominent among which was that of soap. In order to appreciate the influence of chemical investigation upon this, we have to recall to mind the pioneering labours of Chevreul[1] on the subject. The knowledge of the chemical nature of fats to which they led was perfected by later work, particularly by that of Heintz and of Berthelot, which finally proved that the fats were neutral glycerine ethers of various fatty acids.[2] The manufacture of stearine candles and of glycerine, which are important both as commercial and household products, may be regarded as the fruits of the labours just spoken of, in addition to which those of A. de Milly (the originator of the stearine industry), Melsens, and Frémy deserve special mention. Further valuable improvements in these manufactures have been effected by chemical investigation within the last few years (cf. Deite's book upon the manufacture of soap, the article on Soap in Thorpe's *Dictionary*, by the late Alder Wright, Schädler's book on the technology of Fats and Oils, and W. Lant Carpenter's volume on *The*

[1] M. E. Chevreul, born in 1786, lived until 1889. He occupied in his time a number of responsible posts in Paris, the last being that of Director of the Dyeing Department and Professor of Chemistry as applied to dyeing in the world-renouned Gobelins tapestry works. His classical *Récherches sur les corps gras d'origine animale* gave rise to a great amount of work of a physiologico-chemical nature upon dyes, adipocere and other substances.

[2] Cf. p. 441.

Manufacture of Soaps and Candles, etc., 1895). From a commercial point of view, the working out of methods for determining the value of any oil or fat and for detecting adulterations has been of the first importance (see Benedict's admirable work, *Analyse der Fette*, Berlin, 1897; English edition, revised and enlarged by Dr. J. Lewkowitsch, 1895).

Closely connected also with the soda industry stand the manufactures of ultramarine and of glass. The former substance, which is in a special degree a product of chemical research, was discovered in 1828 by Chr. Gmelin, and also at about the same time by Guimet. It has given rise to a large amount of scientific investigation,[1] which has led to material improvements in the manufacture of the various kinds of ultramarine, and has also explained particular parts of the firing process, but from which no final opinion has yet been formed as to the chemical nature of this curious product. The two hypotheses still oppose one another—viz. (1) that ultramarine is a definite chemical compound, and (2) that it is a mixture similar to glass. The recent work of F. Knapp[2] has, however, begun to throw some light upon the cause of the colour of ultramarine.

Although the production of glass reached a high state of development in olden times through pure empiricism, it too has greatly benefited by chemical research. The manufacture of glass with sulphate of soda and the improvements in flint and crystal glasses belong to the present century, while progress has also been made in silvering (by Liebig), and in glass painting, through the discovery of new mineral colours. The investigations of Wöhler, Knapp, Ebell, M. Müller and others resulted in elucidating the chemical reasons for the different colours of different glasses. Lastly, laboratory work has greatly advanced the art of imitating the precious stones, and, generally, of producing new varieties of glass. The chemical reactions which go on during the formation of glass

[1] The work of Leykauf, Büchner, R. Hoffmann, Knapp and Guckelberger may be referred to here.

[2] *Journ. pr. Chem.* (2), vol. xxxviii. p. 48.

have given rise to much experimental work,[1] but the conclusions drawn from this—as to whether glass is a true chemical compound or not—have been very various. Chemical analysis has of late years produced results not merely of scientific interest, but of very great practical importance with regard to the manufacture of glass.[2]

Water glass, which was known to Agricola, Glauber, etc., was made available for technical purposes by Fuchs in 1818, and has since then been used for a great number of different purposes—*e.g.* for impregnating wood, preparing cements, protecting frescoes, etc.

Earthenware and Pottery.—Important practical improvements in this old field of industry are associated with the names of Wedgwood, Littler, Sadler and others. C. Bischof,[3] Richters,[4] and, more recently, Seger[5] have rendered good service in their chemical investigations upon the nature of fireclay, and on the connection between its composition and its behaviour at high temperatures. The labours just cited have also done much to improve the manufacture of pottery, by enabling the proper mixtures of the ingredients to be made. The ceramic art is further greatly indebted to chemistry as regards glazing and the burning-in of colours.

The preparation and application of mortar, especially of hydraulic cement, have likewise been greatly advanced by purely chemical work, whereby a nearer approach has been made to the solution of the much-discussed problem,—how the hardening is to be explained from a chemical point of view. Many investigations have been made with a view of arriving at the explanation of this, the chief property of cements, among others by Winkler, Feichtinger, Michaëlis,[6]

[1] Pélouze, *Ann. Chim. Phys.* (4), vol. x. p. 184; R. Weber, *Wagner's Jahresbericht* for 1863, p. 391; Benrath, *ibid.*, 1871, p. 398; also Benrath's book, *Die Glasfabrikation* ("The Manufacture of Glass," 1875).

[2] See the investigations of Schott, Mylius, R. Weber, Förster, Gray and Dobbie, and others.

[3] *Dingl. Journ.*, vols. clix. cxciv. cxcviii. and cc.

[4] *Ibid.*, vol. cxci. p. 150.

[5] *Ibid.*, vol. ccxxv ii. p. 70.

[6] Cf. his pamphlet, *Die hydraulischen Mörtel*, etc. (Leipzig, 1869).

F. Schott,[1] Fr. Knapp[2] and Michel.[3] The old view of the hardening process, viz. that it consists entirely in the gradual formation of a calcium silicate, had to be abandoned as insufficient; but a complete theory of it still remains to be given.

The advances made in the manufacture of paper can be but partially touched upon here, the more especially since they belong chiefly to the domain of mechanics. The attempts to utilise raw vegetable products, particularly wood and straw, for the production of paper, were first successfully carried out in the year 1846. In caustic soda a reagent was found by means of which cellulose could be prepared from these materials; while of late years a solution of calcium sulphite in sulphurous acid has shown itself especially well adapted for this purpose. The above process for the production of *sulphite cellulose* resulted from the chemical investigations of Tilghman and especially Al. Mitscherlich. The conversion of cellulose into cane-sugar or alcohol is another problem which has been often attacked, and from many different sides, but it still remains to be solved. Should this ultimately be successfully carried out on the large scale, a complete revolution would be effected in agriculture and husbandry generally.[4]

The manufacture of starch and of the products obtained from it has also derived great advantage from chemical investigations. The transformation which starch undergoes upon treatment with acids has only recently been cleared up in some degree by the work of Märcker, Musculus, O'Sullivan, Payen, Brown and Heron, Salomon, Allihn, etc. The earliest observation on the production of starch-sugar was made by Kirchhoff in 1811, and from this an important branch of industry has now arisen; dextrine, which has for long been used as a substitute for natural gum, is obtained as the intermediate product here.

[1] *Dingl. Journ.*, vol. ccii. p. 434; vol. ccix. p. 130.

[2] *Ibid.*, vol. ccii. p. 513.

[3] *Journ. pr. Chem.* (2), vol. xxxiii. p. 548.

[4] For details regarding cellulose and its applications, *vide* Cross, Bevan and Beadle's excellent book on *Cellulose*, etc., (1895).

The beet-sugar industry has developed into something enormous from experiments instituted by chemists on a small scale.[1] Marggraf's discovery, in 1747, that sugar was present in the juice of beet, was not at that time capable of being applied commercially. Achard, a pupil of Marggraf, and, in a lesser degree, Hermbstädt, Lampadius and others, again took up at the end of last century the problem of obtaining sugar from beet on the large scale, and they did succeed in devising a process which was carried out in numerous factories during the years of the Napoleonic wars, when the trade of the Continent was driven in upon itself. But this process was unable to live long, being a very imperfect one, and giving but a small yield of sugar. It is from the year 1825 that the real rise of the beet-sugar industry dates, various factors entering into its growth, not the least of which was the practical application of chemical knowledge. We have but to think, for example, of the development of saccharimetric methods, whose aim was the determination—either by chemical or by physical means—of the percentage of sugar in beet juice; of the improvements in the refining process;[2] of the recovery of the crystallisable sugar in molasses, and so on. The filtration of the refined juice through bone charcoal was first recommended by Figuier in 1811, and then by Derosne in 1812, and has since become an essential part of the process. The use of vacuum pans for evaporating the syrup was introduced by Howard in 1813, since which time many improvements have been made in them. The extremely convenient diffusion process, for obtaining the juice of the beet, was discovered by Roberts (of Seelowitz, Mähren) in 1866, and soon came into general use, at first in Austria. Osmosis, which was first applied on the large scale by Dubrunfaut in 1863 for extracting the crystallisable sugar from molasses, has been developed by

[1] Cf. Stohmann's *Zuckerfabrikation* (1893); E. O. v Lippmann's *Geschichte des Zuckers*; and the article on sugar in Thorpe's *Dictionary of Applied Chemistry* by Newlands Brothers.

[2] The decomposition of saccharate of lime by carbonic acid was introduced by Barruel and Kuhlmann.

researches in physical chemistry,—another instance of the practical utility of scientific investigation.

A passing reference may be made here to the good done to this branch of industry by agricultural chemistry, in the determination of the most favourable conditions for the growth of beet, and the investigation of the composition of the soils and manures employed, etc. Indeed, there is hardly any other branch of technical chemistry so intimately connected with agriculture as the beet-sugar manufacture. The production of artificial manures has received a powerful impulse from the immense quantity of beet now under cultivation. Lastly, pure chemistry itself has benefited in many respects from the careful investigation of beet juice.[1]

The so-called *saccharine*, a compound containing sulphur which is now manufactured from the toluene of coal-tar, and which is used to a certain extent in lieu of sugar, offers an example of the assiduity with which every branch of chemical industry is being exploited with the object of imitating natural products by artificial ones, and even of replacing the former by other more active substances.

Fermentation Processes.[2]

The development of the various manufactures involving fermentation has been immensely advanced by chemical investigation, while at the same time the nature of the processes themselves has been brought into clear relief. In place of the contact theory of Berzelius and Mitscherlich, which was merely a re-statement of the facts in other words and no explanation, we now have Pasteur's vital theory of fermentation. To this also the "mechanical" theory of Liebig had to give way, while Pasteur's opinion with respect to the physiological functions of yeast became in its turn subsequently

[1] E. O. v. Lippmann, *loc. cit.*

[2] For the recent literature on the subject, see Hansen's *Praxis der Gärungsindustrie;* Jörgensen's *Mikroorganismen der Gährungsindustrie;* Märcker's *Spiritusindustrie;* and Thorpe's *Dictionary.*

modified to a material extent through the researches of others.[1]

The labours undertaken with the object of testing or establishing theoretical views have also had a determining influence upon the practical working of fermentation processes, since the knowledge thus gained has rendered it possible to subject these processes to a better control than was formerly the case. Among the more important observations in this branch during the last few years is that of Effront upon the favourable effect of a minute quantity of hydrofluoric acid on the fermentation process.

The good which has been done by the application of analysis to fermented liquors is evident at a glance, since any defects in their mode of preparation thus become apparent. A knowledge of the normal composition of wine and beer has led to rational suggestions for the improvement of those drinks. It would be out of place here to attempt even a bare enumeration of the more important innovations in this branch, many of which are due to Pasteur.[2]

Thé manufacture of spirits may be cited as one of the great branches of industry which has been helped to its present high state of development by chemical work. We have also the enormous production of alcoholic preparations[3] from spirit itself, as well as from the first and last runnings of the still; the manufacture of ordinary and of compound ethers, the latter of which are so largely used in perfumery and for making artificial *liqueurs*; and that of chloroform, iodoform and chloral, whose importance in a medicinal sense is sufficiently well known.

The knowledge that the formation of acetic acid from alcohol depended upon the oxidation of the latter, formed the basis of the *Quick Vinegar Process*,[4] the development of which

[1] Cf. *The History of Physiological Chemistry*, p. 546 *et seq.*

[2] Cf. especially the works of Pasteur, Hansen and Jörgensen.

[3] *E.g.* Ethyl iodide, bromide and nitrite; propyl and isobutyl compounds, etc.

[4] This process was first carried out by Schüzenbach in Freiburg in 1823, and then by Wagenmann in Berlin in 1824.

was the direct consequence of Döbereiner's work; while, on the other hand, the technical production of pyroligneous acid, methyl alcohol, acetone, etc., arose from the chemical investigation of the products of the distillation of wood.

The Aniline Colours and other similar Dyes.[1]

There is no industry which better illustrates the practical good that accrues from scientific chemical researches than that of coal-tar, the working up of this substance and perfecting of the numerous methods involved in so doing having set in motion and continued to permanently occupy the energies of a large army of chemists. It was clearly proved here that pure chemical work was the necessary preliminary to the development of each and every branch of the whole coal-tar industry. In no other section of technical chemistry have there been so many discoveries made by systematic investigation as in that of artificial dyes.

Out of the large number of important investigations by which the industry has been advanced, only the most striking can be mentioned here,[2]—those which have had an undoubted influence in shaping this branch of chemical manufacture. This applies to A. W. Hofmann's classical researches upon aniline and its derivatives, and upon rosaniline, the base of fuchsine (magenta), and its derivatives; and also to the notable work done by E. and O. Fischer upon para-rosaniline and

[1] Cf. especially Nietzki's *Chemie der Organischen Farbstoffe* ("The Chemistry of the Organic Colouring Matters," 1889); G. Schultz's *Chemie des Steinkohlentheers*, etc. ("The Chemistry of Coal-Tar," etc., 1886-90). R. Möhlau's *Organische Farbstoffe welche in der Textilindustrie Verwendung finden* ("Organic Dyes used in the Textile Industry," 1890); and the very valuable report on the Progress of the Colour Industry, etc., published half-yearly by H. Erdmann in the journal, *Chemische Industrie.* The utility of such a report may be gauged from the extraordinary amount of literature continually appearing in this branch of the science.

[2] For the references to special papers, see *The History of Organic Chemistry*, the works cited in note 1, and Caro's lecture on the Development of the Coal-Tar Colour Industry (*Ber.*, vol. xxv. Ref. p. 955).

rosaniline, which established the constitution of these compounds. The deep significance for technical industry which the investigations of Coupier and Rosenstiehl on the toluidines possessed is sufficiently well-known, while important results also accrued from these to the pure science. The beautiful discovery of green dyes from oil of bitter almonds and benzotrichloride by O. Fischer and Döbner (working separately) in 1877 may likewise be recalled, as also the proof that these substances were, like rosaniline and aurine, derivatives of triphenyl-methane. It must not be forgotten that Mansfield's work of fifty years ago laid the necessary foundation for the development of the aniline industry,[1] for it rendered possible the production of benzene and its homologues from coal-tar on the large scale, and also of nitro-benzene.

The first aniline dye which was produced upon a technical scale was the violet prepared by Perkin in 1856, by acting upon aniline with bichromate of potash and sulphuric acid. A. W. Hofmann observed in 1858 the formation of aniline red (magenta), which was shortly afterwards manufactured by another method by Verguin of Lyons, and introduced into commerce under the name of fuchsine. This was quickly followed by Hofmann's discovery of aniline blue, aniline violet and aniline green, which were further proved by that chemist to be derivatives of fuchsine. The discovery of methyl violet by Lauth in 1861[2] and that of aniline black by Lightfoot in 1863 were of great practical importance. While the constitution of this last compound is still enveloped in mystery, that of the other aniline dyes is now for the most part known, thanks especially to the investigations of E. and O. Fischer, mentioned above. In addition to this, new and important methods for the production of rosaniline dyes have been discovered and developed, *e.g.*, formic aldehyde and carbonyl chloride are now used for the synthesis of magenta, methyl violet, and allied compounds.

[1] *Journ. Chem. Soc.*, vol. i. p. 244, vol. viii. p. 110. Mansfield fell a victim to his work, dying of the severe burns which he received as the result of an explosion.

[2] This dye was not, however, prepared on the large scale until 1867.

Chemical research has also borne rich fruit in respect to the alizarine industry. This valuable dye was formerly prepared entirely from the madder root, but is now, practically speaking, obtained only from coal-tar, this revolution having been brought about by Graebe and Liebermann's successful synthesis (in 1869) of alizarine from anthracene, a constituent of coal-tar. In fact, the madder plantations of Alsace, the south of France and Algiers, which were in a flourishing condition twenty-five or thirty years ago, have now almost ceased to exist. In addition to this great practical triumph, the purely scientific results, which consisted in the determination of the chemical constitution of alizarine and similar compounds, must also be borne in mind.

A. Baeyer's successful conversion of phthalic acid into colouring matters (the phthaleïns) was of practical importance, since it led to Caro's discovery of the beautiful eosin dyes, while it also proved itself fruitful from a purely scientific point of view, as the elucidation of the constitution of these phthaleïns threw light upon other branches of the subject.

From the memorable researches of P. Griess upon the diazo-compounds, supplemented by those of Caro, Nietzki, Witt and others, the manufacture of azo-dyes has arisen; the modes of formation and constitution of these were so clearly made out by the above investigators that an endless series of valuable colouring matters can now be produced by certain typical reactions. The first azo-dye was brought into commerce under the name of aniline yellow so long ago as 1864, without, however, its true constitution being known. It is only since 1876 that the enormous development of this industry dates; quickly following upon one another came chrysoïdin, the tropæolines (most of which are yellow and orange dyes), the *Ponceaux* and "Fast Red" of commerce (red dyes distinguished by their purity), together with Biebrich scarlet and croceïn scarlet. The most important discovery of recent years in this direction was that of the "substantive cotton dyes," obtained from benzidine and similar compounds by Bötticher and others in 1884, as examples of which we may cite Congo red and chrysamine. The fact that there are more than 150 azo-colours in the market is sufficient evidence of the immense number of such compounds.

The chemical investigation of methylene blue and the safranines, new dyes of great value, has been of much importance both practically and theoretically, the rational composition of the former having been arrived at by Bernthsen, and that of the latter by Nietzki and Witt. The great aim of so many of the researches upon the organic colouring matters, viz., the elucidation of their relations to other compounds from which they are readily derivable, has in the above cases been attained; methylene blue is derived from thio-diphenylamine, and the safranines from phenazine. Similarly rosaniline, aurine and numerous allied substances have been proved to be derivatives of triphenyl-methane; the azo-dyes to be derivatives of azo-benzene and azo-naphthalene; and alizarine, purpurine, etc., to be derivatives of anthraquinone. It is also now known that the indophenols and indamines, the eurhodols and eurhodines, the rhodamines, etc., are derivatives of definite chemical compounds not in themselves dyes, but which become so by the entrance of certain atomic groups into the molecule. Various attempts have lately been made by Witt, Nietzki and Armstrong among others, to discover definite relations between the chemical constitution of dyes and their colouring properties, but these speculations have as yet no claim to be looked upon as constituting a theory; they are more or less only a re-statement of facts.

The chemical investigation of indigo blue, the most valuable of all blue dyes, has also been ardently prosecuted, with a view of arriving at its constitution, which, however, is not yet definitely settled. Most of our knowledge on the subject is due to v. Baeyer. He succeeded several years ago in preparing indigo artificially from simpler compounds contained in coal-tar, but until quite recently no one was able to convert any one of the known syntheses into a practical commercial process. The new Heumann-Lederer method, however, by which indigo blue is prepared from phenyl-amido-acetic acid, is carried out in a modified form by the well-known Badische Aniline and Soda Manufactory, so that "artificial" indigo is now in the market.

Dyeing and Tanning.

The processes by which colours are fixed upon vegetable or animal fibres have been greatly improved since the chemical nature of dyes came to be known, although there are some cases in which a true explanation is still required of the mode in which the fibres themselves and certain mordants act. The earliest attempt, even if it was an imperfect one, to get clear ideas upon this subject was made by Macquer in 1795. The empiricism which prevailed for so long in the dyeing industry has gradually been done away with, thanks to the efforts of chemists to obtain a truer insight into the reactions which dyeing involves. Attention must be called here to the investigations of Knecht on the subject, according to which the fixation of dyes by wool fibre is dependent on the chemical nature of the latter.

With respect to the application of the more important dyes, previous to the discovery of the coal-tar colours, it may be mentioned that indigo was used in Europe from the first half of last century, and madder red from the second half, while picric acid came into vogue at the beginning of the present one. The use of extract of campechy-wood (which is still very considerable) dates from about the year 1840, and that of the dye from the yellow berries of the Chinese plant, *Sophora japonica*, from about 1848. Reference must also be made to the improvements in the application of metallic colours in dyeing, *e.g.*, Prussian blue, chrome yellow, chrome orange, etc.

Tanning, whose processes up to 1860 or so were almost purely empirical, has been made susceptible of scientific treatment through the investigations of Knapp, Eittner, Böttinger and others. This subject ought to have a great interest for chemists, seeing that, according to Knapp, it constitutes a special case of dyeing, many analogies being apparent between the two. The researches on the various tannic acids have been of value from a theoretical point of view, both to chemistry and to vegetable physiology. Among the important practical innovations, for which this branch of manufacture has to thank chemistry, the mineral tanning introduced by Knapp, Heinzerling and others deserves notice. But there is as yet no general theory of the various tanning processes.

Various Chemical Preparations.

An immense industry—that of so-called chemical preparations—has gradually been developed on scientific lines from apparently insignificant beginnings, which had their origin in the work of the apothecary; such "preparations" belong partly to inorganic, and partly to organic chemistry.[1] As instances of this we may take the great increase in the production of silver salts, bromine and iodine for photographic and other purposes, and the manufacture of numberless other metallic salts, not to speak of newly introduced compounds like the peroxides of hydrogen and sodium. The already imposing list of inorganic preparations is being continually added to. Here again it is scientific investigation which has led to the use of such substances in manufactures generally.

The manufacture of organic preparations is still more extensive. What a multiplicity of compounds, for instance, is comprised under the term *alcoholic preparations!* The various alcohols themselves, their ethers and esters, chloroform, chloral, iodoform, aldehyde, etc., are now all essential to chemical manufactures and to medicine; while among the phenols, carbolic acid and the cresols, hydroquinone, pyrocatechin, resorcin, pyrogallol and others, together with many of their derivatives, have found extended application in medicine, photography, and for disinfecting and other purposes.

[1] For the scientific importance of this branch of industry, and indeed of technical chemistry generally, compare H. Wichelhaus's *Wissenschaftliche Bedeutung chemischer Arbeiten* (1893).

The manufacture of organic acids, of which acetic acid has already been referred to, also shows a continuous development; many of these compounds which occur in nature are now prepared artificially on a large scale, the methods followed being based upon scientific investigation. Thus, salicylic acid is produced from carbolic (Kolbe and R. Schmitt), benzoic acid from toluene, phthalic acid from naphthalene, cinnamic acid from benzoic aldehyde, and oxalic acid from wood by treating the latter with alkali, etc. This last process was discovered by Gay-Lussac in 1829, and its practical application now constitutes an important industry.

The manufacture of aldehydes is likewise the result of scientific work. Some of these, such as benzoic aldehyde, are of importance for the colour industry, while others, like vanillin and piperonal, are flavouring essences or scents. The so-called *ethereal oils* belong to this last category; and much has been done by the researches of Wallach, Semmler, Tiemann, Bertram and others to place their manufacture on a sound basis. Systematic investigation has also brought about great advances with regard to the vegetable alkaloids, as is evident in the strenuous endeavours to prepare artificial compounds which shall have the same physiological actions as quinine, morphine, atropine, etc.

Various other Products from Coal-tar; Illuminants.

Coal-tar is the raw material from which many other organic preparations are obtained,—it is, in fact, a rich mine for numberless useful substances. Formerly a troublesome waste material, it is now of at least equal value with the other products from the distillation of coal. The manufacture of ammonia and salts of ammonia from gas liquor is now a thoroughly rational one, thanks to the careful chemical examination of the latter, and it forms a large and important branch of industry. In consequence of the rapidly increasing consumption of ammonia salts, more and more attention is being paid to the problem of utilising the ammonia which escapes into the air when coal is either converted into coke [1] or is completely burnt. L. Mond [2] has lately set up an ingeniously constructed apparatus on a large scale at Northwich in Cheshire, which serves not merely for heating purposes, but at the same time allows of the condensation of the ammonia produced. The manufacture of coal-gas has developed in a more empirical manner, and has thus been less influenced by recent chemical researches bearing upon the subject than many other branches of industry; but here, too, much good has been done by the application of the methods of gas analysis, and chemical experiments have also borne fruit in the introduction of improved modes of purifying gas. It is interesting to note here that acetylene, which is itself an electro-technical product, appears destined to enter into serious competition as an illuminant with the electric light.

The great influence which chemical investigations have exercised upon the production of other illuminants has already been touched upon, and is shown in the manufacture of stearic acid from animal fats.[3] Attempts have been made upon the large scale to employ (the liquid) oleic acid, which occurs so plentifully in nature, for the production of candles, by making use of the well-known reaction with caustic potash, which converts it into (solid) palmitic acid. The flourishing paraffin manufacture [4] of Scotland and Germany, with its various valuable by-products, also owes much to chemistry. But the latter has still many problems to solve both in this field and in that of the petroleum industry, as is evident from the recent work of Markownikoff, Beilstein and Engler upon the chemical nature of petroleum; these researches are bringing us nearer to a solution of the origin of this useful substance. A short reference has already been made (p. 425) to the theoretical points which bear upon illumination, and to the causes of the luminosity or non-luminosity of different flames.[5]

[1] Tar is now an important by-product in coking.

[2] *Journ. Chem. Ind.* for 1889, p. 505.

[3] Cf. p. 569.

[4] Paraffin, which was discovered in wood-tar by Reichenbach in the year 1830, is obtained practically from lignite or bituminous shale.

[5] Cf. also Hans Bunte on "Recent Developments in Gas-Lighting," *Ber.*, vol. xxxi. p. 5.

Heating Materials.

That the knowledge gained through chemical analysis of the composition of different kinds of fuel, of their products of combustion, and of their chemical behaviour generally, is of the first consequence, requires no demonstration. It is of course impossible to refer here to the large number of important investigations in this field, but reference must be made to the fundamental work of E. Richters and F. Muck;[1] to the improvements in the methods of analysis of furnace gases,[2] which permit of conclusions being drawn with regard to the course of any particular combustion; and to the improvements in heating apparatus which have been brought about by chemical work,—the construction of generators and regenerators, whose history is inseparably connected with the names of Aubertot, Thomas, Laurens and, above all others, Siemens. "Water gas," prepared by passing steam over red-hot coal, is now much used for heating and illuminating purposes, and it will undoubtedly become even more employed in time; this is notably a case in which a great manufacture has arisen from experiments on a very small scale.

Speculations regarding the origin of coal deposits, and the metamorphoses which these undergo, have received much support from the work which has been done upon the composition of coal and of the gases which are found enclosed in it. And it is mainly to chemical research that we owe the means of averting or at least diminishing the great dangers to which coal miners are exposed from explosions of fire-damp, —witness the Davy safety lamp. The subject is still being assiduously worked at from time to time both by chemists and by practical engineers. The zeal which the various recent "Fire-Damp Commissions" of different countries showed in their investigations is still fresh in the public memory.

[1] Cf. Muck, *Grundzüge und Ziele der Steinkohlenchemie* ("The Outlines and Aims of the Chemistry of Coal," 2nd edition, 1891).

[2] Cf. Winkler's *Anleitung zur technischen Gasanalyse* ("Methods of Technical Gas Analysis"); also p. 392 of this book.

The above short sketch is sufficient to indicate how enormous have been the benefits which laboratory research has conferred upon every branch of technical chemistry, and how the latter has been raised to a higher level by a continuous infusion of the scientific spirit. Nowhere can we find a better illustration of Bacon's maxim : *Scientia est potentia.*

THE GROWTH OF CHEMICAL INSTRUCTION IN THE NINETEENTH CENTURY, MORE ESPECIALLY IN GERMANY.[1]

At the beginning of this century there was a marked want of those facilities which, during the last few decades, have been at the command of any one desirous of devoting himself to the study of chemistry. At that time there were practically no laboratories for general instruction. In lectures upon physics, mineralogy and anatomy, chemistry was relegated to a very subordinate place. It is true that there were chairs of chemistry in various universities and colleges, but the lectures on this subject were usually conjoined with those upon one of the others, just named, in such a manner that chemistry was forced into the background. Chemical literature, lastly, was still poor in works which either gave a review of the state of the science at the time, or furnished regular reports of the latest discoveries in it.

In France, where towards the end of the eighteenth century it began to be perceived that instruction in natural science must be fostered by every means at command, a start was made far before any other countries in respect to the development of chemical study. Up till then apothecaries' shops were the only places where work in practical chemistry could be carried on, and there merely after certain prescriptions and not according to scientific methods. Vauquelin was the first to organise a course of instruction in his small laboratory for students anxious to learn, while after the first decade of the century Gay-Lussac and Thénard also taught in their laboratories, which however were exceedingly cramped. Fourcroy had already done an immense deal to raise the standard of scientific instruction, and he contributed greatly by his bril-

[1] In addition to the books referred to in the succeeding pages, compare E. Zöller's book, mentioned on p. 556, and also Wallach's essay in Lexis' *Die deutschen Universitäten*, vol. ii. p. 35 (1893).

liant lectures[1] to ensure to chemistry a worthy position as a course of study. But it was only after Liebig had taken up the subject with his accustomed energy, that chemistry came to be taught in the higher schools in essentially the same manner as that to which we are now accustomed.[2]

The importance of lectures on chemistry, illustrated by experiments, for the proper understanding of chemical reactions, was recognised a long time ago, more especially in France.[3] But during the early decades of the present century this aid to study hardly existed in the higher teaching institutions of Germany, and the so-called natural philosophy of that day was such that it sorely handicapped the development of exact scientific research. Chemistry, in particular, was looked upon by the natural philosophers as being no science at all, and was degraded by them into a mere experimental art.

The efforts made by Davy, however, backed as these were by an exceptional talent for devising and carrying out experiments, and also by Gay-Lussac and Thénard's admirable lectures, resulted from the beginning of this century in an increasing demand for lectures with appropriate experimental illustrations. Liebig has left to us a graphic description of the effect which Gay-Lussac and Thénard's discourses had upon himself, at that time a youth of eighteen. From this account it is evident that these lectures gained an indescribable charm from the "mathematical method, which transformed each problem—wherever possible—into an equation," and by a lucidity of expression which was "conjoined with a wonderful experimental skill."

We know that it was the lectures given by Marcet in London which induced Berzelius in 1812 to abandon the old

[1] Compare Pariset's vivacious *Éloge de Fourcroy*, cited in Höfer's *Histoire*, vol. ii. p. 557.

[2] Cf. below; also O. L. Erdmann's valuable and too little known pamphlet, *Ueber das Studium der Chemie*. Liebig willingly acknowledged the great debt which he owed to Gay-Lussac, with whom he had worked as a student (cf. *Ber.*, vol. xxiii. Ref. p. 824).

[3] Cf. The work of Rouelle, p. 118, note.

method of instruction and to make use of experiments in introducing students to chemical science; and the result of this was conclusive. The subsequent good achieved by Faraday, Liebig, Wöhler, Bunsen, Wurtz, Kolbe and especially A. W. Hofmann, through the new lecture experiments which they devised, requires but to be mentioned. Those experiments and many others have since taken a permanent place in the teaching of chemistry.

Practical instruction in chemical laboratories, as commonly carried out at the present day, was developed by Liebig. The gradual introduction into laboratories, through his example, of teaching methods based upon a strictly scientific foundation, created a wholesome reaction against the still prevailing tendency of the natural philosophy of the day, which was combated by Liebig all the more energetically from his having himself suffered under its pernicious influence.[1] He first emphasised with all the force at his command that the true centre-point of chemical study lay not in lectures but in practical work. With what energy and under what sacrifices he gave personal proof of this is well known.[2] True, Berzelius had already given instruction in his laboratory to a limited number of pupils, mostly elder ones, who in their turn propagated their master's doctrines, but the real development of chemical teaching is due to Liebig. He it was who laid down the order, now classical, in which the various branches of the subject should succeed one another, viz. (1) the systematic study of qualitative and then of quantitative analysis,[3] (2) exercises in the making of preparations, and (3) attempts at independent research.

Liebig's laboratory was the centre from which, after about the end of the twenties, the brightest light radiated. He was the first to enunciate and apply the principle that

[1] Cf. pp. 263–264.

[2] Cf. the Memoir of him by Kolbe in the *Journ. pr. Chem.* (2), vol. viii. p. 435; also Weihrich's essay (already quoted), p. 264, note 2.

[3] The co-operation here of R. Fresenius, who was at one time assistant to Liebig, and the stimulus given by him towards the creation of a systematic course of analytical work, will remain in lasting remembrance (cf. p. 389); the great service rendered by Will must also be emphasised.

his pupils, be they students of pharmacy, technical chemistry, mineralogy or physiology, should learn to treat chemical questions practically. Thanks to the wonderful stimulus which he was able to exert, there was founded in his modest laboratory a school which left its stamp upon the chemistry of the succeeding decades, and whose beneficial influence is still felt all over the world at the present day. The peculiarity of Liebig as a great teacher consisted, according to Kolbe,[1] in his "being able to stimulate his pupils to original thought, and to inoculate them with the scientific spirit while they were working out his own ideas."

The most eminent among the teachers of chemistry on the Continent since the time of Liebig, of whom Wöhler, Bunsen, Erdmann, Kolbe, Kekulé, Wurtz and A. W. Hofmann may be named here, made the essential principles of his method of teaching their own, while each added of course much that was new, with the most beneficial results. The principles on which chemistry is taught are the same both in the German Universities and the Technical High Schools.

Numerous teaching laboratories were in due course founded in the other German universities and colleges on the model of the Giessen one, and about these a few notes may fitly find a place here. How badly off Austria and Prussia were in this respect, even so recently as the year 1840, was vividly depicted by Liebig in his two pamphlets entitled *Ueber den Zustand der Chemie in Österreich,*[2]—*und in Preussen* [3] ("On the State of Chemistry in Austria, and in Prussia"). Even in Berlin there were up to that time no facilities for the study of practical chemistry. H. Rose and Mitscherlich were hardly in a position to give regular laboratory instruction, the space and means generally at their disposal being very insufficient;[4] and the same thing applied to the other "high schools" of Prussia.

[1] In his work, *Das chemische Laboratorium der Universität Marburg,* etc., p. 26. In this the principles of Liebig's method of instruction are described with exceptional clearness.

[2] *Ann. Chem.*, vol. xxv. p. 339. [3] *Ibid.*, vol. xxxiv. pp. 97 and 355.

[4] Cf. A. W. v. Hofmann's *Chemische Erinnerungen aus der Berliner Vergangenheit* (1882)

In the meantime laboratories began to be established elsewhere in Germany, *e.g.* at Göttingen, where Wöhler set up one in the course of the thirties, to be rebuilt and enlarged in 1888; and at Marburg, where Bunsen began a regular practical course in 1840. The chemical laboratory which Erdmann[1] instituted at Leipzig in 1843 remained for a long time the pattern of what a well-organised place of the kind should be. It was only in the course of the fifties that Heidelberg, Karlsruhe, Breslau, Greifswald and Königsberg followed suit with laboratories properly equipped for the purposes in view.

A new era in the history of chemical institutions began about the middle of the sixties, the famous laboratories at Bonn and Berlin,[2] both built according to A. W. Hofmann's plans, being completed in 1867, while the equally well-known Leipzig laboratory, designed by Kolbe, was finished in 1868. The experience gained, both during the erection of these and by their subsequent use, has been applied with good results in the planning of later and even in some respects finer institutes. Of the other new German laboratories, those of Aachen[3] (1870), Dresden (1875), Munich (1877), the Berlin Technical College (1879), Kiel 1880), Strassburg (1885), Göttingen (1888), Heidelberg (1892) and Halle (1894), may be specially named. In Austria, too, various excellent laboratories, have been built during the last two decades, among which those of Graz and Vienna stand out prominent.

[1] Otto Linné Erdmann was born at Dresden in 1804, and died in 1867 while holding the post of Professor of Chemistry at Leipzig, where, since 1827, and especially after the organisation of the laboratory which he had himself founded, he laboured with wonderful energy and with great success. His rich experiences, and the views to which they gave rise, were set forth in the weighty, if short, pamphlet entitled, *Ueber das Studium der Chemie* (1861). That he was also active in a literary sense, his *Lehrbuch der Chemie* and *Grundriss der Waarenkunde* ("Outlines of a Knowledge of Technical Products"), etc., prove. In 1828 he started the *Journal für technische und ökonomische Chemie*, which developed in 1834 into the *Journal für praktische Chemie.* His numerous experimental researches have helped to enrich mineral chemistry, the chemistry of the carbon compounds, and also chemical technology.

[2] Up to that date Berlin was without any large laboratory for general instruction.

[3] *i.e.* Aix la Chapelle.

Among the many German University teachers who have exercised a marked influence during the last forty years, and who have not been already mentioned, von Baeyer takes a foremost place, while in addition there are Claus, Erlenmeyer, E. Fischer, Fittig, Ladenburg, Lothar Meyer, Victor Meyer, Strecker and Wislicenus.

The other countries of Europe have not kept pace with Germany in the establishment of institutes for the teaching of chemistry. There were, it is true, laboratories in France at the beginning of the century in which such men as Gay-Lussac, Thénard, Dulong, Chevreul and others carried out their work, but the opportunities for general chemical instruction were extremely few, the above institutes receiving but trifling support from the State. And the fees which a laboratory student had to pay were exorbitant, being 1500 francs for an eight-months' course. Even the efforts made to establish teaching laboratories during the thirties by Dumas and Pélouze, and later on by Wurtz, Gerhardt and others, were followed with but scant success, because these chemists were thrown entirely on their own resources.

Those conditions were only improved after Wurtz in 1869 presented his report [1] upon the German laboratories to the French Minister of Education, in which he insisted upon the necessity for establishing properly equipped laboratories for practical instruction in chemistry. He stated that at that date there was in France only one chemical institute with the necessary means at command —that of the *École Normale Supérieure*, under the direction of H. St. Claire Deville. E. Frémy, well known by his work in inorganic and technical chemistry, had set up a laboratory in 1864; and in the introduction to his Encyclopedia of Chemistry he gives a detailed account of the principles upon which chemistry was taught in it. Frémy died in Paris in 1894, at the age of eighty.

In Great Britain, too, it is only within the last twenty-five or thirty years that the lack of roomy and well-equipped

[1] *Les hautes Études pratiques dans les Universités allemandes* (1870).

laboratories has been remedied; and to this, especially of late years, the recognition of the fact that the industries of the country would be enormously benefited thereby has greatly contributed. The first laboratory in Britain, small though it was, in which a young man had the opportunity of working practically at the subject, was that of Thomas Thomson[1] in Glasgow, established in 1817. This was therefore the first chemical laboratory for general instruction. After the founding of the College of Chemistry[2] in London in 1845 (which quickly rose into a flourishing condition under the leadership of A. W. Hofmann), the country became by degrees well supplied with suitably equipped laboratories, in which instruction substantially upon the lines of the German school was given. In addition to the Universities and a few of the older institutions for higher education in London, etc., each of the University Colleges now scattered over the country possesses its own chemical laboratory, and the same thing applies in greater or less degree to the colleges and schools for technical instruction which continue to be founded with considerable rapidity. In fact, the mind of the country is now becoming to some extent awakened to the importance of the subject. Among the chemical laboratories, more or less recently erected, those at Manchester, Leeds, Edinburgh, the City and Guilds Institute (South Kensington, London), and Cambridge may be specially named.

In Switzerland, Holland, Belgium, Italy, Russia, Scandinavia and America are now to be found numerous chemical teaching institutes, arranged and fitted up in accordance with the requirements of the age.

The increasing necessity for specialisation in chemistry,

[1] Cf. p. 194.

[2] The College of Chemistry was taken over by Government in 1853, and was made a part of the Royal School of Mines, while at the same time retaining a *quasi*-separate existence under its own name. In 1872 it was moved from its old premises in Oxford Street to South Kensington. The name College of Chemistry was finally merged into that of the Normal School of Science and Royal School of Mines in 1881. In 1890 the N. S. S. and R. S. M. were rechristened the Royal College of Science.

and the consequent resulting division of labour, has made itself evident in the establishment of laboratories for certain definite purposes only. Thus we now find institutions existing solely for researches in chemical physics, agricultural chemistry, technological chemistry, physiological chemistry, pharmaceutical chemistry and hygiene. What a contrast between the present facilities for chemical study and the opportunities of only a few decades back!

Among the more important improvements which have been aimed at and achieved in the construction of laboratories during these last decades, are those which have reference to arrangements for supplying plentiful ventilation and good light. Then the means for carrying out chemical operations have also been both greatly increased and improved, *e.g.* coal and charcoal fires have been superseded by gas, the *Bunsen* burner having played an important part here. The apparatus, too, employed by chemists has undergone many refinements, as is readily seen in the delicate balances and the appliances for filtering, distilling, heating under ordinary and increased pressure, etc., which are now in common use.[1] The making of preparations is at present an easy matter compared with what it used to be, this being in part due to better methods of procedure; by far the greater number of these substances can now in fact be bought pure. Chemists are thus freed from the difficulty which

[1] The following points may be referred to with advantage here:—Water suction pumps were introduced by Bunsen in 1868, and injector pumps a little later by Arzberger, Zulkowsky, etc., to be used for filtering and producing a vacuum. Simple distillation was immensely facilitated by the introduction of the Liebig condenser, while a reflux condenser appears to have been first made use of by Kolbe and Frankland in 1847. Dittmar and Anschütz (independently of one another) were the first to distil under diminished pressure. The water-bath, for which Berzelius devised a convenient form, has since been improved by arrangements, elaborated by Fresenius, Bunsen, Kekulé and others, for keeping the water in it at a constant level. The use of gas regulators for the maintenance of a uniform temperature may also be mentioned, and this again in conjunction with Bunsen's name. Caoutchouc tubing appears to have been first brought into general employment by Berzelius. And the first mention of the use of sealed tubes for carrying out chemical reactions under pressure is to be found in Wöhler and Liebig's research on uric acid derivatives.

was ever present with them sixty or seventy years ago,—of having laboriously to prepare even their most simple reagents. Berzelius had to make his own yellow prussiate of potash, the pure mineral acids, spirits of wine for burning, etc. And how simple were the arrangements generally in his laboratory![1] Many of the aids to practical work which are now accepted as a matter of course had in his day no existence.

Chemical Literature.

The manuals and text-books of chemistry and also the journals have increased to a very large extent of late years, thus greatly facilitating the study of the science. For a long time Lavoisier's *Traité de Chimie* remained the pattern of what such a book should be, and upon it numerous others were modelled, *e.g.* those of Girtanner, Gren, and Thomson. Berzelius' large book on chemistry exercised an extraordinary influence, especially after it had been translated into other languages, and contributed in an exceptional degree to the spread of chemical knowledge.

This great work, great both in its conception and in the manner in which it was carried out, was afterwards taken in many cases as the standard for the arrangement of chemical matter in text-books which appeared later. Of these a few may be mentioned here:—Thénard's *Traité de Chimie Élémentaire*; Mitscherlich's *Lehrbuch der Chemie*; Liebig's *Organische Chemie*; Wöhler's *Grundriss der Chemie* ("Outlines of Chemistry"), from which sprang the well-known and widely-read work of the same title by Fittig; Regnault's *Cours Élémentaire de Chimie*, which formed the basis of Strecker's *Kurzes Lehrbuch der Chemie*; Graham's *Elements of Chemistry*, from which arose Otto's large work, the organic portion of which was written by Kolbe, while H. Kopp wrote the general theoretical part (inorganic and organic), and Buff and Zamminer the physico-chemical. Gerhardt's

[1] Cf. Wöhler's description, *Ber.*, vol. xv. p. 3139.

Traité de Chimie Organique (1853 to 1856), known as the text-book of the type theory, greatly contributed to the propagation of the latter, while Kekulé's book, which began to appear shortly after the last volume of Gerhardt's *Traité* had been published, served to develop the "typical" view, and (in its second volume) strengthened his own assumption as to the mode in which atoms are combined with one another, *i.e.* the structure theory. It is unnecessary to mention here even a few of the numerous text-books of chemistry which have been written since then, for, belonging as they do to the present era, they are already sufficiently well known. A palpable want has recently been supplied by the publication of W. Ostwald's, Nernst's and Horstmann's admirable text-books of general theoretical and physical chemistry, while Lothar Meyer's *Moderne Theorien* has greatly helped to extend the interest felt in questions of theoretical chemistry. Some of the best known text-books on technical and physiological chemistry have been already referred to.

There has likewise been no lack of chemical encyclopædias since the great success of Liebig, in conjunction with Wöhler and Poggendorff, in the *Handwörterbuch der reinen und angewandten Chemie*, which began to appear in 1837. Wurtz's *Dictionnaire de Chimie pure et appliquée*, Watts' *Dictionary of Chemistry*, and Ladenburg's *Handwörterbuch der Chemie* have been written upon a similar plan. The publication of Frémy's *Encyclopédie de Chimie* must also be recalled.

Among the larger treatises of chemistry, which are intermediate between the text-books proper and the dictionaries, that of L. Gmelin justly excited the admiration of his contemporaries by its consistent thoroughness. In Beilstein's *Handbuch der Organischen Chemie*, already in its third edition, the present huge mass of material on the subject has been sifted and arranged in a masterly manner. Dammer's *Handbuch* aims at doing for inorganic chemistry what Beilstein's does for organic.

The periodical journals, whose number has gone on

steadily increasing, have exercised the greatest influence upon the enlargement and spread of chemical knowledge, more especially since the beginning of this century. A short account has already been given[1] of the condition of this class of literature towards the end of last century. In Germany, after the third decade of the present one, all the more important chemical researches were for long published either in Poggendorff's *Annalen der Physik und Chemie* or in the *Annalen der Chemie und Pharmazie*,[2] which was at first edited by Liebig alone, but afterwards in conjunction with Wöhler. The latter journal, more particularly, soon became the medium in which were discussed the experimental and speculative chemical questions of the day. And no one was better qualified to deal with those exhaustively than Liebig himself.

In France the *Annales de Chimie*, founded in 1789 (the year of the Revolution) by Lavoisier, Fourcroy and Berthollet, has always been appreciated and loyally supported. Since 1816 it has appeared as the *Annales de Chimie et de Physique*, its first editors under this new title having been Gay-Lussac and Arago, and it has all along contained the records of pretty nearly all the more important French chemical researches. The *Comptes Rendus*, which has been published weekly by the *Académie Française* since the year 1835, includes among its numerous papers only comparatively few and short accounts of chemical investigations.

In Great Britain, up to the year 1841, papers on chemical subjects were published either in the *Philosophical Transactions*, the *Transactions* of the Royal Society of Edinburgh, etc., or in other more recent journals which have since been superseded, such as Nicholson's *Philosophical Journal*, and Thomson's (later Phillips') *Annals of Philosophy*. Since 1841, or at least since 1848, the *Journal of the Chemical Society* has been the main organ of scientific chemistry in this country. Apart from the original memoirs which it contains, this

[1] Cf. pp. 175 and 179.

[2] Until the year 1839 this journal bore the simpler title, *Annalen der Pharmazie*.

journal has since 1871 greatly extended its usefulness by giving copious abstracts of papers which have appeared in the chemical journals of other countries.

And the other European countries have not been behindhand in the publication of chemical journals; according to the degree in which chemistry has found in them a permanent home, so have journals of every shade and variety sprung up. Most of these were and are still connected with learned corporations—academies and chemical societies—in Austria, Italy, Holland, Belgium, Switzerland, Russia, Roumania and Scandinavia, and the same remark applies to the United States and Canada.

In Germany more particularly, which has now for long been the chief centre for scientific chemical interests, thanks to the favourable conditions for scientific instruction there, a number of new journals for the publication of papers on purely chemical subjects have been added to those older ones just mentioned. Among these are the *Journal für Praktische Chemie*, begun by Erdmann in 1834, continued by Kolbe from 1870 to 1885, and since the latter date edited by E. von Meyer; and, especially, the *Berichte der Deutschen Chemischen Gesellschaft*, which was brought into life with the founding of the German Chemical Society at Berlin in 1868, and in which one finds a record of pretty nearly all that is being done in scientific chemistry, either in the form of original papers or of abstracts [1] from other journals. Mention must also be made here of the *Kritische Zeitschrift*, known later on as the *Zeitschrift für Chemie*, which was supported by such men as Kekulé, Erlenmeyer, Fittig and others, and the critical utterances in which have often helped to throw light upon disputed points in chemistry. The *Chemische Centralblatt* is also a valuable journal of reference for every branch of the science.

Mention still remains to be made of the *Jahresberichte* (" Yearly Reports ") on the progress of chemistry and allied branches of science. The reports which were edited by Berzelius (from 1821 to 1847) are unique, and are abso-

[1] These abstracts are no longer printed in the *Berichte*, but in the *Chemische Centralblatt*.

lutely indispensable to any one who desires to make a detailed study of the progress of chemistry during those years. The continuation of them, which was undertaken by Liebig in conjunction with other chemists, cannot be compared with these earlier volumes, the new *Jahresberichte* having been restricted into mere epitomes of reference with regard to current chemical work. The *Jahrbuch der Chemie*, begun in 1891, and edited by R. Meyer in conjunction with various collaborators, aims at giving a concise statement of the more important advances in pure and applied chemistry.

The critic, whose use as a fermentive and corrective agent will be denied by no one, seems, with but few exceptions, either to have disappeared from the chemical literature of recent years, or at all events to be at present dormant. It is well to remember that the critical acumen which was brought to bear upon the occasional errors of chemical investigation by Berzelius and Liebig, and at a later date by Kolbe, had a consolidating and not a disintegrating effect, even in those cases where the critic's argument had a strongly polemical, and—to the subject of the attack—a personal flavour.

The value of a minute study of good original papers has time and again been insisted upon by the great teachers of chemistry. The records of such experimental labours offer to the student the best means of following out the author's train of thought; they thus strengthen the historical sense, and at the same time strongly incite to criticism and to emulation. They are therefore to be looked upon as among the best literary aids to the study of chemistry. At the same time they possess a high educational value from their style and form alone. As Erdmann well says in his short treatise, already cited, p. 60: "By making use of such sources of information the student learns at one and the same time from a master of the science how and in what form scientific results should be stated, how to distinguish between what is and what is not essential, and how to condense the subject-matter, while at the same time omitting from it nothing of importance, so that no necessary element shall be wanting for its critical examination."

INDEX OF AUTHORS AND SUBJECTS

INDEX OF AUTHORS' NAMES

The figures in thick type refer for the most part to those pages upon which biographical notices occur, although they are also employed in some cases for others on which points of special importance are recorded.

ABEL, 568
Abukases, 30
Achard, 573
Afzelius, 192
Agatharchides, 12
Agricola, 3, 47, 59, **83**, **84**, 88, 93, 95
Aitken, A. P., 535
Albertus Magnus, 30, **31**, 34, 56, 137
Algarotus, 94
Allihn, 572
Ampère, 257, 418
Ammermüller, 495
Anaximenes, 6
Anderson, 478, 484
Andrews, 405, 493
Anschütz, 439, 593
Arago, 499, 596
d'Arcet, 558
Archimedes, 12
Arfvedson, 407
Aristotle, 2, 5, **6** *et seq.*, 15, 20, 38
Armstrong, 452, 579
Arnaldus Villanovanus, **32**, 43
Aronheim, 458
Arppe, 442
Arrhenius, 501, 503, 520
Arzberger, 593
Aubertot, 584
Augustin, 558,
Auwers, 462, 502
Avenzoar, 30
Averrhoes, 30
Avicenna, 30
Avogadro, **215**, 290, 294, 490

BACON, Francis, 101, 585
Bacon, Roger, **30-31**, 43
Baeyer, A. v., 349, 354, 358, **363**, 366, 435, 436, 450, 454, 469, 480, 485, 537, 578, 579, 591
Bährens, 62
Balard, 402, 418, 419, 426, **446**, 566
Balling, 557
Balmer, 495
Bamberger, 363, 436, 466, 468, 472, 473, 477, 481
Bancroft, 147
de Bary, 547
Barrière, 417
Barruel, 573
Basilius Valentinus, **36**, **37**, 41, 46, 47, 48, 51, 52, 53 *et seq.*, 94
Baudrimont, 62
Baumann, 439, 451, 464, 543
Baumé, 139, 154
Bayen, 131, 177
Beadle, 572
Beaumont, 451
Béchamp, 445
Becher, **109**, **110**, 134
Beckmann, E., 341, 359, 454, 462, 502
Beckurts, 551
Becquerel, 498, 528
Behrend, 341, 462
Beilstein, 458, 583, 595
Bence Jones, 228
Benedict, 570
Bergman, **124**, 137, 138 *et seq.*, 247, 384, 386, 417, 512, 523
Berlin, 411
Bernard, Claude, 541, 543
Bernoulli, 494
Bernthsen, 579
Berthelot, 23, 26, 28, 29, 39, 45, 46, 57, 153, 167, 168, 328, 420, 435, **508**, 517, 569

Berthier, 527
Berthollet, 172, **173-174**, 185 *et seq.*, 421, 473, 474, 513-519, 565, 596
Bertram, 582
Berzelius, **203-210**, 210-213, 216 *et seq.*, 223 *et seq.*, 229-237, 240, 241 *et seq.*, 248 *et seq.*, 251-253, 255 *et seq.*, 279, 281 *et seq.*, 292, 313, 388 *et seq.*, 396, 402, 404, 405, 408, 410, 411, 419, 421, 422, 424, 426, 429, 430, 441, 449, 474, 503, 516, 522, 523, 525, 587, 588, 593, 294, 597, 598
Besson, 425
Bettendorf, 410
Beudant, 221, 525
Bevan, 538, 572
Beyer, C., 481
v. Bibra, 539
Bidder, 541, 545
Biot, 499
Biringuiccio, 85
Bischof, C., 571
Bischof, G., 528
Bischoff, 543, 545
Bischoff, C. A., 359
Black, **119-121**, 128, 144, 180
Bladin, 485
Blagden, 166, 502
Blaise de Vigenère, 97
Blochmann, 425
Blomstrand, 237, 329, **340**, 390, 412, 429, 430, 471, 473, 516, 526
Blyth, A. Wynter, 399
Bockorny, 537
Bodländer, 506
Boerhave, 60, 113, **114-115**, 133, 135, 137
Bolley, 556
Boltzmann, 494
v. Bonsdorff, 526
Borcher, 559
Böttger, 63
Böttger, Rud., 568
Bötticher, 578
Böttinger, 580
Boullay, 253, 495
Bourcault, 454
Bourdelin, 118
Boussingault, 274, **533**, 535
Boyle, Robert, 3, 59, 92, 100, **103-107**, 128, 133-136, 140 *et seq.*, 510
Brand, 149
Brandt, 149
Brauner, 410
Bredt, 454
Brefeld, 547
Brewster, 499
Brieger, 549
Brisson, 152
Brodie, 445
Bromeis, 441
Brown, 455, 572, 456
Brücke, 501, 538, 540, 543
Brühl, 349, 498, 499, 509
Brush, 526
Buchholz, 523, 553
Büchner, 548, 570
Buckton, 464
Buff, 328
Buffon, 140
Bunge, 539
Bunsen, **260-261**, 311, 385-386, 391, 392, 407, 413, 418, 424, 426, 433, 469, 493, 494, 497, 511, 526, 528, 557, 567, 589, 593
Bunte, 583
Butlerow, 332, 334, **336**, 344, 435, 438

Cæsalpin, 88
Caëtano, 62
Cagliostro, 62
Cagniard de la Tour, 547
Cahours, 259, 438, **444**, **445**, 468, 469, 492
Cailletet, 492
Calmels, 483
Cannizzaro, **335**, 399, 438
Carlisle, 229
Caro, 470, 576, 578
Carpenter, 569
Carstanjen, 478
Cavendish, **121-122**, 128-130, 145, 150, 169, 415
Champion, 568
Chance, 564
Chancel, 299, 439
Chaucourtois, 371
Chevreul, 438, 441, 530, 542, **569**, 591
Chittenden, 541
Christensen, 403, 427
Christison, 397
Chrustschoff, 417
Ciamician, 484
Claisen, L., 355, 363, 444, 450, 452, 453, 483, 486
Clarke, 423
Classen, 390
Claus, Ad., 345, **348**, 349, 351, 360, 462, 481, 591
Claus, C. E., 413
Clausius, 494, 518
Clément, 426, 561

Cléve, 430
Cloëz, 474, 477
Collie, 363, 349, 416
Combes, 452
Conrad, 453
Cooke, 403
de Coppet, 376, 502
Couper, 329, **333–334**, 337, 345
Coupier, 577
Courtois, 402, 560
Cousin, 168
Crafts, 434, 452
Crell, 180
Croll, 74, 97
Cronstedt, 144, 149, 523
Crookes, 374, 409, 410
Cross, 538, 572
Curtius, 423, 471, 472, 486

Daguerre, 510
Dale, 498
Dalton, 181, **188–195**, 199, 493
Dammer, 595
Daniell, 242
Daubrée, 527
Davy, Humphry, **195–199**, 229 *et seq.*, 238, 239, 241 *et seq.*, 290, 402, 404, 408, 418, 419, 421, 423, 424, 425, 426, 507, 584, 587
Davy, J., 198
Deacon, 565
Debray, 413, 429, 492, 528, 559
Debus, 189, 440, 449, 568
Déherain, 533
Deite, 569
Delitzsch, 474
Democritus, 7, 9
Dennstedt, 484
Derosne, 573
Descroizille, 391
Desormes, 426, 561
Dessaignes, 543
Deville, H. St. Claire, 413, 422, 424, 492, 498, 528, 559, 591
Dewar, 350, 493
Diesbach, 147
Dietrich, 533
Diodorus Siculus, 12
Dioscorides, 5, 11, 15, 16, 18, 50
Dittmar, 418, 593
Dobbie, 571
Döbereiner, 249, 371, 420, 441, 448, 484, 576
Döbner, 481, 577
Dombasle, 531
Dragendorff, 397, 549
Draper, 511
Drebbel, 95
Drechsel, 442, 540
Dschabir, 29
Dschafar or Geber, **29**, 30
Dubrunfaut, 573
Dufet, 506
Duhamel de Monceau, **118**, 136, 146, 147, 149
Dühring, 497
Dulong, **219**, 220, 242, 419, 423, 424, 591
Dulong and Petit, 220, 497
Dumas, 224, 225–227, 253, 258, **272–275**, 276–277, 280 *et seq.*, 323, 370, 371, 396, 404, 407, 433, 437, 438, 460, 475, 490–491, 495, 556, 563, 591
Duppa, 442
Durocher, 527
Dutrochet, 536

Ebell, 570
Ebelmen, 527
Ebert, 485
Effront, 575
Ehrenberg, 478
Eiloart, 357
Einhorn, 483
Eittner, 580
Ekeberg, 181, 204, 412, 523
Elbs, 362, 504
d'Elhujar, 411
Eller, 115
Empedocles, 6, 8
Engelmann, 536
Engeström, 144
Engler, 583, 406
Epicurus, 7
Erasmus of Rotterdam, 62
Erastus, 72
Erdmann, 203, 370, 388, 396, 526, 587, 589, **590**, 597, 598
Erlenmeyer, 334, 336, 338, 470, 591, 597
Étard, 549
Ewan, 559
Eykmann, 502

Faggot, 155
Falck, 545
Faraday, **227–228**, 251, 418, 433, 492, 503, 516
Favre, 507
Faworsky, 435
Fehling, 392, 441, 446, 449, 476, 496
Feichtinger, 571
Fick, 543
Figuier, 430, 573

Fileti, 443
Finkener, 501
Fischer, Emil, 358, **363**, 364, 437, **455**, 457, 462, 472, 485, 537, 576, 577, 591
Fischer, F., 556
Fischer, G. E., 184
Fischer, O., 363, 437, 576, 577
Fittig, **345**, 435, 443, 448, 452, 455, 458, 485, 591, 594, 597
Fitz, 547
Flamel, Nicolas, 35
Fleck, 569
Flügge, 399
Fordos, 420
Forster, 545
Förster, 431, 571
Fouqué, 528
Fourcroy, 163, 168, 171, 172, **174** *et seq.*, 586, 596
Fownes, 485
Frank, A., 557, 566
Franke, 427
Frankland, E., 285, 310, **316** *et seq.*, **322** *et seq.*, 329, 330, 331, 336, 361, 366, 367, 415, 425, 434, 442
Frankland, P., 535
Frémy, 418, 423, 427, 529, 539, 569, 591, 595
Frerichs, 541, 545
Fresenius, R., 385, **389**, 397, 588, 593, 529
Fresnel, 499
Freund, 435
Friedel, 434, 452, 454, 528
Friedheim, 429
Friedländer, 481
Frobenius, 152
Fuchs, 571
Fuchs, N., 524

Gabriel, 481
Gadolin, 409, 523
Gahn, 144, 146, 149, 181, 523
Galen, 48
Gattermann, 424, 449
Gautier, 476, 549
Gay-Lussac, 195, **199-201**, 210, 214 *et seq.*, 239-241, 249, 290, 391, 395-396, 402, 404, 418-420, 422-424, 426, 427, 473, 489, 490, 527, 562, 566, 582, 587
Geber (Dschafar), **29**, 30
Gehlen, 422
Gélis, 420
Gengembre, 177, 421
Genth, 428
Geoffroy the elder, 62, **117**, 138, 154
Geoffroy the younger, 117.
Gerhardt, 283, **286-295**, 300-307, 327, 337, 440, 445, 460, 479, 591, 594
Gerland, 430, 559
Geuther, 366
Gibbs, 390, 428
Gilbert, 533, 535
Gilchrist, 557
Girtanner, 594
Gladstone, 498, 517
Glaser, 108, 155
Glauber, **86**, 90, 92-97, 137
Glover, 562
Gmelin, C. G., 407
Gmelin, Chr., 192, 525
Gmelin, L., 227, 243, 278, 290, 295, 371, 474, 495
Goldschmidt, 483
Gomperz, 7
Gore, 418
Gorup-Besanez, 541
Göttling, 553
Goulard, 155
Graebe, 353, 363, 435, 452, 480, 578
Graham, **244**, 413, 423, 501, 594
Gray, 571
Gren, 180, 594
Grew, 155
Griess, 320, 460, 466, **470-471**, 578
Grimaux, 157, 161, 163, 166, 468
Gros, 430
Groth, 506
Grüneberg, 567
Gruner, 557
Guareschi, 549
Guckelberger, 563, 570
Guimet, 570
Guldberg, 517 *et seq.*
Gustavson, 435
Guye, 501
Guyton de Morveau, 133, 163, 168, 171, **173**, 175

Haarmann, 456.
Hagen, 154, 553
Haitinger, 483
Hales, 128, 129, 131
Hall, 527
Hammarsten, 540, 542
Hampson, 493
Hansen, E. Chr., 547, 574, 575
Hantzsch, 359, 360, 363, 422, 462, 473, 480, 485, 486
Harden, 189
Hardy, 483
Hargreaves, 563
Harnack, 540

Hasenbach, 422
Hatchett, 180, 412
v. Hauer, 430, 526
Hausmann, 385, 525
Hautefeuille, 528
Haüy, 523, 525
Hawksbee, 107
Heeren, 568
Heintz, 441, 539, 542, 569
Heinzerling, 581
Helbig, 563
Hellot, 147
Hellriegel, 535
Helmholtz, v., 543
van Helmont, 59, 61, **75–79**, 128
Helvetius, 61
Hempel, 393, 396
Henneberg, 534
Hennel, 439
Henninger, 541
Henry, 180, 341, 421, 493
Henry, W. C., 188
Heraclitus, 6
Herapath, 494, 495
Hermann, 526
Hermbstädt, 553, 556, 573
Hermes Trismegistos, 24, 25
Heron, 455, 456, 572
Herschel, 494
Herter, 541
Hess, 396
Hess, G. H., 507
Hesse, 496
Heumann, 425, 579
Hiärne, 146
Higgins, 196
Hill, 485
Hinsberg, 481
Hisinger, 204, 229
Hittorff, 406, 506
Hjelm, 411
van 't Hoff, 357, **377**, 406, 489, 499–502, 521
Hoffmann, Friedrich, **113**, 142, 152
Hoffmann, R., 446, 570
Hofmann, A. W. von, 72, 179, 209, 284, 295, 296, **297** *et seq.*, 439, 449, 465 *et seq.*, 468, 469, 470, 476, 477, 480, 490, 576, 577, 589, 592
Hofmann, Frz., 545
Hofmann, K. B., 11, 14, 15, 17, 18, 19
Hofmeister, 541
Holt, 443
Homberg, **107–108**, 144
Hooke, 107, 132
Hope, 408
Hoppe-Seyler, 539–542, 548
Horstmann, 488, 492, 501, 521, 595
Howard, 573
Hüfner, 541, 548
Humboldt, A. von, 214, 264, 273
Hunt, Sterry, 301
Husemann, 397, 549

Ingen-Houss, 531, 536
Irinyi, 569
Isaac Hollandus, 35, 41
Isambert, 492
Ittner, 473

Jacobi, 559
Jacobsen, 455
Jannasch, 526
Janssen, 415
Japp, 309
Jolly, 501
Joly, v., 414
Jörgensen, 428, 430, 574, 575
Jorissen, 406
Joule, 494
Julius Firmicus, 26
Juncker, 133

Kalle, 464
Kane, 257
Kanonikoff, 498
Karmarsch, 556
Karolyi, 568
Karsten, 495
Kaufmann, 422
Kay, 328
Keiser, 403
Kekulé, 307, **308** *et seq.*, 328–330, **332**, 336, 338, 339, 344 *et seq.*, 347 *et seq.*, 352, 353, 433, 440, 442, 446, 458, 463, 470, 471, 589, 593, 597
Kempe, 429
Kerl, 556, 557
Keyser, 414
Kiliani, 455
Kircher, 62
Kirchhoff, G. S. C., 572
Kirchhoff, Gust., 386, 494
Kirwan, 130, 180, 523
Kjeldahl, 397
Klaproth, **178–179**, 385, 386, 388, 403, 410, 411, 523
Klason, 463, 464, 474, 477
Knapp, 406, 556, 570, 572, 580, 581
Knecht, 580
Knop, W., 534
Knorr, 363, 486
v. Knorre, 429

Kobert, 549
Kohlrausch, 501
Kolb, 563
Kolbe, 285, 306, 310, **311–322**, 329, 330, 338, 345, 361, 437, 438, 441, 446, 447, 448, 461, 463, 475, 486, 582, 588, 589, 593, 594, 597, 598
König, 399
Königs, 363, 480, 482
Kopp, E., 568
Kopp, H., 4, **375**, 489, 495, 496, 498, 505, 594
Koppfer, 396
Körner, 350, 353, 479
Kortum, 62
Kraemer, 436
Krafft, 442, 477, 497
Kraus, 537
Kremers, 371
Krönig, 494
Krüger, 341
Krüss, 369, 408, 410, 411, 429, 430
Kühling, 482
Kuhlmann, 573
Kühne, 540, 541
Külz, 543
Kunkel, **109**, 147, 153
Küsser, 503
Kützing, 547

Laar, 355
Labillardière, 484
Ladenburg, **348**, 349, 353, 363, 479, 480, 482, 591, 595
de Laire, 456
Lampadius, 425, 523, 573
Lamy, 409
Landauer, 144
Landolt, 469, 498
Langer, 428
Langlois, 420
Laplace, 168, 488, 508
Lassaigne, 394
Lauraguais, 153
Laurens, 584
Laurent, 278 *et seq.*, **286–292**, 293–295, 301, 440, 457
Lauth, 577
Lavoisier, 4, 129, 158, 159, **160–172**, 247, 386, 392–395, 401, 403, 405, 530, 546, 594, 596
Lawes, 533, 535
Lea, C., 406
Le Bel, 357, 499
Leblanc, 149, 531, **562**
Leclaire, 560
Le Cor, 35
Lecoq de Boisbaudran, 409, 495
Lederer, 579
Lefèvre, 108
Lehmann, 534, 540, 541, 542
Leibniz, 102
Lemery, **107–108**, 134, 154
Lenk, 568
Lenz, 504
Lepsius, 567
Lerch, 453, 542
Leuchs, 541
Lévy, 528
Lewes, 425
Lewkowitsch, 570
Leykauf, 570
Libavius, 57, 59, **74**, 88, 91, 93, 95, 97
Lieben, 438, 441, 483
Liebermann, 351, 363, 435, 443, 452, 578
Liebig, 243, 246, 251, 253, 254 *et seq.*, 258–260, **262–270**, 274, 284, 306, 396, 423, 427, 432, 440, 441, 444, 445, 447–449, 451, 456, 459, 463, 474, 478, 530, 532–533, 535, 536 *et seq.*, 543, 544, 546, 568, 570, 574, 587, **588** *et seq.*, 594, 596, 598
Liechti, 429
Lightfoot, 577
Limpricht, 484
Linck, 568
Linde, 493
Linnemann, 497
v. Lippmann, 455, 573, 574–576
Lippmann, 510
Lister, 550
Littler, 571
Lob, 504
Lockyer, 415
Loew, 537
Lommel, 536
Long, 504
Lossen, 341, 495
Löwig, **402**, 463, 469
Loysel, 163
Lubbock, 180
Lucretius, 7
Ludwig, C., 501, 540, 541, 542
Lunge, 422, 561

Mackenzie, 405
Macquer, 117, **119**, 146, 147, 580
Magnus, 270, 419, 430, 439, 542
Malaguti, 517
Malherbe, 563
Mallet, 409
Malpighi, 530
Maly, 541
Mansfield, 577

Marcet, 587
Marchand, 370, 388, 396, 411
Märcker, 572
Marckwald, 485
Marggraf, 115, **116–117**, 143, 146, 150, 153, 157, 386, 573
Margueritte, 391, 429
Marignac, **369**, 370, 388, 403, 405 *et seq.*, 410, 413, 429, 526
Mariotte, 106, 530
Markownikoff, 436, 583
Marsh, 357
Marsh, 422
Martin, 558
van Marum, 405
Maslema, 30
Mathiessen, 407
Maxwell, Clerk, 494
Mayer, A., 547
Mayow, 107, 129, 130, 135, 164, 165
McGowan, 377
Meineke, 411
Meissner, 543
Melsens, 283, 569
Mendelejeff, **372** *et seq.*, 493
Menschutkin, 518, 521
Merck, 482
Mercurius, 25
v. Mering, 543
Merling, 483
Mersenne, 102
Meslans, 460
Meusnier, 168
Meyer, E. von, 406, 477, 597
Meyer, Lothar, **371**, 372, 458, 591, 595
Meyer, O. E., 494
Meyer, Richd., 598
Meyer, Victor, **350–351**, 355, 359, **363**, 406, 439, 455, 458, 459, 461, 462, 470, 484, 490, 491, 591
Michael, 356, 358
Michaëlis, A., 421, 462, 465, 469, 486, 487
Michaëlis, W., 571
Michel, 572
Miller, 386, 494
v. Miller, 481
Millon, 419
Milly, A. de, 569
Minderer, 93
Minunni, 462
Mitscherlich, Al., 572
Mitscherlich, E., **221–222**, 225, 406, 421, 427, 434, 439, 460, 463, **505–506**, 522, 524, 525, 527, 589, 594
Mohlau, 576
Mohr, 391, 397
Mohs, 522
Moissan, 369, 403, 424, 425, 428, 460, 528, 529
Moitrel d'Élément, 129
Moldenhauer, 569
Mond, 428, 430, 563, 564, 583
Monge, 163, 168
Moraht, 408
Morley, 403
Morris, 456
Morveau. See Guyton de Morveau
Mosander, 204, 410, 525
Mosso, 549
Muck, 584
Mulder, 537, 539
Müller, Fr., 459
Müller, H., 458
Müller, M., 570
Müller, N. J. C., 536
Müller v. Richenstein, 403
Müntz, 535
Musculus, 572
Muspratt, 556, 563
Muthmann, 429
Mylius, 431, 571
van Mynsicht, 73, 97

Nägeli, 538, 548
Naquet, 338, 344
Nasse, O., 541, 548
Naumann, A., 492, 497
Naumann, C. F., 525
Nef, 461, 462, 478
Nencki, 540, 548, 549
Neri, 89
Nernst, 488, 496, 503
Neubauer, 543
Neumann, F. C., 498, 504
Neumann, Kaspar, 115
Neumeister, 541
Newlands, 371
Newlands Brothers, 573
Newton, 140
Nicholson, 229, 596
Nicklès, 418
Nicolas, 569
Nièpce, 510
Nièpce de St. Victor, 510
Nietzki, 452, 453, 578, 579
Nilson, 408, 410, 411, 491, 497
Nobbe, 534
Nobel, 568
Noble, 568
Noelting, 423
Nordenskiöld, 125, 173
Noyes, 403

Odling, 307, 326, 327, 371
v. Oefele, 320, 374, 445, 464

Oettel, 566
Olympiodor, 25, 26, 27
Olzevsky, 492
Ortholph von Baierland, 49
Ost, 448, 483, 556
Ostwald, **377**, 488, 489, 501, 503, 504, 519, 520, 595
O'Sullivan, 455, 456, 572
Otto, J., 397, 568, 594
Otto, R., 464, 477, 487, 549
Overton, 360

Paal, 363, 422, 452, 481, 484, 485
Page, 458
Palissy, 62, 83, **85–86**, 89, 90, 530
Palmer, 488
Paracelsus, 3, 59, **67–72**, 94
Pariset, 587
Parkes, 558
Parmentier, 177
Partridge, 409
Pasteur, **358**, 447, 499, 505, 547, 548, 574, 575
Pattinson, 558
Payen, 556, 572
Péan de St. Gilles, 517
Pebal, 419
Pechiney, 565
v. Pechmann, 363, 453 472, 473, 485
Péligot, **275**, 411, 429, 438, 460, 561
Pelletier, 177, 421
Pélouze, 370, 404, 439, 591
Perkin, W. H., jun., 435
Perkin, W. H., sen., 363, 444, 450, 466, 500, 577
Peters, 534
Petersen, 526
Petit, 210, **219**, 220, 221
Pettenkofer, 371, 542, 545
Pettersson, 408, 491, 497
Pfaff, 385, 422
Pfaundler, 518
Pfeffer, 501, 536
Pfitzinger, 481
Pflüger, 545
v. d. Pfordten, 369, 427
Phillips, 569
Pictet, 483, 492
Pinner, 469, 476 481, 483
Piria, 446, 456
Planck, 501
Plato, 9, 38
Plattner, 390, 407, 558
Playfair, 311, 428, 474, 557
Pliny, 5, 6, 11–20
Plücker, 504
Poggendorff, 596
Ponomareff, 468
Popoff, 341, 452
Porret, 474
Porta, 90
Pott, 116
Prechtl, 556
Preyer, 541
Priestley, **122–124**, 130, 131 *et seq.*, 162, 392, 531
Pringsheim, 536
Proust, 183, **185–188**, 387, 388, 400, 430, 515
Prout, **201–203**, 374
Psellus, Michael, 30
Pseudo-Aristotle, 26
Pseudo-Democritus, 26
Pseudo-Geber, 34, **39–40**, 41, 42, 46, 47, **50–51**, 52–56, 59
Pugh, 535
Pullinger, 431
Pythagoras, 9

Quincke, 428

Rammelsberg, 390, 411, **526**, 528
Ramsay, 130, 374, 393, 402, 408, **415**, 416, 422, 430, 497, 500
Ranke, 543, 545
Raoult, 376, **502**
Raschig, 423, 424
Rathke, 477
Rayleigh, 374, 403, **414–415**
Raymund Lully, 32, **33–34**, 43, 52, 57
Réaumur, 147, 152
Redtenbacher, 441
Rees, 547
Regnault, **258**, 281, 433, 439, 542, 594
Reich, 409, 561
Reichenbach, 583
Reiset, 430, 542
Renault, 504
Renk, 545
Retgers, 506
Rey, 132
Reynolds, 474
Rhazes, 30
Richards, 403
Richter, J. B., **181–184**, 185, 387, 417
Richter, Th., 409, 526
Richters, 571, 584
Riecke, 359
Rinman, 146, 523, 557
del Rio, 412
Ripley, 35
Ritter, 510
Ritthausen, 537

de la Rive, 405, 559
Roberts, 573
Robinson, 563
Robiquet, 249
Rochleder, 441, 537
Romé de l'Isle, 221, 523
Romer, 569
Roscoe, 189, **369**, 412, 418, 429, 511
Röse, 341
Rose, Fr., 428
Rose, G., **389**, 505, 522, 525, 527
Rose, H., 209, 385, **389**, 412, 421, 424, 427, 517, 522, 525, 589
Rose, Valentin, the elder, 389
Rose, Valentin, the younger, 152, 389, 523
Rosenstiehl, 577
Rosetti, 90
Rossi, 438, 442
Rothe, 462
von Rothenburg, 486
Rouelle, 117, **118–119**, 136, 137, 154, 161
Rubner, 545
Rüdorff, 502
Rumford, 507
Runge, 484
Rutherford, 130

Sachs, 536, 538
Sabatier, 425
Sadler, 571
Sala, Angelus, 59, **80**, 95
Salomon, 455, 572
Sandberger, 526
Sandmeyer, 460
Sarasin, 528
Sattler, 560
Saussure, Th. de, 392, 395, 531, 536
Saytzeff, 443, 464
Schädler, 569
Schaffner, 563
Scheele, 124, **125–127**, 129, 130 *et seq.*, 143 *et seq.*, 147, 149, 150–155, 162, 183, 238, 386, 392, 427, 494, 510, 560
Scheerer, 390, 505, 526, 543
Scheibler, 429, 455
Scherer, 180
Schertel, 561
Scheufelen, 458
Schiel, 302
Schiendl, 510
Schiff, R., 495
Schischkoff, 478, 567
Schloesing, 535
Schlossberger, 543
Schmidt, A., 540, 541, 548
Schmidt, C., 541, 545
Schmidt, F. W., 410
Schmiedeberg, 539
Schmieder, 62
Schmitt, R., **319**, 447, 582
Schneider, E. A., 406
Schneider, R., 390, 404, 411
Schönbein, 405, 419, 568
Schöne, 419
Schorlemmer, 341, 497
Schott, 571, 572
Schrader, 532
Schraube, 473
Schrauf, 505
Schroeder, 496
Schrötter, 406
Schürer, 90
Schützenberger, 420, 431, 540
Schützenbach, 575
Schultz, 535
Schulze, H., 422
Schulze, E., 538
Schultze, 510
Schwalb, 458
Schwanert, 484
Schwanhardt, 148
Schwann, 547
Scott, 403
Seebeck, 408, 499
Sefström, 412
Seger, 571
Seignette, 97
Selmi, 549
Semmler, 582
Sénarmont, 527
Sendivogius, 62
Senebier, 531 536
Sennert, 59, **80**
Serullas, 424, 439, 474
Seubert, 414
Shaw, 114
Shenstone, 263, 406
Shields, 430, 500
Siemens, W., 559, 584
Silbermann, 507
Simpson, Maxwell, 442
Skraup, 353, 479, 480
Smith, 526
Smithells, 425
Sobrero, 568
Solon, 9
Solvay, 565
Soret, 405, 495
Soubeiran, 422
Soxhlet, 455, 542
Spencer, 559
Spilker, 436
Sprengel, 532

Stadion, 419
Städeler, 541, 543
Staedel, 495
Stahl, 4, 59, **110–113**, 133
Stahlschmidt, 424
Stas, 275, **370**, 388, 397, 403, 404, 407
Steiner, 478
Stenhouse, 484
Sterry Hunt, 301
Stevenson, 397
Stoehr, 480, 481
Stohmann, **508**, **509**, 534, 556, 573
Stölzel, 557
Stoney, 495
Storer, 531
Strecker, 466, 541, 543, 591, 594
Streng, 526
Stromeyer, 385, 390, 409, 423, 526
Struve, F. A., 555
Suidas, 2
Svanberg, 525
Swab, 144
Swan, 386, 494
Sylvius de le Boë, 59, **80–81**, 82, 92, 93
Synesios, 26, 27

TACHENIUS, 59, 80, **81–82**, 92, 93, 96
Talbot, 386, 510
Taylor, 397
Tennant, 405, 565
Tertullian, 24
Thaer, 531, 532
Thales, 6
v. Than, 425
Thénard, L. J., **200**, 239, 395, 418, 419, 424, 426, 427, 587, 594
Thénard, P., 421
Theophilus Presbyter, 46
Theophrastus, 5, 15
Thom, H., 551
Thomas, 557, 584.
Thomas and Gilchrist, 557
Thomas Aquinas, 31
Thomsen, J., 417, 430, **508**, 517
Thomson, James, 497
Thomson, Th., 189, **194**, 202, 390, 526, 592, 594
Thorpe, 104, 123, 167, 198, 228, 244, 423, 424, 430, 495
Thot, 25
Thurneysser, 62, 72
Tiemann, 454, 456, 476, 582
Tilghman, 572
Tillet, 133
Tollens, 455
Traube, F., 355
Traube, M., 419, 548
Travers, 402, 416
Trommsdorff, 177, 553
Troost, 528
Tunner, 557
Turner, 203, 390
Turquet de Mayerne, **73**, 97
Tutton, 423
Tyndall, 511

VALENTINER, 566
Valerius Cordus, 98
Varrentrapp, 397, 441
Vauquelin, **176–177**, 385, 386, 391, 408, 410, 426, 523, 586.
Verguin, 577
Ville, 531, 533, 534, 535
Vinzenz of Beauvais, 30
Voigt, 477
Voit, 542, 545
Vogel, 543
Vogel, B. H. W., 420, 511
Volhard, 320, 391, 467, 474
Volta, 145

WAAGE, 517 *et seq.*
van der Waals, 494
Wackenroder, 420
Wagenmann, 575
Wagner, 452
Wagner, P., 557
Wagner, R., 556
Walden, 504
Walker, Jas., 377
Wallach, 436, 454, 468, 582, 586
Ward, 148
Warington, 535
Watson, 494
Watson, W., 149
Watt, 167
Watts, H., 595
Weber, 498
Weber, R., 420, 561, 571
Weddige, 477, 481
Wedgwood, 571
Weidel, 353, 479
Weihrich, 264, 588
Weilandt, 497
Weldon, 565
Welter, 420
Wenzel, 186, 387
Werner, A., 341, 359
Werner, A. G., 523, 524
Westrumb, 177, 523, 553
Whetham, 503
Wichelhaus, 581
Widmann, 481
Wiedemann, G., 504

Wiegleb, 62, 177, 523
Wilcke, 120
Wilfarth, 535
Wilhelmy, 520
Will, 397, 453, 456, 588
Willgerodt, 458
Williams, 478
Williamson, 295, **298** *et seq.*, 326, 337, 439, 518
Willis, 107, 134
Winkelmann, 497
Winkler, Cl., 392, 393, 408–410, **411–412**, 526, 559, 561, 571
Winogradsky, 535
Winterl, 417
Wischnegradsky, 479, 482
Wislicenus, J., **356**, 357, 358, 366, 446, 453, 500, 543, 591
Wislicenus, W., 355, 422, 423, 444, 453, 473
Witt, 470, 578, 579
v. Wittich, 541
Wohl, 466
Wöhler, 252, **254**, 262, **270–272**, 389, 396, 406, 408, 424, 427, 432, 444, 456, 463, 474, 525, 528, 543, 570, 589, 594, 596
Wolff, E., 534
Wolff, L., 481
Wolffenstein, 419
Wollaston, **195**, 199, 412, 413
Wray, 153
Wren, 107
Wroblevsky, 492
Wurtz, **295–296**, 298 *et seq.*, 307, 310, 327, 328, 337, 368, 424, 434, 439, 440, 446, 450, 466, 467, 477, 492, 526, 548, 591, 595
Wyrouboff, 506

Young, Jas., 244
Young, Sidney, 497

Zeise, 463
Zamminer, 594
Ziervogel, 558
Zimmermann, Cl., 369, 411, 429
Zincke, 436, 439, 440, 452, 481, 497
Zinin, 466, 470
Zöller, 534, 556, 586
Zosimos of Panopolis, 24, 27
Zulkowsky, 593

INDEX OF SUBJECTS

The figures in thick type refer to those pages upon which subjects are treated in detail or points of special importance are recorded.

ABSORPTION of gases by water, 129
Academia Cæsarea Leopoldina, 102
del Cimento, 102
Académie Française, 176
Royale, 102
Academies and Learned Societies, formation of, 101-102
Academies, Spanish, 28
Acetaldehyde, 449
Acetic acid, 19, 96, 153, 441, 575-576
acid, constitution of, 282-283, **316**, *et seq.*, 441
acid (glacial), 153
acid, synthesis of, 361
aldehyde, polymers of, 449
Aceto-acetic ether, 354, 442, 453
tautomerism of, 354
Acetone, 451
-dicarboxylic acid, 453
Acetyl, 315
theory (Liebig), 258
Acetylene, 435
as an illuminant, 583
Acid amides, 444, 468
anhydrides (Gerhardt), 300, 444, 445
chlorides, organic, 444, 445
nitriles, 314, 475
theory of (Lavoisier), 167
Acides, 172
Acids, 51, 91
constitution of (Berzelius), 233 *et seq.*
constitution of (Davy), 241
constitution of (Liebig), 243-246
nomenclature of (Lavoisier), 172
organic, 153
Acids from plant juices (Scheele), 153
manufacture of organic, 582
always contain oxygen (Lavoisier), 167 *et seq.*; controversion of this view, 197, 238 *et seq.*
Acrylic acid, 442
Actinometry, 511
Adipic acid, 442
Adjective dyes, 147
Aer vitriolicus, 131
Æsculin, 456
Æthal, 275, 438
Æthereum (Kane), 257
Ætherin, 253
theory, the, 253-254
Affinitas, 137
Affinity-coefficients, specific, 519 *et seq.*
Affinity, degrees of, 334
determinations of, 512-521
doctrine of, 512-521
doctrine of (Bergman), 512-513
doctrine of (Berthollet), 174, 185, 513 *et seq.*
doctrine of, its latest development, 519
simple elective, 138
tables of (Geoffroy), 117, 138, 512
units of, 341
views as to its causes, 137-140
Affinity, chemical (Boyle), 105
chemical, views of the Phlogistonists, 137 *et seq.*
doctrine of (Guldberg and Waage), 517, 519
Affinivalenten (Erlenmeyer), 338
Agricultural-chemical experiments at Woburn, 533

Agricultural chemistry, 530 *et seq.*
Liebig's great services, 269, 530 *et seq.*
Air, composition of atmospheric, 122, **130-131**, 145
Albumens, vegetable, 537
animal, 539
Alchemistic period, the, **21-64**
Alchemistic speculations of the 13th and 14th centuries, 41
Alchemists, practical-chemical knowledge of the, 45 *et seq.*
Alchemy among the Arabians, 28 *et seq.*
at the European courts, 35, 61, 62
books on, 23
decay of, 58 *et seq.*
during the last four centuries, 58-64
general notes upon, 2-3
in Egypt, 23, 37
in the Christian countries of the West, 30 *et seq.*
its relation to the Platonist philosophy, 22
origin of, 21, 23 *et seq.*
position of chemists of repute in the 16th and 17th centuries with regard to it, 59 *et seq.*
problems of, 32 *et seq.*
relations of, to astrology, 25
special history of, 37 *et seq.*
theories of, 37 *et seq.*
Alcohol, 57, 98
constitution of (Berzelius), 256
meaning of the word, 98
preparations from, 575, 581
Alcoholometry, beginnings of, 152
Alcohols, 437, 440
constitution of (Kolbe), 318
polyatomic, 438
secondary and tertiary (Kolbe), 318
Aldehydes, 448-451
constitution of (Kolbe), 318
formation of, 449
Aldol, 450
Aldoses, 455
Aldoximes, 462
Alembic Club Reprints, 214
Alexandrian Academy, the, 21, 26, 37, 38
Algaroth, powder of, 94
Alizarine, 578
Alkahest, 52, 96
Alkali, 50
Alkali metals, discovery of by Davy, 197, 238
Alkali metals (Gay Lussac and Thénard), 200, 239
atomic weights of, 407
compounds of, 426
earlier views on their nature, 239
Alkali waste, 563-564
Alkalies, decomposition of, 238
Alkalimetry, 391
Alkaloids, derivatives of pyridine, etc., 482-483
synthesis of, 482
tests for, 397-398
Alkarsin, 261
Alkyl cyanurates, 477
cyanides, 475
Alkyl-pyridines, 479
Alkyls, metallic, 487
Allo-isomerism (Michael), 356
Allotropy, 405, 406, 506
Alloys, 560
Allyl alcohol, 438
Allylamine, 467
Alum, 17, 52, 90, 93
earth, confounding of this with lime, 93
Aluminium, 409, 559
bronze, 560
chloride, syntheses with, 434
Amalgamation processes for obtaining silver, 47, 88
Amalgams, 560
Amides, 444-445
Amidines, 468, 476
Amido-acids, organic, 446
constitution of (Kolbe), 319
Amido-miazines, 478
-pyrimidines, 478
Amidoximes, 476
Amine bases (Wurtz, Hofmann), 295 *et seq.*
Amines, 465 *et seq.*
Ammonia as a type, 296, 300
gas, discovery of, 129
manufacture of, 582-583
salts as medicines, 93
Ammonia soda, 564
Amygdalin, 268, 456
Amyl alcohol, 438
Anæsthetics, 551
Analysis, introduction of the word by Boyle, 106, 141
Analysis, development of, 384 *et seq.*
legal-chemical, 397-398
of articles of food and drink, 398
of gases, 128, 145, 392
of inorganic substances, 384 *et seq.*
of organic substances, 393 *et seq.*

Analysis, qualitative, 82, 96, 140 *et seq.*, **384** *et seq.*
quantitative, 144, 178, **386**, *et seq.*
technical, 398
volumetric, 390
Ancients, practical-chemical knowledge of the, 9 *et seq.*
Anhydrides of organic acids, 444
Aniline, 465, 576
black, 577
blue, 577
colours, 576 *et seq.*
green, 577
red, 576
violet, 577
yellow, 578
Annalen Chemische (Crell), 180
der Chemie und Pharmazie (Liebig), 266, 596
der Physik, 180
der Physik und Chemie (Poggendorff), 180, 596
Annales de Chimie, 175, 596
de Chimie et de Physique, 596
Anthracene, 435
Antimoniuretted hydrogen, 422
Antimony (Basil Valentine), 36, 47, 54
compounds, organic, 469
pentachloride, 424
pills, 94
preparations, 37, 49, 54, 94
Antiphlogistic system, the, 158, 168 *et seq.*
in Germany, 177
in other countries, 180
Antipyrine, 472
Antiseptics, 155, 550
Apothecaries' shops, 49, 91
Apparatus for collecting gases, 129
of the alchemistic age, 34
Aqua fortis, 51
regia, 51-52
vitæ, 57, 98
Arcana, 70
Arabian academies, 28
Arabians, chemistry among the, 28 *et seq.*
Arabite, 439
Archeus, 70, 78
Argon, 122, 393, 414, 416, 419, 493
helium, and other monatomic gases, their position in the Periodic System, 374, 416–417
Aromatic compounds, meaning of the term, 349-351
theory of (Kekulé), 346
(Ladenburg, Claus, and Baeyer), 348-349
Arsenic and its compounds, 55, 94, 404, 422
Arsenic acid, 151
Arsenious acid, 55, 151
Arseniuretted hydrogen, 422
Arsines, etc., 469
Ashes of plants as manure, 17
Asparagine, constitution of, 319
Assimilation in plants, 510, 534, 536 *et seq.*
Asymmetric carbon atom, 355
theory of (van 't Hoff and Le Bel), 356 *et seq.*, 447
Atom (Laurent), 294
Atomic compounds (Kekulé), 339-340
heat, 220, 497-498
hypothesis, the, 181, **188**
theory, (Dalton's), **188**, **194**
theory, further development of, 194 *et seq.*
theory, further development of, by Berzelius, 210 *et seq.*
preparatory work for (Richter), 181, (Proust), 185
volume, 495
weight determinations, 203, 388
weight (Laurent), 293-295
weight system of Berzelius and the opposition to it, 205, 223-225, 227
weight tables (Berzelius), 218, 224
weight tables (Dalton), 193
weights (Berzelius), 211 *et seq.*, 217 *et seq.*, 223 *et seq.*
weights, correction of, 373
weights, deduction of, by Cannizzaro, 335
weights (Dumas), 225
weights, Dumas' opposition to those of Berzelius, 225
weights (Erdmann and Marchand), 370
weights (Gerhardt), 290 *et seq.*
weights, improvements in their determination, 370, 387, 403
weights (Marignac), 370, 388, 403
weights of the metals, 406 *et seq.*
weights of the non-metals, 401 *et seq.*
weights, periodic arrangement of, 370 *et seq.*
weights, ratio of those of hydrogen and oxygen, 403
weights, relative (Berzelius), **210** *et seq.*, **218**, **223**
weights, relative (Dalton), 191 *et seq.*

Atomic weights, relative (Thomson), 195
weights (Stas), 370, 388, 403
weights, uncertainty as to them generally, 227
weights, uncertainty as to those of the metals, 218, 221
Atomicity of the elements, 334, 338
Atoms, conception of, 181
of various orders (Dalton), 191
spacial arrangement of, 252, 344, **356** *et seq.*
Atropine, 482
Aurum potabile, 44
Azo-compounds, 470
Azo-dyes, 470, **578**
Azo-imide, 423
Azoles, 485

Backward substitution, 283-284
Bacteriology, 550
Balance, importance of the (Lavoisier), 161, 169
Barium, 408, 426
Bases, designation of, by Lavoisier, 172
Basic slag, 557-558
Basicity, law of (Gerhardt), 289
of acids, 243 *et seq.*
of acids, criterion of, 245
Beet sugar industry, 275, 573
Beetroot sugar, 116, 149, 573
Benzene, constitution of, as deduced from molecular refraction, 499
constitution of (Baeyer), 349
constitution of (Claus), 348
constitution of (Kekulé), **346**
constitution of (Ladenburg), 348
derivatives, isomerism of, 347, 351 *et seq.*
hexagon formula of, 347
Benzin, 434
Benzoic acid, 97, 256, 443, 582
Benzoic aldehyde, 448
Benzoyl-carboxylic acid, 453
Benzoyl the radical of benzoic acid, 254
Benzyl alcohol, 438
Berichte der Deutschen chemischen Gesellschaft, 597
Berlin Academy, the, 102
blue, 147
Beryllia, 176
Beryllium, 408, 427
Berzelius-Liebig Letters, the, 209, 246, 259, 266, 272, 274, 284
Bessemer process, the, 557
Betaine, 467
Biblical characters as alchemists, 24
Bile, acids of the, 541
Bile, chemistry of the, 541
Bismuth, 47
preparations, 93, 94
Bitter almond oil, 254, 577
Bitter salt, 155
Black oxide of manganese (investigation of by Scheele), 127
Blast-furnace process, the, 557
Blood, chemistry of the, 541-542
gases, 542
Blowpipe, **144**, 385, 390, 523
Boiling point of solutions, 502
Boiling point, laws regulating the, 496-497
Bonds, central, 351
double, 351
Bone charcoal for sugar refining, 573
Bones, constituents of, 539
Boracic acid, 116
Boron, 404, 424
methide, 486
Brandy, distillation of, 90
Brass, 15, 147
Bromine, 402, 566
Bronze, 13, 14
Bunsen burner, the, 593
Butylene, 433, 435

Cacodyl compounds (Bunsen), 261
compounds, constitution of, 315-317
Cadaverine, 549, 480, 483
Cadmia, 15
Cadmium, 409
Cæsium, 407
Caffeïne, 537
Calcination of the metals, 132
Calcination of the metals (Lavoisier), 164 *et seq.*
Calcium, 408, 426
Calomel, 95
Calorimetry, 508-509
Campechy wood, extract of, 580
Capillarity, 500
Carbamines, 476
Carbides, metallic, 428
Carbinols, secondary and tertiary, 438
Carbohydrates, **454—456**
their significance for plant life, 538
Carbolic acid, 550
Carbon, 405, 425
as a constituent of organic compounds, 247
bisulphide, 425
compounds, saturated and unsaturated, 344 *et seq.*

Carbon, determination of, 395-396
assimilation of, by plants, 536 *et seq.*
double linkage of, 358
oxysulphide, 425
Carbonate of ammonia, 53
Carbonic oxide, composition of (Dalton), 190
acid, 79
acid, composition of, 387, 394
acid, composition of (Black), 119-120, 128
acid, composition of (Dalton), 190
Carborundum, 428
"Carboxylic acid," 453
Carboxylic acids, **441-448**
acids, constitution of (Kolbe), 317-318
acids, saturated, 441 *et seq.*
acids, aromatic, 443-444
acids, unsaturated, 442
acids, chlorides, anhydrides and amides of, 444 *et seq.*
Carbures, 172
Cavendish's researches on gases, 122
Cellulose, 572
Cement, 571
copper, 47
Ceramic art, the, 17, 147, 571
Cerium, 179
metals, the, 409-410
Chelidonic acid, 454, 483
Chemia, first use of the word, 26
Chemical compound, different from a mixture, 136
art, the, 9
combination, according to the ancients, 8
combination, according to the Phlogistonists, 135
composition, distinction between empirical and rational (Berzelius), 233
compound, meaning of, 8, 41-42, 134 *et seq.*
compound, old ideas regarding, 8, 41
constitution of organic compounds, methods for investigating this, 361 *et seq.*
equilibrium, statical and dynamical, 518
equivalents (Lavoisier), 169
"tinder," 568
industries, the great, 561
journals, 382, 595 *et seq.*
Chemical nomenclature (Lavoisier), 171
nomenclature (Berzelius), 234
nomenclature of organic compounds (recent), 437
notation (Dalton), 194
periods, the various, 1 *et seq.*
Chemistry, agricultural and physiological, 90, **530** *et seq.*
analytical, in the modern period, **384-399**
analytical, 106, 124, 136, 141
analytical, its development by Boyle, 106
antiphlogistic, 168 *et seq.*
applied, 83 *et seq.*, 146 *et seq.*, **554** *et seq.*
geological, 527 *et seq.*
in ancient Egypt, 9, 12, 13, 16 *et seq.*, 23 *et seq.*
in olden times, **5-20**
inorganic, 367 *et seq.*, **400-431**
its meaning at different periods, 2 *et seq.*
meaning and origin of the word, 2, 23
mineralogical, **522-529**
organic, **246** *et seq.*, **432-487**
pharmaceutical, 48 *et seq.*, 91, 154, 552
physical, **488-521**
pneumatic, founded by van Helmont, 78
pneumatic, its further development, **128** *et seq.*, 145
tasks of, in the various ages, 1
technical, in recent times, **554-585**
technical, in the iatro-chemical age, 87
technical, in the phlogistic period, 146
the aims of, 1 *et seq.*
Chili saltpetre, 566
Chloral, 459, 551
Chloraldehyde, 258
Chloride of lime, 426, 565-566
Chlorimetry, 391
Chlorine, 148, 402
discovery of, 127
its action upon organic substances, 457 *et seq.*
recognised as an element, 197, 239-240
supposed composition of, 240
the name, 239
Chlorophyll, 536
Choline,, synthesis of, 467
Chrome colours, 560

Chromium, 410
discovery of, by Vauquelin, 176
Chrysamine, 579
Chrysene, 436
Chrysoïdine, 578
Cinnabar, 95
Cinnamic acids, isomeric, 443
Circular polarisation, 499
polarisation, magnetic, 500
Citric acid, 153
Classification of organic compounds, 278-280, 289, 302
Coal gas, 583
Coal-tar colour industry, 576 *et seq.*
Coal-tar, products from, 582
Cobalt, 149, 410, 428
-ammonia compounds, 428
blue, 90
Cocaïne, 483
Cohesion, 514
Coins of alchemistic gold, 61
Colcothar, 55
Collidines, 480
Colloids, 501
Colour photography, 510
Combining proportions, proof that these are constant, 185 *et seq.*
weights (Gmelin), 227, 291, 295
Combustion according to Stahl, 110-113
according to Hoffmann, 114
according to Mayow, 107
correct explanation of, by Lavoisier, **164** *et seq.*
-ladder (Gerhardt), 289
phenomena of, 425
theory of (Lavoisier), 158 *et seq.*, **166-167**
Composition of substances according to Becher, 109-110
Compounds, atomic, 339
classification of, at the beginning of the Modern Period, 172
molecular, 339
Comptes Rendus, 596
Condensations, 450
meaning of, 362
of aldehydes, 452
Congo red, 579
Coniferin, 456
Conine, 482
synthesis of, 364
Conservation of matter (Lavoisier), 169
Constitution, chemical (Berzelius), 231, 252
chemical (Gerhardt), 304-307
Constitutional formulæ (Kolbe), 318-319
Copper, 13, 47, 88, 560
oxide for organic analysis, 396
vitriol, 18
Copulæ, 283, 284, 313 *et seq.*
Copulated or conjugated compounds, 283, 288, 313 *et seq.*, 324
Copulation, a consequence of saturation capacity (Frankland), 325
meaning of, 283, 322 *et seq.*
Corpse alkaloids, 549
Corpuscular theory (Berzelius), 216-217
theory (Boyle), 106, 139
Cosmetic, old Egyptian, 18
Creatine, 467
Cresols, 550
Criticism, importance of (Kolbe), 598
Croconic acid, 453
Crotonic acids, 443
Crotonic aldehyde, 450
Crystalline form, its connection with composition, 221-223
Crystallography, 523
Crystalloids, 501
Cultures, dry and water, 534
Cumarone, 485
Cyan-alkines, 477
Cyanic acid, 251, 474
Cyanamide, 474
Cyanogen, 200, 473
compounds of, 473-478
iron compounds of, 427-428
polymers of, 477
Cyanuric acid, 474

Daltonism, 188
Decipium, 417
Decomposition of molecules into atoms, 491
Decomposition of organic compounds, 366
Deduction, significance of (Aristotle), 5, 10
Dephlogisticated air (oxygen), 131
Dephosphorisation of iron, 557
Desmotropism, 355
Destillatio per decensum, 19
Dextrine, 572
Diagonal formula of benzene, 349
Di-aldehydes, 449
Di-amines, 465
Diamond, artificial production of, 529
Di-azines, 481
Diazo-acetic ether, 471-472

Diazo-compounds, 466, 470 *et seq.*
oxidation of, 472
Didymium, 410
Diffusion, 500
-process for sugar, 573
Digestion, 540
Di-ketones, 452
Dimorphism, 505
Dissociation, 492
electrolytic, 501 *et seq.*
Distillation, 19, 49, 57, 90, 593
Distilleries, 90
Di-sulphones, 464
Docimacy, beginnings of, 47
Docimacy of the noble metals, 88, 390
Double atoms (Berzelius), 236
Drinks, analysis of, 399
Dualism (Berzelius), 233, 243 *et seq.*
fight against, 237 *et seq.*, 275 *et seq*
overthrow by Unitarism, 282 *et seq.*
Duplication of the metals, 26
Dye character and chemical constitution, supposed connection between, 470
Dyeing, 17, 48, 90, 147, 580
Dyes, 147, 576
Dyes, distinction between *adjective* and *substantive*, 147
synthesis of, 364
Dynamic hypothesis, the, 359
Dynamite, 568

EARTHENWARE, 17, 47-48, 85, 89, 571
Earths (Becher's), 110
Eau de Javelle, 565
Ecgonine, 482
Échelle de combustion (Gerhardt), 289
Effect, chemical, 514
Elasticity, 514
Electric conductivity, 503
Electro-chemistry, 504, 516, 565
-chemical equivalents (Faraday), 228
-chemical theory (Berzelius), 230 *et seq.*, 516
-chemical theory (Davy), 229
-metallurgy, 559
Electrolysis, 228 *et seq.*, 242, 503
Faraday's law of, 228, 503, 516
of salts of fatty acids, 314
Electrolytic determination of metals, 390
Electrolytic law (Faraday), 228
Element, meaning of the term (Boyle), 105, 134
Element, meaning of (Lavoisier), 170
Elements, Aristotle's four, 7
classification of, 161
discovery of in recent times, 401 *et seq.*
discovery of in the phlogistic period, 150 *et seq.*
discovery of supposed new, 417
natural families of, 372
of the alchemistic period, 37 *et seq.*
of the Phlogistonists, 134-135
old views regarding, 6 *et seq.*
Elixir, for transmuting metals, 41
Encyclopedias of chemistry, 595
Energy, 508 *et seq.*, 521
Enzymes, 548
Eosin dyes, 578
Equilibrium, dynamical, 518
statical, 518
Equivalents, electro-chemical, 228, 503
of the elements (Gerhardt), 290 *et seq.*
of the elements (Gmelin), 290
of the elements (Laurent), 294
first table of (Richter-Fischer), 184
of the elements (Wollaston), 199
Equivalents instead of atomic weights, 199, 227
Erbium, 410
Erucic acid, 443
Esters, 439
Ether as a fifth element, 8
-acids, 439
constitution of, 256
from alcohol, 98, 152
Ethereal oils, 20, 436
Etherin theory, the, 253-254
Ethers, compound, 439
mixed, 298, 439
simple, 439
varieties of, 152
(Williamson), 298
Ethionic acid, 439
Ethyl, 256
Ethyl ether, 439
ether, formation of (Williamson), 298
sulphide, 463
-sulphuric acid, 439
theory, the, 256
Ethylene, composition of (Dalton), 189-190
oxide, 439
Eurhodines, 579
Eurhodols, 579

Experimental lectures, 265, 586 *et seq.*
methods, development of, by Boyle, 103-104
Explosives, 567-568

Fats, 19, 98, 153-154, 540
Fatty acid series, structure of compounds of the, 344-346
Fatty acids, 441
acids, constitution of (Kolbe), 314 *et seq.*
Fatty oils, &c., 19
Fermentation, 79, 546 *et seq.*
former views regarding, 152
processes, 546 *et seq.*, 574
significance of (v. Helmont)) 77-78
theories, 546 *et seq.*
Ferments, organised and unorganised, 547-548
Ferric acid, 427
Ferricyanogen, 474
Ferrocyanogen, 473
Ferrocyanogen compounds (Berzelius), 206
Filter papers, incineration of, 388
Filtering appliances, 593
"Fire air" (oxygen), 131
"Fire-damp Commissions," 584
Fixation of carbonic acid by alkalies (Black), 119-120, 128
of mercury, 27
"Fixed air" (Black), 120
Flame colourations (Marggraf, Scheele), 143
reactions (Bunsen), 385
Flesh, chemistry of, 543
Fluorene, 436
Fluorine, 403
its analogy to chlorine, 418
compounds, 206, 418, 424
compounds, organic, 460
oxygen compounds of, 460
Foods, analysis of, 399
Formazyl compounds, 472
Formic acid, 153
aldehyde, 449, 450, 537
production of, in plants, 537
Formulæ (Gerhardt's), 304-305
graphical (Kekulé), 343
rational (Kolbe), 321
Formyl-acetic ester, 453
Four-volume formulæ, 293
Freezing point of solutions, 502
Friction, fluid, 501
Fuchsine, 576
Fulminate of mercury (Kekulé), 308
Fulminic acid, 478
Fulminic acid, isomerism with cyanic acid, 251
Fulminuric acid, 478
Fumaric acid, 443
Furfurane, 350, 484
Furfurol, 484
Furnace gases, 557

Gall apples, juice of, 98
Gallium, 374, 409
Galmei, 15
Galvanic current used in analysis, 390
Galvano-plastic process, the, 559
Gas analysis, **392**
analysis, beginnings of, 145
analysis, technical, 392
Gas regulators, 593
Gas sylvestre, 79
"Gas," the generic term, 79
Gases, absorption of, 493
critical pressure of, 493
critical temperature of, 493
discovery of many by Priestley and Scheele, 123-124, 127, **128** *et seq.*
kinetic theory of, 493
liquefaction of, 492-493
their first collection over mercury, 129
the chemistry of gases in the phlogistic period, 128 *et seq.*
van Helmont's researches on, 78-79
Gastric juice, 541
Geber's writings and doctrines, 29
Geometrical isomerism (J. Wislicenus), 356 *et seq.*
Generators, 584
Germanium, 374, 411
German silver, 558
Glass, history and manufacture of, 16, 47-48, 89, 147, 570
Glauber's salt, 92
Glucoses, 454-455
constitution and synthesis of, 454 *et seq.*
Glucosides, 456
Glycerine, 154, 328, 438-439
Glyceryl, 328
Glycocoll, 319
Glycollic acid, 319
Glycogen, 543
Glycol, 328
Glycols, 439
Glyoxal, 449
Glyoxaline, 485
Gold, 11-12, 46-47, 88, 558

Gold, amalgamation of, 12
compounds of, 430
determination of its atomic weight, 430
separation from silver, 12, 88
Goulard's lotion, 155
Gradverwandtschaft, 333
Great chemical industries, the, 561
Groups of elements, 367, **370** *et seq.*
Guanidine, 468
Guanamines, 468
Gun-cotton, 568
Gunpowder, 567
Gypsum, 93

HALOGEN carriers, 458
derivatives of hydrocarbons, 457 *et seq.*
Halogens, the, 239-240, 402, 403
compounds of the, 239, 418, 424
hydrides of, 418
their action on the unsaturated hydrocarbons, 459
Handwörterbuch de Chemie, 267, 595
Heat, latent (Black), 120
latent (Lavoisier and Laplace), 162
nature of (Lavoisier), 162, 170
of combustion, 507 *et seq.*
of formation, constancy of (G. H. Hess), 507
specific, its relation to atomic weight (Dulong and Petit), 220
Heat-capacity of atoms, 220
Heating materials, 584
Helium, 393, 415, 416, 419
Hermetic Society, 62
Hermetic, 24-25
art, 24-25
Heterologous compounds, 303
Hexagon formula of benzene, 347
Hexoses, 454
Historia naturalis of Pliny, 5
History of chemistry, alchemistic period, 21-64
of chemistry, from Lavoisier till now, 158-598
of chemistry, iatro-chemical period, 65-99
of chemistry in early times, 5-20
of chemistry, phlogistic period, 100-157
Hoffmann's drops, 98
Homologous compounds, 303
Humus theory, the, 531 *et seq.*
Hydracides, 241
Hydrates of the metallic oxides, discovery of, 187
Hydrazine, 423
Hydrazines, 472
Hydrazoic acid, 423
Hydrazones, 451, 455, 472
Hydrides of aromatic hydrocarbons, 436
Hydrocarbons, **433-437**
synthesis of, 361, 434
aromatic, 435
researches on, 433 *et seq.*
unsaturated, 433 *et seq.*
Hydrochloric acid, 51, 91
acid gas, 129
acid, manufacture of, 565
acid as a type, 302
Hydrocyanic acid, 153, 474
Hydrofluoric acid, 418
its first use for etching glass, 148
Hydrogen, 78, 121, 402
acids, 241 *et seq.*
a constituent of organic compounds, 247
as a type, 302 *et seq.*
as the primary material (Prout), 202
as the unit in the determination of atomic weights (Dalton), 192
compounds of the halogens, 418
determination of, 395-396
properties of liquid (Dewar), 493
replacement of in organic compounds, 458-459
Hydrogen acids, theory of the (Davy; Dulong), **241** *et seq.*, 245
Hydro-phthalic acids, 349, 358
Hydroxylamine, 422
as a specific reagent, 365-366, 451, 462
Hygiene, relations of, to chemistry, 397-399, 545-546
Hyponitrous acid, 422

IATRO-CHEMICAL doctrines of Paracelsus, 69 *et seq.*
doctrines of Sylvius, 81
doctrines of van Helmont, 76-78
period, the, **65-99**
Iatro-chemistry, general notes upon 3
problems of, 66
Iatro-chemists, practical-chemical knowledge of the, 87 *et seq.*
Illuminants, 583
Indamines, 579
Indigo blue (Baeyer), 579
blue, artificial production of, 579
Indium, 409

Indole, 485
derivatives of, 472
Indo-phenols, **579**
Induction, photo-chemical, 511
Inductive methods, the gradual appreciation of, 32, 65, 101
Industrial gases, 393
Industries, the great chemical, 561
Inflammable air, 121
Inorganic compounds, structure of, 343, 367 *et seq.*
Inorganic compounds, systematising of, 367, 371, *et seq.*
Institut National, 176
Instruction, growth of chemical, **586-598**
systematic chemical, 207, 265, 271, 312, **586** *et seq.*
technico-chemical, 556
chemical, in laboratories, 176, 194, 204, 207, 265, 588 *et seq.*
International Commission on chemical nomenclature of organic compounds, 437
Iodine, 402, 566
recognised as an element, 240
Iodine (Gay-Lussac), 200
"Iodo-benzene," 459
Iodo- and iodoso-compounds, 351, 458–459
Iodonium bases, 459
Iodoso-benzene, 459
Ions, 503
Iridin, 457
Iridium, 413
Iron, 13-14, 47, 89, 147, 557-558
industry, the, 557
ores used in olden times, 13-14
Iron-carbonyl, 428, 487
Iron, chloride of, 95
compounds of, 427
Isatoic acid, 320
Iso-butyric acid, 441
Iso-cyanides, the, 476
Iso-cyanuric acid, 477
Iso-diazo-compounds, 473
Isogonism, 506
Isologous compounds, 303
Isomerisation of hydrocarbons, 435
Isomerism, **250-253**
geometrical, 356 *et seq.*, 500
of position, 352-353
physical, 447
in spite of identity in structure, 356
Isomers, **250-253**
structural-chemical, interpretation of, 351 *et seq.*
Isomers among the unsaturated acids, 443
Isomorphism (Mitscherlich), **221** *et seq.*, 505 *et seq.*, 524
appreciation of its value by Berzelius, 222-223
polymeric, 505
Iso-nitroso compounds, 461
Iso-propyl alcohol, 438
Iso-pyrazolone, 484

Jahrbuch der Chemie, 598
Jahresberichte der Chemie (Berzelius), 208, 597
der Chemie (Liebig), 267, 598
Journal, Allgemeines J. der Chemie, 180
de Physique, 175
für praktische Chemie, 312, 597
Journal of the Chemical Society, 596
of the Society of Chemical Industry, 556
Journals, chemical, 595-598
Journals, old German chemical, 179
Juices, the animal, van Helmont's views upon, 77-78

Kermes mineral, the, 94, 155
Ketones, **451-454**
constitution of (Kolbe), 318
Ketones, fatty aromatic, 452
Ketonic acids, **451-454**
Ketoses, 455
Ketoximes, 462
Krypton, 416

Laboratories, establishment of, 556, 586 *et seq.*
for students, 207, 265, 312, 586 *et seq.*
improvement of, 590
instruction in, 586 *et seq.*
recent, 590 *et seq.*
technical, 593
Laboratories, Egyptian, 9
Libavius' effort to establish chemical laboratories three centuries ago, 75
Laboratory fittings and apparatus, 593
Lactic acid, 153
acid, constitution of, 319
acids, the, 319, 446
Lactones, 448
Lactonic acids, 448
Lana philosophica, 55
Lanthanum, 410
Lapis infernalis, 95

Latent heat (Black), 120
Law of Boyle and Mariotte, 106
Lead, 14
acetates of, 57, 96
sugar of, 57, 96
Leblanc soda, 562 *et seq.*
Lecture experiments, 587
Lehrbuch der Chemie (Berzelius), 207
Levulinic acid, 453
Leyden papyrus, the, 23, 24, 26
Liebig-Wöhler Letters, the, 254, 262, 263, 268, 272
Liebig's *Autobiography*, 263
"Life air" (oxygen), 131
Light, chemical action of, 509-511, 536
refraction of, 499
Lime used in ancient times, 19
Linkage, interchange of, 351, 355
Linking bars (Couper), 334
of atoms, 332 *et seq.*, 343
Literature, the more recent chemical, 594 *et seq.*
technico-chemical, 556
Lithium, 407
Lucium, 417
Lutidine, 480

MADDER red, 578
Magenta, 576
Magisterium, 41
Magnesia alba, 155
Magnesium, 408, 559
Magnetism of chemical compounds, 504
Maleic acid, 443
Malic acid, 153
Malonic acid, 442
Manganese, 149
black oxide of (Scheele), 127
compounds, 427
Mannite, 439
Manuals of chemistry, 74, 108, 114, 119, 154, 266, 271, 273-274, 295, 312, 594 *et seq.*
Manures, artificial, 567, 574
Marcasitae, 56
Marsh gas, 79
gas as a type, 309
Martin process, the, 558
Masrium, 417
Mass-action, 513, 514, 517 *et seq.*
Matches, 568
Materia Medica (*Dioscorides*), 5
prima, 42, 63
Matière de chaleur (Lavoisier), 170
Meconic acid, etc., 483
Medicines of various orders for the transmutation of metals, 40
of ancient times, 18 *et seq.*
of Paracelsus, 70-71
of the iatro-chemists, 73 *et seq.*
of the phlogistic period, 154-155
recent, 550-551
Melame, 474, 477
Melamine, 474
Meleme, 477
Mellitic acid, 444
Mellone, 474, 477
Melting points, 497
Mercaptals, 464
Mercaptans, 462-463
Mercaptols, 464
Mercurius philosophorum, 27, 30 *et seq.*, 40 *et seq.*
Mercury, 15, 47
as a constituent of metals, 39 *et seq.*
salts of, 53, 95
Mesdem, the old Egyptian cosmetic, 18
Mesitylene, 353, 435
Metabolism, animal, 542, 545
vegetable, 530 *et seq.*
Metalepsy, 277
Metallic calces, 59, 164, 165, 112
chlorides, 91-92
compounds, recent work with, 426 *et seq.*
oxides, 55
salts in the alchemistic age, 52-53
Metallo-organic compounds, 322 *et seq.*, 486-487
Metallurgy, furtherance of, by Agricola, 84, 88
in the alchemistic period, 46
in the phlogistic period, 146
of the ancients, 11
of recent times, 557-560
Metals, colouring of, 38
derivation and meaning of the word, 11
duplication of, 26
ennobling of, 21, 25 *et seq.*, 37 *et seq.*
increase in weight on calcination, 132, 164.
nature of (Boyle), 134
nature of (Stahl), 134
old chemical theory of the, 38 *et seq.*
oldest knowledge of the, 11 *et seq.*
supposed composition of, in the alchemistic age, 39, 41

Metals, transmutation of, 2, 8, 21 *et seq.*, 37 *et seq.*
Metamerism, 252
Methane, composition of (Dalton), 190
Metargon, 416
Methods, analytical, 386 *et seq.*
technico-chemical, 398
synthetic, in organic chemistry, **361** *et seq.*
Methyl violet, 577
Methylene blue, 579
Microscope, its application to chemical research (Marggraff), 117, 143
Milk, chemistry of, 542
Mineral potash, 567
Mineral system, the chemical (Berzelius), 205
pigments, 560
pigments of ancient times, 18
tanning, 581
waters, artificial, 555
waters, analysis of, in the Phlogistic period, 142
Mineralogical chemistry, **522-529**
systems (Berzelius and others), 525
Minerals, analysis of, 176, 178, 205, 384 *et seq.*, 389, 525
artificial production of, 527-529
classification of, 523 *et seq.*
classification of (Bergman), 124
Klaproth's researches on, 178
nomenclature of, 525
Mixture-weights of the elements (L. Gmelin), 203
Modern chemical period, the (from Lavoisier), **158-598**
Molasses, crystallizable sugar from, 573
Molecular compounds (Kekulé), 339
weight, determination of, by vapour density, 376, 490
weight, determination of, in solutions, 376, 502
heat, 509
weight (Laurent), 294
Molecule, definition of the term, by Laurent, 294
Molecules, liquid, complexity of, 500–501
Molybdenum, 410, 429
Molybdic acid (Scheele), 151
Monochlor-acetic acid, 459
Mono-saccharides, 456
Mordants, 17, 47, 90, 147, 580
Morphotropism, 506
Mortar, 571
Mosaic gold, 147
Mosandrium, 417
Mucic acid, 153
Multiple types, 302
Muscular power, sources of, 545
Mustard oils, the, 467, 476
Myronic acid, 456

NAPHTHALENE, 435
Naphthenes, 436
Narcotics, 551
Natural Philosophy of the early part of this century, 263
Neon, 416
Nestorians, the, 28
Neurine, synthesis of, 467
Neutralisation, law of (Richter), 183
Nickel, 149, 410, 558
Nickel-carbonyl, 428, 487
Nickolanum, 417
Nicotine, 483
Niobium (Marignac), 369, 410, 430
Nitragins, 535
Nitric acid, 51-52, 148, 566
acid, composition of, 122, 150
Nitric oxide, 130
Nitrification, 534
Nitriles, 475
dimolecular, 478
Nitro-benzene, 460
reduction of, 466
Nitro-ethane, 461
-glycerine, 568
-methane, 461
-prussides, 428
Nitro-compounds, organic, 460 *et seq.*
Nitrogen chloride, 424
compounds, inorganic, 422 *et seq.*
compounds, organic, 465 *et seq.*
discovery of, 130
estimation of, 275, 396-397
diffusion of (Ramsay and Travers), 402
direct assimilation of, by plants, 535
density of, 415
group of elements, atomic weights of, 404
group of elements, compounds of, 422 *et seq.*
oxides of, 422
iodide of, 424
Nitrolic acids, 461
Nitrols, 461
Nitroso-compounds, 461
Nitrous acid, 150
Nitrous oxide (Davy), 196

Nitrum, 16, 52
Nomenclature, chemical (Lavoisier, etc.), **171-172**
chemical (Berzelius), 234
Nordenskiöld's *Life* and *Journals* of Scheele, 125, 173
Notation, chemical (Dalton), 194
chemical (Berzelius), **235-237**
Nuclei, original and derived (Laurent), 278
Nucleus theory (Laurent), 278-280
Nutrients, 544-545
Nutrition of animals, 544 *et seq.*
of plants, 531 *et seq.*, 536 *et seq.*

Oelsüss, 154
Oils, ethereal, 20, 582
Oils (fatty) known in ancient times, 19
Optical activity, its connection with chemical constitution, 357
Organic chemistry, development of, up to 1811, 246 *et seq.*
special history of, **432-487**
compounds, chemical behaviour of, 365
compounds, constitution of, 317, 251 *et seq.*, 303 *et seq.*, 332 *et seq.*, 344 *et seq.*
compounds, modes of decomposing, 366
compounds, structure of, 336, **344**
substances, qualitative composition of, 248, 393
substances, quantitative composition of, 248, 393 *et seq.*
Organic compounds, knowledge of, in the phlogistic age. 151
classification of, 278-279, 289, 302
compounds, distinction from inorganic, 246-247
Organo-metallic compounds (Frankland), 322 *et seq.*, 486-487
Orthrin, 255
Osazones, 455, 472
Osmium, 414
Osmose, 501
Osmosis of sugar solutions, 573
Osmotic pressure, 501-502
Ostwald's *Klassiker*, 214
Oxalic acid, 153, 582
acid, synthesis of (Drechsel), 442
Oxalines, 468
Oxalo-acetic ester, 453
Oxaluric acid, 468
Oxazoles, 486
Oxidation theory (Lavoisier), 166-167
Oxy-acids, constitution of (Kolbe), 319
organic, 446
Oxydes, 172
Oxygen, absorption of, by palladium and platinum, 430
"rendering active" of, 406
atomic weight of, 403
acids, theory of the, 167, 197, 237-238
law, the (Berzelius), 212
as the unit in atomic weight determinations (Berzelius), 217
importance of, for the antiphlogistic system, 171
importance of, for the atomic theory, 210 *et seq.*
discovery of, 124, 127, **131** *et seq.*, 165
compounds of the metals, 426 *et seq.*
compounds of nitrogen, phosphorus, etc., 422-423
compounds of the halogens, 418
compounds of sulphur, 420
Oxy-muriatic acid, 239
Ozone, 405-406

PALLADIUM, 413
"Panacea" of the ancients, 27, 44
Paper manufacture, 572
Parabanic acid, 468
Paraffin industry, the, 583
"Parallelosterism," 496
Para-oxybenzoic acid, 448
Para-rosaniline, 576
Pathology, its relations to chemistry, 549 *et seq.*
Pattinson process, the, 558
Penta-methylene-diamine, 480
Pentite, 439
Pepsine, 541
Peptones, 541
Periodic system of the elements, **370** *et seq.*
Periods of the elements, 371-372
Periods, the various chemical, 1 *et seq.*
the various chemical, their characteristics, 1-4
Perkin's violet, 577
Peroxides of organic acid radicals, 445
Per-sulphuric acid, 420
Petroleum industry, the, 583
Pharmaceutical preparations of ancient times, 18
Pharmacy, development of, 71

Pharmacy in the iatro-chemical age, 91 *et seq.*
its relations to chemistry, 552
Pharmacy, text books of, 553
instruction in, 553
Phenacetine, 551
Phenanthrene, 435
Phenols, 440
manufacture of, 581
Phenyl-hydrazine, 472
as a specific reagent, 365–366, 451, 455, 462
Phenyl-propiolic acid, 444
Philippium, 417
Philosopher's stone, the, 30 *et seq.*, **40** *et seq.*
Philosophical Transactions, the, 102, 596
Phlogistic period, general history of the, 100–128
period, merits of the, 155–157
period, special history of the, 128–155
period, the, **100–157**
system, fall of the, 168
Phlogisticated air (nitrogen), 131
Phlogiston an element, 135
assumption of the hypothetical, 110 *et seq.*
its identification with hydrogen, 122, 130, 159
theory, beginnings of (Becher), 110
theory, development of (Stahl), 111 *et seq.*
theory, its development after Stahl's time, 115 *et seq.*
theory, general notes on the, 4
theory, its value, 113
Phlogistonists, practical-chemical knowledge of the, 140 *et seq.*
Phloroglucin, tautomerism of, 354
Phosgene, 425
Phosphines, etc., 469
Phosphonium bases, 469
Phosphoric acid, 117, 150
acids, basicity of, 244
Phosphorous oxide, 423
Phosphorus, 149, 404
manufacture of, 569
pentafluoride, 340
allotropic modifications of, 406
hydrides of, 421
oxygen compounds of, 423–424
Phosphyl compounds, 462
Photo-chemistry, 509–511
Photography, historical notes on, 510
Phthaleïns, 578
Phthalic acid, 578
Physical chemistry, special history of, **488–521**
chemistry, its general significance, 375 *et seq.*
methods, application of, to chemistry, 156, 376
Physics, influence upon chemistry at the beginning of the phlogistic period, 101
Phyto-chemistry, 536 *et seq.*
Picric acid, 460, 580
Pigments of antiquity, the, 17–18
Piperidine, 408
Pitocarpine, 483
Planets, their relation to the metals, 25
Plant juices first used as *indicators* by Boyle, 131
Plant-nutrients, 532 *et seq.*
Plastic compounds (nutritive), 544
Platinum, 149, 413, 558–559
bases, 430
chloride, compounds of with carbon monoxide, 431
compounds, 430
metals, the, 413–414
their atomic weights, 414
Poisons, methods for detecting, 397–398
Polarity, electric, of atoms, 230-231
Poly-azines, 350, 353
Poly-azoles, 353
Polybasic acids, doctrine of the, 243 *et seq.*
Poly-carboxylic acids, 442
-sulphonic acids, 464
Polymerism, 252
Polymorphism, 505, 525
Poly-saccharides, 456
Porcelain, 17, 63, 147, 571
Position, determination of chemical, 352–353
Position-isomers, 353
Position of elements in the periodic system, 373
Potash, distinction of, from soda, 92, 118
industry, the, 567
salts, deposits of, 567
salts as medicines, 92
Potashes, 16, 17
Potassium, 238, 407, 426
carboxide, acids from, 453
Potter's art, the, 17, 47, 89, 571
Pottery, 17, 47, 89, 571
Powder, smokeless, 568
Precious stones, artificial, 16, 570

Precious stones, old imitations of, 16
Precipitates, 96
Prediction of alcohols (Kolbe), 318
of new elements (Mendelejeff), 373–374
Preparations, chemical, their manufacture, 581–582
old chemical, 18
old medicinal, 18, 19
officinal, in the phlogistic age, 154
technico-chemical, in the phlogistic age, 147
Pressure, critical, 493
Priestley's researches on gases, 123–124
Primary material, assumption of a, 374
Princes as patrons of alchemy, 61
Principe oxygine, 167
Prism-formula of benzene, 348–349
Progression, law of (Richter), 183–184
Proïn, 255
Propiolic acid, 443
Proportion-numbers of the elements (Davy), 199
Proportions, doctrine of chemical (Richter), 182 *et seq.*
law of constant (Proust), 181, **185–188**
law of multiple (Dalton), **189** *et seq.*
law of multiple, further developed by Berzelius, etc., 210 *et seq.*
Protyle (Crookes), 374
Prout's hypothesis, 201 *et seq.*, 370
Prussian blue, 147
Prussic acid, 153, 473
Pseudo-forms of compounds, 354
Ptomaïnes, 398, 549
Ptyalin, 541
Putrefaction, 548 *et seq.*
bases, 398, 549
Putrescine, 549
Pyrazine, 485
Pyrazines, 481
Pyrazole, 483
Pyrazolone, 484
derivatives of, 472
Pyrene, 436
Pyridine, constitution of, 350, 479
bases, 478 *et seq.*
Pyrimidine, 481, 485
Pyrimidines, 481
Pyroligneous acid, 153
Pyromucic acid, 484
Pyrrol, 350, **484**
Pyrrolidine, 484
Pyrroline, 484

QUALITATIVE tests for substances, 141–144
Quantitative researches, period of, 158 *et seq.*
Quick vinegar process, the, 575
Quinazoline, 485
Quinazolines, 481
Quinoline, 350, 480, 485
derivatives of, 481
Quinoline, synthesis of, 479 *et seq.*
Quinones, 452
Quinoxalines, 481
Quinta essentia, 8
Quintessence, 96

RACEMIC acid (discovery of), 206
acid, isomeric with tartaric, 252
Radical theory, first steps towards the, 246, 249
theory, the newer, 313–320
the older, **253–262**
theory, supersession of the older, 285
theory, fusion of the older radical theory with the type theory by Laurent and Gerhardt, 286
Radical vinegar, 96
Radicals, oxygenated, 255, 257
chemistry of compound, 259
compound, 247, 249
polyatomic, 302
the clearer definition of, 260
variability of, 257, 279
Reaction, time-rate of, 518–519
Reactions, specific, of organic compounds, 365
Reagents, introduction into analysis, 141 *et seq.*
Reciprocal reactions, 518
Reform of chemistry by Lavoisier, 166, **170**
Refraction-equivalents, 498
Regenerators, 584
Replaceable value of elements, 323, 326-327
Residues, theory of (Gerhardt), 287
Respiratory compounds (nutritive), 545
Retene, 436
Rhamnite, 439
Rhodamines, 579
Rhodium, 414
Rosaniline, 576
Royal Society, the, 102
Rubidium, 407, 426

Ruby, artificial production of, 529
glass, 89, 147
Russium, 417
Ruthenium, 413

SACCHARIDES, 456
Saccharimetry, 573
Saccharine, 574
Safety lamp, the *Davy*, 197, 584
Safranines, 579
Sal, 50
ammoniacum, 53
mirabile, 92
Salicin, 456
Salicylic acid, 319, 447, 550, 582
Saliva, chemistry of the, 541
Salmiac, 53, 93, 149
Sal nitri, 52
Sal petræ, 52
Sal polychrestum (Glaser's), 155
Salt as a constituent of the metals, 41
its meaning in the alchemistic and iatro-chemical periods, 50, 82–83
Rouelle's definition of a, **119**, **136**
Salts, constitution of (Berzelius), 233
constitution of (Liebig), 245
nomenclature of (Lavoisier), 172
Saltpetre, 52, 92
as a manure, 17
Sarcosine, 467
Saturation-capacity, assumption of a constant (Kekulé), 338 *et seq.*
-capacity of the elements (Frankland), **322** *et seq.*, **325**, 337
of carbon, 327 *et seq.*
Scandium, 369, 410
Scheele's *Letters and Journals*, 125
researches on gases, 129
Schweinfurt green, 560
Secretions, animal, 541
Seignette salt, 97
Selenium (Berzelius), 404
compounds, inorganic, 421
compounds, organic, 463
Sheep's-wool grease, 20
Siderum, 417
Silicates, fusion with alkaline carbonates (Bergman), 388
Silicon, 406
alkyl compounds of, 486
compounds of, 424
carbide, 428
Silver, 12, 46, 88, 558
mirrors (Liebig), 269
nitrate of, 53, 95, 560
allotropic modifications of, 406
Silver, oxides of, 427
extraction of, from ores, 558
Smalt, 90
Soap, 17, 148
manufacture, 569
Societies, learned, 101–102
Soda, 17, 18
artificial preparation of, 118, 148, 149, 569
industry, the, 562–564, 565
salts as medicines in the iatro-chemical age, 92
Soda-waste, working up of, 563-564
Sodium, 238, 407, 426, 559
Sodium peroxide, 426, 559
Soils from a chemical-agricultural point of view, 531 *et seq.*
Solar spectrum, chemical action of the, 510
Solder, 14
Solidification, Raoult's law of, 502
Solution, theory of, 376, 501 *et seq.*
Soporifics, 551
Spagiric art, the, 25
Special history of modern chemistry, **381** *et seq.*
Specific gravity of gases, 129
heat (Lavoisier ; Laplace), 162
heat of solids, 497
heats of the metals, relation to their atomic weights (Dulong and Petit), **220**
volume, 495
Spectrum analysis, 385–386, **494**
Spirit-lamp (Berzelius), 388
Spirit of wine, 49, 57, 98, 152
sweetening of, 57
Spirits, manufacture of, 575
Spiritus, 50
fumans Libavii, 95
igno-aëreus (Mayow), 130
Mindereri, 93
salis, 50-51
tartari, 97
Starch, 19, 149, 456, 538, 572
sugar, 572
Stassfurt salts, 567
Statique Chimique (Berthollet), 174, **185**, 514
Stearine candles, 569
Steel, 89, 146, 557
Steren, **496**
Stereo-chemistry, 355 *et seq.*
of nitrogen, 359 *et seq*, 462
Stereo-isomerism, 355 *et seq*, 443, 459, 473
Stibines, &c., 469

Stöchiometry, 387
founding of (Richter), 185
Strontium, 197, 408
Structural formulæ (Couper), 334
Structure, chemical, 332, **336** *et seq.*
theory, beginnings of the, **332** *et seq.*
theory, development of the, 342 *et seq.*
Sublimate, 53
"Substantive cotton dyes," 578
dyes, 147
Substitution, first observations upon, 276
-form (Gerhardt), 288
laws of (Dumas), 276
theory (Laurent), 278
partial admission of, by Berzelius, 284
Succinic acid, 97, 442
Sugar from beet juice, 116, 149, 573
estimation of, 392
refining of, 573
Sugar, known by the ancients, 19
from sugar cane, 99
Sugars, synthesis and constitution of, 455-456
Sulphines, 464
Sulphinic acids, 465
Sulphite cellulose, 562, 572
Sulphone-ketones, 464
Sulphones, 464
Sulphonic acids, 463-464
Sulphonal, 464
Sulphoxides, 464
Sulphur, allotropic modifications of, 406
atomic weight of, 404
balsam, 56
as a constituent of the metals, 39, 55-56
Sulphur auratum, 94
Sulphur, compounds of, 420
compounds of the alchemistic period, 55-56
compounds, organic, 460, 465
Sulphur ether, 152
Sulphur, milk of, 56
Sulphur philosophorum, 52
Sulphur, recovery of, from alkali waste, 563-564
Sulphures, 172
Sulphuretted hydrogen, discovery of, 129
Sulphuric acid, 51, 91
acid, anhydrous, 562
acid, fuming, 148
acid, manufacture of, 148, 561
Sulphurous acid gas, 19, 129
acid, practical utilisation of, 562
Symbols, chemical (Berzelius), 235
(Dalton), 194
with bar across them (Berzelius), 236, 283
Synthesis by *condensations*, 362
of organic compounds, 361 *et seq.*
Syrians, culture among the, 28
System, chemical, of the minerals (Berzelius), 524
natural, of the elements, 370 *et seq.*
periodic, 370 *et seq.*
Systematization of inorganic compounds, 367
of organic compounds, 433
Système unitaire (Gerhardt), 303-304

Tables des Rapports (Geoffroy), 117
Tables of affinity (Geoffroy), 117
Tannic acids, 537
Tanning, 580
Tantalum (Marignac), 369, 412-413, 430.
Tar, products from, 582
Tartar, 96-97
emetic, 73, 97
Tartaric acid, 153
acids (optically isomeric), 447
Tartarus, 70, 97
Taurine, constitution of, 319
Tautomerism, 354-355
Technical Schools and Colleges, 556
Tellurium, 403-404
compounds, inorganic, 420-421
compounds, organic, 462-463
Temperature, critical, 493
Tension-series of the elements (Berzelius), 231-232
Terbium, 410
Terpenes, 436
Terra pinguis (Becher), 110
Tetrolic acid, 443
Text-books of chemistry. See Manuals
Thallium, 409, 427
Therapeutics, relation of, to chemistry, 494-495
Thermo-chemistry, **507-509**, 517
Thiacetic acid, 463
Thiamides, 468
Thiazoles, 486
Thio-aldehydes, 451
Thiocyanogen, 475
Thionylamines, 465
Thiophene, 350, 484
Thomas-Gilchrist process, the, 557

Thorium, 411
Thymol, 550
Tin, 14, 89
compounds of, 429
dioxides, isomeric, 251
perchloride of, 95
Tincturæ, 38
Tincture for changing silver into gold, 27
Tinder, chemical, 568
Titanium, 179, 411
Titrimetry, 201, 390
Toluidines, 577
Traité Élémentaire de Chimie (Lavosier), 164
Tri- and Tetra-methylene derivatives, 435
Tri-amines, 465
Tri-azines, 481
Triazole, 485
Tri-methylamine, 467
Trichloracetic acid, 275, 280, 283, 459
Trimethyl-carbinol, 438
Trimethylene, 435
Triphenyl-methane, 436
-phosphine oxides, isomeric, 340
Tropæolines, the, 578
Tungsten, 410, 429
Tungstic acid (Scheele), 151
Turpentine, oil of (known to the ancients), 19
Turpeth mineral, 95
Tutty, 15
Two-volume formulæ, 293
Type metal, 560
Type theory, the newer (Gerhardt), 300 *et seq.*
theory, the newer (Kekulé), 307 *et seq.*
theory, the newer (Sterry Hunt), 301
the newer (Williamson), 298 *et seq.*
the newer, preparatory work for, 295 *et seq.*
theory, the older (Dumas), 280-281
Types, chemical, 281
condensed, 302, 307
duplicated, 302, 307
(Gerhardt's), 300 *et seq.*
mechanical, 281
mixed, 308-309
real, as opposed to formal, 307, 321
auxiliary, 303

ULTRAMARINE, 570
Unitarism, beginnings of, 237 *et seq.*
Unitarism, development of, **275**, **281** *et seq.*
Universal medicine, 44
Universities, establishment of, 65
Unsaturated compounds, 345
Uranium, 179, 410, 429
Urea, estimation of, 392
synthesis of, 252, 361
Ureas, substituted, 477
Uric acid, 153, 268
derivatives of, 468
synthesis of, 468
Urine, 542
analysis of, 543
chemistry of the, 542-543

VALENCY of the elements. See Saturation-capacity
constant or varying, 337-342
doctrine of, its influence on the development of chemistry, 331 *et seq.*
maximum, 338
of radicals, 326-328
of carbon, **327-330**
definite, 331
varying, **325**, 336
its application to inorganic compounds, 367
Vanadium (Roscoe), 369, 412, 429
Vanillin, 449
Vapour density determination, methods of, &c. (Dumas, Gay-Lussac and Hofmann, Victor Meyer), 225, 376, **490** *et seq.*
densities, abnormal, 492
pressure of liquids, 497
pressure of solutions, 502
Vegetable physiology, 530 *et seq.*
Versuch über die Theorie der chemischen Proportionen (Berzelius), 213, 235
Vitriols, 52
"Volume-atoms" (Berzelius), 216
Volume theory (Berzelius), 216, 229
Volumes, law of, extension by Avogadro, **215**
law of (Gay-Lussac), 200, **214** *et seq.*, 490
law of, its appreciation by Berzelius, 216 *et seq.*
specific, 495
Volumetric analysis, 390
analysis, beginnings of, 201

Wahlverwandtschaft, 138, 333
Water as a type, 298 *et seq.*

Water as the primary element (van Helmont), 76
analysis of, 555
composition of, 121, 167-168, 387, 394
its supposed transformation into earth and air respectively, 20, 161
Water-baths, 593
culture, 534
"Water gas," 584
glass, 93, 571
White precipitate, 95
Wood spirit, 275, 437

Yttrium earths, 409

ZAFFRE, 87
Zeitschrift für analytische Chemie, 390
für Chemie, 597
für physikalische Chemie, 377
Zinc, 15, 47, 89, 146, 559
dust as a reducing agent, 435
preparations in the iatro-chemical age, 95
alkyls, 486
chloride, 95
salts, 95
Zirconia (Klaproth), 179
Zirconium, 411
Zoo-chemistry, 530, **539** *et seq.*

THE END

Richard Clay and Sons, Limited,
london and bungay.

Bei Fragen zur Produktsicherheit wenden Sie sich bitte an:
If you have any questions regarding product safety,
please contact:

Walter de Gruyter GmbH
Genthiner Straße 13
10785 Berlin
productsafety@degruyterbrill.com